THE ASCENT OF JOHN TYNDALL

Roland Jackson is a northerner by background. As a scientist, educator and science communicator, he was Chief Executive of the British Association for the Advancement of Science (now the British Science Association), of which Tyndall was President in 1874, and Head of the Science Museum, London. A member of the Alpine Club, like Tyndall, he has ascended many of the same peaks as his subject. His interests now centre on the history, ethics, and policy issues of science and technology. He is a Visiting Fellow of the Royal Institution of Great Britain, where Tyndall was a central figure in the mid-nineteenth century, and a Research Associate in the Department of Science and Technology Studies at University College London. Roland is a General Editor of *The Correspondence of John Tyndall*, being published in 20 volumes by the University of Pittsburgh Press.

Praise for *The Ascent of John Tyndall*

'Lively and fluent, mirroring his subject's incessant activity'

Jan Golinski, *ISIS*

'It was not until 1945, more than half a century after his death, that a semi-authorised biography of Tyndall was published. Now Jackson has authoritatively redressed this injustice.'

Jules Stewart, *Geographical*

'Excellent biography ... *The Ascent of John Tyndall* is a long-overdue, magnificent tribute to an important, but largely under-appreciated scientist. Highly recommended.'

Richard Carter, *Friends of Charles Darwin*

'Mr. Jackson amasses a wealth of detail to give a fuller picture of this extraordinary man ... [He] has done a great service in his detailed and careful presentation of John Tyndall's life at a time when science is under attack, neglected and misunderstood, especially by those in government.'

Peter Pesic, *Wall Street Journal*

'Splendid monument of a biography'

Barbara Kiser, *Nature*

'This story reveals much about Tyndall . . . [this biography] is immensely long and devotedly successful at unearthing the facts of Tyndall's life . . . '

Jonathan Parry, *London Review of Books*

'One of the most important mountaineering biographies to have been published in recent years . . . Roland Jackson's biography of John Tyndall is not only a tour de force of scholarship, it's also an eminently readable book ... It's a magnificent piece of work and a must-read for every scholar of Alpine history.'

Alex Roddie, *The Great Outdoors*

'Jackson's book is as comprehensive as it is overdue ... Jackson at once recounts the important events of Tyndall's life and uses Tyndall himself to build a richly textured picture of the social and scientific world in which he lived. The book favours a rigorous attention to detail ... Jackson's impressive facility with the scientific and political contexts of Tyndall's late-nineteenth-century world enables him to weave together a series of themes that define both the man and the period, providing a useful and comprehensive launching pad for a wide variety of forays in to the social and scientific worlds of Victorian England.'

Joshue Howe, *Annals of Science*

the ASCENT *of* JOHN TYNDALL

Victorian scientist, mountaineer,
and public intellectual

ROLAND JACKSON

OXFORD
UNIVERSITY PRESS

OXFORD
UNIVERSITY PRESS

Great Clarendon Street, Oxford, OX2 6DP,
United Kingdom

Oxford University Press is a department of the University of Oxford.
It furthers the University's objective of excellence in research, scholarship,
and education by publishing worldwide. Oxford is a registered trade mark of
Oxford University Press in the UK and in certain other countries

© Roland Jackson 2018

The moral rights of the author have been asserted

First published 2018
First published in paperback 2020

Impression: 1

Published in the United States of America by Oxford University Press
198 Madison Avenue, New York, NY 10016, United States of America

British Library Cataloguing in Publication Data
Data available

Library of Congress Cataloging in Publication Data
Data available

ISBN 978-0-19-878895-9 (Hbk.)
ISBN 978-0-19-878894-2 (Pbk.)

Printed and bound in Great Britain by
Clays Ltd, Elcograf S.p.A.

To Nicola, with all my love and gratitude,
and remembering our ascent of Monte Rosa

'Physics no one reads, History every one reads'

—John William Draper[1]

'All history becomes subjective; in other words, there is properly no History; only Biography'

—Ralph Waldo Emerson[2]

ACKNOWLEDGEMENTS

This book has been at least seven years in the making. I am particularly grateful to Nicola, who has had to live with John Tyndall and me for that period. To her I dedicate this book, with love and enormous thanks.

I used to wonder why the list of acknowledgements in biographies is generally so long. Now I know. A legion of people has helped me with information, advice, and critique. Without their assistance and knowledge I could not have written this work. I should like to blame them for any remaining errors, but such are of course all mine.

I have benefited hugely from interaction with the many scholars working on the Tyndall Correspondence Project, which is publishing his letters in nineteen volumes between 2015 and 2023. I have been welcomed into this community, which combines leading-edge scholarship on Tyndall with knowledge and expertise in so many associated domains. Particular thanks to the project's director, Bernard Lightman, and to my other fellow general editors, James Elwick and Michael Reidy. Special thanks also to Melinda Baldwin, Michael Barton, Ruth Barton, William Brock, Janet Browne, Geoffrey Belknap, Geoffrey Cantor, Gowan Dawson, Graeme Gooday, Diarmid Finnegan, Nanna Kaalund, Elizabeth Neswald, and many others behind the scenes.

John Tyndall is one of the Royal Institution's most famous figures. It has been a privilege to sit at a desk there as a Visiting Fellow opposite a portrait of Michael Faraday, at the invitation of Frank James, and occasionally to speak in the hallowed Faraday Theatre. Frank's support and encouragement has been crucial, combining commentary on several draft chapters with innumerable informal tutorials and access to the Royal Institution's archive, which holds so much of Tyndall's personal material, and I offer my heartfelt thanks. Thanks also to Charlotte New and Jane Harrison for advice and access to the collections and archive—including their discovery or rediscovery of letters and other material—and to others who work or have worked in the collections: Wahida Amin, Rupert Cole, David Coombes, Katherine Doyle, Harriet Lloyd, and Laurence Scales.

The Royal Institution is associated with University College London (UCL). I am grateful to Joe Cain and the Department of Science and Technology Studies at UCL for a Research Associateship.

The Athenaeum was Tyndall's home from home, strategically placed between the Royal Institution, the Royal Society, and Westminster. The library and archive staff, including Kay Walters, Jennie de Protani, and Laura Doran, have been of great help, as has Michael Wheeler, authority on the Club's history and much besides.

In the world of mountaineering, I am indebted to the staff and many volunteers of the Alpine Club Library, including Hywel Lloyd, Glyn Hughes, and Tadeusz Hudowski, and to others with knowledge of the history of mountaineering: Clare Roche, Sir Brian Smith, Ian Smith, and, in St Moritz/Pontresina, Diane Conrad.

Graham Farmelo and Helen Haste gave me much wise advice at an early stage, and others have offered valuable information and perspectives: Michael Bailey, Geoffrey Boulton, Robert Bud, Aileen Christianson, Miguel DeArce, Norman Denison, Martina Dürkop, Julia Elton, Gabriel Finkelstein, Aileen Fyfe, Angus Hawkins, Bruce Hunt, Philip Hunt, Fredrik Jonsson, Paul Kerry, Philip Meredith, Charles Mollan, Sir John Rowlinson, Lin Skippings, Nick Snook, Gregory Tate, James Ungureanu, and a mystery reader.

The social network of Twitter is particularly useful to the historical researcher. @ProfTyndall is on Twitter, as is @Roland_Jackson. Many have offered help, whether they know it or not, through their levels of activity and interaction, including (in addition to many of the above): Thony Christie, Vanessa Heggie, Rebekah Higgitt, James Sumner, Ed Hawkins, @EduBoisReymond, and @GuyCallendar.

It was a pleasure to visit Elton Hall, home of Louisa's mother's family, at the invitation of Sir William Proby, Louisa's great-great-nephew. I thank both him and the archivist, Jenny Burt.

My visit to Carlow and Leighlinbridge was also a delight. I thank Dermot Mulligan, John Shortall, and their staff for access to the Carlow County Museum and Carlow County Library. I am grateful to Norman McMillan, Martin Nevin, and Kevin Higgins for their local knowledge, insights, and company by the banks of the River Barrow.

While so much material can now be found online, my thanks to the following for specific information or access to physical archives: Lord Avebury, John Lubbock's great-grandson; Anna Sander, Balliol College, Oxford; Colin Harris, Bodleian Libraries, Oxford University; staff of the British Library; Dave Shawyer, BT Archives; Adam Perkins, Frank Bowles and staff of Cambridge University Library; Steven Leclair, Canada Institute for Scientific and Technical Information, National Research Council; Natalie Midgley, Central Library, Halifax; Helen Harrison, Harrow School; Stephanie Arias, Huntington Library, San Marino, California; Anne Barrett, Imperial College Library; Lord Inglewood, Louisa Tyndall's great-great-nephew; Ourania Karapasia and James Peters, John Rylands Library, Manchester; staff of the London Metropolitan Archives; staff of the National Archives, Kew; David McClay, National Library of

Scotland; Carsten Lind and Katharina Schaal, Archiv der Philipps-Universität Marburg; Sir David Pollock, descendant of Frederick and Juliet Pollock; Charlotte Murray and Verity Andrews, Reading University; Elizabeth Harper and John Willans, Royal Albert Hall; Rupert Baker, Joanna Corden, Katherine Ford, Katherine Harrington, Joanna Hopkins, Fiona Keates, Keith Moore, and staff of the Royal Society Library; Bob Richardson, St Bride Foundation; Grace Timmins, Tennyson Research Centre, Lincoln; Jonathan Smith and James Kirwan, Trinity College, Cambridge; Christopher Hilton and staff, Wellcome Library; Daniel Sudron, West Yorkshire Archive Service, Calderdale.

Many individuals and organizations have given permission to quote from manuscript material or provided illustrations, for which I gratefully thank them. Credits are given in the Publisher's Acknowledgements and List of Picture Credits.

It has been a pleasure to work with OUP, and those associated with it, on the development of this book. Particular thanks to my editor, Latha Menon, for her interest and support from the beginning, to Jenny Nugee, Kate Farquhar-Thomson, Phil Henderson, Katrina Baillie-Smith, Carrie Hickman, Dan Gill, Pam Birkby, Elizabeth Stone, Denise Baggett, and to any others of whom I am personally unaware.

CONTENTS

LIST OF PLATES

A NOTE ON WORDS

Names, spellings, and the meanings of words change over time. In this book I have chosen to reflect the language of the period to an extent, as I believe that helps one better to comprehend the ideas in their time, and to empathize with their originators.

So, I use the word 'scientist' only twice (and on the cover). Though it was coined in 1834 (by William Whewell), it did not catch on until much later in the century.[1] Tyndall and his circle never used it. Tyndall regarded himself as a 'natural philosopher' and a 'physicist'.

Ideas of matter, force, and energy were in flux in this period. Many chemists (including Michael Faraday) did not accept the existence of material atoms, at least until the last quarter of the nineteenth century. Tyndall was a firm believer in physical atoms, and in molecules as groups of atoms combined. For him, *something* had to vibrate to create the waves of radiation that could be detected as light or heat. That something was an atom or molecule, and the waves were carried through space by an invisible substance called the ether (a substance in which physicists no longer believe) to a receiving atom or molecule, which was then put in motion. 'Radiant heat', meaning thermal radiation, the study of which Tyndall made his own, covered visible as well as invisible radiation. Tyndall termed the invisible radiation 'obscure heat', or 'black heat'. We would now call it infrared, as the radiation lies beyond the visible red end of the spectrum of light (as seen in a rainbow).

There are some contemporary names for substances that will be met in this book for which there are different modern names: 'carbonic acid' is now called carbon dioxide, and 'olefiant gas' is the hydrocarbon ethene. I have used the contemporary names for clarity, except in direct quotes. The same has been done for place names: for example, Leighlin Bridge is Leighlinbridge today; Bel Alp is Belalp; and Chamouni is Chamonix.

ABBREVIATIONS

BL	British Library
CP	Louis Pasteur. Papiers. V—Correspondence. CLXXXVI Suzor–Zévort, Bibliothèque nationale de France
CUL	Cambridge University Library
Hirst Journal	Typescript journals, RI MS JT/2/32a–e
ICHC	Imperial College Huxley Collection
NA	National Archives
NHM	Natural History Museum
PP	Parliamentary Publications
RA	Royal Archives
RDS	Royal Dublin Society Library & Archives
RI MS	Royal Institution of Great Britain manuscript
RS	Royal Society of London
SB	Staatsbibliotek zu Berlin
TC1	Tyndall Correspondence volume 1: Cantor and Dawson (2016)
TC2	Tyndall Correspondence volume 2: Baldwin and Browne (2016)
TCC	Trinity College, Cambridge
TCD	Trinity College, Dublin
Tyndall Journal	Typescript journals, RI MS JT/2/13a–c
UCL	University College, London
WT	Wellcome Trust Library
YU	York University, Toronto

INTRODUCTION

A pure white cone of snow rose above John Tyndall and his two Swiss companions. They took the last few steps in the rarefied air, and stood on the topmost point. It was the early afternoon of Monday 19th August, 1861. The summit of the Weisshorn, most beautiful and majestic of Alpine peaks, had been trodden for the first time.

Tyndall was about forty years old, at the height of his powers, and not only as a mountaineer. A few months before, he had stood before the Royal Society, Britain's elite scientific body, to deliver its prestigious Bakerian Lecture. He had given the assembled Fellows a detailed account of his seminal research on the absorption of heat by gases. In the process, he set the foundation for our modern understanding of the greenhouse effect, weather, and climate change. On that basis alone he would deserve a major biography, but there is far more to John Tyndall than this.

Tyndall is one of the most intriguing and significant figures of the mid-nineteenth century. This outspoken Irish-born physicist and mountaineer, who rose from a humble background to move in the highest reaches of Victorian science and society, and marry into the aristocracy, is central to the development of science and its place in cultural discourse. He was one of the most visible public intellectuals of his time. Tyndall is best known in scientific circles for his research on the absorption of heat by gases in the atmosphere. But among much else, he explained why the sky is blue, how glaciers move, and helped establish the germ theory of disease. Beyond that, he was a pivotal figure in placing science on the cultural and educational map in mid-century, and in negotiating its ever-problematic relationship to religion and theology. His iconoclastic speeches and writings sent shockwaves through society, not least in religious circles. With friends such as Michael Faraday, Thomas Huxley, Charles Darwin, Rudolf Clausius, Hermann Helmholtz, Emil du Bois-Reymond, Thomas Carlyle, and Alfred Tennyson, he was at the heart of nineteenth-century thought.

There are many ways to tell the story of a life and its influence. I have let John Tyndall speak extensively for himself, to capture his voice and to create a sense of the man and his time. His values and outlook were shaped by the social, political, and religious environment of his upbringing in rural southern Ireland. While he broke eventually with some of those influences, others retained, even enhanced, their power.

What emerges is a picture of an individual at the heart of science, and of debates about science and religion, in mid-Victorian Britain. Tyndall was a man of contrasts: driven by a sense of duty and a Protestant work-ethic, but a romantic with a deep regard for poetry; a frequent sufferer from ill-health and insomnia, but with immense physical stamina; freely generous and affectionate, but a stickler for any right, great or small; a staunch friend to many, but a combative enemy in disputes; in religion, eventually a pantheist—if one can classify him at all—but a good friend of many clerics; a champion of underdogs, but no democrat; a tantalizing mixture of the Liberal and the Conservative; a hybrid of the inductive and empirical worldview of the British with the romanticism and idealism of the Germans; a loner at times, but a skilled social networker; a superb conversationalist and lecturer, but often cantankerous. He did not suffer gladly those he thought fools, and had a limited sense of humour. Satire and irony were foreign to him; he spoke his mind directly. To the modern view, his attitudes to women and to race in particular are striking, indeed offensive. That they were typical of the time does not reduce their impact today. He aroused strong opinions, strong loyalties, and strong enmities throughout his life.

We meet Tyndall in this account only briefly as a boy. As a man, we can picture him of middle height, sparely built, but tough and athletic-looking. His face was rather stern and strongly marked, his eyes grey-blue, and his fine hair light brown in youth (Plate 1). He often wore a hat 'on account of the unusual length of his head'.[1] In his middle and later years, he sported one of those typical Victorian beards, a luxuriant growth that faded to white whiskers as the years advanced. His voice carried a hint or more of Irish tones.

This is not simply the story of a physicist, although one would be warranted on that basis alone. Tyndall had such wide interests that many strands weave throughout this book. All have resonances today: scientific discovery; the politics of science; inventions and their commercialization; government service; science communication; education; the development of mountaineering; political upheaval in Britain, Ireland, Europe, and America; religion and theology; poetry; attitudes to women and sex; and attitudes to race and slavery. Lives can look logical and inevitable in retrospect. But as lived, they are not. The future is unknown. In this biography I have let Tyndall's life unfold as it did. I hope that reveals the contingency and unpredictability of his scientific and private lives, and enables you to feel your way into his being.

PART I

c.1822–1850

FROM CARLOW
TO MARBURG

1

IRISH BEGINNINGS

c. 1822–1844

The River Barrow flows from Glenbarrow in the Slieve Bloom Mountains to Waterford Bay in south-east Ireland. Just downstream from the county town of Carlow, amid fertile fields, lies the village of Leighlinbridge, birthplace of John Tyndall. We know the day of his birth—2nd August—but not the exact year; the parish records were destroyed in a fire at the Public Records Office in Dublin during the Irish Civil War of 1922. Tyndall gave conflicting information at different times. He settled on 1820, but it was probably a year or two later.

The Ireland into which Tyndall was born inspired two of his great passions: a love of the outdoors and a keen interest in religion and politics. The landscape around the village where he grew up instilled in him a deep emotional bond with the natural environment. And a sense of power and menace; as a young teenager, he capsized while trying to retrieve a small boat and had to swim for his life. This was a formative incident he often recalled. But the influence of the environment was normally calming. As an adult, if the pressures of work became too much, he would flee to the countryside whenever he could.

The pastoral beauty of the river contrasted sharply with the social, religious, and political tensions of the time. Ireland in the early nineteenth century was a ferment of change and unease. Not only, like England, was it emerging from the challenging economic consequences of the French Revolution and Napoleon's eventual defeat at Waterloo, it was gripped by revolutionary fervour. Still reverberating in collective memory was the United Irishmen Rebellion of 1798, an uprising against British rule. Though this was a popular revolt rather than an organized movement, it resulted in thousands of deaths.[1]

The creation of the United Kingdom of Great Britain and Ireland in 1801 had produced a new Westminster Parliament that included 100 Irish members. In Ireland, the Protestant Ascendancy—the ruling landowner class, whether Whig or Tory—was socially and politically dominant, with a powerful administration based at Dublin Castle. It was not until 1829 that Catholics could serve as Members of Parliament. Agrarian violence was widespread by the time John started school, and in the 1830s, the Catholics' refusal to pay tithes for the upkeep of the established Anglican Church of Ireland led to the Tithe War. Over a five-year period there were widespread riots and attacks on property. Hundreds of people were killed, including police and those resisting payment. The burden of tithes was reduced in 1838, ending the violence, but not abolished until Gladstone's government disestablished the Church of Ireland in 1869. By the 1840s, the Nationalist movement in Ireland consisted almost entirely of Catholics, while Unionism was primarily Protestant, whether Anglican or Presbyterian.

Both John's parents came from relatively prosperous landowning families. His paternal grandfather had owned 110 acres at Coolcullen in Kilkenny, and lived in the relative luxury of Prospect Hall, a three-storey house in the town. While his father's upbringing was Protestant, his mother's family were Quakers, and owned large tracts of land near Ballybromell.[2] However, disapproval of the union on both sides of the family left the new couple in relatively impecunious circumstances. John's father, John Tyndall senior, set up as a shoemaker and leather dealer in Leighlinbridge, and served in the Irish Constabulary.

Given his background, it is not surprising that Tyndall developed a strong interest in religion and politics. Like his father—an educated man, an avid reader, and a staunch Protestant—John developed a forensic skill in religious and political argument, and an intense antipathy to the Church of Rome. He and his father corresponded frequently, and many letters exist from those exchanges. We have fewer glimpses of his relationship with his mother, Sarah, and his older sister Emma.[3] Nevertheless, later in life he spoke of his mother with great affection, declaring that he had rarely seen such gentleness combined with so much courage.[4] It may have been his mother who kindled in him a lifelong interest in poetry. She often recited it to him on their walks together, and from childhood onwards, John loved to commit long pieces to memory to recite to himself while walking.[5] In addition to this maternal influence, John's upbringing was infused with a strong Protestant work ethic and a pervading ethos of manly independence. His escapade with the boat as a teenager is characteristic of a childhood of frequent scrapes. A slight lad who grew into a lean man, John was an agile tree-climber, and was unafraid to box. He developed a strong moral sense as a boy, showing remorse at losing his temper with a donkey, as he thought of its humility and helplessness.[6]

In about 1828 the Tyndall family moved to nearby Nurney, on the slopes above the Barrow valley, and five years later much further afield to Castlebellingham. But by

1836 they were back in Leighlinbridge, and John began attending the National School in Ballinabranagh, 4 miles away. There he was fortunate to encounter John Conwill, a dedicated teacher who, ironically, was a Catholic who came from the hedge school tradition. These offered the rural population in Ireland an education outside the Anglican system, and were held in the open air, or in barns or houses. Despite their religious differences, John's father recognized Conwill's ability. When Richard Bernard, Dean of Leighlin—the Protestant Minister who later became a supportive mentor to Tyndall—objected to Conwill's influence on the boy, Tyndall senior reputedly exclaimed: 'Reverend Sir, if Conwill taught upon the altar steps, I would send my son to him, as I have no doubt but he will receive from Conwill a sound secular education that will fit him for life'.[7]

Conwill, an able mathematician, lived halfway between Leighlinbridge and Ballinabranagh. He and John—who was soon his star pupil—would often walk along the Barrow towpath, solving mathematical problems as they went. Under Conwill's tuition, John studied grammar, algebra, geometry, and trigonometry, building on his own reading through access to his father's books. One formative volume was Blair's *Lectures on Rhetoric and Belles Lettres*, which Tyndall reported later as useful to his writing, giving him what he described as a natural liking for good style.[8] Literature was a common topic of discussion; subsequent letters between the two men are full of literary and classical allusions. Crucially, Conwill also taught Tyndall the rudiments of surveying. They became his passport out of Ireland.

* * *

A Division of the Ordnance Survey of Ireland was based in Carlow.[9] Thomas Colby of the Royal Engineers had established the Irish Survey in 1824. The aim was to map the country at 6 inches to the mile, to help the government assess people's tax liability and their eligibility to vote, which were based on the value of their property. Completed in 1846, the Survey was a large undertaking, with about 2,000 staff by 1840. It proceeded broadly from north to south, organized into five Districts, each split into Divisions. By 1838 the Fifth Division of C District was headquartered at Leighlinbridge, with Irish-born Lieutenant George Wynne in command. Though the Survey was commanded by officers of the Royal Engineers, who reported to the Master-General of the Ordnance in London, much of the work was done by Irish civilian assistants. The widespread education provided by the hedge schools for the Irish 'lower orders' meant that they were generally viewed as better surveyors than the English and Scottish assistants.[10]

Tyndall joined the Survey in April 1839, in his late teens, on a wage of 1s 6d per day (about £36 per annum). It was a higher initial rate than many of his contemporaries, perhaps because of his mathematical abilities and his knowledge of surveying imparted by Conwill. Tyndall started by printing and writing plans for the Castlecomer area of Kilkenny, before working as a draughtsman. As his employers recognized his speed and accuracy, his pay quickly rose, reaching 2s 6d by the end of the year; possibly an incentive to prevent him leaving for the English Tithe Survey once he had trained. In May 1840, as work progressed, Wynne moved his headquarters south-west to the port town of Youghal in County Cork.

In Youghal, a town of about 10,000 and the first place in which Tyndall lived away from home, he shared lodgings with Phillip Evans, a fellow Carlovian, and John Tidmarsh, a Methodist lad from Cork. The three men secured two rooms for 5s a week with McGrath, a Roman Catholic who enjoyed religious arguments. Tyndall's rhetorical abilities, honed by his father and John Conwill, were such that he managed to trounce his colleague William Ginty in a public debate, despite taking the Catholic side. Ginty had left the Irish Survey in Leighlinbridge for the English Tithe Survey in 1839. He had returned that October, joining Tyndall in his lodgings, and the two men formed one of Tyndall's several friendships that long outlasted the Survey. Tyndall was contemptuous of those who were unversed in theology, declaring: 'A man may be a Newton in either the political or mathematical world and still be a child in the ways of religion'.[11] Yet he claimed no deliberate confrontational intent. He assured his father, after a dispute with a Catholic 'bigot': 'I neither seek the combat nor shun it when it comes'.[12] A few years later he went further, writing: 'I have a constitutional hatred of contention'.[13] That seems genuine. His later feuds, and there were many, arose more by accident or self-righteousness than by deliberate intent. But he did not shun the combat, even though it often unsettled him.

Away from home, the young Irishmen found plenty of opportunities for romance. Tyndall revealed little directly in his letters or journal, but others teased him or made occasional references to his love life. In July he had a romantic encounter with a Miss Stedman.[14] His cousin Maria Payne, who wrote quite intimately to him, tried unsuccessfully to extract information about other romances, and asked about a certain Ellen Wall: 'was she not a favourite?'.[15]

By August 1840 Tyndall was out in the field, extending his surveying expertise, though unable to understand the Irish of the farmers he stayed with; the Celtic language was barely spoken around Carlow at the time. A month later, when Wynne left to care for his sick wife, the regime at the Survey changed. Tyndall tried to extract a testimonial from Wynne. But to his disappointment Wynne refused, on the grounds that if he did it for Tyndall, everyone would ask for one. Nevertheless, he told Tyndall that if he were approached there was no one he would more happily support.[16] He relented in February 1842, and provided a reference when Tyndall was considering alternatives to surveying, writing from the Greek island of Zante, where he was now back in service.[17]

Work at the Survey was unremitting. Tyndall was denied leave at Christmas, which he spent with Ginty at the Tidmarsh family house in Cork.[18] Early in 1841 he informed his father that he expected a transfer to England in the summer,[19] but he heard a few days later that additional work would delay the move by a year. The assistants were told that only those who exerted themselves would be selected for the English Survey, said to be more lucrative, even if that rumour would prove illusory. Seen as hard working and responsible, Tyndall earned the nickname 'Corporal Tyndall' in his office, and became the Survey's highest-paid assistant. Nevertheless, he considered asking Dean Bernard and Captain Steuart (a local landowner who had employed his father) to put in a word for him at the Inland Revenue, or at the Bank of Ireland.[20]

In January 1841, in a sign of things to come, Tyndall had his first significant brush with authority. He led a petition to Lieutenant Whittingam, Wynne's successor, asking him not to transfer their immediate boss Corporal Mulligen, who was seen as firm but fair by the civil assistants. Whittingham was not amused, and made it clear that he would not countenance such insubordination again.[21] The next month, as restructuring continued, Tyndall was sent unexpectedly on a field expedition to Mount Uniacke, about 7 miles from Youghal, where he struggled again with the locals' Irish and found it hard to get food in their squalid mud and thatched huts; perhaps a reflection of their intense poverty. Although his understanding was that he would be moved to Cork in April, he was sent instead three months later to Kinsale, a port in County Cork, during the General Election of 1841. There he lodged once more with Ginty, Evans, and Tidmarsh.

While he was away from Leighlinbridge, Tyndall missed two episodes of electoral excitement. In the first, in October 1840, three Protestants, including Captain Steuart (standing as Conservatives), faced three Catholics (standing as Liberals) in elections for the Office of Poor Law Guardians. The roles had been created under the Irish Poor Relief Act of 1838, and gave the incumbents the authority to create workhouses. As Tyndall's father put it: 'the Priests are using every art to have their pets returned. May they be disappointed is my sincere wish'.[22] Steuart (who won the poll) and two of the Liberals were elected. Tyndall was rarely understated in his views on Catholics, or on Irish politics, and declared his political allegiances floridly to his father. Equating democracy with mob rule, he invoked the Terror of the French Revolution: 'When I see such stumps of upstart dignity standing forward and with brazen impudence opposing men who are as much superior to them as the midnight moon to the yelping of the shephard's cur, I already see in anticipation the tragedy of Robespierre acted over again when everything great and noble was made to bend before the withering Simoom of democratic tyranny'.[23]

The second political episode, the General Election of July 1841, was of greater significance, as the charismatic Irish nationalist and champion of the Repeal movement, Daniel O'Connell, campaigned to have the Act of Union repealed. County Carlow had been a political battleground for years, and its two seats regularly changed hands between Conservatives and Liberals. At the recent by-election in October 1840—while Tyndall was in Youghal—to his relief and that of his father the Conservative Colonel Henry Bruen had beaten the Liberal Frederick Ponsonby. Ponsonby was not a Repealer, which Daniel O'Connell blamed for his defeat.[24] Carlow became a focal point of the 1841 General Election when O'Connell's son contested a seat, bringing the full support of his father with him. Tyndall's father, a committed Orangeman— the Protestant Orange Order had been founded in 1796—claimed that O'Connell's people were intimidating Bruen's tenants to leave the county during the election. Bruen expected his tenants to vote for him, and voting at the time was in public. A large mob of O'Connell's supporters invaded Leighlinbridge in early July. According to Protestant testimony, they attacked the house of Tyndall's uncle Caleb. Caleb fired a gun in self-defence, slightly wounding Mary McAssey, a relation of Tyndall's mother, in the thigh. Under a watchful army presence (Colonel Jackson had deployed cannon at a key bridge in Carlow), Colonel Bruen and fellow Conservative Thomas Bunbury were returned in the election, beating the Liberals with majorities of just eight votes. Daniel O'Connell himself lost in Dublin. Nationally, Robert Peel's Conservatives replaced Lord Melbourne's Liberals.

This tense election took its toll on relations between Protestants and Catholics in Leighlinbridge. After Tyndall's father sacked his apprentice, who supported the Repealers, Catholics refused to patronize his shop. Worried about his father's income, Tyndall sent him money.[25] Caleb Tyndall was tried for his part in the disturbance, but a jury of ten Protestants and two Catholics found him not guilty, and in early 1842, he was appointed High Constable for the Barony of Idrone West.[26] Tyndall was attacked in Kinsale during the election by a party of fishermen carrying green boughs, the emblem of the Nationalists, but he managed to buy them off with some beer.[27] His father narrowly escaped a severe beating when a mob of several hundred Repealers from Kilkenny assaulted a friend in Leighlinbridge whom he had tried to protect.[28]

Given his father's politics and contacts, the arrival of a Conservative government increased Tyndall's likelihood of securing a position. He had been studying French in Youghal, and took drawing lessons in September while he still explored if there might be a job at the Inland Revenue.[29] There was also America—at that time the destination for many Irish people, even before the Great Famine. It was a possibility that Tyndall actively contemplated. Not only did he have relatives in America, but William Wright from Leighlinbridge, one of his close childhood friends, had emigrated to Cincinnati after marrying in 1841, and encouraged him to follow suit.[30] Others went south. Several of Maria Payne's family emigrated to Van Diemen's Land—now Tasmania—during this period.

It was in Kinsale that Tyndall began his ventures into print, under one of his many pseudonyms, Walter Snooks. His poem about the conflict in Leighlinbridge—*The Repeal Meeting at Leighlin*—was published in the *Carlow Sentinel* on 16th October. *Carlow*, a tirade against O'Connell, followed on 30th October. The third offering, *The Testimonial*, a heroic take on Bruen's victory, appeared in November. In December came the powerful *Landlord and Tenant*. Poetry was important to Tyndall, and remained so throughout his life. Far more than today, it was a part of cultural discourse and a means of communication between friends. Letters from Tyndall during this period contain much verse, original and quoted, which Ginty and others reciprocated. About seventy-five of Tyndall's poems survive, mostly from his early adulthood. Many are hackneyed and formulaic, yet they reveal Tyndall's values and beliefs. The early ones give a glimpse of a juvenile sense of humour, in stark contrast to his later writings, both private and public, from which humour is noticeably absent. And among his lovelorn friends he was in demand as a versifier to create acrostics. Nevertheless, there are some potent pieces. *Landlord and Tenant* steers an intriguing line between good and bad landlords and happy (Protestant) or wretched (Catholic) tenants.[31] While critical of the slovenly ways of many of the peasantry (who were implicitly Catholic), Tyndall was adamant about the responsibilities, as well as the rights, of the predominantly Protestant landlords, and critical of the behaviour of many of the absentees.

Tyndall was away from home at Christmas 1841 for the second year in succession. At the last minute, he was told he could not be spared; a particular disappointment, as he was hoping to explore possible jobs. On Christmas Day he went to church, which he attended regularly while away. Then with a friend, he took a gun to forage for food to supplement their fare. Failing to obtain anything, and finding themselves hungry and far from home, they begged their way back to their lodgings as house after house refused to give them even so much as a cold potato, a likely indication of continuing widespread poverty as much as lack of charity.[32] In January 1842 Tyndall transferred to Cork and was put to colouring the plan of the city.[33] He told his father that he would be able to visit home after the Irish Survey finished in March, and eventually obtained leave from 1st May to 10th June, having seen about 150 of his colleagues set off on the steamer for England.[34] Ginty, Evans, and Tidmarsh had by now left Cork for England, and finally Tyndall got his own orders. In August 1842, leaving Ireland for the first time, he sailed for Liverpool and the English Survey.

* * *

Tyndall spent much of his first sea crossing on the deck of the steamship *Ocean*, watching the sunset and star gazing. Next morning, he admired the Welsh coast and the Snowdon hills before the ship put in to Liverpool's Clarence Dock. Ginty and William Latimer met him on the quay, welcoming him and Latimer's brother George, who had travelled over too. Ginty had written to warn Tyndall of the impact of the 'modern Babel' that was Liverpool. This second largest port after London was a stunning vision to rural Irish eyes: 'just imagine every cabbage stalk from where you stand, into Carlow and 6 miles round you transubstantiated into masts for ships—every kettle...in the Barony into a steamer...—all the sparrows that you have seen for the last 7 years into Policemen, Excisemen, &c., &c., &c. and every corn-creak in Whyte's toy shop in Cork into a windmill. —these, with the assistance of a fertile imagination...may enable you to form a distant idea of "Liverpool" alias confusion ad infinitum'.[35]

Soon after arriving, Tyndall was sent on to the English Survey office at Preston, where plans were produced of Lancashire and Yorkshire. He found immediately that though living costs were similar to those in Cork, the prospect of earning more money was limited. There had been no increase to his 2s 10d per day since early 1841. Promises and recommendations for increases towards the end of his time in Ireland had proved illusory. Now he was in the hands of Captain Henry Tucker, known for his aversion to increasing men's pay.[36] That stimulated him to seek better paid work. He plotted with his father to obtain a position as a map-maker at the Map Office of the Tithe Commissioners in Somerset House, using the reference from Wynne and identifying all the people who might put in a word for him, particularly with Colonel Bruen.[37] He even found two local landowners who might be prepared to speak to Tucker, but it came to nought.[38]

Tyndall lodged with fellow assistants Phillip Evans and George Latimer. His working day started at 6.30am and finished at 5pm, and in the autumn, he took French lessons for a further 2 hours each evening, so that he could read the work of many 'clever Engineers'.[39] At the Mechanics' Institute in Preston, founded in 1828 and aimed at both working men and professionals, he attended a lecture each week during the winter. This was his first real exposure to a wide range of sciences: mechanics, astronomy, general physics, chemistry, botany, and physiology. He was also struck by the range of cheap publications available in England, including the graphic *Illustrated London News*, first published in 1842.[40] They offered opportunities to a man like Tyndall, who sought to educate himself.

Tyndall's romantic attachments in England are hard to pin down. To one of his colleagues, John Chadwick, he wrote that the 'daughters of merry England' did not please him as much as the Irish lasses. Chadwick responded that though 'the iced wine of Ireland is more palatable than the hot lascivious vintage of England', Tyndall ought to explore the latter. He hinted at Tyndall's sexual experience by calling him 'not much burdened with precise morals—in short not quite a Joseph'.[41] One friend reminded him that he had been 'the centre of attraction of some 5 or 6 of the belles of Cork', and he may have had a soft spot for Christina Tidmarsh, John Tidmarsh's sister. Ginty teased him about Ellen, the 'lady of the raven plume' in Kinsale, who featured in several poems that passed between them.[42] Ginty himself was on the receiving end of a tremendous prank when Tyndall published in the *Preston Chronicle*, under the name of W. Ginty, the poem *On Leaving Westmorland*, narrating Ginty's supposed discovery of his 'false' Mary, a girl he knew.[43] He sent a copy anonymously to Mary, after which there was some concern that her family might see it, especially since her engagement to someone else had just been announced.[44]

By the spring of 1843, dissatisfaction with the English Survey was widespread among the assistants. The mostly Irish surveyors were increasingly unhappy with their treatment: the lack of pay increases, and the manner in which the Royal Engineers were using their public funding to undercut the civilian surveyors working on the Tithe Survey. Despite knowing that he would be dismissed if he was found out, Ginty spoke to the editor of the *Liverpool Mercury*. He told Tyndall that the editor was interested in their grievances and needed the facts. Would Tyndall help?

Tyndall did. Much of the inside information came from a colleague, Archie McLachlan, working in the Survey's Leeds Office. He wrote a long letter to Tyndall, discussing whether a newspaper or pamphlet campaign would be a better way to bring the issue to public attention.[45] Clandestine exchanges followed, principally between 'X' (Tyndall), 'Y' (Ginty), and 'B' (McLachlan). Their initial plan was to encourage those with experience on the Survey to provide evidence of mismanagement. McLachlan invited them to comment on a draft letter to the Master-General of the Ordnance, George Murray, which laid out their complaints: misdirection of public money, and mismanagement leading to discontent, demoralization and unsatisfactory work by the staff. He suggested that Tyndall should lead the process, taking care to ensure that the letter's authorship could not be inferred from its context.[46] Over the next months, Tyndall worked with them to pull the information together.

Meanwhile, Tyndall's attempts at poetry continued. His poem *An Hibernian's Song to ____ ____*, which mocked Daniel O'Connell, was published in the *Preston Chronicle* in May 1843. But he was not so successful when he tried to publish in the *Carlow Sentinel*. The editor, Thomas Carroll, told him that his last submission was 'unsuitable', and suggested he write graphic sketches of public men instead.[47] Undeterred, Tyndall tried again, also unsuccessfully, in July.[48] Friends admired his poetry, and one even suggested he become a full-time poet, but that was never his talent or intention.[49] He restricted himself largely to political and personal matters. Few poems exist from his mature years, at a time when several men of science, such as the physicist James Clerk Maxwell and the mathematician James Sylvester, were more prolific. They used poetry, both in private among friends and occasionally published, to explore and present their scientific ideas.[50]

Tyndall spent a week out of the office at the end of June, checking surveyors' measurements, followed by a fortnight in the village of Goosnargh, about 7 miles from Preston, with his colleague Billy Marquis. The whole experience was invigorating; in the country alone Tyndall found 'the mind unshackled'.[51] The visit was further enlivened by the attractions of Lizzy Barton, the youngest daughter of his host, keeper of the General Elliot pub, for whom Tyndall acquired a passion, in competition with Billy. But his future was still uncertain. Nothing had come of his Irish contacts and he was still considering his options in America.

Tyndall's political ideals were severely challenged by conditions in England. A few days after his arrival in Preston (Figure 1)—where the average age at death was just 19 in 1844[52]—he witnessed political violence on the streets. Following a mass meeting, a strike had been called in Preston's cotton factories, to demand better pay and the democratic rights outlined by the Chartists. A huge crowd, preventing some workers breaking the strike, confronted police and soldiers, who opened fire after the crowd refused to yield, killing four demonstrators. Nearly fifty years later Tyndall recalled this vivid moment in his recollections of Thomas Carlyle and the misery that drove Carlyle's writing at the time, as Carlyle railed against the treatment of workers, which he compared to the conditions of slaves.[53]

Although Tyndall approved of the stand the authorities had taken during the Preston riots, the conditions of workers in England, and the treatment of civilians on the Survey, who were mostly Irish, soon hit home.[54] In June 1843 he read Carlyle's *Past and Present*, finding in it 'a morality so righteous, a radicalism so high, reasonable and humane, as to make it clear to me that without truckling to the ape and tiger of the mob, a man might hold the views of a radical'.[55] Tyndall flirted with radicalism. In 1846, commenting on the political position of a friend's father that he noted was radical, he thought it 'the natural result of an observant nature'.[56] But he never went the whole way. That he did not become a thorough-going radical (though he termed himself instinctively a liberal) may be partly due to the lasting influence of his father, but the stiffening of his sinews must have been heavily influenced by Carlyle.

Fig. 1. Preston in the mid-nineteenth century.

Thomas Carlyle, historian, polemicist, and social commentator, is almost forgotten today. Yet in his time he was one of the most visible and influential public intellectuals, provocative to a fault and lionized at dinner parties. Tyndall was profoundly inspired by him. Carlyle was no democrat, nor was Tyndall at any stage of his life, but the inequities in society and the abuse of power raised his moral ire. Ginty wrote to Tyndall that he himself was almost a Repealer, and wondered if Tyndall and Evans— had they been born Catholics—would have been too.[57]

Tyndall's sense of injustice about the Survey of London in particular spurred him to write the first of what became five open letters to the Prime Minister, Sir Robert Peel, based on the information that he, Ginty and McLachlan were gathering. It was published in the *Liverpool Mercury* in September 1843, under the pseudonym 'Spectator'.[58] The letter started by deprecating the fact that the Survey was directly allocated to the Royal Engineers, rather than put out to open competition. It continued by illustrating their huge cost overruns and unfair competition against the civil surveyors, and finished by laying bare the miserly wages and poor conditions of service of the Irish assistants. The stakes were increased when Tyndall and twenty-four of his colleagues wrote directly to George Murray, the Master-General of the Ordnance, explaining their refusal to sign a form about terms of employment presented to them.[59] To save money, the Survey was reducing staff numbers by seeking voluntary repatriation of some to Ireland, with the rest to remain at current rates of pay. This situation Tyndall and others considered inadequate and unfair. The letter was handed to Captain Tucker on 23rd September. But it was not sent on, and after a meeting between the men and Tucker in October, they wrote to him with a further missive for the Master-General, reiterating their case.[60] This letter reached Colonel Colby, the Superintendent of the Survey, who prevented it landing on the Master-General's desk. After this further rejection, Tyndall published four more letters as 'Spectator' in the Liverpool Mercury in October and November, incorporating detailed advice and calculations from McLachlan. The blow fell on 7th November, when Tucker gave notice that nineteen assistants and fifteen labourers would be dismissed a fortnight later. The dismissals were not simply of the poor performers. Tucker took the opportunity to remove the trouble-makers—Tyndall, Evans, Ginty, and George Latimer among them.[61]

Tyndall now needed work, and looked for references. Dean Bernard provided one, but Colonel Bruen would do nothing for him, considering him 'the tool of some designing villain'.[62] His father was equally critical, telling him to submit to authority and return to work. Tyndall vigorously defended his actions, arguing: 'the cause which I espoused was worthy and the course which I adopted was open, manly and respectful'.[63] He disabused his father of any notion that he was acting against the government. Taking his last week's wages, and a little money from a fund the assistants had created for themselves, he set off by train to London to find work, while entertaining the ambitious hope of laying his case personally before the Master-General.

The sights of London impressed the country boy, who stayed at the Robin Hood Tavern in Holborn: 'Well, surely extremes meet in London—wealth and poverty, beauty and deformity, florid health and pale disease, all congregate together in this mighty city'.[64] Tyndall did not manage to see the Master-General, but was advised instead to write to him. He did so, enclosing testimonials from Wynne and Dean Bernard. Even Tucker, to his credit, had provided a good reference, giving his reason for dismissing Tyndall as the need to reduce numbers, even though he had dismissed many of the best assistants while retaining poor performers.[65] Tyndall wrote to the Master-General again a week later, documenting the skills possessed by himself and his fellow sacked colleagues.[66] After the dismissal had been confirmed by Colby, Tyndall wrote once more, explaining why he had disobeyed the order to sign the paper drawn up by Tucker, and protesting that he had at no time attempted to force an increase in pay.[67] He had merely sought to draw the attention of his superiors to the unsatisfactory conditions, naïvely imagining that they would see the error of their ways.

Tyndall spent a fortnight in London. Contacts were friendly and supportive, but there was no work. He was advised that there would be better prospects in Ireland given the unemployment among English surveyors. He even placed an advertisement for himself in the *Times*, which failed to produce any responses, nor did a speculative letter to the engineer Sir John Rennie. Finding himself without enough money for the fare back to Liverpool, he borrowed 10s and returned third class, using what money was left over to buy a surveying chain and drawing materials to take back to Ireland. Tyndall made the crossing to Cork with Phillip Evans, and by the time they had paid for the coach to Carlow and Leighlinbridge, they had only a halfpenny between them. They couldn't even afford breakfast.

*   *   *

Back in Ireland in December 1843, Tyndall looked for work as a surveyor or draughtsman. He gained small commissions through friends, including Elizabeth Steuart, wife of Captain Steuart, who corresponded with Tyndall for the next fifty years. He did not let go either of poetry or political agitation, though the *Preston Chronicle* now shared the view of the *Carlow Sentinel*; it published just a stanza from Wat Ripton (another nom-de-plume) in January, commenting that his poems 'discover considerable poetic feeling, and no small skill in the art of weaving verse...But they have one capital fault—they are too high flown: sense being, in many instances, sacrificed to sound'.[68] The fifth letter to Sir Robert Peel appeared in the *Liverpool Mercury* in February, restating the arguments about the unjustifiable low pay and poor conditions on the English Survey.[69] A sixth was planned but never published.

Tyndall's awareness of inequalities of birth had been honed in England. In March 1844, while helping his old teacher John Conwill at Ballinabranagh School, he was impressed by the academic abilities of the boys. But he remarked: 'I could scarcely repress a sigh when I thought of the extreme probability of each and every one of the boys who composed the class dropping unheeded into the grave and leaving no name behind—it might be safely predicted that such would not be the case if they got fair play'.[70] He now laid firm plans for America, the land of opportunity, telling his father: 'The thought of going to America has more or less occupied my mind for years, but within the last nine months it has become a settled determination on my part'.[71] His cousin William Tyndall, already in America, advised against going because so many surveyors were unemployed.[72] Wright reinforced this, saying that there were thirty applicants in Ohio for every job, although he hoped that Tyndall would come out in any case, because there were other opportunities for enterprising men.[73] Tyndall was not disheartened, especially as Wright had mentioned a shortage of draughtsmen, a position for which he considered himself well qualified.[74] He determined to go out in April 1844, against the advice of the Dean, sailing on the *Thomas P Cope* from Liverpool. He hoped that George Latimer would travel with him, but at the end of March, Latimer wrote apologetically to say that he had changed his mind.[75]

The situation altered when Tyndall's uncle Caleb suggested that he make a proposal to survey County Carlow for a land valuation. With a day's notice, Tyndall submitted sample drawings, bid £231, and found himself up against the County Engineer in front of the Grand Jury. To his surprise, he won. The work had to be completed by November, and financial security was immediately promised by Captain Steuart. But a few minutes later, a juror announced that the election was informal, and would not stand. Shortly afterwards Tyndall had an interview with Colonel Bruen, who informed him that sending the letter to the Master-General had been a most injudicious act. He also urged him not to go to America.

By the end of April, given the continuing opportunity of the Carlow survey, Tyndall had decided against crossing the Atlantic. The Carlow contract was due to come before the July Assizes, and Tyndall wrote to Tucker to ask for a recommendation for his experience, steadiness, and fidelity in discharging his duties, telling him that 'the great majority of the Grand Jury are favourable to me personally, but they consider experience indispensable towards the proper execution of the applotment, and measuring my experience by my years, they infer that it must be limited'.[76] Tucker replied that he could not alter a word of his previous reference. Unfortunately, Tyndall was unable to use it. When he thought he was going to America, he had given it to Evans, forging the name, because it was a better reference than the one Tucker had written for Evans.

In early June Tyndall did the rounds of influential people to canvass support for his bid, and lined up Ginty and Latimer to work with him. There were ten other competitors. But the Grand Jury received legal advice that it was not competent to appoint, and Tyndall's chance was gone. Captain Steuart thought the Jury had made a mistake, but Tyndall suspected villainy and the desire of some people to prevent the survey from proceeding. He had not been impressed by the jurors: 'what an admixture of meanness and cunning, aye and self-complacency arising from the consciousness of power'.[77]

He continued to find sporadic work, while maintaining his drive and desire for self-improvement. Still studying French, he also delved into logarithms, and read widely, including Chaucer, Shakespeare, Spenser, and Sheridan. He was a regular reader of the *Mechanics Magazine*, a popular scientific weekly directed at working men, and wrote a letter to it giving intriguing, if naïve, objections to Newton's particle theory of light. For example, he suggested that if a candle were lit in a dark room, emitting particles of light, then light must gradually increase: 'it will be sixty times as light in an hour as it was in a minute'.[78] One article, describing Newton's commitment to 'patient investigation', influenced him strongly. Tyndall commented on the impact of real study in developing proper understanding, and resolved to continue like that himself. He drew, painted, and spent many days writing a 'tale' that seems not to have been published. In the garden, meanwhile, he sowed onions, turnips, and lettuce.

At last his luck changed. With the help of a colleague, Bob Martin, he received an invitation from Thomas Wren to run a survey office in Preston. He wrote to suggest a trial employment at a salary of £1 10s per week—about £75 per annum—and Wren agreed.[79] So at the end of August, after an extensive visit to the engineering marvel that was the new atmospheric railway at Dalkey, Tyndall sailed from Dublin, wondering when he would next return.

2

RAILWAY MANIA

1844–1847

For the second time in his life, Tyndall arrived by sea at Clarence Dock in Liverpool. As before, he was soon on the train to Preston; and on his arrival he was welcomed by his friends Bob Martin and Bob Allen. Now in his early twenties, Tyndall had a spare, athletic figure; as an adult he weighed about 10 stone, and stood under 6ft tall. In his lean, pale face was set a 'prominent nasal organ very largely developed'.[1] His brown hair may have had a tinge of Celtic red,[2] but there was no sign yet of the Victorian beard that became a later feature.

Britain in the mid-1840s was on the cusp of 'railway mania', a period of boom and bust in railway construction that changed not only the face of the British countryside but also patterns of trade and travel. It was a cutthroat, largely unregulated business. Before rail companies could acquire the necessary land for their routes, they had to obtain parliamentary approval, but there were few checks on the financial viability of their proposals. In 1846, at the height of the mania—and of British railway share prices—Parliament passed 272 Bills, authorizing 4,500 miles of railway. Much was never built, as companies merged or collapsed.[3]

Thomas Wren was the agent for the proposed Liverpool, Ormskirk, and Preston Railway. Business was slack in his office when Tyndall arrived, and the newcomer experienced some jealousy from the existing staff at his appointment. Perhaps for those reasons the arrangement did not turn out as expected. Tyndall parted company amicably after just a week, and travelled to Manchester to hustle for work. Within days he was hired at 3 guineas per week (4 in the field) by the surveying firm Nevins and Lawton, where Billy Marquis also had a job. He tried to secure positions for Ginty and George Latimer. Ginty had just found a position in northern Ireland, but Latimer came over to Manchester.

Tyndall was first sent to survey in Staffordshire near Tutbury, and then around Bedford for the proposed Bedford and Ely Railway. His task, with theodolite and surveying chain, was to measure the 'levels', or height of the land, to find the best route for a railway, for example by minimizing gradients or avoiding expensive embankments. He was there until mid-October, out in the countryside most of the time, where the landscape was cut by hedges, brooks, and lanes, making surveying difficult. The fields and woods were alive with game. He startled a great number of partridges and hares, and sympathized with the plight of the coursed hare, asking himself: 'What must the poor little thing's feelings be when she felt the very breath of the murderous hounds behind her?'.[4]

Back in Manchester, Tyndall was put to work on the Chester, Stockport, and Manchester Railway, sharing the load with Latimer. The period leading up to 30th November, the deadline for railway companies to submit their plans for parliamentary approval, was a hectic one. Tyndall found himself caught in the crossfire as the various railway companies fought to outdo one another. Unwell, and plied with 'coffee, brandy, soda water, meat and fowl', he experienced his first all-night panic as the documents neared completion.[5] Nevertheless, amid the surveying, there was time to admire the local girls. In Cheshire one stood out. His journal reports: 'Among the Lancashire witches I have not met one in her station to match this fair Cheshire girl—her forehead was white and ample, her brow like the dark streak of an artist's pencil, her eyes soft black with an intellectual expression approaching almost to pensiveness, her cheeks rather pale but slightly tinted with the sun's influence, her lips contained the sweetness of a cherub's—indeed she was passing fair'.[6]

But work at Nevins and Lawton was diminishing. Sensing this, Tyndall applied, unsuccessfully, for the surveyorship of the Borough of Salford. His fortune turned in mid-December, when he received instructions to work with Richard Carter, surveyor of the proposed West Yorkshire Railway, based in Halifax. For most of the next three years, the textile town of Halifax would be his home.

After a short break for Christmas Day (which he spent in Manchester), Tyndall was put to work, surveying at Low Moor, south of Bradford. The beautiful moors around Halifax and Bradford were rent by the immense iron works and coal mines of this rapidly industrializing landscape: 'The whole extent was covered with huge mounds of cinders and studded with coal pits, cinder ovens and furnaces. At night the latter were particularly imposing—for miles round their belching might be heard and their red jaws seen pouring forth volumes of flame'.[7] Occasionally, the buildings themselves went up in flames. In July 1847 Acroyd's Mill caught fire, illuminating the whole valley, and putting about 2,000 people out of work. Tyndall tried to marshal the water carriers, but the task proved too much for him.

A surveyor's life was mobile, travelling from place to place and finding lodgings where he could. Nights were spent in public houses that ranged from the delightful to the vile. In the new year of 1845, Tyndall came upon the Moravian settlement at Fulneck, near Leeds, and stayed there in peaceful, clean surroundings while he levelled the countryside nearby. The frequent biting cold often brought freezing fog that deposited a layer of ice on his instrument's eyeglass each time he made a measurement, and froze his leggings solid. Problems with the weather could be compounded by hostile landowners. On Colonel Tempest's estate, at Tong Hall near Bradford, the gamekeeper appeared with a shotgun and ordered him off the land. But Tyndall had a knack of establishing friendly relations with the landed gentry, in England as he had in Ireland. He later smoothed any remaining discord by giving the mollified Colonel a lesson in levelling. His style stood him in good stead later as he made his way in Society. As he described it, 'frankness and fearlessness, provided no trace of vulgarity or rudeness appear, goes a great way with such people'.[8]

In February, he worked through deep snow along the line between Low Moor and Copley Mill, crossing the frozen summit of Beacon Hill above Halifax: 'far beneath you, in the deep darkness, a thousand lamps twinkle like stars, while the illuminated factories, with their hundreds of windows each, appeared like so many constellations'.[9] A poem about the landscape, *On Beacon Hill*, probably dates from this period. Once the surveying was complete, office jobs followed. First for Carter, then back with Nevins and Lawton, where he worked on plans for the Churnet Valley Railway and tested levels for the proposed Manchester and Salford waterworks. Visiting the wild Cheshire moors in frosty weather, he admired the nine-arch Ashton viaduct.[10] He took the opportunity to visit the Mechanics' Exhibition, where he encountered a wide-ranging presentation of paintings, armour, 'Eastern splendour', and phrenology. One exhibit, a medallion of Wellington, elicited both loud clapping and hisses from the assembled visitors—eventually 'ancient prejudice at last prevailed', and Tyndall cheered too, revealing an instinctive politics.

On his arrival in Preston, Tyndall had resumed his religious and intellectual explorations. He attended the Wesleyan Chapel, and in complete contrast, heard Emma Martin, the radical feminist, speak twice, first on 'Christianity Detrimental to Human Happiness', and subsequently on 'The Inutility of Divine Worship'. He was impressed by Martin's fiery style: 'never have I heard a bolder specimen of the infidel tribe…one poor fellow made an attempt at opposition but was soon floored'.[11] Although he thought her arguments could be successfully opposed, he didn't try. In Manchester he sampled two approaches to religion, the Irish Presbyterian and the Wesleyan, and one Sunday in Halifax he visited both the Anglican Trinity Church and the South Parade Methodist Chapel, noting their diametrically opposed doctrines on the baptism and apostolic succession.[12] He thought the Methodists too literal in their interpretation of scripture, commenting with a latent physicist's amusement on their concept of hell as a place of the blackest darkness filled with devouring flame. Later, in the Trinity Church, he 'could not subscribe to the preacher's doctrine of human depravity'.[13] In a lengthy response to a letter from Bob Martin, who had 'got religion', and who claimed to prove from Scripture the natural depravity of man, Tyndall gave his uncompromising opinion of those who 'bind men on pain of eternal damnation to believe in them'.[14] Setting out his own beliefs, he felt that principles within him were constraining him to a love of order, beauty, and goodness: 'A noble sacrifice of self in the cause of humanity is, I am persuaded, the principle on which depends the final triumph of justice and truth'. This he saw embodied by Jesus, who circumvented the 'formulas' of religion.[15]

The work with Nevins and Lawton dried up at the end of March, but Tyndall was now in a good position to find alternatives. Ten minutes before closing on an engagement with the surveyor of the Manchester and Salford Waterworks he received a note from Carter asking that he join him. They agreed on £175 a year plus expenses, and a few months later Tyndall was given authority to agree terms for Ginty's employment. For much of the time he was with Carter, he was checking levels as part of a dispute with their competitor, the Manchester and Leeds Railway. He was even arrested once, after crossing the rival's line, though he wriggled free from his captor.

A visit to London with Carter in May for preparatory work for a parliamentary submission gave Tyndall an opportunity to sample the cultural life of the capital. He took in Bellini's opera *La Somnambula* at Drury Lane, visited the SS *Great Britain* three days after Queen Victoria, and tasted the high life at the Star and Garter Hotel in the 'paradise' of Richmond, enjoying a sumptuous meal that included salmon, stewed eels, roast mutton, and copious draughts of champagne, port, and sherry.[16] He visited spectacular exhibitions at the Colosseum (on the east side of Regent's Park), the venue designed by the architect Decimus Burton, who also designed Marble Arch and the Athenaeum Club. An acre of canvas in the dome of the Colosseum gave a panorama of London as seen from the top of St Paul's cathedral. Tyndall the tourist was suitably impressed by the Swiss cottage and alpine scene, complete with chamois, and the chill Cavern of Adelsberg with its crystals and stalactites. He visited St Paul's and scratched his name on top of hundreds of others in the Whispering Gallery.

In June he was back in London to help give evidence to a House of Lords Committee.[17] In the Committee Room, the contending engineers included two of the most famous of the period, Robert Stephenson and John Hawkshaw. Hawkshaw, chief engineer of the Manchester and Leeds Railway, argued that steep gradients were possible with modern locomotives and, in opposition to Stephenson, supported the line between Low Moor and Bradford, despite its 1/70 incline and enormous quantity of traffic. In the evenings Tyndall had more time for sightseeing: to Covent Garden to hear the

Brussels Company; to the Lyceum for *Cinderella*; to the Royal Academy to view Landseer's *Peace and War*; and to St James's Park for a stroll, during which he saw goats pulling little carriages containing infants. After three days of committee hearings, Tyndall noted that 'the West York had defeated its Titan foe—both schemes were now extinct and the promoters stood on an equal footing'. They had fought each other to a standstill. Nothing was yet settled. Carter was now asked by the promoters of the two enterprises to re-survey and lay out a proposed West Riding Union Railway, which aimed to connect Leeds, Bradford, Halifax, Huddersfield, and Dewsbury.[18] Caught up in the railway fever, Tyndall applied for twenty shares in the Huddersfield, Halifax and Bradford Junction Railway, and in September, he acquired ten shares in the Chelmsford and Bury Railway. Later, he sold them on, unable to live with the financial risk.

Meanwhile, Tyndall was breaking away from his strict Protestant and conservative roots. In politics, his observation of the poverty and exploitation of factory workers in Lancashire and Yorkshire challenged his initial support for the suppression of the Chartist-inspired protests he had seen in Preston. Realizing that poverty, rather than Roman Catholicism, seemed to underlie the antipathy of the rural Irish to their Protestant landlords, he became less partisan. On 12th July, he no longer celebrated Orangeman's Day, which marked the Glorious Revolution of 1688 and the victory of the Protestant William of Orange over the deposed Catholic King James II at the Battle of the Boyne in 1690, when once 'the clamour of gunpowder boomed like celestial music on my ear'.[19] In religious matters he challenged the unquestioning acceptance of doctrine and literal interpretations of the Bible, advocating mixed scientific and religious education to two Catholic priests, who argued with him that all education should be based on religion.

He was eager for intellectual stimulation, and for knowledge as a means to his betterment through self-education. In July 1845 he joined the Halifax Mechanics' Institute, having benefited so much from the corresponding institution in Preston. He read George Combe's *Outlines of Phrenology*, that widely discussed 'science' of the time. His topics of conversation ranged from politics and religion to science and the beauty of nature. With Carter's daughter he discussed the intellectual power of men and women, recording: 'Miss Carter assumed their equality—I was so ungallant as to support the negative'.[20] His lack of gallantry was not contrived. He often exhibited surprise at women's intellectual capabilities, and though he imagined that women could understand anything revealed by the savants, he did not believe they had the same powers of imagination and discovery.

At the end of July he had a few days' leave. Though he claimed to have had no intentions towards Lizzy Barton, he was so devastated to hear of her marriage to an apothecary that he changed his plan of going to Goosnargh.[21] Instead he went to

Preston, then on to Lancaster, and by canal to Kendal. After exploring the ruins of Kendal Castle, with its glorious view of the fells, Tyndall walked the 17 miles to Wordsworth's house in the pouring rain—though it was something of a pilgrimage, he did not seek a meeting. Returning from Ambleside to Lancaster the next morning, he took a seat in the coach just vacated by the social theorist Harriet Martineau, whose reputation he already knew.

* * *

In late August a tall young man arrived in the Halifax office to take up a job as an assistant. Thomas Hirst, who would become Tyndall's closest friend for nearly fifty years, was 'a youth upwards of 6ft high and 16 years of age—an immense development of the brain which is in true keeping to his extraordinary power of thinking' (Plate 2).[22] Gradually, over the following months, they developed a close relationship within the wider group of surveyors. From his circle of old friends, Tyndall had persuaded Carter to hire Jack Tidmarsh; Ginty was nearby in Ripon; and Latimer was back with Nevins and Lawton. For the rest of August and into November, Tyndall often worked 10–13-hour days, six and sometimes seven days a week, all around Leeds, Bradford, and Halifax. One Sunday in November, he worked 19½ hours, and on Friday 28th November, he worked round the clock as the parliamentary deadline approached.

In the New Year of 1846, there was still no respite. As he corrected and coloured plans, occasionally having to leave the office in foul weather to check levels, Tyndall filled his diary with short entries. The reason for the burst of activity was a further parliamentary hearing in London in February, which was preceded by the customary theatrical delights and some heavy drinking. Almost 200 allegations had been made against the West Riding Union Railway, and although Tyndall's levels were corroborated and 'our opponents were vanquished', the respite did not last long.[23] As Tyndall learned from a report in the *Times*, their petition had not been approved.[24] Although he was incensed ('such a flagrant case of mislegislation was never known... Our cause has been defeated and private influence may be thanked for our disaster'), officially the mishap was the result of a clerical error. The petition was sent back to the Committee, giving Tyndall and his colleagues another chance. Before returning north, Tyndall glimpsed Queen Victoria, who from her coach 'fixed her clear blue eyes right upon Mr Carter, Mr Ellershaw and myself, and acknowledged our bows with a gracious nod'.[25] He took in a concert at the Adelaide Gallery and, like many journalists of the time, was transfixed by a woman who posed to recreate celebrated statuary. According to Tyndall she was 'the brightest specimen of human beauty that my eyes had seen or my imagination formed'—or at least she was until he encountered the Duchess of Sutherland a month later, 'one of the most beautiful women I ever saw'.[26]

Over the next few months, Tyndall—now clearly well fed, as he weighed 10st 4lbs, 11lbs more than during his recent visit to Preston—was often in London, trips that he interspersed with visits north to check levels. His evening theatre entertainments became less highbrow; at the Lyceum he saw the 'curious freak of nature General Tom Thumb, 15 inches high' (actually 25).[27] Science featured more in his activities. He took in lectures at the London Polytechnic, debated the implications of the latest geological findings for the Mosaic account of creation, discussed the nebular hypothesis for the formation of the solar system, and made notes on matter and forces. He also revealed in his journal some early thoughts on the relationship of the brain to the mind. He argued that 'the brain was the organ which manifested the will and working of the immaterial mind'; that this was its sole medium of communication with external nature; and that if this communication were destroyed the manifestation of the mind ceased, though the mind itself might still live.[28]

Finally the time had come to face the Select Committee again in Parliament. Tyndall arrived in London on 29th June, the same day that Prime Minister Peel resigned, following his defeat over the Irish Coercion Bill, a piece of legislation that would have allowed rule by force in Ireland during times of unrest. Peel had fared better when his earlier Bill to repeal the Corn Laws had been passed, despite opposition from members of his own party who had defended the interests of agricultural landowners who wanted to keep the price of corn high.

Tyndall and his colleagues unfurled their large maps in the Committee Room 'like flags of defiance'.[29] Hawkshaw, now in charge of the venture, was examined at length. The stakes were high. The Committee's decision could sanction an expenditure of £2,000,000. The next day, as the Leeds and Bradford Railway opened, the Committee found in favour of Hawkshaw's team, and against the claims of the Huddersfield and Manchester Railway.[30] George Latimer was on the losing side. Tyndall sympathized, but celebrated by watching the fireworks in Vauxhall Gardens. The furious work was over, and the West Riding Union Railway Bill received Royal Assent on 18th August. In November the company was absorbed into the Manchester and Leeds Railway, which subsequently changed its name to the Lancashire and Yorkshire Railway Company in 1847, finally becoming a component of the giant London, Midland, and Scottish Railway in 1923.[31] Tyndall helped with the office work following the decision, and did some re-measuring in the field.

Politically, Tyndall continued to engage with more radical ideas. He bought the first number of Douglas Jerrold's *Weekly Newspaper*, describing its radical anti-Tory agenda as 'a rallying point and rock of defence to those who most need both'.[32] He was particularly impressed by an article entitled 'The Last Hours of the League'—written by Elihu Burritt, founder of the League of Universal Brotherhood in 1846 after witnessing the suffering of the Irish peasantry—and the moral crusade it advocated, which called for the non-violent overturning of current structures to enable freedom, peace, and fair prosperity for all. He was also intrigued by articles about the Whittington Club, which was soon to be formed on Jerrold's initiative. This club, with an educational character, was aimed at the lower middle classes, and offered full membership and participation to women. Keen to join, Tyndall wrote to the secretary to ask if geography precluded his membership, and later discovered that non-residents could pay a fee of half a guinea followed by an annual subscription of the same amount—which he thought very low. He joined towards the end of the year, and renewed his membership in February 1848 from Queenwood College. The experience stimulated him to write four letters to the *Halifax Guardian* in April 1847—under the pseudonym 'A Progressionist'—addressed to the clerks, assistants, mill-owners, and employees of Halifax, urging the formation of a literary institution like the Whittington Club.[33] He was in tune with a broadly radical agenda of human progress through education for the masses. That agenda never left him.

While in Halifax Tyndall heard a series of speeches by Henry Vincent, the radical promoter of the cause of universal suffrage, who had already been imprisoned twice for his inflammatory speeches. Vincent spoke over several evenings to huge audiences, composed of people from all social classes, about temperance, moral education, and the education of women. According to Tyndall: 'Miss Carter must have been delighted—his reference to the natural position of women must have tickled her fancy'.[34] Tyndall approved of Vincent's moral exhortations: 'Two navvies sat near me…their unsophisticated souls could be seen shining through their countenances as they hung with the apparent helplessness of infancy upon the speaker's lips'.[35]

As Tyndall explored the intersection of education, religion, and science, his reading widened. He absorbed two of George Combe's works: *Remarks on National Education*, which argued that the State should take an active responsibility in ensuring secular and moral education for all, but should not interfere in the personal matter of religion; and *On the Relation between Religion and Science*, which maintained that the natural laws governing God's Creation on Earth were revealed through science rather than religion. As Tyndall declared: 'I am no churchman—I would break the connexion of Church and State tomorrow'.[36] He read Combe's 'common sense' philosophy with interest, though he does not seem to have given his phrenology much credence.[37]

For some time, Tyndall had been receiving news from Ireland that his father was not well. Obtaining leave from Carter, and making the trip back with Jack Tidmarsh, he passed first through the huge Free Trade celebrations in Manchester,[38] then on to Liverpool, from where he made a wet crossing to Dundalk. He appears to have chosen Dundalk so that he could make a nostalgic trip to Castlebellingham, where he visited the school and church, unrecognized by anyone from ten years before. From the coach he saw signs of the potato blight that led eventually to the Great Famine: 'The crops looked very luxuriant but the plague was visible among the potatoes. Some fields were quite black and crisped'.[39] He travelled by train from Drogheda to Dublin and on to Carlow, from where he took a coach to Leighlinbridge. Arriving at his father's bedside, he found it surrounded by 'lugubrious countenances'.[40] He spent two weeks with his father, procuring medicine and trying to make him comfortable. But eventually news came from Carter that the Bill for the railway had been passed, and Tyndall was forced to return. Sailing on the *Duchess of Kent* from Dublin to Liverpool, he was able to view old Snowdon's craggy summit from the ship's deck.

* * *

Back in England, Carter's position was precarious. In mid-September Tyndall observed that Hawkshaw 'had snatched the survey of the Huddersfield line from us'.[41] The next day he heard that Carter had been removed as surveyor, according to one colleague because of 'the influence of deep and long working treachery'.[42] This meant that Tyndall was likely to be looking for work himself. At the end of the month the crisis arrived, but while Jack Tidmarsh had to leave, Tyndall was able to stay, and on 21st October he witnessed the grand ceremony that accompanied the raising of the first sod of the West Riding Union Railway in Halifax. The sod was cut by Sir Charles Wood, MP for Halifax and Chancellor of the Exchequer in the Whig administration formed by Lord John Russell following Peel's resignation. The freshly cut earth was placed in a barrow held by Lord Morpeth, MP for West Riding, who ceremonially tipped it out again.[43] Tyndall was thoroughly impressed by Lord Morpeth, whom he heard speak the next evening at the Manchester Athenaeum. Morpeth, who had welcomed the very first meeting of the British Association to York in 1831, had supported Tyndall and his colleagues in Parliament. Talking of the findings of modern science, Morpeth expressed his hope that the spirits of the men who made these discoveries recoiled 'with modest awe, instead of swelling with self-sufficient pride', and gave 'fresh reasons to be reverent, acquiescent and lowly'—sentiments of which Tyndall approved.[44]

In Ireland the famine, and economic situation, were becoming dire. For Tyndall, part of the blame lay with the indolent peasant Irish. But it was the often-absentee landlords that he felt should shoulder the lion's share: 'In prosperity he was a tyrant, ignorant of the duties of his position...in his adversity he has shrunk into the whining mendicant, abandoning all self-reliance and imploring the aid of government'.[45] Tyndall was clear that Repeal of the Union would not solve Ireland's woes. The poor man, he argued, 'would be compelled to fight the title aggressor at home even as he has been battled against in England',[46] although in a cryptic note added to his journal on 24th February 1848—that year of European revolution—he implied that he had

26

modified his convictions. If so, it seems his was a short-lived radicalization. Tyndall's view was that the government should arm the peasants with mattock and spade and let them loose on the uncultivated land. For him, property ownership came with duties as well as rights, and if landlords could not supply the capital for cultivation, then the government should. Nevertheless, he thought that the government's plan to relieve Ireland's distress was bold and comprehensive, with a Board of Works to employ people in work of public utility.

Tyndall was not idle in the face of the trouble. Conscious of the need to alleviate poverty, he addressed letters to the surveyors of Wakefield, York, and Liverpool, urging them to give 8 hours wages for their starving countrymen—but the idea did not catch on. Then the Repeal movement lost its leader when 'The Liberator', Daniel O'Connell, died unexpectedly at Genoa in May 1847. Although Tyndall did not share O'Connell's politics, he did recognize his achievements: 'Though my sympathies have not lately connected me with O'Connell, I cannot contemplate his extinction without a feeling of exalted sorrow—almost akin to awe'.[47] Regarding O'Connell as a superb agitator, Tyndall believed it unlikely that Ireland would ever have the same devotion for another such figure.

Work became patchy. Before Christmas—which Tyndall enjoyed with the Gintys despite an invitation to Preston—he took on a job in north-east England, surveying from Ripon towards Durham. For the first half of 1847, he had a different commission, working around Pye Nest, the estate of Captain Henry Edwards near Halifax. Tyndall was impressed by Edwards and the way he treated his staff despite being a 'true blue'.[48] It was while working at Pye Nest that Tyndall received a communication from Conwill urging him to come to Ireland quickly, as his father was sinking fast.

On his arrival in Dublin he met 'starving shivering beings; sitting on the door steps and clinging to the railings . . . woeful indeed are the privations endured by the people . . . it is most unnatural to see men idle and land wanting cultivation'.[49] He was incensed that land was uncultivated while people were starving, for what he saw as the lack of loans from the Bank of England, and he wrote a letter to the *Sligo Champion* castigating the officials in Sligo for their response to the crisis.[50] He was equally enraged at seeing foxhunters dashing through poverty-stricken streets: 'these red-coated gentry drive me to despair . . . they are cursed with a moral leprosy'.[51] Having spent a fortnight in Leighlinbridge, watching his father become weaker and weaker, he eventually had to leave, fearing that he would never see him alive again. His father died on 27th March, three days after Tyndall left: 'one of the strongest links between me and life is broken . . . I knew it would be thus, tears, bitter, hopeless tears'.[52]

At the end of April, possibly because of his father's death, Tyndall suddenly became unwell—'completely broken down', in his own words—and asked for leave.[53] This may have been the first of several depressive episodes that struck throughout his life. Carter's daughter advised him to go to Ben Rhydding, the famous hydropathic establishment that her father patronized, but Tyndall chose instead to go to his old haunts at Preston and then on to Goosnargh. In Preston, spending a night with the Allens, he admired the magnificent New Literary and Philosophical Society, and passed judgment on the new Mechanics' Institute, which he thought would be a superb building. He then travelled, with Marquis, to Goosnargh, where '"our peer-less Lizzy" opened her dark fringes in almost incredulous wonder'. Reports of her marriage had been false.[54]

Back in the welcoming atmosphere of Goosnargh, Tyndall started to feel better. He read extensively, including radical poets and writers: Shelley's *Queen Mab*, which drew parallels for him with Byron's *Cain*; and Carlyle's *Chartism*, which Tyndall described as 'A noble production. I...thank the gods for having flung him as a beacon to guide me amid life's entanglements'.[55] He spent nearly three weeks recuperating, reading more Shelley—*Helen and Rosalind*, the *Revolt of Islam*, and *Prometheus Unbound*—and often walking the 6 miles between Goosnargh and Preston, to visit the circus or theatre, and to talk with the Allens. Leaving Goosnargh, he was in good heart: 'I came broken down; I return re-edified—I came with hope; I return with regret'.[56]

In Halifax his reading became more scientific. In addition to books about engineering and mechanics, and his frequent perusal of the *Mechanics Magazine*, he devoured Gideon Mantell's *Wonders of Geology*, as well as the current sensation *Natural History of the Vestiges of Creation*, which he bought on 2nd July. This book, published anonymously in 1844 (the authorship of Robert Chambers was not publically revealed until 1884), was a major pre-Darwinian theory of evolution, and brought the idea of the transmutation of species before the public eye. Darwin himself acknowledged it, writing in the *Origin of Species* that 'in my opinion it has done excellent service in this country in calling attention to the subject, in removing prejudice, and in thus preparing the ground for the reception of analogous views'.[57] Though Tyndall judged it '*quoique peut-être faux dans ses déductions*',[58] without specifying what he thought those errors were, he made pages of notes. He also continued to read Carlyle. Of Carlyle's *Heroes and Hero-worship* he declared: 'the writer must be a true hero. My feelings towards him are those of worship "transcendent wonder" as he defines it'.

During his railway years, Tyndall enjoyed a variety of female company. He was smitten by Fanny Smith (a relative of Margaret Allen), whom he termed 'The Witch', although he seemed to feel that he had been deceived by her.[59] This relationship, such as it was, may have stimulated him to write the poem 'Society'—'though the lovely spell which works in woman's eyes points not to me'—published in the *Preston*

Chronicle.[60] On one day in early 1847, his journal records notes on the attractions of Miss Tatum, Miss Wright, the Misses Shaw, Miss Piercy—'that type of beauty which the young Mahommedan expects to meet in heaven'—and Miss Fanny Piercy (at eighteen, about a year younger than her sister)—'an excellent piece of physiological workmanship'.[61] As in Ireland, his romantic attachments are hard to fathom, but he did receive an unexpected proposal of marriage in Halifax via his landlady. The offer came from his fellow lodger, Miss Latham, whose playing on the harp was much admired. The proposal caused great merriment. Tyndall commented: 'Why the poor lady is old enough to be my mother'.[62] This age difference in reverse would subsequently prove no barrier to him.

In July Tyndall received a life-changing letter from Bob Martin, who asked him to contact George Edmondson about a teaching job at a new school. Edmondson, a Quaker and innovative educator, was taking over a building in the southern county of Hampshire, to run a school with a novel curriculum that included surveying and railway engineering. Marquis had been invited, though he declined. Tyndall hedged his bets, and with Carter unable to employ him he tried to switch to Hawkshaw, who said there was no place for him, an answer Tyndall attributed to antipathy towards Carter. A few days later Tyndall went to nearby Tulketh Hall, Edmondson's current school, and was impressed by the educator's progressive vision.[63] He was offered £100 per annum and a share of profits. Not being a financial risk-taker, he asked for a flat £150, which Edmondson accepted.

Tyndall's preparations to leave Halifax took place against the background of the 1847 General Election. Five candidates, vying for two places, plastered the town with handbills. In the event the electorate returned the Whig Sir Charles Wood, and the Tory Captain Edwards (of Pye Nest). Although the Conservatives won the most seats nationally, a split in the Tory vote between the protectionists, led by Lord Stanley, and the free-traders, led by Peel, kept Lord John Russell's Whig government in power. Immediately after the election, Tyndall wrote an open letter to Edwards, but the *Halifax Guardian* refused to publish it—perhaps because of its political stance, or perhaps because of the lengthy and florid epistle Tyndall supplied. Stripped of its verbiage it urged the need for education, and the founding of a Literary Institution in Halifax to rival Preston's Mechanics' Institute and Literary and Philosophical Hall. At this time Tyndall's politics were still verging on the radical. He quoted Mazzini, the Italian political activist, admiringly: 'On the very day when democracy shall raise itself to the power of a religious party it will carry away the victory—not before'.[64] But he added in a critical note that he leant more towards Carlyle. He believed strongly in human progress, but full democracy was a step too far.

Tyndall now terminated his employment with Carter, turning down his offer of a ticket to the post-election dinner: 'I declined seeing a blue banner bearing the words "Conservative Association", I thought I should have no business there'.[65] Tyndall believed Carter owed him £257, but offered to settle for £200, of which £180 was left on account. Carter hosted a farewell dinner on 16th August, inviting Hirst and their colleague Jemmy Craven too. A few days later, Tyndall left for Manchester and from thence the south.

3

QUEENWOOD COLLEGE
1847–1848

Queenwood College stood in an undulating chalk district of Hampshire in south-ern England; a rural landscape of knolls and little valleys, interspersed with rich woods of beech, yew, and elm. The imposing building, H-shaped and turreted in Italianate style, started life in 1842 as Harmony Hall, designed to house a pioneering project in community living. The architect of this new-age lifestyle was the socialist Robert Owen. But his grandiose ambition outstripped his resources, as only about 100 people joined his envisaged community of 700. The project was bankrupted. This was where, as Hirst put it, 'the socialists first practically tried to live by the law of love, and of course miserably failed'.[1] Tyndall saw the downfall of this institution run on 'Communist principles' as a consequence of their embracing universal suffrage, which he thought voted out those who could have run it effectively.

Following the demise of Owen's community, the building and surrounding estate were let to the Society of Friends, who set up Queenwood College in 1847 under George Edmondson (Figure 2). It was a revolutionary school, one of the first in England to introduce the practical or laboratory teaching of science.[2] Science was not taught in any of the major public schools at the time, except on a limited, voluntary basis. Edmondson was able to introduce agriculture to the school, a mission of his, creating a class of 'farmers', who were also taught surveying and engineering. They were older than the other pupils, and were an unruly element that many teachers—but not Tyndall—struggled to control. Such a progressive school attracted forward-thinking parents and guardians, drawn from the ranks of professional men, army officers, doctors, farmers, and manufacturers in the north of England. The first pupil to arrive, a fortnight before Tyndall, was thirteen-year-old Henry Fawcett, who became a reforming political economist, serving in Gladstone's Liberal government in the 1880s; he was blinded in a shooting accident at the age of twenty-five. James Mansergh, later President of the Institution of Civil Engineers, was another. Not for the first time, nor the last, Tyndall had landed on his feet.

Fig. 2. Queenwood College, Hampshire.

Arriving at Queenwood on 20th August by carriage from the nearby railway station of Dunbridge, Tyndall was welcomed by Edmondson, his wife Anne, and his 'fair daughter' Jane. He was promptly issued with Erasmus Wilson's *Treatise on the Skin* and Liebig's *Animal Chemistry*,[3] so he could prepare his first lectures. He settled easily into life at Queenwood, began to teach surveying, and happily took on his additional responsibilities for managing the school's accounts and acting as its secretary. The rural environment suited his temperament, and the historical interest of the region offered tempting opportunities for exploration. At the beginning of September, Tyndall made an early visit to Salisbury Cathedral and Stonehenge, while his free afternoons on Wednesdays enabled trips to places such as Romsey's Abbey Church and Deanbury Hill. Tyndall and his friends would also wander and relax in the grounds of Norman Court, a nearby country house. Later he went on an expedition into the New Forest to find the 'Rufus Stone', which marks the point, according to legend, where King William II was killed by an arrow in 1100 while hunting.

A fortnight after Tyndall started at Queenwood, Edward Frankland arrived as a teacher, a chemist, and, according to Tyndall, 'a deeply thoughtful and instructive young man'.[4] He had recently been the assistant at the Putney College of Civil Engineering for Lyon Playfair, who soon became one of the key figures in British science and science education, later a member of Parliament and ennobled as Baron Playfair. As with Tom Hirst, this meeting was the start of a lifelong friendship for Tyndall. Frankland found that Tyndall's mind—despite his reading of scientific books—was 'almost a complete blank as far as chemistry or physical science in any form was concerned'.[5] Tyndall attended Frankland's chemistry and botany lectures, making extensive notes in his journal, and took individual lessons, in exchange for teaching Frankland mathematics.[6] He also took phonography (shorthand) lessons from Jane Edmondson, although he never used the technique in any of his surviving journals. He followed up the sessions with Frankland by reading Liebig's work on atomic theory, and Faraday's objections. The scientific community was split in the mid-nineteenth century between those, following John Dalton, who believed in the physical reality of atoms, and those like Faraday, who considered atoms as 'atmospheres of force' grouped around a point. Expressing an opinion he would later alter completely, he wrote: 'I like the views of the latter best, they are much more refined than the mechanical conception of Dalton'.[7]

As Tyndall became more familiar with practical chemistry, both men were often in the laboratory. Tyndall nicknamed Frankland 'Hermes' or 'Tris', after Hermes Trismegistus, the fictional alchemist. On one occasion the laboratory was the scene of a massive explosion, when Frankland 'blew the apparatus to atoms, cracking 3 windows', fortunately causing no serious injury.[8] Having access to the laboratory enabled Tyndall to experiment with inhaling ether, chloroform, and nitrous oxide (laughing gas), the latter with his students, including Henry Fawcett, who 'lay down and snuffed the dust like a donkey intending to roll'.[9] Tyndall was so entranced by the ether experience, as 'the brick walls began to shine more brightly and to twist and quiver in sundry funny contortions', that he passed out.[10] Fortunately Frankland was conscious enough to remove the tube from his mouth.

Tyndall's first lecture at Queenwood, on the skin, was attentively received, and praised by the Edmondsons. He followed it with a series on the same topic over several weeks, after which he lectured on the steam engine. The experiential and practical emphasis of Edmondson's educational vision at Queenwood influenced Tyndall greatly, and carried through into evidence he gave to Select Committees decades later. Edmondson had been influenced by the Swiss educationalists Johann Pestalozzi and Philipp von Fellenberg, who advocated discovery-based and child-centred learning. During his time at Queenwood, Tyndall was fascinated to hear about Pestalozzi from their fellow teacher John Yeats, who taught geography, history, and commerce. Pestalozzi had worked at Fellenberg's Institution at Hofwyl near Bern and was said to be responsible for eradicating illiteracy in Switzerland by the 1830s. Both Yeats and another colleague, John Haas, who taught German and French, had worked there.

Queenwood's reputation attracted high-profile visitors. Until the formation of the Department of Science and Art under the Board of Trade in 1853, education was the responsibility of the Committee of Council on Education. The Marquis of Lansdowne—Lord President of the Council under Lord John Russell—visited in October, finding Tyndall in his shirt sleeves labouring under a pile of railway sections. Lord Morpeth, First Commissioner of Woods, Forests, and Land Revenues, and a man Tyndall admired, was expected in November but did not appear, perhaps owing to the death of his father, whom he succeeded as seventh Earl of Carlisle. Elihu Burritt also visited, and had a long conversation with Tyndall. Two visitors who arrived together in early 1848 became particularly significant for Tyndall:[11] the Reverend Henry Moseley, mathematician, Fellow of the Royal Society and one of the first schools inspectors, and the Reverend Richard Dawes, educationalist, rector, and headmaster at King's Somborne nearby, until he became Dean of Hereford in 1850.

At the end of 1847, Tyndall went on leave to Ireland over Christmas and the New Year, travelling by coach and train to Bristol to catch the boat to Cork. He bought a copy of *Vestiges of Creation* on the way, an indication of the book's significance for him, though he recorded no further comments on it. Cork, in the grip of the advancing economic malaise, he found depressed and depressing. Passing through the wretched alleys of the town, Tyndall was convinced 'of the necessity of the whole of Ireland serving an apprenticeship to industry and cleanliness', and surer than ever that land-lords should shoulder their responsibilities.[12] On the boat trip over he had a long after-dinner discussion with Lord Mount Cashel, who took the side of the landlords, while others argued for the tenants. Tyndall regretted that neither side could see that 'man's duty to man extends beyond mere legal contracts, that his duty towards his neighbour is to love him, to cheer him when distressed, to help him when in difficulty and to teach him when ignorant'.[13]

Moving on from Cork, Tyndall passed through Youghal for old time's sake. Catching up with a few friends, he heard of the dire conditions in Ireland; no rent was being paid, and the landlords themselves were unable to make the Poor Law payments. He arrived in Leighlinbridge two days before Christmas, and remained for a week, working on algebra every morning, including Christmas Day, from 5 or 6am onwards. After Christmas he finally brought himself to visit his father's grave in Old Leighlin church: 'on coming away I turned instinctively to bid him goodbye but there was no answer'.[14]

Returning to England via Dublin and Liverpool, Tyndall went to Preston on a whim to see Bob Allen, Bob Martin, and Billy Marquis, before stopping in Manchester to see Ginty and Tom Hirst. He arrived back in Halifax just in time to catch the end of Ralph Waldo Emerson's lecture on Napoleon, and afterwards went straight out and bought some of his works. After reading Emerson's aphorism 'Trust men and they will be true to you', he decided not to confront Carter about some £100 that was still outstanding, although he was concerned about when he would secure it (Carter eventually settled just before Tyndall left Queenwood for the continent the following year). As he passed through London, he ran into Frankland, who introduced him to the chemist August Hofmann—first Director, at the instigation of Prince Albert, of the Royal College of Chemistry, which had been set up in Oxford Street in 1845—and took him round Putney College. They ended with a visit to Lyon Playfair's laboratory in Duke Street, though in Playfair's absence.

In the New Year, Tyndall lectured at Queenwood on mechanics, algebra, and geometry, in addition to practical surveying. Having heard Emerson speak in Halifax, Tyndall was now reading his work avidly. He wrote to Hirst: 'every time I rise from his book I find a new vigour in my heart—…There are many parts of his writings very difficult, especially some portions of the Transcendentalist, and Idealism. The rule he lays down will I believe make all clear—let us by enacting our best insight by doing that which we feel to be right, strengthen our powers and purify our vision, and all will be understandable'.[15] The ideas of transcendentalism, of which Emerson was a founding exponent, originated on the east coast of the United States in the 1820s and 1830s; in essence a liberal Protestantism, influenced by English and German Romanticism. They emphasized the intuitive and spiritual above the material and empirical, and the importance of individual expression and self-reliance. But unlike the Romantics, the Transcendentalists retained a respect for the empiricism of science. The transcendental belief in the essential goodness and relatedness of man and nature resonated with Tyndall's awareness of nature and its importance to him. He absorbed transcendental perspectives further by subscribing to the *Truth Seeker*, a publication from Dr Lees and George Phillips (known as January Searle). The *Truth Seeker*, Lees's third temperance journal, published by John Chapman, operated between 1845 and 1849, publishing readings in biblical criticism, reports on the revolutions in Europe, articles examining the health risks of alcohol, literary criticism, and much original poetry.

Emerson became one of Tyndall's formative influences, and helped to anchor his ideas about using the imagination to transcend experimental findings and look for the connections and explanations behind them.[16] In April he wrote down an early statement of the value of reflection and thought, speaking of the need to unite the practical man and the theorist: 'Doubtless the properties of the lever were known and taken advantage of long before its laws were investigated, but how limited would have been its application had not these properties come within the dominion of thought'.[17]

Formal religion now diminished in importance for Tyndall. To begin with, during his time at Queenwood, he was a regular attendee at nearby Broughton church. But he increasingly found the sermons insufferable: 'Intolerable sermon in church, from someone enabled to extract 7 or 8 hundred a year from his wretched congregation'.[18] In early February 1848, he wrote: 'No church, which enables me to spend the day better'.[19] A week later he cut his little toe, 'and therefore went not to church'.[20] He remarked on 'some pretty girls' in the congregation, but he remained unimpressed by the theatrical histrionics of the preacher at Broughton, preferring the 'poor stammering old servant of God' at East Tytherley church.[21] It was clear to Tyndall from his arrival at Queenwood that his religious views differed from those of Edmondson and his brother-in-law Josiah Singleton, but that did not concern him. Religious discussion was frequent, between the traditional churchgoer Singleton, Frankland, who called himself an independent, and Tyndall, who declared himself neither. By Christmas, Tyndall noticed that Frankland had moderated his views, though he did not specify their nature: 'Opinions which six months ago he would have shrunk from with abhorrence and dread are now beginning to make inroads on him'.[22]

It seemed that Tyndall, Frankland, and, finally, Hirst, with whom he continued to exchange letters, were all losing their traditional faith. But despite the loss of faith, Tyndall's intrinsic morality was in no danger of wavering. Already his educational philosophy, an extension of his own moral beliefs, was being conveyed forcefully to the boys he taught. In his classes and lectures, inspired by Carlyle and Emerson, he emphasized the moral worth of those who work for humanity rather than themselves, and the need to be truthful and self-reliant rather than imitate others. In another lecture, at a time when notions of the wisdom of the crowd and radical ideas of democracy were in the air, he urged the importance of individual action and responsibility, and the need for self-denial and courage to overcome difficulty.

However, all was not well between the Edmondsons and their staff. The Quakers' expectations of sobriety, Sabbath observance, and 'solemn silence in the dining room' did not sit easily with their employees. Tyndall would often escape with Frankland and others for a gin or a brandy and cigar which, being prohibited, possessed 'a zest which in a land of plenty would be a desideratum'.[23] In May, returning from the pub 'in a very undesirable condition', they caused a scene.[24] Tyndall and Edmondson had a different attitude to rules, as Tyndall commented: 'He would have a rule observed strictly. I would have the maker of a rule superior to it and would have him infringe it whenever by virtue of his supremacy he proved it inconvenient'.[25] Following a rebuke from Anne and Jane Edmondson, after he made an indiscreet remark, he was also aware that his attitude could work against him: 'Kindness and courtesy however I can never reduce to a commercial standard. I will choose a rough tongue before a hypocritical one'.[26] As time went on, Frankland and Tyndall became increasingly unhappy with the regime at Queenwood, especially the weakness they perceived in Edmondson and what they saw as the inappropriate influence of his wife. They planned to hand in their notice.

Matters came to a head at the end of May. According to Tyndall and Frankland, Edmondson had asked their fellow teacher John Yeats to spy on Josiah Singleton, Edmondson's brother-in-law, in a manner they saw as being intended to force Singleton out. Despite Tyndall's and Frankland's objections, Singleton did leave, much to the dismay of his pupils, many of whom signed a touching leaving note and made an emotional presentation of a pair of world globes. Frankland and Tyndall now told Edmondson of their intention to go. Anne Edmondson wrote them a long letter, without her husband's knowledge, saying it was quite wrong of them not to have confronted Yeats directly, and fearing that that they might leave in circumstances which she thought they would later regret.[27] Tyndall wrote a verbose and sententious essay in response (which Frankland hand-delivered), explaining that it was Edmondson himself who had drawn views out of them that they would otherwise have kept to themselves.[28] He made clear his view of the appropriate roles of men and women in an unsent draft to Jane Edmondson after a later dispute: 'I would earnestly repeat the advice which I urged upon you this morning—that is to keep yourself wholly unconnected with school matters. You cannot imagine what a repugnance some men feel to the interference of ladies; and moreover, these are the very men who will offer to ladies the most loyal service as long as they remain in their own sphere'.[29]

This demarcation of roles was an attitude he retained. Rejecting a request in 1872 from the editor of *Woman* magazine to subscribe, he wrote: 'I should pause before declining ... if I felt any lively sympathy with our latter day movement regarding "the rights of women". But there are many of us who would go any length with you in the endeavours to right the real wrongs of women, and who, in common with what they regard as the true womanhood of England, consider much of the present movement as leading to mischievous results. In fact I sometimes think that it is because the men of the age have become to some extent women, that a few of the women aspire to be men'.[30] Unlike contemporaries such as John Stuart Mill, Tyndall was no advocate of women's emancipation. In 1873 he wrote to Martha Somerville, daughter of the scientific author Mary Somerville: 'I would open all doors to the education of women; but had I known your mother's views on the suffrage question I should have argued the point with her when I had the pleasure and the privilege of seeing her at Naples'.[31]

The letter to Anne Edmondson is typical of his direct approach. On giving a stern message to a past surveying colleague he had written: 'my letters are intended to cut deep. I speak my mind fully without exaggeration or diminution'.[32] Later, he claimed: 'I have one single guide in all my letters, and that is, to write as I feel'.[33] Journal entries are even more private than letters, though both might occasionally be shared with other close friends, and in at least one emotionally charged episode, Tyndall tore pages from his journal to send to the object of his desire. There is no evidence that Tyndall wrote with the intention to make either letters or journal entries public, though he occasionally drew on them, so they offer us the closest insights into his personal life that we have, even if they do not reveal all.

A few days after the exchanges with Anne Edmondson, Tyndall left for London to see Carter, who had asked him to help prepare for a hearing against the Lancashire and Yorkshire Railway. It fell to Tyndall to show that the work had been done as stated, which he did. As Tyndall explored alternative employment, Carter promised that he would notify him of any vacancy in Robert Stephenson's establishment. Later, Tyndall met William Yolland in Southampton, home to an Ordnance Survey office to which both Archy McLachlan and Bob Martin had moved. After the meeting, Yolland said he would offer Tyndall employment, if he thought he would accept it. Tyndall had admitted that he was the author of the 'Spectator' letters, which Yolland told him they had suspected.

* * *

Tyndall and Frankland had been planning a trip to Paris in the holidays. Their goal was to stay for five weeks so that they could attend some lectures and Tyndall could improve his French (he had also started teaching himself German, as he had earlier planned a trip up the Rhine in the summer).[34] They practised French enthusiastically with John Haas, who had arrived in May from Switzerland after a narrow escape at Fribourg during a fight between the military and the local populace. Immediately after Tyndall's return from London, and after Tyndall and Frankland had confirmed to Edmondson their intention to leave, Haas saw them off from Southampton.

Passing the Isle of Wight and half a dozen tremendous men-of-war slumbering around Spithead, they landed in the heavily fortified town of Le Havre, which was festooned with banners proclaiming 'Liberté, Egalité, Fraternité'. In their hotel, a dark-eyed French girl gave them a bed candle and complimented them on their French. They then took the train to Rouen, in an open-class carriage to see the view. Tyndall was struck by the scenery, the friendliness of the French, the lack of beggars, and 'the finest viaduct I have ever seen, higher than that of the Manchester and Birmingham at Stockport'.[35] In the cathedral in Rouen singing and organ music 'filled the air with melody and all this in honour of a crucified carpenter's son!'.[36] After a theatre visit the two men looked into—but appear not to have entered—one or two licensed brothels, where four or five girls sat with arms, breasts, and necks bare: 'Some of them were truly beautiful girls and so young that it was pitiful to see them.[37] They also inspected the sites of barricades, observing that people seemed to harmonize with the military: 'the soldier is not isolated and treated with contempt as in England'.[38] In the museum,

Tyndall remarked: 'The skeleton of the monkey occupies the box next to that of the man'.[39]

When they arrived in Paris on 19th June, all appeared tranquil, though the military was patrolling the streets every hour. It was the calm before the storm. The two men had chosen a particularly interesting time to visit France. The February 1848 revolution had ended the Orléans monarchy and established the Second Republic, and Tyndall and Frankland arrived in Paris during a further revolution against the conservative government. Undeterred, they went sightseeing in the Tuileries and to the theatre. Deciding to visit Amiens, Tyndall left Paris on 23rd June, missing the start of the June Days Uprising—in which the citizens revolted against the conservative turn of the Second Republic—by mere hours. Instead he saw 5,000 National Guards leave Amiens for Paris to help quell the rebellion, with their ration of a loaf of bread stuck on each bayonet. Hearing rumours of 20,000 killed,[40] Tyndall fled for Boulogne, weeping with anxiety over the safety of Frankland, who was still in Paris.

On his return to Queenwood, news came that the revolution had been crushed, and he planned to go back to Paris to look for his friend. Just before he set out, he heard that Frankland was safe. He had clambered over a barricade to get back to his hotel, survived 2 hours' artillery fire, and witnessed the carnage first-hand. Three democrats and two *gardes* had been killed in his street. On 2nd December 1848, Louis Napoleon was elected President of the Second Republic. Four years later he suspended the elected assembly, establishing the Second French Empire, which the Franco-Prussian War would bring to an end in 1871.

Tyndall retraced his steps via Le Havre to Paris, passing bivouacked dragoons on the Champs Elysées, to find Frankland in his room in the Rue Serpente. This time the sightseeing took in smashed houses and widespread damage in the Rue St Jacques and around the Pantheon and the Bastille. He described the aftermath of the revolt in letters to the *Carlow Sentinel*.[41] The insurgents had shot many of the wild animals in the Jardin des Plantes, although the elephant had been saved by its thick hide. Sporadic killings continued, but Tyndall rather relished the danger: 'the very frequency of the deaths deprives death of terror'.[42] They went for an evening walk along the Champs Elysées, which was thronged with crowds enjoying themselves as though the revolt had never happened. Tyndall reflected on the rights of men—and women, whom he noted were demanding their rights in France—and on the strong nature of a government that allowed a society to flourish. For him, the ideal society was one that afforded scope for individuals to succeed, while encouraging the noble and the good and holding the base and wicked in check, all within a general plan 'overruled by a God'.[43] Whether the government was despotic or constitutional was immaterial. But compared with what he had witnessed in France, which had renewed order even at the cost of many lives, he held that Ireland, despite its social injustices, was not ready for Repeal. It would not be capable of self-governing in the same way. Tyndall thought that Ireland needed the help of England, and his experience told him that England, despite its history of past sins, sincerely wanted to help.

Although Tyndall and Frankland failed to hear the chemist Joseph Gay-Lussac lecture in the Jardin des Plantes (he didn't turn up), over the next few days, they did hear Jean-Baptiste Dumas lecture on organic chemistry, as well as several other notable speakers. More sightseeing followed, this time to the park and chateau at Meudon—mostly destroyed by fire in January 1871 after being bombarded during the Franco-Prussian War. At St Cloud—also later destroyed in the War—they feasted their eyes on the art and tapestry. They attempted to visit the National Assembly but were turned away, prompting them to write to Lamartine (who had been instrumental in founding the Second Republic) for a ticket. Lamartine's wife told them regretfully that her husband could not help; he had been forced to leave the government after the revolt.[44]

After more than a fortnight in Paris, both men set off for Brussels to visit the battlefield at Waterloo. During a trip to the city's Natural History Museum, Tyndall commented—more than ten years before the publication of Darwin's *On the Origin of Species*—'On looking at those biped animals it is scarcely possible to repress the idea of their analogy to man and the possibility of his having sprung from the same root. The idea is not flattering'.[45] They declined the frequent invitations from brothel keepers to have 'the finest girls in Europe'—'one feature the moral man will dislike'—attending instead a glorious performance of Verdi's opera *Jérusalem*, which had been premiered in Paris the previous year.[46] Taking a Tilbury, a light carriage, to Waterloo, they toured the field with a guide, drank beer in La Belle Alliance, the house where Wellington and Blücher embraced after the battle, and bought two or three rusty souvenir grape shot. Travelling back to London via Antwerp, Ghent, and Ostende, they once again resisted the temptations of a brothel to which their guide took them. Tyndall may have found sexually aware young women unsettling, remarking once of a group of peasant girls: 'one with a face positively beautiful but still with those terrible "understandings"'.[47]

As he passed through London, Tyndall visited the British Museum, followed by the Polytechnic, where he heard lectures on patents and the sources of heat. While there, Frankland finally introduced him to Lyon Playfair, in time an important contact. Back at Queenwood Tyndall's routine continued, with the addition of some physics experiments in the laboratory. At the end of July, he gave his first lecture on physics (which he saw as synonymous with the widely used term 'natural philosophy'), coupling it to the human desire for knowledge and the continuous progression of ideas it produces.[48] This, he argued, demanded action, not simply the reflection and dreams of the poet.

For some time, Tyndall and Frankland had been discussing the idea of going to Germany to study. Frankland had already visited, and knew the charismatic chemist Robert Bunsen (of the eponymous burner), who wrote placing his laboratory in Marburg at Frankland's disposal and invited him to join his investigation into ozone. Tyndall would also be able to work there for a PhD, which did not require a previous degree. He was aware that the decision to study in Marburg was both potentially life-changing and, because he was about to put his entire savings into funding his education, risky. The university was barely known in England, and they would be the first two men from Britain to graduate there.

In his private life, matters were also unsettled. Bob Allen had promised to find him a wife—as so many of his friends were now getting married—but Tyndall recognized his reality: 'Who would join her fortunes to such a Will O the Wisp? I don't expect to meet one, and if I should I would rather think of putting her in a glass case like a beautiful Egyptian Ibis to be worshipped in due season than of causing her fate to oscillate with mine'.[49]

Tyndall's last lecture to his students at Queenwood before setting off for Germany was on 25th September. It was an emotional affair, and charged with religious feeling. He urged his students to independent thought and action, and justified his desire to go to Germany, in words clearly inspired by Emerson: 'the higher you climb the fewer examples you have above you and the more you are dependent on yourself . . . What are sun, stars, science, chemistry, geology, mathematics but pages of a book whose author is God! . . . I want to know the meaning of this book'.[50] According to Frankland, Anne Edmondson, Haas, and some of the boys were left in tears. Tyndall was owed £168 in wages and took £58, leaving the balance with Edmondson; either prudence on his part, or the inability of Edmondson to pay it all immediately.

Before his departure for Germany, there was time for a fortnight in Preston, Manchester, and Halifax, visiting old friends (in Preston, he even spied the Witch in the Lune Street Chapel). At Spring Bank in Over Darwen, Josiah Singleton's new house and school, Tyndall taught a lesson on mechanics, algebra, and Euclid. A further attraction for the German visit was Tyndall's increasing interest in German philosophy, perhaps kindled by reading Carlyle. He bought a copy of Fichte's *Characteristics of the Present Age*, with its view of the historical progress of the human race from the rule of instinct to the rule of reason. He also read Schlegel, who argued that it was not the function of philosophy to prove the existence of God. In London, having secured a passport from the Prussian Consul, he prepared for his departure. He bought a copy of Ollendorff's language teaching methodology, and works by

Locke and Paley. Paley's *Natural Theology or Evidences of the Existence and Attributes of the Deity*, published in 1802, which put forward the 'watchmaker' analogy as proof for the existence of God, was required reading for anyone studying natural philosophy, and a compulsory text at Cambridge University. On 18th October 1848, Tyndall and Frankland set off down the Thames on board the *Rainbow*, bound for Rotterdam, breathing in the fumes of chloride of lime used to combat the cholera that had descended on the river. It was a rough crossing, and Tyndall was violently seasick. On their arrival, spurning the railway from Rotterdam to Bonn, they took the scenic route, travelling up the Rhine all the way to Mainz. While Frankland went on ahead, Tyndall stopped in Frankfurt for a night, alongside hordes of Prussian soldiers, for a little sightseeing before the start of his new life in Marburg.

4

MARBURG

1848–1850

In 1848, before its annexation by Prussia in the Austro-Prussian War of 1866, the medieval town of Marburg lay within the Electorate of Hesse, one of many small states which eventually became part of unified Germany in 1871. Dominated by its castle, the town sits in a wooded, gentle valley astride the River Lahn (Figure 3). The only way to reach Marburg was by horse and carriage (the railway did not reach the town until 1850), so Tyndall travelled the 60 miles from Frankfurt in 11 hours, arriving in Marburg in the evening of 25th October. He was met by Frankland, and they supped at the Hotel zum Ritter, a regular dining place for academics from the university. The following day, after a breakfast of black bread, butter, and coffee,

Fig. 3. Marburg in 1843. Steel engraving by Joh. Poppel after a drawing by L. Lange.

Fig. 4. Robert Bunsen (1811–1899).

Frankland introduced Tyndall to Heinrich Debus, chemistry demonstrator in Bunsen's laboratory, and a man destined to become a lifelong friend of both Tyndall and Frankland. At this point Debus could not speak English nor Tyndall much German, so Frankland translated. Later, over dinner at the Ritter, Tyndall was introduced to Robert Bunsen himself (Figure 4), and afterwards he went for a walk with the philosopher Theodore Waitz. The next day he called on Christian Gerling, professor of physics, and Friedrich Stegmann, professor of mathematics.

Tyndall and Frankland lodged with landlady Frau Baum, on neighbouring floors of a house situated conveniently opposite the Ritter. From his window, Tyndall could see the castle, with the apple trees clustered below, and the observatory with its tower and white walls. Outside the door ran the Ketzerbach, a stream fringed with acacias. The woods of Spiegelslust bounded the view to the left, and the heights of the Dammelsberg to the right.

The university at Marburg was established in 1527, one of the oldest Protestant foundations in Germany. Tyndall settled into a routine of hard study: up at 5am; German until breakfast at 7.30; a lecture on physics from 8 to 9; Bunsen's lecture on chemistry from 9 to 10; and then into the laboratory, where his goal was to master qualitative and quantitative analysis. In the afternoon, from 4 to 5, he had a private mathematics lesson from Stegmann—this involved uphill work on calculus, which he cracked by the end of November—followed by tea at 6, then more mathematics until bed at 10.[1] He thought the teaching at the university impressive, although Stegmann's and Gerling's rooms were stiflingly hot. According to Tyndall it would take 'years of devoted effort to bring England up to the same standard'.[2] He found Bunsen's lectures superb, and he had some private chemistry tuition from Debus, as well as supplementary mathematics from an advanced student, Christian Menges. In the new year, he planned to instruct Hirst and Jemmy Craven in chemistry from afar.[3]

In between the science and mathematics, Tyndall was an avid reader of philosophy and theology, including Locke, Fichte, Paley, and Kant, and he took lessons from Waitz. He also read Volney's *Ruins of Empire*, with its vision of the unification of religions through the common truth underlying them all, although he found Fichte much deeper, writing: 'Volney describes the face of nature, while Fichte exhausts her mind'.[4] Comparing Fichte with Volney and Locke, Tyndall saw 'a direct antithesis between him and them; he had got upon a higher cliff and seen further than either'.[5] Although Tyndall did not record exactly which works of the German philosophers he was reading, he must have found personal resonances in German ideas: aspects of idealism; ideas of consciousness and will; the importance of the individual; moral duty; and the role of intellectual intuition or inspiration, tempered by practical experience.

Tyndall was unimpressed by Paley's *Natural Theology*, however: 'I am willing enough to be convinced upon the subject of which Paley writes, but what he says is insufficient to this end. The Great Spirit is not to be come at in this way; if so, his cognition would only be accessible to the scientific and to very little purpose even here'.[6] He copied some of Schlegel's words into his journal, which argued that the concept of 'natural theology', which saw knowledge of the divine essence as a branch of natural science, was wholly false. Tyndall and Frankland sat up late on New Year's Eve discussing the matter. Frankland recorded that Tyndall believed God and the Universe to be a unity, and compared it to a life gradually developing itself.[7] This was a forerunner of the position Tyndall took years later; that matter contains within itself 'the promise and potency of all terrestrial life'.[8] Already, aspects of transcendentalism and pantheism were crystallizing in his mind. He explained Kant's philosophy to Hirst, and its influence on Emerson: 'Kant proves, that Space and Time, which are the forms of all ideas, flow from the mind—Space is the form of all sensible appearances, and knowledge of space and time is not a knowledge of objects from without but only a knowledge of our mental organization. Hence Emerson's expression "know thyself" and "study nature" become at last one maxim'.[9]

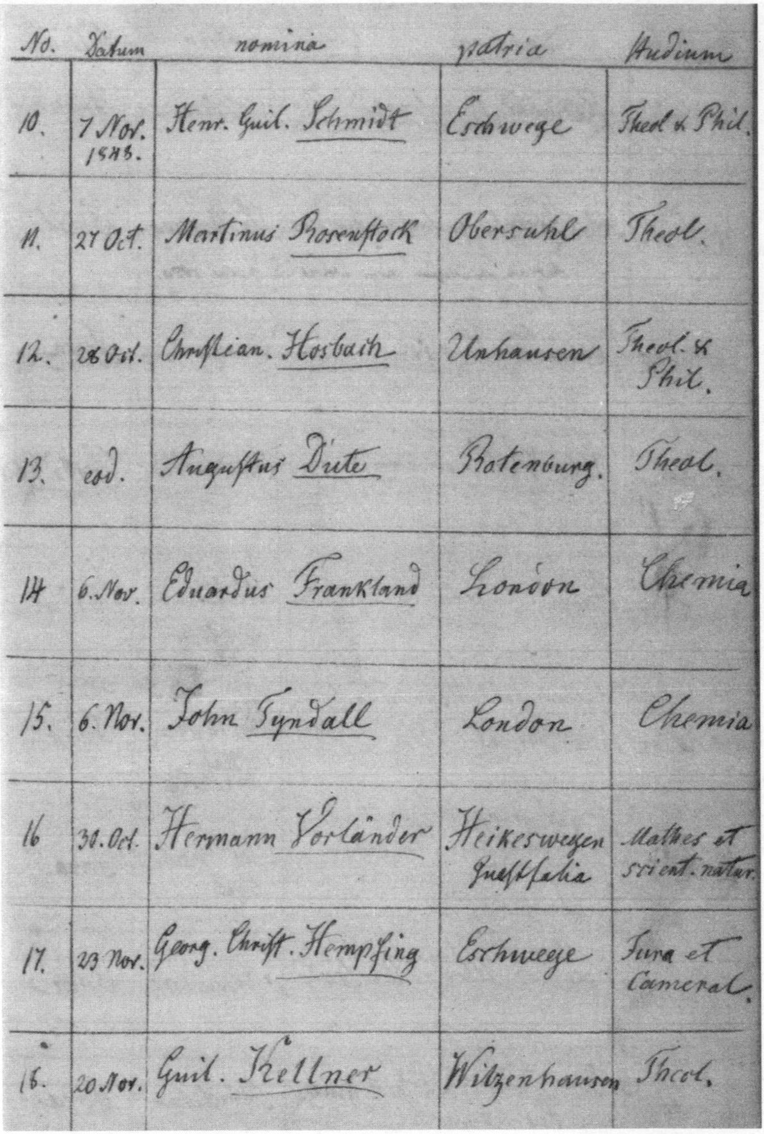

No.	Datum	nomina	patria	studium
10.	7 Nov. 1848.	Henr. Guil. Schmidt	Eschwege	Theol. & Phil.
11.	27 Oct.	Martinus Rosenstock	Obersuhl	Theol.
12.	28 Oct.	Christian. Hosbach	Unhausen	Theol. & Phil.
13.	eod.	Augustus Dute	Rotenburg.	Theol.
14	6. Nov.	Eduardus Frankland	London	Chemia
15.	6. Nov.	John Tyndall	London	Chemia
16	30. Oct.	Hermann Vorländer	Heikeswegen Westfalia	Mathes et scient. natur.
17.	23 Nov.	Georg. Christ. Hempfing	Eschwege	Iura et cameral.
18.	20 Nov.	Guil. Kellner	Witzenhausen	Theol.

Fig. 5. Tyndall's matriculation entry at the University of Marburg on 6th November 1848.

Tyndall visited Justus Liebig's laboratory in Giessen with Frankland and others, and witnessed the civilian protest following the execution of the politician and revolutionary, Robert Blum, in Vienna. As in France, Tyndall had arrived in Marburg at a time of great political unrest. The Frankfurt National Assembly—the first freely elected parliament in the emerging German state—had been elected in May. Although Blum was a member, such status did not save him from execution when he joined the revolutionaries fighting in Vienna in October. Tyndall himself was no revolutionary, as he wrote to Hirst: 'I was once obliged to leave Halifax on account of ill health if you remember...I took "Chartism" along with me and read it in the green fields between Preston and Goosnargh...I wish heartily that we had a good translation of it for German students, they are brave young fellows but mad with the phantasm of democracy—a Republic they imagine is the panacea for all earthly ills—This is a natural reaction against the irrational King-government which has so long pressed them down, but like other reactions it swings just as far beyond the true centre'.[10]

On Christmas Day 1848, Tyndall encountered the German Christmas tree, the tradition Prince Albert's example had encouraged in England: 'In the centre of a room stood a pine, hung round with wax lights and having numberless figures of cocks, dogs and men, in fancy confectionery, suspended from the branches. On the top was perched a little angel with outspread wings'.[11] Nevertheless, though taken with local customs, he did not forget his countrymen at Christmas. Informed by letters from Ireland of the suffering caused by several years of potato crop failure, Tyndall gave a donation to the poor, so it could be 'arranged that 12 poor people from Leighlin should spend Christmas Day comfortably'.[12]

The social life in Marburg was a revelation. Tyndall commented that everyone knew everyone else, people cultivated their social instincts, and the class differences were negligible compared to those in Ireland. But unlike his behaviour in England, he did not throw himself into the fray, which provoked comments from some that he might be married. As he told Hirst, he had a reputation for being 'cold and zurück-haltend' (reticent) among women, and he suggested that Hirst, who was planning to visit, should 'buckle up your heart against the seductions of our Marburg beauties'.[13] His poem *Alone*, written into his journal at this time, expresses his preference for freedom in the countryside rather than the social throng.[14] And in February, while attending a large ball with Frankland that included 'some lovely girls', he declined invitations to dance and refused to be introduced to any ladies until he was forced to at dinner, after which he enjoyed conversation with them.[15] He was committed to the hard work and discipline necessary to succeed, but he also lacked confidence in his German, although he was learning quickly. At a ball a month later, though he still declined to dance, he enjoyed talking with the fair girl who sat opposite. He gave an amusing account in the *Preston Chronicle* of the life of German students, who, he claimed, were duelling one minute and sitting gently with their sweethearts the next.[16] He witnessed one duel in which the cut 'seemed to sever one half of his antagonist's face from the other; his left face was cut through from the ear to the mouth, it was a ghastly wound'.[17]

The university holidays started in early April: 'the students are all away and I devote the whole of my time in the laboratory to Electro Chemistry—it is deeply interesting, I intend to follow it up as far as it has been developed'.[18] At the end of the month, taking a break, Tyndall spent several days exploring the countryside from Marburg to Kerstenhausen, Cassel, Hameln, and Münden, walking 50km on the first day (not an exceptional distance in Victorian times). He reflected that great events were passing yet leaving little trace in his journal. In March he had noted troops of Bavarians passing through Marburg daily on their way to Schleswig Holstein, some mounted on cattle. And in April a notice in the Ritter had caught his attention; the Germans had captured two Danish ships in a raid led by the Prince of Coburg, Prince Albert's brother. Germany at this point was almost in rebellion, 'with many places in active revolt. Dresden revolutionized, bombarded, and subdued. The Hungarians thrash the Austrians and declare their independence. Russia sends 50,000 troops to aid her ally. At Marburg we are purchasing arms and forming a corps of armed students; waiting "like greyhounds on the slip" until an opportunity offers to strike for freedom'.[19] Though in Hesse the Schleswig Holstein question—a territorial dispute between Denmark and the German Federation—seemed to absorb public attention, this was swamped in Hannover by the refusal of the king to acknowledge a central ruling power. Tyndall wrote: 'The people are exasperated ... Even grave men with gray hairs submit with a sigh to the conclusion that the princes must be driven from the land'.[20] Talking to Tyndall, people pointed to Cromwell's hard-won and bloody victory over despotism and saw the same future for the German lands.

Tyndall decided to visit Dresden a few weeks later. On the way, he was stunned by the beauty of the countryside, especially that around Eisenach. At the castle of Wartburg, on the precipice overlooking Eisenach, he visited the room in which Luther, the challenger of the Roman Church, found shelter from the Pope while translating the Bible. Having read Carlyle's *On Heroes, Hero Worship, and the Heroic in History* Tyndall had felt compelled to visit the scene and view the mark on the wall indicating the spot where Luther had flung his inkstand to repel his vision of the devil. He wrote a poem about Luther and even sketched a letter to Carlyle—which was never sent—during the long ramble. After travelling from Eisenach to Leipzig (where he took in the opera), he passed through Weimar, 'where sleep Goethe and Schiller'.[21] En route a gentleman pointed out the Napoleonic battlefields of Jena and Lützen. At the end of May, in Dresden, with a map of the barricades in hand, he saw the recent destruction in the city during the uprising of 3–9 May against the repression of the democratic movement. Parts of many of the houses had been torn away by cannon balls when the Prussians had arrived to aid the Saxon military. Fighting had continued until some 1,600 people had died, and eventually towels and tablecloths were hung out in surrender. Traces of blood remained on streets and carpets. After Dresden, Tyndall travelled on to Wittenburg to see Luther's grave, his house, and his old furniture: 'Upon the spot where he burnt the Pope's bull is planted an oak tree protected by railing and surrounded by flowers'.[22] Two vignettes of his travels appeared in the *Preston Chronicle*.[23]

Back in Marburg, Tyndall heard the physicist Hermann Knoblauch for the first time, and was impressed: 'young, speaks quickly and understands his subject'.[24] He sent his first letter to Carlyle on 6th June, ostensibly commending him to take an interest in Hirst, who had recently written to Tyndall of the impact of Carlyle's *On Heroes*: 'I have got a glimpse of that religion that exists apart from logic…I was trying to see the truth there was in different creeds and could only see a huge jumble of useless forms with no life, no meaning in them. I could not see—blind as I was—that there was a religion apart from them'.[25] Carlyle deftly passed the letter on to George Phillips, a journalist who had recently returned from America and who was heading up the *Leeds Times*—he was 'January Searle' of the *Truth Seeker* and Secretary of the Huddersfield Mechanics' Institution—who contacted Hirst directly and offered to make his acquaintance.[26]

Hirst now told Tyndall that he was planning to come out to Marburg. Tyndall warned him not to expect too much: 'Our town is nichts besonders, nay it is positively old and ugly and were it not lifted up into heaven's sunshine by the hill on which it stands would be a semiplace of perdition. Our University is not grand, it is broken into parts and presents no imposing front. Our laboratory presents rather a scoundrel-like appearance, but…it holds a man whose superior as a chemist is not to be found within a radius of 8,000 miles from the Piece Hall of Halifax'.[27] Among the items Tyndall asked Hirst to bring with him were postage stamps, a second series of Emerson's essays, and copies of *Punch*.[28]

Towards the end of July, Tyndall began to socialize more. Frankland had been far more active on the social scene, and, according to Tyndall, was observed to 'sun himself in the eyes of Miss Fick' on 29th July.[29] Tyndall added in his journal his later discovery that this was the day on which Frankland became engaged to Sophie Fick. Tyndall rarely commented on his birthday, but this time, on 2nd August 1849, he wrote in his journal: 'determined to do my duty; still willing to work as hard as ever; to fear God, honour a true king and love my friend'.[30] Hirst arrived in Marburg at the end of August, and Tyndall 'felt a kind of pride as his enormous frame erected itself before me in being able to present to Germans such an example of English bone and muscle'.[31] They had planned a walking trip to Switzerland, but Tyndall heard from Jemmy Craven, the day before his major examination with Professors Gerling, Hessel, Bunsen, and Stegmann, that Hirst's mother had died. Able to tell him only that she was very ill, it was not until Hirst was in the carriage to return home that Tyndall could bring himself to voice the truth. Knowing that Hirst had been cheated out of an inheritance from his father (while Jemmy Craven had inherited a fortune), he offered, while sending his condolences, to provide any help he could in future.[32] However, to Hirst's surprise, it transpired that his mother had left him an income of £150 per annum, more than enough for his needs. Knowing Tyndall's desire to stay on the continent he immediately offered him half his new income.[33] Tyndall was stunned by the offer, one he was tempted to accept if it allowed him to spend three or four months in France. He asked Hirst how much he could really lend him, and established a formal agreement to pay the money back in two years' time.[34]

On 17th September, his examination safely behind him and just the thesis to complete, Tyndall left for a long walking holiday, vowing to discover more about German literature. He travelled first with Frankland to Giessen and then walked on his own, dodging heavy showers, to Wetzlar on the River Lahn, which he followed for some 45 km to Limburg, arriving with 'toes smashed and heels blistered'.[35] He limped on to Nassau—with its enormous crags, and a section of the river bordered by a ruined castle on one side and lush green meadows on the other—and the next day on to Koblenz. Koblenz had fallen victim to cholera (as had England, according to alarming accounts from home). Local reports claimed that several people had died the previous day. In the hope of escaping the scourge, Tyndall found as clean an inn as possible, thinking he had 'doubtless unwittingly passed through many infected streets'.[36] From Koblenz he took a boat up the Rhine to Biebrich, and then travelled by rail to Heidelberg to visit its ruined castle. Nearing Rastatt there were many visible signs of fighting, including railway offices destroyed by fire. But there was beauty too. Passing Freiburg, before crossing the border to Basle, he found the maidens 'exceedingly beautiful...the soft black eyes transacting witchcraft and their cheeks like sunned peaches inviting your lips to test their sweetness'.[37]

In Zurich he was equally struck by the beauty of the city's daughters: 'were I not a predestinated old bachelor I should hardly venture to dwell so long upon these thoughts'.[38] He made his first significant mountain ascent here, reaching the summit of the Rigi at 1,800m, and then bartered with the proprietor of the Staffel hotel to stay the night on the mountain.[39] Heading into the Bernese Oberland, passing the fabled site where William Tell was said to have killed Gesler with his crossbow, he made his way over several days to Grindelwald. From there, under the great north walls of the

Eiger, Mönch, and Jungfrau, Tyndall set off over the Kleine Scheidegg to Lauterbrunnen. It was a typical mountain day—cloud swirling around the peaks, and avalanches thundering from the Jungfrau. Passing the famous Staubbach Falls, Tyndall the nascent physicist calculated the speed of water at the bottom as 3 miles per minute. But though the mountains had impressed him, there was no sign of the deep love he was to have for them in later years. On the road out to Basle he took a look back, thinking, 'this is the last I saw, or perhaps will see, of the Swiss mountains'.[40]

Writing from a distance, the two friends Tyndall and Hirst shared their religious struggles. Tyndall declared that for two years he would have given anything to be a Christian: 'this want within myself was near driving me to join the Methodists, expecting that their prayings and groanings and religious excitements would arrest the dry-rot of my soul'.[41] But what emerged now, reinforced by his reading of Carlyle, was his Protestant-inspired work ethic: 'as Carlyle said, it is work that is the key: "Consider how even in the meanest sorts of labour the whole soul of man is composed into a kind of real harmony the instant he sets himself to work!"'. He wrote to a friend: 'I think if I were asked what is the greatest sin to which humanity is subject I should point to idleness'.[42] Tyndall saw work as 'the most practical act of worship, the best heart purifier that man possesses'.[43] Carlyle's writings had a powerful effect on the language, beliefs, and social commitment of many young people in the 1840s and 1850s. These often self-educated readers liked the 'vigorous rhetoric which broke with conservative tradition, and his demand for freedom of thought and self-dependence'.[44] Tyndall was one of them. Of his time studying for a PhD, he later said that it was reading the works of Carlyle, along with Fichte and Emerson, that held him to his work 'in the long cold mornings of the German winter'.[45] It was Carlyle who also inspired him to read Tennyson, of whom he wrote: 'I am acquainted with no soul so energetic and beautiful'.[46]

As far back as April, Tyndall had noted that he had a 'capital subject' for the dissertation for his doctorate. He outlined his thesis topic to Hirst: 'Conceive a screw with a triangular thread, conceive one of the bevelled surfaces of this thread produced indefinitely. I am to investigate the properties of this surface and to determine the conditions of equilibrium upon it'.[47] This was a mathematical problem, supervised by Stegmann, which required trigonometry and algebra but no calculus, and by early November, Tyndall thought he might already have discovered the equation for the surface. The new semester, which had started at the beginning of November, included sessions with Knoblauch, who Tyndall found 'exceedingly kind, he is as simple and unassuming as a child'.[48]

Tyndall's philosophy of science was now taking shape. In an article on 'Goethe and Faust', published in the *Preston Chronicle*, he revealed his thinking, and the need for explanatory models or physical pictures: 'In explaining many of the phenomena of nature we are obliged to use a sensible image as a satisfaction to the intellect; magnet-ism and electricity demands a hypothetical fluid; chemical combination demands the atomic hypothesis; polarization the hypothesis of ether particles and vibratory motion'. He went on to draw a parallel with moral nature: 'The moral experiences also have their imagery; and Goethe has given them under angels, devils, warlocks, and witches'.[49] In a five-part series entitled 'Sisters of the Rhine', which was again published in the *Preston Chronicle*—for which Hirst thought he could have got him payment—Tyndall mused on wider matters of philosophy, politics, and relationships.[50]

* * *

It was Hermann Knoblauch (Figure 6) who gave Tyndall what proved to be his big break in science. Knoblauch was one of a strong group of German savants, including Emil Du Bois-Reymond, Rudolf Clausius, Hermann Helmholtz, and Wilhelm Siemens, who had studied in Gustav Magnus's laboratory in Berlin. Like many others, Knoblauch was intrigued by Michael Faraday's recent discovery of diamagnetism, the weak repulsion of substances by a magnetic pole. This was in stark contrast to the strong attractive magnetic force exhibited by a few substances such as iron. Knoblauch

Fig. 6. Hermann Knoblauch (1820–1895).

invited Tyndall to join him in his researches. To follow up Faraday's investigations, Knoblauch had apparatus made in Berlin, although his primary interest was in radiant heat, a subject to which Tyndall himself later turned with great success. Knoblauch's time was limited, however, and the work fell largely to Tyndall. He started his experiments at the end of November, just as he 'hewed the last difficulty of [his] dissertation to pieces'.[51] His work over the next six years on this subject shaped his whole career, and Knoblauch's exploration of the relationship between crystal structure and the transmission of heat in different directions may well have influenced Tyndall's own efforts to relate observed properties to underlying structures. These ideas suffused Tyndall's work on diamagnetism, and guided all his later research.

Taking just one break, when he met Liebig at his laboratory in Giessen, Tyndall worked right through December (including Christmas Day and into New Year's Day) on his diamagnetism experiments. He was frequently up at 5am and working with Knoblauch in the laboratory to 7.30 or 8pm. During this time he made some attempt at a life outside work, attending one evening ball, although he refused two female advances, for which he 'got laughed at of course and censured and scolded'.[52] He got to know Heinrich Debus: 'an extraordinary fellow—very narrow in many of his views, but powerful and original in the advocacy of all'.[53] Tyndall defended his doctorate in the new year (Figure 7), and was able to devote his full attention to his research.

The aim of Tyndall and Knoblauch was to test the findings of Julius Plücker, a German mathematician turned physicist, who had also been following up Faraday's discovery of diamagnetism. Plücker thought that the optic axis of the crystal determined its orientation when repelled by a magnet. By the beginning of 1850, Tyndall had disproved this, using in one case a piece of Iceland Spar he had chipped six years earlier from a crystal on the mantelpiece of the inn at Frognall in Staffordshire, where he had stayed while surveying. It had become clear to him that the cleavage plane of the crystal (the direction in which it splits when sharply struck) was a determining feature, an idea that developed into his belief in the importance of molecular structure. However, the way he established this is revealing. Rather than using variably-shaped crystals, he developed a carefully controlled series of experiments using cubes, discs, and thin bars of different materials, thus simplifying the geometric complexity of the problem. He also paid meticulous attention to chemical composition. Even at this early stage, his consummate skill in experimental design and execution was evident. On 22nd January Tyndall's and Knoblauch's paper was on its way to Giessen, from where Frankland took it to England the next day.[54] It was published in the *Philosophical Magazine*,[55] the most important journal for papers in natural philosophy at the time in Britain, apart from the Royal Society's *Philosophical Transactions*.

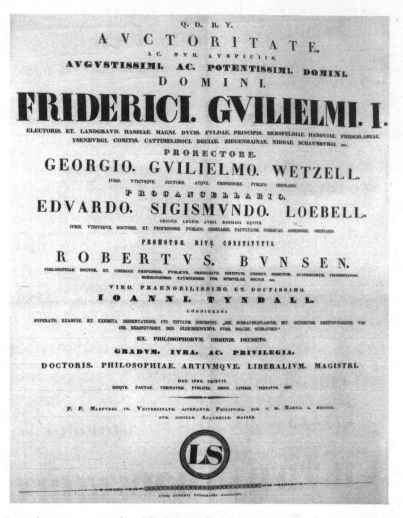

Fig. 7. The announcement of Tyndall's doctorate from the University of Marburg in March 1850.

The success of these initial experiments encouraged Tyndall to continue, and to try to find a way to stay on in Germany. He asked Hirst to help him find someone who would pay for a series of sketches on German university life to fund his stay. So far none of his published poetry or prose writing had earned him a penny. As he explained to Hirst, he had obtained clear new results, 'and this threatens a very beautiful theory which has obtained considerable celebrity for its propounder. He will be dreadfully angry no doubt and will do all in his power to overturn us, but in this he will assuredly fail Tom, for Nature is on our side, and we shall limit ourselves strictly to declaring what she has confided to us'.[56] These new findings disproved Plücker's conclusions about optically positive and negative crystals, and his law that the diamagnetic force decreases more slowly than the magnetic, which Tyndall and Knoblauch showed was an artefact of his experimental arrangement. Their second paper, which Tyndall later referred to as the First Memoir, was published in the German journal *Poggendorf's Annalen*, and then in the *Philosophical Magazine* in July.[57]

This paper was noteworthy not only for its demolition of Plücker but also for the challenge it laid down to Faraday, widely regarded as the greatest physicist of his generation. Faraday had proposed that a 'magnecrystallic' force determined the orientation of a diamagnetic substance in a magnetic field, which ran counter to Plücker's identification of the optic axis. In a clever series of experiments with powdered substances subjected to pressure, Tyndall showed that it was not necessary to posit any such new force. All Plücker's and Faraday's results could be explained on the basis of Tyndall's concept of a 'line of elective polarity', dependent on underlying structure, and of the geometries of attraction and repulsion. The behaviour of a substance between magnetic poles was determined by its chemical composition and its density in different directions, a clue to which was given by the plane of cleavage. Already Tyndall's two defining characteristics as a physicist were evident: his desire to understand and explain phenomena in terms of structure, and his experimental skill, which allied a dexterity of execution with the ability to create simple physical models and systematically control variables.

In mid-February Hirst sent Tyndall £20, which enabled him to complete his experiments in Germany. Hirst also told Tyndall in passing of an 'authoress an anonymous one though privately known as a young lady (28) living at Haworth near Keighley... She has written a novel called "Jane Eyre"...that would do you good to read. By the law, I did not think it possible that we could possess such a precious heap of feminine flesh, & such a heart in it, too! I have fallen positively in love with her. Philips & I are going to prostrate ourselves at her feet some day soon'.[58] Tyndall replied, 'Do present my vows also at the shrine of "Jane Eyre"—and be sure to tell her that I am the nicest young man of the three!'.[59]

The *People's Journal* turned down Tyndall's idea for sketches of German university life, but he was able to continue some (unremunerated) writing. In March he sent a poem on the death of Dean Bernard to the *Sentinel*,[60] and the *Preston Chronicle* published 'Man and Magnetism', in which the Carlyle-inspired Tyndall praised the 'man of steel' who holds to his resolve on all accounts, influencing others around him.

Such people act on each other, he said, their strength rebounding, rendering 'all tory-isms and quack conservatisms in the long run impossible'.[61] In 'Zig-Zag', stimulated by the arrival on 20th March of the first locomotive in Marburg, Tyndall took the opportunity to remind readers of the early and ridiculed work of Denis Papin on the steam engine in this very town.[62] He also showed his understanding of the development and provisionality of scientific knowledge, and the danger of using it as evidence for theology: 'Science is valuable…but we must beware of making it the foundation of moral or religious convictions'.[63] He ended with a ringing criticism of manufacturers who exploit their workers by not fairly sharing the profits. Such people needed fettering, he said, and he thought 'noble-hearted men' of their own class would see to this.[64] He did not believe revolution was necessary.

Tyndall continued to read Carlyle's work as it was published. In June 1850 he described buying 'Parliament', the latest of the *Latter Day Pamphlets*, and reading it in Trafalgar Square.[65] He bought others as they appeared. Tyndall's instinctive belief in the 'right' of the man of 'might' to rule reasserted itself against any radical tendencies. His religious sense was equally influenced or reinforced. By July 1850, if not before, he was reading *Sartor Resartus*, along with Emerson's poems (and Dickens's *Nicholas Nickleby*). They made an immediate impact. Reading them, Tyndall wrote: 'Emerson is a pantheist in the highest sense and so is Carlyle. I dropped an hour ago upon a very significant passage in the Sartor. "Is there no God then, but at best an absentee God sitting idle ever since the first Sabbath at the outside of his Universe and seeing it go?" At the "outside" of his universe. I imagine Carlyle's untrue creed is folded in this Sentence. And here the difference between his faith and that of Paley's is very distinct. According to the latter God bears the same relation to the Universe that a clockmaker does to the clock. He is an omnipotent mechanic detached from his work. With Carlyle the universe is the blood and bones of Jehovah'.[66]

Thomas Huxley wrote to Charles Kingsley in 1860 that '*Sartor Resartus* led me to know that a deep sense of religion was compatible with the entire absence of theology'.[67] The message for Tyndall was the same. What Carlyle also gave was moral impetus, self-discipline, and the commitment to practical action in the world. In his 'Address to Students' much later, Tyndall said that without Carlyle and Emerson 'I never should have become a physical investigator, and hence without them I should not have been here today. They told me what I ought to do in a way that told me how to do it, and all my consequent intellectual action is to be traced to this purely moral force'.[68]

Tyndall had intended to leave Marburg at the end of June, spend a short time in Berlin and then return to England.[69] In the event he left earlier, postponing the Berlin visit until the following year. He told Hirst: 'My last investigation has been of infinite value to me, nature has stood here and there at the gaps and corners of the enquiry pointing in this direction and in that, and seeming to smile down upon me the promise: "I will reward you my son, if thou persevere". To speak plainly, I see a rich unexplored field before me and my object now is to fix myself in a position to cultivate it'.[70] Nevertheless he needed a job, and finding one in science would be difficult. His plan was to work to finance a return to Germany.

Hirst now aimed to come to Marburg himself. He asked Tyndall about his plans for his return to England; whether he might seek a situation as a tutor or surveyor, perhaps working again with Carter. Hirst told Tyndall that Phillips wanted to talk to him about a scheme to establish 'a People's College…It is a favourite notion of his for giving first rate instruction to classes who could not afford to go to Cambridge…establishing an University something after the German fashion (with regard to expense especially)'.[71] Tyndall, however, found the idea of spending eight hours a day measuring and plotting uninviting, thinking instead that he might throw himself into a private school till he had time to look around elsewhere, and while Edmondson had asked if he would return to Queenwood, Tyndall preferred to avoid it if possible. He thought that Wynne, now Government Inspector of Railways, might help him find a position, and at Wynne's request, sent a statement of his qualifications. In writing, Wynne had also asked Tyndall if he still adhered to the pure doctrine of the Church of England. Without specifically answering no, but determined to give a correct impression of himself, Tyndall fell back on Carlyle and Emerson: 'Doubtless you are acquainted with the writings of both, and if so will have observed that though they never attack the Church of England nor any other church, their writings in many places breathe a spirit to which the Church of England would scarcely subscribe'.[72]

Just before leaving Marburg, Tyndall attended both the Lutheran and the reformed churches. While he was singularly unimpressed by the sermon in the former, he did note the lack of social segregation. Perhaps it was his growing confidence with the language, but he also attended a spree of balls. He promised Hirst a welcome in Marburg, writing: 'it will be impossible to remain until you come, but I will so arrange matters in England that I may accompany you and see you comfortably installed'.[73] On 12th June Tyndall took a carriage to Giessen, and then went on to Frankfurt. As he left, a 'fair head and face leaned through one of the high windows' in the Elizabeth Strasse. Tyndall looked up, raised his hat, 'and so the matter ended'.[74] We know no more than that.

In Frankfurt he visited the art gallery, paid homage at the statue of Goethe, and went to the theatre, although he was unimpressed by the audience's enthusiastic reaction to a presumably scantily clad dancer: 'I am inclined to think that the stage as a moral agent would be improved by the entire abolition of ballet dancing'.[75] In the cathedral he regarded the organ, the architecture, and the memories of ages as 'the present spells of the Romish faith', holding the people in their power.[76] He compared the characters of the French and the Germans to the detriment of the French: 'the Frenchman all sparkle and glare and fanfaronade; the Prussians with their sunshiny gemütlichkeit; honest, unequivocal, sincere'.[77] Gliding down the Rhine from Mainz to Rotterdam, past the stunning scenery and castles, he mused on beauty and knowledge: 'Beauty comes not at the nodding of the will, it will not be coerced but flows freely into the open heart...Human knowledge seems to me to reduce itself into the question of what are laws and what are not, this once discovered content is the consequence'.[78]

Having crossed the Channel, Tyndall's luggage was examined at Catherine's Wharf, and he was forced to pay duty for a battery he had brought over. He set off to Hammond's Hotel in Covent Garden. The first person he saw, 'tall and symmetrical, with a face as beautiful as Ariadne', was a prostitute.[79] Walking along the Strand to Temple Bar a stream of girls flowed past, 'some with the soil of sin upon their beautiful cheeks, the taint deepening as the years advanced, until finally it was written in brazen characters legible to all the world...Such sweetness, such beauty, abolished for ever'.[80]

Tyndall was back in the vibrant metropolis. Now he had to find a way to break into the world of science.

PART II

1850–1860
BREAKING IN

5

MAKING A NAME
1850–1853

The opportunities for making a living out of scientific research in the 1850s were limited. The independently wealthy, like Charles Darwin, could pursue their interest. But for most people, it was a case of finding time while holding other positions: in the church, medicine, or military service. From the outset, Tyndall's desire was for fundamental research. He left its exploitation, throughout his life, mostly to others. With this mindset, his most likely option was to secure a post in a university or similar institution. Competition was fierce, and success relied more on patronage than merit. Even then, any lecturing commitment and the need to build up income from student fees would restrict research time.

Tyndall spent ten days in London after arriving from Marburg. He visited the offices of the *Philosophical Magazine*, one of the oldest scientific journals published in English. Its driving force was Richard Taylor, whose son William Francis (though not publicly acknowledged as such) joined him as editor in 1852 as the firm Taylor & Francis was established.[1] Tyndall quickly came to know both Taylor and Francis, and the latter became a crucial figure in his entry into the scientific world. So did Faraday, Director of the Laboratory at the Royal Institution, whom he visited for the first time, and 'felt warmly welcome'.[2]

Tyndall realized that making his way would be difficult. But he heard from John Phillips, assistant secretary of the British Association for the Advancement of Science, that his paper on diamagnetism was of interest.[3] So after a couple of weeks with Singleton at Spring Bank, and a few days with Hirst in Halifax, he took the train from Manchester to Edinburgh at the end of July for his first major scientific event, the 1850 Annual Meeting of the British Association.

The British Association had been founded in 1831, in response to concerns about the decline of science in Britain and the perceived inability of the Royal Society to address them.[4] It arranged a meeting in a different city each year, at which scientific papers were read and discussed, new initiatives and lobbies planned, and reports commissioned to advance scientific activity. More inclusive than the Royal Society, Britain's elite academy of science, the British Association allowed women to attend and, though this was uncommon, to speak, the Royal Society did not admit women as Fellows until 1945. Nevertheless, at the British Association the role of women was primarily passive and social. Their presence helped create a public supportive of science, but was largely unchallenging to the position of the men of science.[5] Few attended Section A, Mathematical and Physical Science, in which Tyndall's paper would be given. The meeting in Edinburgh was Tyndall's introduction to the big guns of science. James Forbes was President of Section A. The Professor of Natural Philosophy at Edinburgh, and a Tory Episcopalian, Forbes had discovered the polarization of radiant heat and carried out extensive work on the motion of glaciers. Also attending Tyndall's session was Sir David Brewster, like Forbes, one of the founders of the Association. Brewster, a devout Presbyterian, concentrated on optics. He was one of the four editors of the *Philosophical Magazine*.

Tyndall's paper, on 2nd August, was the second of the day.[6] He read a poem by Emerson to calm his nerves. Flanked by Lord Wrottesley, who became President of the Royal Society in 1854, Tyndall gave his paper faced by 'a bunch of ladies with mild brown eyes and every time I raised mine I found theirs fixed on me as if I had been reading the story of Jack and the beanstalk or something else equally interesting'.[7] According to the *Athenaeum*, the paper gave rise to 'a very animated discussion'. Tyndall described it as 'a hand to hand fight'.[8]

It was the first occasion, but by no means the last, on which Tyndall took on the Glasgow-based William Thomson, later Lord Kelvin (Plate 4). He was a few years younger than Tyndall but already had a high reputation. Thomson was a Vice-President of Section A along with the Cambridge physicist George Stokes, who was also present. Forbes, who had seen Plücker's experiments in Bonn, was impressed. He remarked: 'Here we have a memoir which tends directly to invalidate the views of Faraday and Plücker'.[9] Thomson praised the 'beauty and ingenuity' of the experiments, challenged some of the details, and then defended Faraday's view of the 'directive force' of magnetism, as opposed to the combination of attractive and repulsive forces for which Tyndall argued. Tyndall had seen Faraday a few weeks before, telling him he felt compelled to differ. Faraday had replied, 'No matter, you differ not as a partisan, but because your convictions compel you'.[10] No one else tackled Tyndall, and the discussion continued between him and Thomson while people thronged into the room.

Tyndall had made his mark, but was unable to capitalize on it. Despite staying in a cheap temperance hotel, he did not have enough money to attend the rest of the meeting, and missed Thomson's paper. He wrote ingratiatingly afterwards for some references from Thomson, who replied at length, suggesting possible experiments.[11]

Tyndall was not sure that he could afford to return to Germany. While based with Singleton at Spring Bank, he explored alternatives to an uncertain scientific career. Captain Wynne, his former boss at the Ordnance Survey, invited him to stay and offered advice. Tyndall was conscious of Wynne's social status but confident of his own intellectual equality: 'Rank offers no barrier to the intercourse of related minds; it is a shield and fence at times and must be dropped when men of natural quality meet each other'.[12] He wrote at Wynne's suggestion to Captain Yolland, who had offered him surveying work two years before, and even to Edmondson about the possibility of returning to Queenwood. Yolland, the past revolt forgiven and forgotten, offered him a position at £90 per annum,[13] and Edmondson jumped at the chance of employing him. But Tyndall wanted to return to Germany. He asked Hirst to let him know if he could make £40 available, which might enable him to postpone returning to Queenwood, to exploit his success at Edinburgh and respond to Thomson's suggestions.[14] Hirst was happy to oblige. Edmondson agreed to a deferral for six months, and Tyndall made plans to travel out with Hirst. The plans were cemented when Francis offered to pay for translations of scientific papers for the *Philosophical Magazine*. Wynne—who had also offered him money at any time— warned him that it was a utilitarian age, and advised him to look seriously for work.[15]

Singleton and the boys bade him tearful farewells at the end of September, and he stayed with the Gintys overnight in Manchester before heading to London. Hirst and Jemmy Craven were waiting for him. Hirst stood clutching £2 from the *Leader* for Tyndall's article on 'The Propensities and their Equivalents', about the Carlyle-inspired moral duty to meet challenges with hard work.[16] They were his first literary earnings. Tyndall's choice of the *Leader*, initially edited by George Lewes, is illuminating. Founded that year, 1850, this weekly newspaper lasted until 1860. Radical in its politics, it supported a universal religious sense unfettered by dogmatism, and it was Hirst who had suggested he try it. Coming back from the Haymarket Theatre that evening, accosted frequently by prostitutes, Hirst felt Tyndall's moral strength beside him and the shame of temptation, to which he appears to have yielded three years previously.[17] It seems unlikely that Tyndall ever succumbed. They went on a quick visit to Queenwood, where Tyndall gave a lecture on light. Back in London he called unsuccessfully on Carlyle (who was out), visited Francis to arrange about the translations, dropped in on Frankland, and stayed overnight with the Wynnes before taking ship with Hirst for the continent.

* * *

Tyndall returned to Marburg in mid-October 1850 to a country in turmoil. Hesse Cassel was in a state of internal warfare, caused by the unpopularity of the Elector and his ministry, who were attempting to levy taxes without the assent of representative government. Prussian soldiers were billeted in Tyndall's house, rushing out on one occasion as rumour spread that the Bavarians were approaching. That was a false alarm, but on 12th December about 3,000 Bavarians and Austrians did enter Marburg, 'owing to agreement between Prussia and Austria...In every house are billeted Bavarians and Austrians who pay nothing'.[18] By the end of the year, the Bavarians were still in Marburg, and the spirits of the people low. Tyndall worked away against this background of political unrest, spending the first week on translations for Francis and another article for the *Leader*.[19] Knoblauch brought him up to date on Plücker's progress, and he soon restarted his experiments.[20] He did not neglect to cultivate Faraday, sending him crystals of calcareous spar but carefully restating his opposition to Faraday's theory.[21] Faraday was unperturbed, welcoming the work of others: 'Where science is a republic, there it gains; and though I am no republican in other matters, I am in that'.[22] But Tyndall's collaboration with Knoblauch now drew to a close. He had noticed in November that Knoblauch was taking credit in front of Bunsen for his work, and resolved to dissolve their 'curious partnership', though they remained friends over many years.[23]

In the period up to Christmas, Tyndall spotted an opportunity to establish the laws of magnetism for bodies in contact or separated by small distances. Lenz and Jacobi had described the law for bodies at a distance, when the magnetic force is proportional to the square of the magnetic strength. Tyndall's extensive investigation, showing that the mutual attraction of a magnet and a sphere of soft iron in contact is, by contrast, directly proportional to the strength of the magnet, demonstrated his careful experimentation. Using spheres of material, he teased out with great precision the changing relationship as the distance between the sphere and the magnet became extremely small. Faraday wrote of the results, published in the *Philosophical Magazine* in April: 'it appears to me that they are exceedingly well established and of very great consequence'.[24]

Looking for other avenues to explore, Tyndall investigated the behaviour of water jets, stimulated by the work of Magnus which he did not think fully explained them. He published his results in February 1851, explaining how air bubbles are formed, and how their bursting creates the sound of moving water, or the 'murmur of a brook'.[25] This required irregular motion. He wrote, in a typically visual phrase: 'Were Niagara continuous and without lateral vibrations, it would be as silent as a cataract of ice'. An effusive letter from Francis on 31st December enclosed the proof of the water jet article and a proposal for a monthly 'Report on the Progress of Physics' in the *Philosophical Magazine*, with translations of French and German papers. Tyndall started the Reports at the beginning of 1851. The payments for translations and reviews enabled him to spend longer in Germany than he had thought possible. The work had the additional benefit of familiarizing him with a wide range of the latest researches in physics, and of bringing him to the grateful notice of many of the continentals, who now found their papers published in England. In time, Tyndall became perhaps the most significant networker of Continental and British physicists.

At the end of February, De Haas wrote to say that Edmondson would like him back at Queenwood as a teacher.[26] Tyndall had calculations to make. He strongly wanted to remain with his researches, but needed money. He asked Francis how much he could rely on from translations, even offering to write 'rhymes, romances, & metaphysics' if Francis could get him a literary position.[27] Francis regretted he could not provide more than £30 per annum.[28] He advised Tyndall against emigrating to Australia—an indication of Tyndall's desperation for work, though it is unclear what he planned—and promised to look out for posts if he stayed longer in Germany. Perhaps he might pick up some public lectures at the Royal Institution or London Institution, or a place at one of the medical schools or other colleges; just at this time, Frankland was appointed to the new Owens College, forerunner of the University of Manchester. Tyndall again asked Wynne's advice. Though Wynne had earlier warned him of the difficulties of his chosen course, this time he saw that Tyndall sought 'fame rather than money', and was more optimistic: 'I do not think these are the days when a man who has with ability & industry devoted himself to science and added to our stock of knowledge will be suffered to linger in obscurity or feel want'.[29]

So Tyndall decided to stay in Germany. He proposed to the Edmondsons that they invite Debus to Queenwood, while he planned to spend time in Gustav Magnus's private laboratory in Berlin, where many of the foremost German physicists had worked.

Taking a break beforehand, Tyndall left Marburg on 6th April for a long walk with Hirst through the thickly forested landscape around the Lahn valley. Over a week they walked about 100 miles, ending up in Wetzlar, where Tyndall imagined that Goethe had pondered over *The Sorrows of Young Werther*, set in a fictional village nearby. They talked as they walked of philosophy and life, of religion and its relation to science. Effectively rejecting the religious background of his mother, and of Edmondson, Tyndall said of the Quakers: 'They have quelled enthusiasm and shorn the heart in a great degree of its natural impulses'. A few months before, he had outlined his beliefs to Hirst, as his attraction to transcendentalism became evident.[30] He thought the universe best illustrated by a human body:

> All are but parts of one stupendous whole
> Whose body nature is and God the soul.[31]

Like Carlyle, Tyndall did not believe in a detached God, Paley's maker of the watch, sitting outside the universe. Nor did he believe in a personal or personified God, yet he still felt some pervasive power in nature. He thought Harriet Martineau's book—*Letters on the Laws of Man's Nature and Development*—missed the point, criticizing its atheistic and positivist or 'mechanist' stance: 'The transcendentalists understand the mechanics but the mechanics cannot understand the transcendentalists'.[32]

Tyndall found lodgings in Berlin with Herr de Baux, a painter, at 95 Dorotheen Strasse near the Brandenburg Gate. The room was noisy, always a problem for Tyndall, but he managed to get used to it and drafted his paper for the British Association. Soon after his arrival, Knoblauch took him to the meeting of the Physikalische Gesellschaft (Physical Society), which had been established in 1845 by Emil Du Bois-Reymond (Figure 8), Knoblauch and four other young researchers. He found he was to be elected, and hurriedly prepared a short lecture at the request of Du Bois-Reymond, the President. Doing the rounds a few days later he met Du Bois-Reymond properly—'athletic, with a neck like an ox and a chest like Hercules'.[33] Under his tutelage he repeated Du Bois-Reymond's celebrated experiment of creat-

Fig. 8. Emil Du Bois-Reymond (1818–1896).

ing an electric current with the muscles of his arm (boasting competitively that Knoblauch could barely cause a deflection). He liked Du Bois-Reymond immediately: 'there is a healthy straightforwardness about him. He may be a little vain of his production, but that is external and will fade away'.[34]

Moving from Marburg to Berlin, Tyndall was now at a great centre of German physics. His visit established relationships that lasted for life. He was introduced to Gustav Magnus (Figure 9), 'a friendly looking little man', who would make a place for him in his laboratory.[35] He met Heinrich Dove, Peter Riess, and Johann Poggendorff, editor of the main German journal of physics, finding the same kind reception everywhere. Another physicist he met briefly was Rudolf Clausius (Figure 10), whose paper critiquing the work of Thomson he had translated. He wrote: 'I should like to have known more of Clausius', a wish that was granted later and developed into his closest friendship on the continent.[36] At the house of Eilhard Mitscherlich, Tyndall was introduced to his son-in-law, the physicist Gustav Wiedemann, and Wiedemann's young wife Clara, with whom he spoke English. She would work later on English translations of his books. Even the great Alexander von Humboldt, author of *Kosmos*,

Fig. 9. Gustav Magnus (1802–1870).

Fig. 10. Rudolf Clausius (1822–1888).

gave him an audience, declaring that: 'The astute and, at the same time, thorough works of Mr John Tyndall are well-known to me'.[37] He found that they all looked 'with wonder—almost worship—upon Faraday'.[38] Knoblauch's father gave him a ticket to Weber's opera *Oberon*: 'Theatre beautiful, scenery beautiful, music beautiful, and I sat there most of the time a dumb, gratified listener'.[39]

Tyndall was soon at work on diamagnetism in Magnus's laboratory. In a few days, he recorded that he had been 'lucky beyond anticipation. Diamagnetism behaves exactly like magnetism; a double current excites a double amount'.[40] He wrote to Faraday, reporting: 'it has been again my misfortune to arrive at conclusions very divergent from those of Prof. Plücker'.[41] To his disappointment he found that Edmond Becquerel had published similar results first.[42] But to Hirst he was sanguine: 'He has not exhausted the matter however, and my method of experimenting is better than his'.[43] Hirst's concerns lay elsewhere, teasing Tyndall about his lack of romantic experience: 'In last week's *Leader* is your "Forester's Grave"; it is a strange one for you...there is evident intellectual rust about the love machinery'.[44]

Francis was proving enormously supportive. Tyndall wrote to Hirst: 'He has sent copies of my Memoir to Wheatstone, Sabine and many others—on the whole his bearing towards me is most good natured'.[45] Francis obtained for Tyndall the position of Berlin correspondent for the *Literary Gazette*, though Tyndall declined it because his experiments took priority. He also tried to arrange for Tyndall to lecture at the British Association, a third general lecture in addition to the usual two. It would have been a remarkable coup—the planned lectures were from the big names of Richard Owen, the biologist, and George Airy, the Astronomer Royal—but it did not come off.[46]

Tyndall left Berlin towards the end of June, having seen little of the city while working so hard. Now aged about thirty, he was conscious of the value of three years in Germany, of what he had achieved in difficult circumstances and what he yet might achieve in science.[47] Reflecting Hirst's strictures on his romantic senses, he wrote: 'I lack the warm aspirations which I once felt, and I believe this is a necessary consequence of my pursuits:

> Love is exiled from the heart
> When knowledge enters in.[48]

Though he continued: 'the presence of a good child reassures me that a considerable portion of heat is still within me... But the love of maidens is always damaged, almost annihilated, by the conviction of their imperfections... The best that I have known are only blest with casual irradiations, and this is not enough'.[49] A few months later he declared: 'for it is to preserve the wild independence of mine that I have forsworn marriage up to the present hour'.[50] His romantic idealization of women cannot have helped.

Leaving Germany, Tyndall headed for the 1851 British Association meeting in Ipswich, where he lodged with Francis and came into contact again with Faraday. Here also he met, probably in the train to the meeting, the naturalist Thomas Huxley, later 'Darwin's Bulldog', who became a close friend (Plate 5). Faraday, whom Tyndall encountered in the street, wanted to hear the paper on diamagnetism. As he could only stay until the next day, it was duly brought forward, though several of those whom Tyndall would have liked to hear it were occupied with the visiting Prince Albert 'and his train of asinine flunkeys'.[51] In this paper, the 'Second Memoir', Tyndall used a torsion balance to measure diamagnetism in bismuth and disprove Plücker's proposition that the laws governing magnetism and diamagnetism were different.[52] Faraday was impressed. He 'felt prepared to admit that that some of Dr Tyndall's results seemed to promise an explanation of Plücker's perplexing results and conclusions'.[53] In all, Tyndall had four papers accepted for this meeting, as he sought to make his presence felt, but administrative blunders meant that only three of them were given; the other two tackled air-bubbles in water and thermo-electricity.[54]

Tyndall's disagreement with Faraday, and with Thomson, who had put Faraday's ideas into mathematical form, centred on their different ways of explaining the same experimental results. Faraday believed that diamagnetic substances simply moved from places of stronger to weaker magnetic force. There was no attraction or repulsion, just a directing force. This is the essence of Faraday's field theory, and of his extraordinary contribution to magnetism. Tyndall, by contrast, saw diamagnetism in terms of attraction and repulsion of particles with polarity, analogous to the attraction and repulsion of electric charges. This view was shared by many other physicists, including the continentals such as Plücker and Wilhelm Weber. Faraday was in a minority. So, for Tyndall, diamagnetic substances had to be polar. Following the Ipswich meeting he set out to demonstrate that diamagnetic polarity exists.

After a gap of a fortnight, back at Queenwood, he remarked with relief: 'This night finished my memoir on "Diamagnetic polarity". I never laid down my pen in greater physical prostration'.[55] This paper became the 'Third Memoir', out of six eventual major publications on the subject.[56] Tyndall's skill at experimental design and execution was again evident, as he prepared a sample of the diamagnetic metal bismuth so that it would show larger effects in a magnetic field, and devised experiments with a small sphere of carbonate of iron as a sensitive means of testing the relative force at various places in the field. He showed that the movement of this sphere could indeed be explained in Faraday's terms. Nevertheless, the voice of the believer in diamagnetic polarity asserted itself. He claimed that Ferdinand Reich's experiments, showing that matter evoked by one pole is not repelled by an unlike pole, 'compels us to assume the existence of *two kinds* of matter, and this, if I understand the term aright, is polarity'.

* * *

This proved to be the end of the first phase of Tyndall's work on diamagnetism. It had brought him to the attention of many of the foremost physicists of the day, but he was no nearer to finding a position from which to continue his researches. He determined to live abstemiously and to whittle down his debt. He settled at Queenwood in company with Debus, with a day out to visit the Great Exhibition of 1851. The College was at a low ebb, with dwindling numbers. Tyndall read the riot act to the 'farmer' students, who were like 'untamed colts…I took them together and…got them all to promise that they would not enter a public house nor visit any of the neighbouring villages without permission'.[57] When they returned again one evening in 'outrageous disorder', he called their bluff by threatening to resign if he could not have them expelled. From then on, all was good behaviour.

Tyndall was earning £150 a year plus board and lodging at Queenwood. But he was concerned, as he wrote to Faraday, that the teaching of elementary mathematics and physics would lead to 'retrogression'. He had become aware in Ipswich of a professorial post in Toronto, offering £350 per annum. He asked Faraday's advice, and whether he had a hope of finding a scientific position in England, where he would prefer to stay.[58] Faraday's response was that in his place he would choose Toronto, and he offered to be helpful if asked for a reference.[59] Tyndall boldly sought testimonials from more than a dozen contacts, including Bunsen, Magnus, Becquerel, Thomson, Forbes, James Joule, and Edward Sabine. The sheer number of significant people he could ask, some of whom he barely knew, and who were positively responsive in return, shows how quickly he had made his mark in Britain and Germany. Hirst advised caution about Toronto, suggesting that Faraday did not know what kind of place it was.[60] Bunsen likewise queried the choice.[61] But the testimonials for Toronto, to which Huxley was also applying for a post, were sent off by Francis 'with a sad heart' at the end of October. Taylor and Francis attempted to influence matters by writing to Sir Robert Kane—an editor of the *Philosophical Magazine* and the first President of Queen's College, Cork—about a professorship there. But this turned out not to be vacant after all when the incumbent withdrew his resignation.

Just as Francis sent off the testimonials for Toronto, he told Tyndall of another post in faraway Sydney, for which the professors were to be chosen by the astronomer and mathematician Sir John Herschel, George Airy, and two others. Tyndall sent Herschel the testimonials already collected for Toronto, and a supportive letter from Faraday.[62] Huxley wished him well. He envied Tyndall's chance of going to Sydney where his fiancée, whom he had met when on the *Rattlesnake* expedition, was still living, while Huxley sought sufficient financial security to enable them to marry.[63] Tyndall recollected that by coincidence he had met Huxley's Captain Stanley—who died in Huxley's arms on the *Rattlesnake*—in Preston nine years before. When he heard from Herschel in early 1852 that he had not been appointed, he took the decision calmly; his preference was for a position in Britain.[64]

But he had caught the eye of many, not least that of Edward Sabine (Figure 11), the reforming Treasurer of the Royal Society. Sabine was one of a group who sought to reduce the influence of aristocrats with a dilettante knowledge of science, and elect as Fellows men of scientific ability. His interest in Tyndall may have been stimulated by the value he placed on the study of magnetism; he had instigated the magnetic survey of Ireland in 1834. On hearing that Tyndall might go to Toronto, Sabine put in motion his election as a Fellow of the Royal Society.[65] Faraday offered, unprompted, to sign the certificate, and Tyndall secured the mathematician James Sylvester and Huxley as two further signatories. William Grove, the physicist, and John Phillips

Fig. 11. Edward Sabine (1788–1883).

signed from personal knowledge, with Airy, Playfair, Charles Wheatstone, the physicist and inventor of the electric telegraph, Edward Forbes, the biologist, and Thomas Henry, the brewer, from general knowledge (Plate 7). Tyndall, who frequently alleged in his journal that he did not push himself forward, noted with satisfaction that: 'the movements have come spontaneously from others, which is of course all the more gratifying'.[66]

The Royal Society was, and remains, the powerbase of the scientific elite. Tyndall became a Fellow at a time of transition, from an organization encompassing aristocratic, political, and scientific membership to one in which election is almost entirely on scientific merit. This move, instigated in 1847 by limiting election to fifteen Fellows each year was led by a group of reformers, including Sabine and Grove. Of Tyndall's close friendship circle, Huxley had been elected in 1851, and Frankland would be elected in 1853. But there is a big difference from today. Given the explosive growth in the number of scientists, the proportion who are Fellows of the Royal Society now is vanishingly small. In the 1850s this was not the case. A significant proportion of the active researchers were Fellows, and their meetings were interdisciplinary.

Based at Queenwood, Tyndall no longer had access to a major laboratory. Instead, convinced by his work on magnetism of the importance of the structure of substances in determining their properties, he carried out experiments on the conduction of heat through wood. He aimed to probe the relationship between the structure of the wood and its ability to conduct heat in different directions. Nor did he have the latest scientific literature. Discovering that he did not have access to the *Philosophical Transactions*, Faraday lent him his personal copy of his recent paper on lines of magnetic force. In the middle of January 1852, after struggling for weeks with his experiments on the conduction of heat, a new experimental design struck Tyndall just as he went to sleep. Though apparently a chance thought, he felt that all his hard work had cleared his intellect, which was now 'sensitive like a daguerreotype plate to the ray of light which then entered'.[67] His skill in designing delicate instruments, often using the best German apparatus, enabled him to obtain a precision that few could match. He aimed to write up the work as a paper on molecular influences for the *Philosophical Transactions*, the journal of the Royal Society.[68]

While pushing forward with this research, Tyndall kept up his wider involvement by continuing to provide translations and overviews for Francis, for publication in the *Philosophical Magazine* and Taylor's *Scientific Memoirs*. During 1852 he published four more 'Reports on the Progress of the Physical Sciences' in the *Philosophical Magazine*, and other translations and responses, covering work by people such as Magnus, Riess, and Dove. Francis also asked him to review a text book on *Elementary Physics* by Robert Hunt, Professor of Mechanical Science at the Government School of Mines. Tyndall, writing anonymously, gave an excoriating verdict: 'every page of the book is a commentary on the incompetency of its author'.[69] This friendship with Francis offered Tyndall an unrivalled position, able to select and recommend work for issuing in England. It meant that his own papers had a hotline to publication, bolstered when he became an editor of the *Philosophical Magazine* in 1854. Tyndall may not have publicly 'pushed himself forward', but he knew how to play the game behind the scenes. He also understood the demands of journalism, asking Francis with respect to one article: 'Who is Woodhead? ... Does he cut any figure in the world? If so it would perhaps be worthwhile to demolish him…If he was a man of any mark…the review might be made a spicy article'.[70]

Tyndall's teaching ability had given him a reputation around Queenwood as a lecturer. He spoke in February in Warminster, and then at the Southampton Polytechnic Institution, on 'Light and Electro-Magnetism'. Despite mismanagement by the organizers, which put him under great pressure, the *Hampshire Advertiser* reported that the lecture was 'of a class…which we should wish to see frequently introduced'.[71] The newspaper had been equally impressed by his lecture on 'Magnetism' after a concert by Queenwood pupils in December. In a percipient remark, the writer commented that Tyndall's 'brilliant experiments reminded us of attendances at the lecture table of the Royal Institution'.[72] His immediate success there, at his first opportunity a year later, was no fluke.

Tyndall came down with 'flu' at the end of March. He managed to stagger to Romsey to give a lecture on magnetism, just as Hirst arrived unexpectedly at Queenwood from the continent to stay for a couple of weeks before Easter (having found his vocation in mathematics, a study he pursued for the rest of his life). Hirst was impressed by Tyndall's command of his classes. They had freedom but were ever obedient. Not only was Tyndall a superb teacher, he had a deep love for children, which he conveyed to Hirst: 'I feel a sweet joy on looking at our boys sometimes they are so free, healthy and happy. It's a pity that I am not a father, surely nature intended me to be one'.[73] He recognized his debt to them too, writing later after a joyous time with the children of a friend: 'Without them I should be a cold stern hardhearted fellow, but they fuse the hardness & banish the chill'.[74] Tyndall's educational philosophy was based on this sensitivity: 'the longer I live the more I am convinced of the necessity of gaining a boy's heart if you wish him and you to progress together'.[75] Hirst perceived Tyndall's powerful influence on the whole institution: 'Everybody in the place, from the principal down to the youngest pupil, is led by him more or less'.[76] Tyndall frequently voiced his dissatisfaction with Edmondson's leadership, which he considered weak and ineffectual. He thought it might make a good object of social research: 'Laws are best studied by examples which do not bewilder us by their vastness, and the science of government may I imagine be successfully studied in a school'.[77]

By early May, neither Tyndall nor Huxley had heard about Toronto. But Tyndall's path into the Royal Society was open, primarily owing to his work on magnetism. He was elected on 3rd June, one of fifteen recommended by the Council from an original thirty-four,[78] and received a letter instructing him when to attend for formal admission. He was taken suddenly ill and could not go, but was able to travel to London a few days later, after the end-of-term concert at Queenwood. He sought out

Francis and called at the Royal Society, where he paid his £10 fee and £4 for the *Philosophical Transactions*. That evening Huxley introduced him to the President, Lord Rosse, and he was admitted to the scientific elite. At the end of the meeting, he was given Thomson's report on his paper on molecular influences, with some remarks from Sabine. Tyndall commented: 'The report on the whole was a flattering one, but Professor Thomson, as is very natural to a young man, wishes to shew that he knows something about the matter'.[79] It was some years before Tyndall's jealous spikiness towards Thomson dissipated. To give him his due, Thomson highlighted Tyndall's 'results on the conduction of heat in wood which I believe to be new, and which are certainly very interesting', and praised his experimental method.[80] Indeed, Tyndall saw himself first and foremost as an experimentalist rather than a theoretician or 'analyst', even if Thomson would qualify as both. Faraday had written in his paper 'On Lines of Magnetic Force': 'As an experimentalist, I feel bound to let experiment guide me into any train of thought which it may justify'.[81] Tyndall heartily agreed: 'our knowledge is a thing flowing and not a thing fixed, and as long as this is the case the widest generalization which man has yet made is in danger of being swallowed up by one still more comprehensive. The analyst is the exponent of the centripetal tendency of the human mind, while the experimentalist is centrifugal, and seeks continually for still wider expansions'.[82]

Just after his election, Tyndall was awarded £50 from the Government Grant for his experiments on the conduction of heat. The Grant fund, initially £1,000, had been established in 1850 to support scientific research. It was administered by a Government Grant Committee of about forty Fellows of the Royal Society (Tyndall was a member from 1861 to 1883, though active only until 1878). Between 1852 and 1864, Tyndall received six grants, totalling £565, to support his work on the conduction of heat, diamagnetism, glaciers, and radiant heat. This money could only be spent on equipment. As the then President, Lord Wrottesley, was keen to make clear to the government when there was a threat of withdrawing the grant under the financial stringencies of the Crimean War, the grant obtained for the government 'the gratuitous services of men of first-rate eminence'.[83] Tyndall gave other service to the Royal Society, including serving four times on its Council. He was also listed in 1859, with Darwin, Huxley, and others, as a leader of the Scientific Relief Fund, initiated by John Peter Gassiot to support men of science and their families in financial distress.[84] He was appointed to the Committee itself later that year.

Tyndall was able to spend a couple of weeks in London, including a day with Edmondson interviewing candidates for teaching. It gave him the opportunity to develop and cement relationships with significant people in science. He sought out Sabine at his military office in Woolwich to discuss terrestrial magnetism, talked to

Thomson and Faraday, breakfasted with Sylvester, and socialized with Huxley and the botanist Arthur Henfrey. At the School of Mines, with Francis, he saw Edward Forbes and the metallurgist John Percy. He visited Wheatstone at King's College; called on the instrument-maker Darker, who was behindhand in building him some conduction apparatus; and took in a popular evening lecture at the Polytechnic from the famous John Pepper (inventor of Pepper's Ghost).

Another possible position now appeared. At the end of June, Tyndall heard from Francis that the Chair of Natural Philosophy at Queen's College Galway was vacant, and went straight to Wynne for advice. Wynne wrote to his relation John Wynne, the Under-Secretary for Ireland, and advised Tyndall to ask Sabine to write to the influential Lord Dunraven, the politician and archaeologist. Bringing all the guns he could to bear, Tyndall wrote also to Kane. Meeting Tyndall by chance in the street, Sabine reported that he had written to Lord Dunraven, who he said was surprised to be asked and could not help, not being a supporter of the current Government. In passing, Sabine remarked to Tyndall: 'I learn that you are an Irishman, well I should never have found it out'.[85]

There is an implication here that Tyndall had developed a less 'Irish' accent, which he perhaps believed would more quickly lead to acceptance in England. This encounter led to an ingratiating letter from Tyndall to Sabine, in which he poured out all his social insecurities: 'I believe a couple of generations back the people from whom I am sprung belonged to the higher portion of the middle classes. But my father was a poor man, who made a livelihood by selling leather and shoes, during a portion of his life he was a policeman. In these few words I sum up all his shortcomings...it now remains for me to make the experiment whether a man with nothing but naked character to recommend him, may not...find the doors to an honourable activity open to him'.[86] Sabine assured Tyndall that his concerns were misplaced, and that no one would judge him for his lowly birth. He mentioned that he was also an Irishman, 'as far as being born there gives that title'.[87] Three years later, Sabine wrote offering the possibility of the new Chair of Meteorology in Toronto.[88] In a typically long-winded manner, Tyndall rejected that opportunity, because the responsibilities were primarily observational rather than experimental. He sensed on meeting Sabine afterwards that the letter had not gone down well. It reminded him of Sabine's response to his earlier letter about his background. Tyndall supposed that 'a cold mechanical feeling is not the nature on which the warm feelings of the human heart should be showered'. With more self-awareness, he might have reflected on the impact of his loquacious and servile mode of writing to Sabine. Nevertheless, his modest Irish background, two barriers to social advancement in one, was a significant challenge to overcome. It is indicative that when he was well established, in 1874, he refused to fill out the questionnaire on which Francis Galton based his book *English Men of Science*, an exploration of the hereditary and environmental factors disposing to scientific success.[89] Perhaps it is because he still felt some embarrassment, or saw himself as Irish.[90]

For the next few weeks at Queenwood, after a break with Debus on the Isle of Wight, Tyndall was not well, his sleep disturbed. He applied for the Galway post, and asked Hirst if he would take his place at Queenwood if he were successful, to which Hirst agreed.[91] A little later, on a visit to Berlin, Hirst called on Magnus, who he noted 'like everybody else…had a great liking to Tyndall'.[92] He told Magnus about Galway. Magnus immediately promised: 'I will write then to Sir John Kane, for I consider Tyndall to be one of the ablest of your young scientific men, and have to him also a great personal friendship'.[93] None of this support was enough. Tyndall heard in November that George Stoney had been appointed, who 'had Lord Rosse's interest and that was too powerful to contend against'.[94] He was not particularly disappointed.

The British Association was due to meet in Belfast in September 1852. Tyndall had not intended to go, a surprising decision given his need to establish himself. But he heard from Francis that Sabine, President that year, had again intervened by asking if he would accept a Secretaryship of the Physical Section. Tyndall arrived in Belfast on the afternoon of Wednesday 1st September, commenting: 'Nearly 5 years have elapsed since my foot rested on Irish ground…Little did I imagine 5 years ago that I should mingle among the magnates of the British Association, and mingle with them by a fairly won right, at my next visit'.[95] He put up in Joy Street, 'not a very aristocratic place, but suited to my exchequer',[96] hearing Sabine's Presidential Address that evening. The next morning he was at his post as Secretary.

For his own presentation he procured a battery and electromagnet from the Catholic Bishop Denvir, while he noted that 'a skirmishing fire was kept up over every paper'.[97] Tyndall had just started to respond to Thomson's paper on magnetic lines of force when the Lord Lieutenant arrived, accompanied by Lord Naas—the Chief Secretary of Ireland, who had formally acknowledged Tyndall's application for the Galway post—and their wives. Tyndall claimed he 'managed the matter very coolly', as Brewster participated, and Sabine sat behind. On the Friday, Tyndall demonstrated his experiments on the conductivity of heat in fifty-seven different kinds of wood, showing its dependence on the direction of the wood fibre and ligneous growth layers.[98] Late the next day, he gave his paper on diamagnetism, using crystals of calcareous spar, carbonate of iron, and pieces of white wax of the same shape and size as the spar to show that the magnetic effect was due to the manner of arrangement of the molecules, not their shape.[99] Thomson delivered 'some very eulogistic remarks', to which Tyndall opined: 'Thomson I believe is a decent soul at bottom but he is greatly afraid of his fame. I think it will never be extraordinary'.[100] The *Athenaeum* reported Thomson as saying that Tyndall's discoveries 'had cleared away a mass of rubbish, and set things in their true light'.[101] Tyndall had his eyes on Stokes too, who had just revealed his discovery of fluorescence: 'Stokes has been greatly praised...and he is a proud fellow...the time will come when he can't afford to be proud to me'.[102] He had the grace to add a note in his journal in October 1853: 'I will let the record stand to prove what an egotist thou wert'.[103]

After the meeting, Tyndall was able to pay a quick visit to Leighlinbridge, where he found his mother, whom he had not seen in five years, 'as brisk as a bee'.[104] He gazed at the vista of his boyhood, the Barrow flowing through the green meadows, the fine trees of the Lodge, the distant hills of Dunleckney, and Mount Leinster beyond. Finding the Steuarts out, he left a card at the Lodge and set off for Dublin, and thence to Holyhead and London.

Much of Tyndall's time towards the end of the year was taken up by work for Francis, translating papers for the *Philosophical Magazine* and the new volume of Taylor's *Scientific Memoirs*. Tyndall had been appointed one of the four editors of the *Scientific Memoirs*, in which translations of foreign papers were published. He was responsible for natural philosophy (physics), while Francis looked after chemistry, Huxley the zoology, and Henfrey the botany. Completing the translation of a paper of seventy-two pages by Helmholtz, he recognized the value to his scientific education, but bewailed the diversion of his mind from original thought. On occasion, he would use his position to challenge papers with which he disagreed. He took issue in the *Philosophical Magazine* with the work of Richard Adie, who argued that thermo-electric currents could not cause a diminution of temperature. It became Tyndall's first scientific dispute in print. Hirst took him to task for the manner of the attack, which Tyndall claimed contained 'nothing personal that could be avoided'.[105] It was not to be the only time that Tyndall misjudged the impact of his language. Hirst was also concerned that success was going to Tyndall's head. He wrote an admonition that Tyndall took to heart and copied into his journal.[106]

Then as now, the Royal Society's Anniversary Meeting, at which medals are awarded and elected positions filled, took place at the end of November. The 1852 meeting was Tyndall's first. He was preoccupied with his own position, and perturbed by a seemingly cold shake of hands from Sabine. Nevertheless, he was gratified that Humboldt received the premier Copley Medal, prompted, he fondly thought, by a review he had written in the *Philosophical Magazine*. Joule and Huxley received the Royal Medals.

At Christmas Tyndall turned to old friends, but he did not overlook scientific opportunities. He was in Manchester with the Gintys and Tidmarshes, dined with the Franklands, and paid a visit to Joule in Salford, where he found William Thomson.

He spent New Year's Eve at Spring Bank, then a night in Preston to visit Margaret Allen and her dying three-year-old son: 'one of the saddest, solemnest, and most beautiful scenes that have yet crossed my experience... You could scarcely believe it, he appeared to be whispered away by angels'.[107]

* * *

Tyndall was lucky with his patrons and supporters. Sabine had eased him into the Royal Society. Du Bois-Reymond prompted his next break, by recommending him to Henry Bence Jones (Plate 8), a physician and friend of Faraday's, and soon to be a Manager of the Royal Institution. Tyndall had received a letter from Bence Jones in October, saying he had mentioned him to John Barlow, Secretary of the Royal Institution, 'as one likely to give a good course of lectures; and invited me to deliver a lecture on some Friday evening either before or after Easter'.[108] This was an invitation to give a prestigious Friday Evening Discourse, a formal lecture to the assembled Society audience of Royal Institution members and their guests.

Tyndall chose to speak on diamagnetism, a topic of which Barlow approved, mentioning the availability of 'the same electric magnet with which Dr. Faraday rotated a ray of light & established diamagnetism'.[109] He went with Barlow to the Royal Institution in January to make arrangements, meeting Faraday 'like an alchemist at work beside the fire'. Faraday told him that he would be obliged to say 'no' to some of his results on diamagnetism in his first lecture.[110] Bence Jones invited Tyndall to this lecture, Faraday's Discourse 'On the Magnetic Force', and to dine with him. Being summoned instead to Sir James Clark's, Bence Jones ensured that Tyndall too was bidden. He found himself 'perfectly at home' with Lyon Playfair; August Hofmann, the first director of the Royal College of Chemistry; and Warren de la Rue. De la Rue, involved in the successful stationery business established by his father, became best known for his astronomical photography, and had a share in the invention of the first envelope-making machine, one of the most remarked-upon exhibits at the Great Exhibition. Tyndall had visited de la Rue's factory with Edmondson in 1847, commenting that the machine 'performed the function of the human hand with perfect exactitude, had a savage looked upon the instrument his first impulse would doubtless be to fall down and worship it'.[111] After dinner, Tyndall had the pleasure of signing Frankland's certificate for Fellowship of the Royal Society. His more experienced colleague was a year behind him.

As Tyndall's big day approached he contemplated the task ahead: 'success there is more a matter of the heart than the head... courage is more needed than knowledge'. He was understandably nervous, as the stakes were high. Indeed, throughout his life, Tyndall had to master his nerves before lecturing. It was a test of character, and one reason he put so much preparation into his lectures. It was a control of the will that he later exercised equally during his challenging mountaineering exploits.

The evening before the lecture, when he demonstrated his experiments first at the Royal Society, 'the whole affair passed off most agreeably'.[112] Then on the morning of Friday 11th February, Tyndall repeated many of the experiments with Faraday. That evening, after dinner at Barlow's where he sat between John Phillips and Edward Forbes, Tyndall gave his first Discourse at the Royal Institution 'On the influence of material aggregation upon the manifestations of force'.[113]

The lecture was an immediate success. Frankland, in his obituary of Tyndall, remembered it well: 'The lecture, although of such an abstruse character, took his audience—mostly popular as it was—by storm'.[114] Tyndall restated his ideas about the influence of molecular arrangement on diamagnetism, instead of relying on the posited 'optic axis' force (by Plücker) or 'magnecrystallic' force (by Faraday). He paid warm tribute to Faraday, though he challenged his interpretation of the phenomenon. At the end, Faraday crossed the floor and shook his hand, as did the presiding Duke of Northumberland. Samuel Whitbread, the brewer, waited behind to say: 'anything to surpass your lecture tonight I never heard anywhere'.[115] People thronged to be introduced, until Faraday finally took him upstairs, where his wife and niece (Jane Barnard), 'a sweet intelligent girl', were at supper, and stuffed him with sherry and sandwiches. He relaxed over the weekend at his friend James Bevington's, discharging his energy in a snowball fight in Richmond Park, before returning to Queenwood.

The possibility of Toronto faded into the background as invitations poured in. Bence Jones offered him four lectures at the London Institution, and suggested that he would be offered a Professorship at £200 a year. The London Institution, located in the City of London and modelled on the fashionable Royal Institution in Mayfair, combined lectures to a relatively well-off business audience, with research and other amenities.[116] Tyndall thought it might be safer to try a single lecture first, and gave it on 16th March, 'On certain points connected with the transmission of heat through bodies'. Bence Jones also mentioned a vacancy at the Royal Institution itself at £150 a year, plus two-thirds of the lecture fees amounting to a further £100. This was a Professorship of Chemistry, following the resignation of Thomas Brande in 1852; Bence Jones considered it unlikely that the College of Chemistry would let their director, Hofmann, the obvious candidate, accept the appointment. Tyndall was unsure of the attraction of a chemistry position, since his interest and expertise was in experimental physics, and sought Huxley's advice.[117] Huxley told him not to be modest, suggesting that he should look ahead to taking Faraday's place at the Royal Institution, but that the London Institution was also tempting, since it too would give him laboratory time. He intimated that Playfair, Edward Forbes, John Percy, and Sir Henry de la Beche at the School of Mines were interested in his going there if they could remove Hunt, whose recent book Tyndall had so mercilessly dissected. Huxley added that Tyndall should cultivate Playfair, who was well placed with the government and about to become joint-secretary of a new Department of Science and Art. He passed on compliments, saying that Gassiot, a wealthy businessman with his own laboratory, had told him that Tyndall's lecture had made his scientific fortune, and that his name had been mentioned with great praise at the elite Philosophical Club of the Royal Society, 'the most influential scientific body in London'.[118]

Barlow invited him to give a second Discourse in June, an unusual honour, and offered him four lectures at the Royal Institution, which would qualify him for consideration as a professor.[119] Tyndall proposed that the lectures should be 'On Air and Water', and agreed to give them in May and June.[120] He wrote to Bunsen, who suggested the complete structure and experimental options for a lecture on water.[121]

The offer from the London Institution, which Tyndall discussed at the City Club with Gassiot and De la Rue, was tempting: a course of six lectures, an evening lecture, a monthly meeting of the Laboratory Committee, and the rest free for experimental work. Tyndall determined to accept if the arrangement did not interfere with his connection with the Royal Institution. Faraday's view was that he could hold one professorship and one lectureship; he said that he would prefer the professorship to be at the Royal Institution, giving nineteen lectures, though he was concerned that they could not offer the £300 per annum they thought might be necessary to secure him. In the end, he took both.

In April Bence Jones proposed terms for the Royal Institution, with the formal duties to start in January 1854.[122] The offer was £200 for nineteen lectures, plus six Christmas lectures in alternate years at £50. There was an intimation that the salary would increase over the next few years by £100 per annum. The rules were also bent. The appointment could not be made officially until the following April, but Tyndall could start his duties and draw a salary. Though delighted and grateful, Tyndall went into the arrangement with his eyes open. Referring to the requirement to 'superintend all experiments ordered by the Committee of Managers', he asked to be granted 'freedom of action' beyond the requirement to lecture, as he believed an investigator needed to follow his researches wherever they led. It was a brassy request that appears to have been granted. Tyndall mentioned that he thought he would be offered a professorship at the Museum of Practical Geology (the School of Mines), which he would also like to take up, but only if the Royal Institution agreed.[123] Bence Jones wrote to confirm the election date of 6th July, with the duties to start in January but the salary immediately.[124]

The four lectures on 'Air and Water' were well received, to Tyndall's self-satisfaction. After the second he was introduced by Faraday to Sir Henry Holland, later the president of the Royal Institution, who invited him to breakfast to see his picture of a geyser—he had been in Iceland. This was the theme of Tyndall's second Discourse, 'On some of the eruptive phenomena of Iceland'.[125] He had prepared over Easter, experimenting with artificial geysers in Frankland's laboratory in Manchester, and again receiving advice from Bunsen, who had also visited Iceland.[126] He was not entirely happy with the lecture—though it stimulated the directors of the Crystal Palace to have a geyser built—but congratulated himself on his position.[127] He wrote: 'I have stood at no man's door craving admittance ... I never sought the Royal Society, still it came. I never sought the Royal Institution, but it has come. I never sought the society of the great and eminent, still I have got into such society'.[128] Tyndall was not disposed to force himself too obviously on his social superiors, a reflection of his insecurity about his relatively poor Irish background and his sense of personal dignity.

Tyndall's thoughts as he left Queenwood reveal his attitude towards children, and the importance he gave to their moral education. His last day there as a teacher was on 6th July 1853, though he was to return often to visit friends, particularly while Hirst or Debus remained. He was sad to leave the boys: 'But I carry away with me little blossoms of affection, watered by the dew of the heart and which to me are sweetest of all, affection, not for women, but for little boys'.[129] A few days before leaving, he wrote a long letter to one pupil, Alexander Cumming, then aged about thirteen, opening his heart and offering advice: 'I know that you are ignorant of most of the vices and follies of the world—ignorant of many practices which are destructive even to boys ... Recollect the first temptation to sin of any kind is always the most easily strangled—strangle the first my boy and preserve your freedom ... Do your duty my son—that's the grand maxim'.[130] Tyndall's time there led to several friendships with parents and guardians of pupils: with James and Mary Coxe (uncle and aunt of the Cumming brothers), who became significant to him in Scotland, with the poet Charles Mackay, and with James Bevington. Tyndall found Bevington a dweller 'on the same mental plane'.[131] He owned Neckinger Mills, a firm that started leather manufacturing in Bermondsey in the early 1800s.[132] Like Hirst and Wynne, he offered Tyndall money as a loan or gift.

* * *

With a secure post to look forward to, Tyndall left London on 11th July, and remained abroad until late September. He would have left the previous day had the Prussian Consul not required a letter from a banker to issue a passport. The brash Tyndall went straight to the top, asking Edward Cardwell—President of the Board of Trade in Aberdeen's coalition government, and Wynne's boss—to certify his identity. Cardwell told him he had heard of Tyndall's standing, but asked: 'how do I know you are Mr Tyndall?'.[133] Tyndall rushed to Playfair for a note, and back to Cardwell, who then vouched for him.

This period in Berlin, lodging at Universitäts Strasse 4, and subsequently in Paris, further extended his relationships with German and French natural philosophers. He immediately reconnected with his German friends, calling on Du Bois-Reymond, who was shortly to marry in England,[134] dining with Magnus and going with both of them to the Sozietät der Wissenschaften (Prussian Academy). The next evening he was with Clausius at the Physikalische Gesellschaft. He focused on his work, writing to Faraday: 'Society is the enemy of work, and here I see a danger which lowers upon the foreground of my own future. I must ... set my face against visiting if I would get any work done'.[135] While in Berlin, he investigated Peltier's discovery of the generation of cold in a circuit of two dissimilar conductors, and explored the idea of making Trevelyan's rockers the subject of a Discourse. Arthur Trevelyan had shown these to the Royal Society of Edinburgh in 1829; a brass bar, with two parallel knife-edges on the bottom surface, can be made to move from side to side and emit a musical note when heated and placed on a cold lead block, perfect material for a demonstration lecture. James Forbes had himself followed up these experiments.

Tyndall's mental state in Berlin was poor, occasioning a mini breakdown that he often recalled. He was suffering from headaches and 'nervous attacks' that made him think of jumping out of the window, and led to his barricading the bedroom door in case he sleepwalked. The pressures under which he put himself, and which he never fully mastered all his life, had taken him close to a severe breakdown. Rather than returning to Britain for the Hull meeting of the British Association, he decided to go to the Harz region for the mountain air, then to Marburg, before going on to Paris. He thereby missed Plücker's reiteration of views that Tyndall believed he had laid to rest.[136] Walking through the Harz mountains, Tyndall absorbed the scenery; the forests, green valleys, and crags, and the legends that sprang around them. Though he travelled part of the way with a student he met on the road, his preference was for solitary contemplation: 'I like ... to have no human presence or speech between me and nature'.[137] The week did him a power of good, and he resolved to retreat to the country in future if he could, whenever his nerves were again attacked. Staying at the Ritter in Marburg, he felt as though he had never been away. He dined with Knoblauch, who showed him his experimental arrangement for radiant heat, perhaps stimulating thoughts for later investigations.

He took the train to Frankfurt with Dove, the boat down river to Cologne, and the train to Brussels and Paris.

In Paris, a meeting with Magnus brought him into easy contact with French physicists. Taking lodgings at Cité d'Antin, Tyndall met Magnus at the instrument-maker Rhumkorff's. They went on to the Collège de France with Jean-Baptiste Biot and Henri Regnault, where Tyndall realized that he would need to master conversational French better to profit from these interactions. At the Académie des Sciences, where he was interested to find the meetings open to the public (unlike those of the Royal Society), he met Émile Verdet, heard Owen read a paper, and saw Stokes at a distance. Visiting the Conservatoire he valued the demonstrations of models and apparatus, observing that there were 'a good deal of the minor adjuncts of science which never find their way into books'.[138] He met the Abbé Moigno, spent much time with Guillaume Wertheim and Magnus, and dined with César Despretz, who became a good friend. Both Moigno and Despretz were devout Catholics; Tyndall's antipathy was not to the individuals but to the institution of the Church of Rome. Bence Jones arrived, and invited Tyndall to his house in Folkestone when he returned to England.

In Folkestone, Tyndall's class expectations were disabused. He had expected Lady Millicent, Bence Jones's wife, to be haughty, 'but she met me so frankly and kindly as to make me forget she was the daughter of an Earl'.[139] Back in London, he took over lodgings from Francis in 9 Chadwell Street, Islington, some 2½ miles from the Institution, and with the support of Charles Anderson, Faraday's assistant of twenty-five years, started familiarizing himself with the facilities. By the end of October, he had enough material, with some to spare, for nineteen lectures on heat. He continued writing them while wrestling with translations for Francis. In the laboratory he revisited the experiments on rockers that he had started in Berlin, quickly producing tones from bismuth, which Forbes had found inert. In subsequent days he succeeded with other metals, with non-metals, and even with glass, commenting that 'Forbes's experiments must have been coarsely made'.[140] His experiments on heat conduction continued into November as he compared rock salt and fluorspar. The former permitted more 'obscure' heat to pass, the latter heat of high incandescence. This would be significant later.

Tyndall was approached by De la Beche in November, to lecture at the School of Mines.[141] Though Bence Jones was against it, Faraday was supportive of the idea. De la Beche suggested he might look to earn £1,000; he said Hofmann was earning that, and James Forbes £1,200 in Edinburgh. In late December, Edward Forbes, knowing De la Beche's difficulty in coming to a proposition, suggested a workload of about thirty lectures and an occasional commitment to the evening lectures for working men, as well as attendance at Council. The salary would be £200 plus 'spoils' (fees). An additional attraction was that, as a government post, there might be a pension attached. Tyndall felt he could accept the lecture load. But he thought that the commitment to serving on Council would cut across his obligations to the Royal Institution, his 'wedded spouse'.[142] He declined the offer. A few months later, Percy told him that Stokes had been chosen as Professor of Physical Science, replacing Robert Hunt. At the same time, Huxley succeeded Edward Forbes, who had moved to Edinburgh as Professor of Natural History.

As if election to the Royal Society and appointment at the Royal Institution were not enough, Tyndall heard in November from Bence Jones that he was to be the recipient of a Royal Medal, one of the premier awards of the Royal Society; Francis had told him in July that he was in the running.[143] Two medals were awarded annually, in the physical and biological sciences. The designated recipient of the other medal was Charles Darwin. Gassiot had proposed Tyndall for his work on diamagnetism, which he considered would help solve 'the true cause of the variation of the magnetic needle', an indication of the practical significance given to the study of magnetism at the time.[144] To Hirst, Tyndall wrote: 'I think the best thing I could do would be to take a little prussic acid and kill myself quietly here while the halo is still around my head. People would then say "if he had lived &c."'.[145] The decision to award the medal to Tyndall was close. He was initially tied with the more experienced Hofmann—who had been proposed by Bence Jones— at six votes each, but then chosen. It is an indication of the speed with which he had developed his reputation that he outcompeted Hofmann at this point. But matters quickly became complicated. Gassiot told Tyndall that there were objections. As Tyndall reported it, when people 'set about scrutinizing the affair; first they found that Thomson was opposed to me, second that my two first papers were joint papers with Knoblauch, third that my third paper had been got up in the private laboratory of Prof. Magnus, fourth that this same paper was published in German before it appeared in England, fifthly that one or two men who have got a name in England thought that I had not quite made out my point'.[146] After consultation with Gassiot and Faraday, Tyndall determined not to accept this singular honour, the only time that the Royal Medal has been declined, and wrote to the Secretary Samuel Christie and to Gassiot to do so. Gassiot approved, writing: 'I would sooner hold the position you have so promptly taken than be the recipient of 20 medals'.[147]

The medals were presented by the President, Lord Rosse, on 30th November. The Copley Medal went to Dove, received on his behalf by Sabine, and the single

Royal Medal to Darwin. Huxley then stood up to ask 'why his intimate friend whom he would not name was not beside Mr Darwin'.[148] Tyndall's letters to Christie and Gassiot were read out, 'and received with great applause'.[149] So ended an unsavoury process. It did not significantly damage Tyndall's relationships among the Fellows—though it was a sign of his prickly character—nor affect his career in any substantial way. Indeed, the following year he was nominated again, for his researches in Physics,[150] but lost out directly this time to Hofmann. At the end of the meeting, Wheatstone introduced him to Charles Babbage, and he 'felt some interest in scanning his bitter countenance and thin cynical lips'.[151] He noted with approval the constitution of the new Council of the Royal Society: 'Grove, Stokes, Wheatstone, Baden Powell are new men in my line—Huxley is also one of them'.[152] In a short space of time, with the help of people like Sabine, Faraday, Gassiot, and Bence Jones, he had become a significant new figure in British science.

6

⊹⇒◎ ◎⇐⊹

CLASH OF THEORIES
1854–1856

New Year's Day of 1854 found Tyndall in pensive mood, as he confided to his journal: 'the old year is gone forever. I have not done much during the last year but much has been done with me'.[1] Little could he have imagined his current situation: a Fellow of the Royal Society and the designated recipient of its Royal Medal, Professor of Natural Philosophy at the Royal Institution, and sought after by many other organizations. Yet he professed that this public recognition was of no great moment, writing: 'I deliberately cultivate a principle within me which enables me to look upon many of the attractions of society with indifference'.[2] He claimed also a dislike of pushing himself forward in 'social or friendly, or warmer things', asserting: 'I have when younger thought how delightful it would be to play an influential part in a retired way, leaving the responsibility and the noisier glory of coming forward, to others. In Preston this was my place. I believe I moved the whole but was never prominent. This was my character when a boy, and though greatly mollified it is my character still'.[3]

This was disingenuous. Tyndall did enjoy society. Dining at Barlow's he felt at home, meeting Colonel Philip Yorke, a Manager of the Royal Institution, to whom he provided information from Magnus on equipment for experiments on projectiles.[4] He was equally at ease in the riotous environment of the Red Lions—an informal scientific dining club that was the metropolitan outgrowth of a convivial group that met at British Association meetings—declaring: 'An evening in the month so spent refreshes a man's social interests'.[5] Arthur Henfrey was the secretary and Edward Forbes the ebullient chairman, who proposed Tyndall's health as the youngest lion. Tyndall took Hirst as his guest, who reported that Forbes sat on a lion's skin, waving its paws and tail, and that the diners would shake their coat tails and roar. A couple of days later Hirst was introduced to Faraday for the first time. Faraday's unexpected arrival at Tyndall's lodgings caused a flap, since they had been smoking and Tyndall knew that

Faraday hated the smell. He sprayed Eau de Cologne on the carpet and burnt himself with candle wax in his haste, telling Faraday that Hirst was a smoker but 'in other respects he is all right and is my friend'.[6] Tyndall remained an inveterate smoker.

In the early part of January, Tyndall's headaches returned while he worked on his Royal Institution lectures. In the twelve weeks before Easter, he gave weekly lectures on the topic of heat, the subject he would make his own in the following decade. In February and March, these were interspersed with six evening lectures on 'Electricity and Magnetism' at the London Institution. Tyndall soon gained confidence, telling Hirst: 'My lectures thus far have gone on pretty prosperously...I of course am the chief learner on such occasions. If I can get over a year or two I have no fears of ultimate success for I find my power of utterance becoming more and more unshackled'.[7]

Tyndall's paper on rockers—bodies at different temperatures in contact, which vibrated and emitted sounds—was read at the Royal Society in January.[8] He had started this work in Berlin, not only because he thought it might make a good Discourse, but also because if James Forbes's explanation was correct, that heat could exercise a repulsive force (Forbes called it 'a new mechanical agency for heat'), then the implication for understanding the nature of heat would be profound. Forbes had contradicted Faraday's original explanation of the phenomenon, which was based on rapid expansion and contraction. Tyndall disproved experimentally each of Forbes's three laws governing the behaviour of rockers, validating Faraday's explanation in the process. He remarked in his journal: 'They all seemed amused at the manner in which I have "demolished Forbes" as they express it. It is just what he would like to do himself!'.[9] Perhaps presciently, Grove wrote in his report on the paper that 'some inconvenience may result from the introduction into the Phil Trans of a paper of a controversial character'.[10]

On the evening following the Royal Society meeting this new research became Tyndall's third Discourse; after a dinner with Bence Jones. Dinners at Bence Jones's or Barlow's were now regular occasions for him. In February, at Bence Jones's, Tyndall found Holland opposite him, and they talked about Carlyle and Tennyson. In March, at Barlow's, he noted that he and Frederick Pollock (Figure 12a), a lawyer and a Manager of the Royal Institution, often gravitated towards each other. It was the start of a lifelong friendship which quickly extended to his wife, Juliet (Figure 12b), and their family. She was one of several Society women with whom Tyndall developed mutually admiring and often flirtatious relationships. As Hirst later described her: 'Mrs Pollock is exceedingly lady-like, her conversation is spirited and full of intelligence. She has also imitative powers of no common order which impart to her a graphic and racy character. Tyndall is clearly a great favourite of hers'.[11] He might also have mentioned that she was a novelist, an accomplished watercolourist, and an expert on French drama and contemporary European literature, which enabled her to break the masculine barrier in the literary reviews, even if anonymously.[12]

Tyndall's lectures progressed well. The audience was increasing each time, and the theatre nearly full. He found the assistant Charles Anderson's knowledge of experiments extremely useful, but was surprised by his lack of understanding; Anderson had 'mere mechanical knowledge, he never appears to reason upon the experiments'.[13] Tyndall was not sleeping soundly though, resorting to brandy and water as an 'opiate', signs of the future. Hirst sent Francis over one evening to stop him 'pondering', and help him get to sleep. They talked of Turks and Russians as crisis approached in the Crimea.

Tyndall was now in demand as a writer too. In March, at the instigation of Playfair, he was approached by Edward Hughes to contribute to a series of reading books for schools.[14] Hughes offered £50, later increased to £75 for the equivalent of eighty pages of the *Philosophical Magazine*. Playfair told him that the physician Neil Arnott made £1,300 a year by writing, and Tyndall thought the money would be easily earned. Faraday was supportive, though he advised him not to make his name too cheap, and he decided to accept.

Fig. 12a. Frederick Pollock (1815–1888).

Fig. 12b. Juliet Pollock (1819–1899).

Work was now piling up in front of him for the two months after Easter, including a Discourse and a new course of seven lectures on 'Ebullition, Combustion and other Phenomena of Heat'. He was also committed to a significant lecture on education, and its impending delivery caused him much anxiety.[15]

Tyndall's first course of Royal Institution lectures had ended to great satisfaction, but he was in need of rest. Determined to take a complete break from thoughts of science, he went to Queenwood and spent two days walking, reading, climbing, and playing cricket. The stay recharged his batteries, and he was glad to find that this sort of break would do so. Back in London he found a letter from an American cousin, Hector Tyndale, and tracked him down at the radical publisher John Chapman's in the Strand.[16] He was delighted to meet this relative who had made contact a year and a half before, finding a man of the same age and height but more muscular, with 'a wonderful elasticity of thought', a democrat 'and I think rather hard on the English'.[17] Perhaps Tyndall was starting to identify now with the ruling English elite. They spent every evening together until Tyndale left for Paris, including a Red Lions dinner. Tyndall then returned to Queenwood for a few days, going to Southampton on 9th

April to see Ginty sail for Rio de Janeiro on the *Great Western*, to manage Rio's gas works, leaving his wife and five children for the moment in England.

Tyndall opened his 'second campaign' after Easter, finishing the first of seven weekly lectures spectacularly with a 30-foot geyser. His social horizons were expanding too, with an evening in May at the Pollocks in Montagu Square, a house that was eventually to become almost a second home to him. Juliet Pollock he saw 'like a domestic moon shedding brightness and happiness through us all'.[18] The guests included the artist William Boxall, who became director of the National Gallery in 1866, and the poet Catherine Macready.[19] All admired Tennyson.

Before his lecture on education, Tyndall had one more Discourse to give for the season. The subject was 'On some phenomena connected with the motion of liquids',[20] for which he offered a series of visual and auditory treats as he expounded on the cohesive properties of water. The lecture ended with the deflection of a bright sheaf of light into a falling jet of water, the light seemingly washed down by the descending liquid. It was a beautiful demonstration, seen by some in retrospect as a precursor of fibre optics, and the applause 'rather exceeded the bounds permitted in the Royal Institution'.[21]

The Lectures on Education, given on Saturday evenings, were a consequence of Faraday's intense concern about the popularity of spiritualism and table-turning.[22] He criticized the system of education that could produce such credulity, and commissioned with Bence Jones a series of seven Lectures to highlight the wider issues.[23] William Whewell, Master of Trinity College Cambridge, gave the first, speaking on the 'History of Science'. He argued that every major intellectual advancement had been preceded by some great scientific discovery. Tyndall was underwhelmed, writing to Hirst: 'He did not by any means fill the measure which would have satisfied either your thirst or mine upon the subject'.[24] The following week it was Faraday's turn. He gave a tour de force on 'Mental Education', pointing out how easily one could be deceived by the senses unless proper care and judgement were exercised. He was explicit that the prevalent belief in spirit-rapping and table-turning was one example of such deficiency in reasoning.

Tyndall's lecture, on 27th May, was entitled 'On the Study of Physics as a Branch of Education', and he used it to lay out his already fully-formed philosophy of education. In the process, he effectively defined our modern understanding of the scope of physics.[25] The title is misleading. Tyndall spoke of physics as a means, not a branch, of education, a line he took with the encouragement of Faraday and Hirst. He saw the inquisitiveness of the human intellect as 'an imperative command' from the Creator to explore the physical universe, increasing our understanding. The process was bidirectional: from observations to the cause which unites them, and from that conception to the prediction and experimental testing of its consequences. The first part—induction—required a willingness to abandon preconceived notions, a surrender to Nature, a humility. The second—deduction—required a logical approach and the testing of conclusions. Nevertheless, the ultimate purpose of this knowledge was power. The humility of the approach to understanding was not extended to the knowledge so gained. He declared: 'The subjugation of Nature is only to be accomplished by the penetration of her secrets and the mastery of her laws'. Recognizing that 'the implements of human culture change with the times', and giving as an example the recent development of steam power, it followed that one could not unconditionally accept the systems of the past. As he expressed it: 'I do not think that it is the mission of this age, or of any other particular age, to lay down a system of education which shall hold good for all ages'.

Bence Jones had urged Tyndall to promote the study of science at the expense of language. But he rejected this position. He argued that the study of Greek and Latin remained vital to enable us to access the ancient mind, but that we now knew things through physics quite hidden from the ancients, making its study essential for modern education. Apart from the intrinsic value of its study, physics offered a means of education that developed the ability to abstract and generalize as much as to deploy logic and sharpen investigation. Too rigid an approach might impair the imagination, but even then, he thought, the 'cold unimaginative reckonings' of the investigator could excite wonder and furnish themes for the poet. And while the investigator might be inspired by the pursuit of knowledge alone, his findings might have important practical applications. So it was vital for those seeking practical outcomes not to constrain the freedom of the researcher. Tyndall even argued that the admission of Members of Parliament should be contingent on a knowledge of the principles of natural philosophy, and that the study of physics might be a moral influence to draw people away from drunkenness, by offering new enjoyments. His vision for its significance was not restricted to class. In his view: 'there is one mind common to us all' ... 'Who can say what intellectual Samsons are at the present moment toiling with closed eyes in the mills and forges of Manchester and Birmingham? Grant those Samsons sight, give them some knowledge of physics, and you multiply the chances of discovery, and with them the prospects of national advancement'. Such education might be a stimulus to miners to develop for themselves means to combat dreadful accidents, serving two ends at the same time, 'the elevation of the men and the diminution of the calamity'.

It was not just education in physics that Tyndall promoted. He stressed the value of giving students the freedom and responsibility to explore questions for themselves. He saw the profession of schoolmaster as of paramount importance, and ended with a call for science to be seen not just as an instrument of material prosperity, but as the means of lifting the mind to the contemplation of the Creator's workmanship.[26] He was well satisfied when the ordeal was over: 'never was lecture more silently listened to in the spaces which stretched between one rumble of applause and the succeeding one'.[27] One listener told him he was two centuries ahead of his time, to which he cheekily responded that so was Galileo.[28]

Tyndall's final scientific lecture of the season was on 6th June. He went to the Pollocks that evening, and was struck by 'a face of surpassing beauty placed as summit to a figure of corresponding grace'.[29] His journal is full of entries commenting on the physical beauty of women, invariably younger than him, but his ideal was divorced from a robust sexual attraction: 'I thought of the immense benefit diffused from a beautiful person, and that while in that girl's presence or even permeated by the thought of her, I could not do or think anything ignoble'.[30] Intellectual attraction was important to him. He wrote to Hirst, on hearing that he might become engaged: 'For I hold that a man who cannot love is an incomplete man, but I also hold that to place an alliance of the kind upon a firm basis the understanding ought to be satisfied as well as the heart'.[31] When Hirst revealed that he had become engaged to Anna Martin, Tyndall wrote her a lovely letter, suggesting that he had suspected the depth of the relationship for some time.[32]

Three days after Tyndall's last lecture, Faraday gave the final Discourse of the season, 'On Magnetic Hypotheses'.[33] It proved a turning point. Tyndall went straight to Queenwood for a few days, to refresh himself and attend the annual end-of-term concert, but the lecture had struck home. It would shape his work for the next two years.

* * *

Faraday had taken issue with the atomic and molecular theories of electricity and magnetism expressed by physicists such as André-Marie Ampère, Wilhelm Weber, and Auguste De la Rive. For Faraday, an atom did not have a material existence in the sense that Tyndall and many others understood it. Faraday, in his field theory, imagined an atom as a point with 'an atmosphere of force grouped around it'.[34] The field was not to be explained in terms of matter, matter was rather a particular modification of the field.[35] By contrast, Tyndall saw the structure of matter at the molecular level as critical to the mediation of force (or energy, as we would now denote it). They were viewing the problem from different vantage points: Faraday from the concept of a field and Tyndall from an idea of matter. So Tyndall set to work again, to make a case for understanding and visualizing magnetic phenomena in his terms. He had done little systematic work on diamagnetism since the end of 1851, but he started again as soon as he was back from Queenwood, free from lecturing demands.

He was on the back foot from the start, getting himself into 'a labyrinth of difficulties'.[36] Needing money for more sensitive apparatus, he applied to the Royal

Society for a Government Grant of £50 or £100, hearing the next day from Sabine at a dinner at Colonel Yorke's that he had been granted the full £100.[37] He wrote to Weber in Göttingen, whose work on diamagnetism he admired, to ask if he could arrange for a copy of his apparatus for measuring diamagnetic polarity to be made. Weber realized that the design would not be sensitive enough for Tyndall's requirements, but gave instructions to the instrument-maker Leyser, in Leipzig, to make a more delicate one.[38]

Hirst returned from Ireland in mid-July and stayed with Tyndall in Islington—taking him to hear Grisi in *La Favourita*—before they went to Queenwood. Tyndall spent a quiet week there, writing the first instalment of his book for Hughes (he eventually wrote chapters on 'Natural Philosophy' for all four books of Hughes's *Reading Lessons*, published in 1855, 1856, and 1858). Hector Tyndale arrived back from Paris, and after a trip to Sydenham and an evening with James and Mary Coxe, they set off for the Isle of Wight. They spent nearly a week on the island, though Tyndall became quite ill during the trip, 'invaded by symptoms like cholera',[39] to the extent that he thought he might die, and was unable to make his usual long walks. He did have enough energy to pay his homage at John Sterling's grave, having read the biography by Carlyle. He lay down and stretched out, 'just to see whether the same grave would answer for me'.[40] Two years earlier, on a visit to the *Victory*, he had laid himself where Nelson fell.[41]

In London throughout August and into early September, Tyndall patiently worked away at his experiments on diamagnetism. Sundays were his only days off, but he found to his disgust that the passing of a Sunday Beer Act, after Sabbatarian pressure and temperance support, meant he could not buy beer between 2.30 and 6pm. He knew that Faraday was exploring the idea of diamagnetic polarity again; they would occasionally demonstrate experiments to each other, but worked independently. He had a long discussion with him, noting: 'we differed, but differed so cordially, that it was pleasanter than agreement'.[42] Eventually, with new and expensive apparatus, Tyndall repeated experiments he had previously made on the polarity of bismuth, 'and found them all correct'.[43] By September, he was able to write 'diamagnetic polarity is secure',[44] and start writing his paper on the subject. As he explained to Hirst: 'I have been engaged in a tolerably exhaustive comparison of magnetism and diamagnetism, and have I trust established a complete antithesis throughout, and proved beyond a doubt that diamagnetic bodies possess polarity the reverse of that of iron'.[45] There were still questions to check. He was side-tracked for a week, carrying out experiments on the conductivity of cubes of metal that William Hopkins had asked him to do, but he could now prepare papers for the meeting of the British Association at the end of the month. Dove was staying with the Sabines, and Tyndall was invited to dine, glad that he had seen them in their own home and cordial, after the letter to Sabine about his past that he now thought was a mistake to have written. He spent time with Dove and with Faraday, paying a visit with both to the Panopticon in Leicester Square, which had opened on 8th March as a showcase for achievements of the sciences and arts.[46]

The British Association meeting of 1854 took place in Liverpool. Discovering that he was planning to travel second class from Euston, Faraday bought tickets for Tyndall and himself, while other grandees travelled in first. On the platform Tyndall was introduced to Sir Roderick Murchison, the Scottish geologist who became Director of the School of Mines after Sir Henry de la Beche died in 1855. With his renewed interest in magnetism, he gave two papers on the subject.[47] The first concentrated on the nature of the magnetic field, attacking Plücker's law of the different behaviours of the magnetic and diamagnetic forces, and stimulating a discussion in which Whewell, Thomson, and Faraday joined.[48] On the following day, Tyndall gave his second paper, on the diamagnetic force.

The crucial question for Tyndall was whether diamagnetic bodies possessed a polarity opposite to iron (as Weber believed), the same as iron (as von Feilitzsch claimed), or had no polarity (the position of Faraday and Thomson). Tyndall demonstrated that the diamagnetic bismuth behaved the opposite to iron in all situations, and claimed that this proved that the diamagnetic force is a polar force the reverse of magnetic polarity, agreeing with Weber. Thomson emphasized his understanding that an ordinary diamagnetic is simply a substance less magnetizable than air, but agreed with Tyndall that the resultant polarity of bismuth, however caused, was the reverse of iron. Tyndall viewed this (unfairly) as Thomson 'backing out of the position he had assumed with regard to diamagnetic polarity'.[49] He wrote to Hirst: 'Thomson has in fact backed out of almost every position he has assumed in regard to the phenomena of diamagnetism and magnecrystallic action. And he has done so leaving the public to suppose that he had been misconstrued or misapprehended—which tact may possibly increase his reputation with the general public, but in the private opinion of me at least does not add a whit to his nobleness'.[50] He was nevertheless amazed at Thomson's output of papers, which far exceeded his.

The Red Lions dinner—which to Tyndall's dismay cost twice as much as the President's dinner—was presided over by Edward Forbes, President of the Geological Section that year. Léon Foucault was Francis's guest, and Dove Tyndall's. With his lecturing skills in demand, Tyndall starred in a public demonstration in St George's Hall of Foucault's experiments to show the rotation of the Earth, shouting at the top of his voice because of the size of the crowd. After the meeting he set off for North Wales with Francis, spending a day walking in the Vale of Llanwrst before viewing the

Abergele falls and going on to Bangor by rail. Apart from a desire to see the scenery and to climb Snowdon, his work on the effect of the compression of substances on their cleavage and their magnetic properties stimulated him to visit the Penryn slate quarries. Taking the train to Caernarfon, he walked on to Llanberis, where he met the geologist Woodward and afflicted him with 'questions on stratification and cleavage'.[51]

Climbing Snowdon with Francis and Woodward was a delight. They were treated to a cloud inversion: 'The morn was glorious—blue sky and bright sunshine. A couple of hours toil brought us above the region of the clouds which belted the hills and through which the black summits protruded, like volcanic islands through the sea...We rested two hours at the summit pondering upon the sublime doings around us and then descended'.[52] The succeeding days were spent walking, through Capel Curig, Dolgellau—where the church bells were ringing in honour of the fall of Sebastopol—and Llangollen, passing the bonfire of a straw effigy of the Emperor of Russia. There he parted from Francis, to take the *Princess* from the Clarence Dock in Liverpool to Dublin.

Tyndall travelled third class to Bagenalstown, meeting the recently promoted Major Wynne on the station, and drove to Leighlinbridge. There he saw Elizabeth Steuart, and spent the days writing and rambling along the Barrow, observing: 'were the people cleaner all would be delightful'. Fellow passengers in the boat back were Maxwell and Mary Simpson, travelling to see John Martin in Paris. Martin, the brother of Mary Simpson and of Anna Martin, and soon to be Hirst's brother-in-law, was a Young Irelander recently liberated from banishment in Australia. The Young Ireland movement, an offshoot of Daniel O'Connell's Repeal Association, had launched an abortive rebellion in 1848, resulting in the penal transportation of many members to Van Diemen's Land. Tyndall took the train to Blackburn and walked to Spring Bank, marching out after dinner to inspect the quarries and the stratification of the shale. Then he was off to Frankland in Manchester for a laboratory visit, and back to London.

*   *   *

Tyndall's next significant introduction came from Huxley. In October, Huxley told him that he had suggested to John Chapman that he split the 'science' reports (primarily reviews of new books) in the quarterly *Westminster Review*, with Tyndall taking the Physico-Chemical and Huxley the Natural History.[53] The pay would be six guineas per issue. Tyndall had not met Chapman. He was surprised to hear he was alive, as he had read in Llanrwst that he had been taken away by cholera; the victim proved to be a distant relative. He was attracted by the idea, and asked for six months so he could complete his paper on diamagnetism, finish the final 30 pages of the 120 due for Hughes, and prepare his lectures for the early part of 1855.[54] Huxley persuaded him by saying it was not too arduous and that he would be able to control the level of work, so Tyndall undertook to start with a piece for the April issue.[55] Tyndall's income and expenses are hard to pin down—no accounts exist—but he felt insecure for decades, despite his later growing income. That anxiety, a function of his background and the awareness of his many impoverished relatives in Ireland, may have impelled him to take on work that distracted him from research, adding further to his pressures.

The larger-than-life Edward Forbes died suddenly in November, aged thirty-nine. The coincidence of name (though they were unrelated) was perhaps a prompt for James Coxe to ask if Tyndall might be interested in James Forbes's post at Edinburgh, since he was known to be in ill-health.[56] The position was worth £800–£1,000 per annum, with six months for research. Tyndall replied that he was not sure he was ready to assume the responsibilities of such people as Forbes, and would need to discuss it with Faraday, to whom he felt committed. He added: 'I hope there are no religious tests in the University—though born in the Church of England I should pause before swearing to the 39 Articles'.[57] Perhaps a slip of the pen—he was born into the Church of Ireland—or perhaps a sign that he was increasingly identifying with England. Faraday said he would be sorry to see him go but would not seek to stop him. Tyndall told Coxe that though he would not want to contest an election, if the Town Council were to offer the post 'I do not at present see any sufficient cause to prevent me from accepting it'.[58] Hirst approved, since contesting an election might imply that he wished to leave the Royal Institution. He thought that Edinburgh could offer more scope, even if it reduced contact with Faraday.[59]

Tyndall sat between Huxley and Percy at the Anniversary Meeting of the Royal Society. The Royal Medals went to Hofmann and the botanist Joseph Hooker, to whom Tyndall was warming. There was: 'an evident sincerity about Hooker that I like'.[60] He was getting to know others too, coming into contact with the surgeon and zoologist George Busk who, like Hooker, became one of the influential group of nine forming the X-Club, a society of like-minded men of science, a decade later. Grove asked if he would like to become a member of the Philosophical Club, as he proposed to nominate him. Tyndall responded that he wanted to be pulled rather than pushing himself forward, somewhat ungraciously since Grove was doing just that.

The turn of the year brought both introspection and public recognition. While recognizing the need to concentrate on his science as he wrote up his next paper on diamagnetism, he was conscious of his wider intellectual and emotional needs. He found that reading Shakespeare and Goethe helped dim the scientific imagery that was so deeply stamped on his brain. Disgusted by his idleness he wrote: 'There are principles in the human heart that cannot be roused by science—principles on which the culture of science and all other duties depend'.[61] He found it 'perhaps a humiliating thought that a man's soul should be so much at the mercy of his body—that his aspirations should rise with the peristaltic action of his bowels'.[62]

Further public recognition came his way: his appointment as Secretary to the Committee for Magnetism and Electricity for the 1855 Paris Exhibition, and an invitation to give the prestigious Bakerian Lecture at the Royal Society on his recent work on diamagnetism. Given these commitments, his general health, and his feeling that he could not leave the Royal Institution while Faraday was giving the Christmas lectures, he did not attend the marriage in Ireland on 27th December of Hirst, his closest friend.[63] He gave the Bakerian Lecture 'On the Nature of the Force by Which Bodies Are Repelled from the Poles of a Magnet' on 25th January 1855, and repeated it in a Discourse the following evening. It was well received. Grove did not see how the arguments could be overcome.[64] Tyndall proudly noted that the Astronomer Royal was there, and 'many others of that calibre'.[65] He had a discussion afterwards with Stokes and Sylvester, who said, 'I believe that you and I will live when Whewell, Macaulay and such men are resolved into their primitive elements. We add to the stock of human knowledge—they expending their force in learning what is known by others'.[66] Tyndall had been worried about Stokes's attitude towards him, writing: 'I fear that Stokes is not my friend and I know not why...At Edinburgh he looked coldly at me. At Ipswich in reply to a hasty and unconsidered remark of mine on an optical experiment he looked and spoke cuttingly in reply'.[67] He may have been oversensitive. Stokes was a very different character, and a devout Christian, but their relationship over many years was mutually respectful. Tyndall frequently sought Stokes's advice on theoretical points while Stokes, in his role as editor of the *Philosophical Transactions*, valued Tyndall's contributions and reports on papers.

The Bakerian Lecture became Tyndall's 'Fourth Memoir' on diamagnetism, published later in the year in the *Philosophical Transactions*.[68] It followed an exchange of ideas with Faraday about the idea of a 'magnetic medium', with which Tyndall disagreed.[69] This work was Tyndall's major experimental contribution to diamagnetism, though the opinion of the two referees was split on its merits. W. H. Miller called it 'a large and very important addition to the knowledge of diamagnetism'.[70] Thomson was not so convinced, writing: 'Still I think that...Mr Tyndall is frequently contending against an imaginary adversary...all Mr Tyndall's experiments and views are in perfect accordance with those indicated by Faraday from the beginning and advocated by myself as early as 1846'.[71]

Effectively, two different models could explain the same experimental results: Faraday's field theory and 'directive force', and Tyndall's mechanical approach based on attraction and repulsion. In the lecture, Tyndall set out his view of the importance of structure, drawing on his experiments on the influence of the structure of wood on its magnetic behaviour. He had extended this work in a series of systematic and controlled experiments with bars and spheres of different substances. The conclusion was that the position taken up by spheres depended on their structure, which Tyndall created by compressing them in specific directions. In all cases diamagnetic and magnetic substances behaved as opposites. He was explicit in his belief in polarity: 'The magnetic force, we know, embraces both attraction and repulsion, thus exhibiting that wonderful dual action which we are accustomed to denote by the term polarity'. Nevertheless, there was uncertainty. Since his and Weber's experiments had only been made with the diamagnetic bismuth, he felt the need to establish the evidence for diamagnetic polarity with a wider range of substances. This he did in experiments published in his next paper, the 'Fifth Memoir'.

Tyndall was gradually meeting the major figures in science, whom he cultivated with consummate skill. In early February, at a breakfast at Barlow's, Faraday introduced him to Airy, who remembered their exchange about a testimonial. Tyndall was impressed: 'He is a proud man they say and I was led to deem him a dry crust—but the man is full of humour and he and Barlow spouted parodies during half of breakfast time'.[72] At a dinner at Percy's house in Bayswater, he met the geologist Andrew Ramsay, who promptly invited him to dinner in ordinary clothes, 'a great privilege I intend to use'.[73] Tyndall did not enjoy formality, and proudly stated much later that he never owned Court dress. At Barlow's again he was introduced to Mary White-Cooper, wife of the oculist William White-Cooper, who became surgeon-oculist to the Queen in 1859. She invited him the next day to meet Owen at breakfast; they had seen each other across a room at the meeting Owen chaired in Jermyn Street for a memorial to Edward Forbes, but not been formally introduced. Tyndall took to Owen immediately, impressed by his scientific 'acquirements' and his love of poetry. He soon discovered that Huxley was not so taken, describing Owen as 'antithetically mixed, one half great and the other half the opposite'.[74] On Valentine's Day, which passed unusually without a reference, Tyndall gave a soirée lecture at the London Institution on 'Boiling and Boiling Springs', an opportunity for dramatic geyser demonstrations.[75] Warren de la Rue, one of the Managers of the London Institution, and soon to be a Manager at the Royal Institution, was present. Another Manager told Tyndall that he had never seen the audience so delighted.

The annual commitment of nineteen lectures at the Royal Institution, one each week, was confined to the first half of the year. In 1855 Tyndall gave eleven before Easter on 'Magnetism and Frictional Electricity', and eight after Easter on 'Voltaic Electricity', ending in early June. His preparation of demonstrations led to a paper on the polymagnet, a device for exhibiting before a class of pupils many of the phenomena of electromagnetism and diamagnetism.[76] He reflected on his experience:

'Success in lecturing is made up of two components, knowledge, and vigour to impart that knowledge to others. The former is obtained by study, but beyond a certain point the latter is weakened by study...and the grand problem is so to arrange both that the effect shall be a maximum'.[77]

With the lectures, the writing up of his diamagnetism paper for publication, the exchanges in the *Philosophical Magazine* on the nature of the magnetic medium, and the start of his reviews of science for the *Westminster Review*,[78] there was no time for original research. He must have felt that Hunt was a target, as he sent Chapman a critique of the revised edition of Hunt's *Elementary Physics* for the *Westminster Review*, which was as unfavourable as his first.[79] He was not well at times, suffering from frequent debilitating headaches from which he recuperated at Queenwood, but there was time for socializing: at Mrs Powell's conversazione; at White-Cooper's, impressed by Miss Poyser's earnest and beautiful eyes, and fine figure; with the Pollocks where the only other guests were one of the Spedding family, his wife, and 'fair daughter'; and at the Royal Society, where he reported: 'A conversation with Mr H Spencer, a good mind I believe'.[80] This was an early encounter with the philosopher Herbert Spencer, to whom he was introduced by Huxley.

At Easter, Tyndall spent further time at Queenwood, arriving later than he would have liked, as he was completing and packing apparatus for the Paris Exhibition. Earlier in the year he had been invited by Playfair to be a juror, judging exhibits. In addition, as Secretary to the Committee for Electricity and Magnetism, with Wheatstone as chairman, he supervised the production of exhibits to illustrate magnetic and diamagnetic phenomena. On Easter Monday he went with Hirst and Anna to Southampton to see Mrs Ginty and her children sail in the *Great Western* to Rio de Janeiro to join Ginty.

Du Bois-Reymond came over after Easter to give a series of lectures on Animal Electricity. It was the first time he had spoken at the Royal Institution, and Tyndall was often with him during his stay. In Berlin in 1853, Tyndall had spent some time reading, making an apparatus, and repeating Du Bois-Reymond's experiments with a view to giving a future Discourse and writing an article about the fierce dispute between Du Bois-Reymond and the Italian Carlo Matteucci.[81] He had not been impressed then, and after Du Bois-Reymond's last lecture he wrote: 'I do sincerely believe that a week's earnest work would overthrow the edifice which it has taken so many years to raise'.[82] Whatever the relative merits of the German and the Italian on 'animal electricity', Tyndall was sceptical of the whole business.

While arranging publication of his paper in *Philosophical Transactions* and making drawings for it, Tyndall heard that he had been elected to the Philosophical Club,[83] along with Stokes and Huxley.[84] To his surprise, Warren de la Rue was blackballed. He dined there on 24th May, sitting between Huxley and Stokes, before going on to the Royal Society. He had also been feeling better, which he put down to the experiment of several glasses a day of bitter ale: 'Mr Bass has proved my apothecary…further trial is needed'.[85] Such a trial would not have been a great tribulation to Tyndall. He was partial to alcohol, and wrote at times to friends when manifestly drunk.

At the end of the month, successive social evenings gave Tyndall his first glimpses of two people he already admired, who would become good friends: Thomas Carlyle and Alfred Tennyson. The first was at Lord Ashburton's, where Carlyle was a regular guest. The stellar crowd included Sir Robert Peel (son of the former Prime Minister), 'a fine young fellow', Lord Elgin, and the Duc D'Aumale. Cardwell, ex-President of the Board of Trade, came up and squeezed his hand: 'But the man who interested me most was a hairy faced individual who was pointed out to me just before we separated…It was Thomas Carlyle—I looked deeply into his countenance as he was putting on his coat to go away. His eyes appeared sad and beautiful to me—…the sadness of one who while he mourned the follies of men had his own high trust and consolations'.[86] The next evening, after spending the day with Wheatstone and being shown improvements to the electric telegraph, he glimpsed Tennyson at a Royal Society soirée at Somerset House. His head was bad again, and though he managed to dine with Bence Jones, in the company of Wheatstone, Gassiot, and Jane Marcet, he cried off invitations to Lord Londesborough, Juliet Pollock, and Sylvester, where he would in fact have met Carlyle. He was well enough again on 1st June to give a successful Discourse 'On the currents of the Leyden battery', which he introduced by describing the usefulness of analogies such as positive and negative fluids to explain electrical phenomena even if one did not believe in their existence. He finished with striking demonstrations of electrical activity, projected onto a big screen.[87] The positive reception relieved his mind and gave him the tranquillity to attack his other work.

Tyndall now had a fortnight before he was due to depart for France for the Paris Exhibition, and a Sunday with Bevington relaxed him for the last of his lectures on Voltaic Electricity. On the eve of his departure from London, he dined with Bence Jones and went on to Faraday's Discourse, where he was introduced to Kane, fellow editor of the *Philosophical Magazine* and the recipient of his enquiries about the posts at Cork and Galway. In the first mention of a femme fatale, he helped Miss Drummond to repeat an experiment after a lecture. She was one of three daughters of the redoubtable social hostess of 18 Hyde Park Gardens, Maria Drummond, widow of Thomas Drummond, inventor of a powerful lime-light, and the well-respected Under-Secretary of Ireland until his early death in 1840.

* * *

The Exposition Universelle in Paris, the second international exhibition after that of 1851 in England, ran from May to November 1855, attracting some five million visitors.

It was an opportunity for France to stake its claim to the superiority of its industrial and artistic capabilities. The exhibition succeeded in conveying French achievements but, unlike the British 1851 Exhibition, failed to make a profit. Tyndall was due in July for his duties as an exhibition juror, with Sir David Brewster, for Class VIII, 'Arts relating to the exact sciences, and instruction'. He took the opportunity for a two-week sightseeing visit to northern France with Debus, crossing to St Malo. Before setting out from there, they saw the tomb of Chateaubriand, who had died in Paris during the 1848 revolution. This visit caused Tyndall to dwell at length on thought and consciousness: 'the man in this world, as well as the taking down of the structure after death, are certainly the work of molecular forces...but how do thought and consciousness spring from these molecular combinations?... Experience shows them to be twin phenomena, but is the association necessary? Can conscious thought exist apart from matter or can it not?'.[88] There followed a lengthy walk over several days to Mont St Michel, as he encountered continental Catholicism. Comparing women and men prostrating themselves before an icon, he thought 'a woman in some degree beautiful...a man contemptible'.[89] For him, the woman was 'as the child who listens with wondering faith to the recital of a ghost story'.[90] Later, at the cathedral in Rouen, observing a congregation at Mass consisting predominantly of women, Tyndall quoted in his journal from Tennyson's *Locksley Hall*: 'Nature made them blinder motions bounded in a narrower brain'. With respect to Catholicism, he wrote to Faraday (a devout Sandemanian): 'Women are its chief supporters at the present day and I suppose the reason is that they have more feeling and less intellect than men'.[91] It takes a certain insensitivity to write to a religious man in such a way.

As they walked, Tyndall and Debus pondered the nature of life. Tyndall saw the growth and development of living things as the result of molecular interactions, ultimately of the light and heat of the Sun transmitted to individual molecules through the 'ether' (an invisible substance presumed by many, though not by Faraday, to suffuse space). Life he defined as 'an incessant effort to restore a disturbed equilibrium'.[92] But reducing in some measure the mysteries of life to mechanics would in no way impair the miraculous beauty of nature. Tyndall rejected a 'vital force', arguing that the processes of life must be due to the operation of the forces with which the particles of matter are endowed, just as much as for crystallization. There was no essential distinction between organic and inorganic. No external force or architect was needed. He was clear about the limits of science, arguing: 'We have regarded the phenomena of life objectively, but we must not confound this with the subjective phenomena of thought and consciousness'.[93] But, he asked, 'who endowed the molecules thus—who gave to these forces their particular directions?...Do we not in these processes trace with dim eyes the operation of an intelligence which transcends our own?'.[94] For him, the physical problem marked the limit of understanding 'and beyond this we must substitute for it the power of Faith'.

In Paris, many hotels were full. He eventually found a berth in the Hotel Manchester. In the post was an invitation to lecture at the London Institution, which he accepted, and he heard to his delight that Huxley had been appointed to the Fullerian Professorship at the Royal Institution.[95] A letter from Henry Moseley asked if he would accept an invitation from Lord Panmure, Minister in the War Office in the new Palmerston Government, to act as an examiner in Physics under the Board of Ordnance for appointments to Cadetships in the Royal Artillery, for a sum not less than £50.[96] The examination was to be held at Kings College London, and he was required to prepare questions on Experimental Science and attend a meeting in London on 23rd July.[97] This experience of examining highlighted his view of the paucity of understanding of the natural sciences by those in authority, and its poor weighting relative to the classical languages. The marks given for all the sciences were equivalent to those for a single classical language.[98]

The British jurors met in the Rue de Cirque on 2nd July, chaired by Lord Ashburton. Tyndall's time in Paris gave him the opportunity to meet many eminent men for the first time, among them Auguste de la Rive and Jules Jamin, and to observe at first hand the rivalries within the French scientific community. He spent much time with Elie Wartmann and Wertheim among the philosophical instruments in the exhibition, including clockwork, measuring instruments, astronomical instruments, and microscopes, and dined regularly with Wertheim and Wheatstone, often at the Palais Royale, and at Despretz's. While they examined the exhibits, Brewster told him of the intention to offer him the Chair of Natural Philosophy at St Andrews, but the incumbent had not moved as expected. After a fortnight of judging, Tyndall had to leave Paris for his examiners' meeting. The burden fell particularly on Wartmann, who asked plaintively in early August if Tyndall could return; Brewster had left, as well as his replacement, and Dove had not even arrived. He was left with 'four French, two Swiss it is not enough'.[99] In the event, Dove and Magnus did arrive, and Wartmann thought it no longer necessary for Tyndall to travel back.[100] The judges did their work. Out of 24,000 exhibitors in the whole exhibition about 11,000 won prizes.[101]

On his return from Paris, which he had found noisy compared with London, Tyndall was able to take up his researches again, working on Weber's experiments on diamagnetic polarity. He succeeded 'in reducing all to certainty', and demonstrated his results to De la Rive and Faraday, the latter expressing 'his perfect satisfaction'.[102] De la Rive was keen to see the polar action of a diamagnetic body that was also an insulator.

Tyndall proved this with the 'heavy glass' Faraday had used to show the rotation of polarized light by a magnet.

Given his commitments to lecturing and research, Tyndall decided that his lodgings in Chadwell Street would now be too far from the Royal Institution for the next season. He vacated them on 18th August, leaving the maid Sarah with tears in her eyes, to base himself until November at Queenwood, where he had a bedroom in the house and a space in the lodge for working. He spent the time until early September working, walking, reading Tennyson's *Maud* and *The Princess*, and talking with Hirst, Debus, and Francis, who visited as Hirst's guest.

The autumn meeting of the British Association beckoned again. It was in Glasgow in 1855, and Tyndall left Queenwood on 6th September for a five-day walking trip through the Lake District with Frankland, *en route* to the meeting. Frankland had a house with a small laboratory on the shore of Windermere, and a boat on the lake. Tyndall joined him, reading *Maud* on the way. Their expedition started with an ascent of Orrest Head just after sunrise, from the summit of which Tyndall declaimed Tennyson's poem on Will: 'Oh well for him whose will is strong'. After a row on the lake they walked past Coniston to spend the night at a farm at the base of the Langdale Pikes. For the next few days they suffered typical Lake District weather: pouring rain interspersed with beautiful views of the landscape. The imposing 'Donjon Gill' (Dungeon Ghyll) set the sublime, rocky, romantic scene as they toiled out of Langdale in terrible rain, aiming for Wasdale. A map-reading error by Frankland led to a huge detour as they mistakenly descended the long Langstrath valley to Stonethwaite before climbing the Styhead pass over to Wasdale, past chasms crammed with vapour, precipices, and mountain torrents, to collapse into a welcoming house for the night. From Wasdale, the passes took them past the head of Ennerdale, in the rain again, and over to Buttermere, where Tyndall 'had not witnessed a more glorious scene in all the lake district. Mountain, wood and water conspired to form a noble picture'.[103] From the inn the next day they walked up to the torrent of Scale Force and past Honister Crag into Borrowdale with its famous Bowder Stone. They were rowed across Derwentwater to Keswick, visiting Southey's tomb and house before taking the coach to Penrith and the train to Glasgow.

Tyndall was now on the Council of the British Association, of which the thirty-two-year-old Duke of Argyll was President, a man he liked 'with his rough bearskin coat and rough red hair'. He took lodgings with Frankland. He was again a Secretary to Section A, although with the illness of Stevelly, the old Secretary, a heavy share of the duties fell on him. The pressures meant that he had to turn down social invitations from White-Cooper, Brewster, and several others. Richard Monckton Milnes, later Lord Houghton, the poet, literary patron, and politician, was in the chair for the

Red Lions. A 'fine old Swedish Count', sitting beside him, was 'rather amazed when first he heard the Lions roar'.[104] Thomson had looked forward to Tyndall's presence in Glasgow, his home city. In a long letter in June, hoping he would attend, he asked the well-connected Tyndall for news of Helmholtz and Du Bois-Reymond, and for what he knew of Werner Siemens's plan for measuring the times occupied by military projectiles in passing over short distances in various parts of their paths.[105]

On the Saturday, Tyndall gave his paper on the polarity of diamagnetic bodies.[106] The work he presented, using the sensitive apparatus produced by Leyser according to Weber's plan, removed, according to Tyndall, the last remaining doubt that diamagnetic bodies under magnetic excitement possess a polarity the reverse of magnetic ones. Thomson challenged him, and when Whewell rose in support 'throwing the weight of his great authority in favour of Thomson', Tyndall unwisely 'said something to the effect that Thomson had acted a safe part in suffering himself to be guided by Faraday'.[107] Whewell glowered, and Thomson's supporters seemed offended, though the audience appeared to be on Tyndall's side. He apologized if he had caused offence, but the words hung in the air. Thomson was not perturbed, inviting him to dine afterwards to meet Liebig, but it left a bad taste in Tyndall's mouth, and an antagonism with Thomson's friends that would last long. He gave one further paper 'On the comparison of magnetic induction, and calorific conduction in crystalline bodies'.[108] He had shown that the line of best conduction of heat in gypsum is that of the least magnetic induction. There was, therefore, no unity of agency, a finding relevant to his thoughts about the relationship of structure to properties.

At the end of the meeting, Tyndall took off for the Trossachs and Stirling with Frankland, visiting Loch Lomond and Loch Katrine, though Frankland refused to climb Ben Lomond, claiming a sore foot. Passing the field of Bannockburn they took the train south, separating at Carlisle for Tyndall to return to London and on to Queenwood, where he aimed to spend the autumn away from the noise of London. He was there for several weeks, in a reflective mood after Glasgow. He had been ambivalent about going to Glasgow at all, since Faraday would not be present, and did not attend a British Association meeting again until 1858. Frankland was also unhappy, writing: 'The more I think of that British Association the more am I convinced of the deep injury it annually inflicts upon science … When we go to the Tyrol next year we must contrive to take in the Naturforscher Versammlung in Vienna as I should much like to see how they manage those things in Germany'.[109] Concerned at the impact of his remarks about Thomson, Tyndall confessed to Faraday. Faraday replied in a letter full of sensible advice, advising him not to jump to conclusions about people's motives and to be more diplomatic, gently chiding him: 'it is better to be blind to the results of partizanship and quick to see goodwill'.[110] Nevertheless Tyndall was content with his achievements, and excited about a new discovery made at Queenwood as a direct consequence of his work on the compression of substances for his diamagnetism experiments.[111] It led to a Discourse the following year on the cleavage of crystals and slates.

At the end of October, Tyndall returned to London, sleeping for the night in a small room in the Royal Institution before looking at new lodgings in 96 Mount Street, Grosvenor Square, 10 minutes' walk from Albemarle Street. He had left behind at Queenwood what he hoped was a better regime to manage the school. Declaring that the position for which Nature intended Edmondson was 'that of a waiter in a hotel', he had organized a weekly meeting of Hirst and fellow teachers De Haas and Wright to manage the place better.[112] His finances were improving to the extent that he went with Francis to deposit his first £100 in the Joint Stock Bank.

Without lecturing commitments before Christmas, Tyndall could concentrate on his researches, even if his health remained a continual subject of complaint. Suffering from constipation—movements of the bowels were a common feature of letters between Tyndall and Hirst, and occasionally with Faraday—Tyndall experimented with his diet. Oatmeal porridge without tea or coffee had helped at Queenwood, and he started half-heartedly recording his diet and its effects. He put himself under the direction of Bence Jones for advice, and dined several times with him or with Barlow in November, in the company of people like Hofmann and Grove. At the end of the month, he spent a couple of hours with Foucault, due to be presented with the Copley Medal at the Royal Society's Anniversary Meeting, and dined afterwards at the Huxleys with Hooker and the mining geologist Warington Smyth, meeting Huxley's 'excellent wife' Netty for the first time. They had recently married now that Huxley was financially secure enough to bring her over from Australia.

A long letter to Grove in December reveals Tyndall's views about the nature of research, his perception of constraints at the Royal Institution and the significance of his latest findings.[113] The letter was stimulated by a request from Grove, acting for the Royal Society Government Grant Committee, for Tyndall to justify his expenditure. It is not clear what instigated Grove's approach. Tyndall argued that the grant gave him the freedom to respond quickly when the management of the Royal Institution might not allow it, and especially now when the issue of diamagnetic polarity was still disputed, even after his Bakerian Lecture. He offered, if requested, to return to the Royal Society the instrument he had bought with the grant, and challenged the view that his application ought to be more definite in the statement of his objectives. To him that was unreasonable, since he was working at the fringes of science, where the outcomes and directions could not be predicted. He bridled at what he took to be slurs on his character, writing that if his record and character were not deemed sufficient he 'would beg to withdraw from all participation in the government grant for the promotion of science'. This was a purely professional letter; on a personal level, his relationship with Grove was close.

Tyndall was now in a position to reveal his final thoughts on diamagnetic polarity. On 20th December, dinner at the Philosophical Club preceded a Royal Society meeting at which his new paper was read. This, the 'Fifth Memoir', entitled 'Further Researches on the Polarity of the Diamagnetic Force', dealt with criticisms that the previous experiments of his and Weber's, which they claimed to show diamagnetic polarity, might instead be due to induced currents and should be repeated with insulators.[114] Von Feilitzsch had tried this and been unable to detect any effect, but Tyndall succeeded. Again, his ability as an experimentalist revealed itself. He made a series of extremely sensitive experiments with copper, antimony, and, crucially, with insulators. He found deflections to be permanent rather than temporary, which would have been the case if there were a momentary induced current. This proved to him that insulators, as well as conductors, showed diamagnetic polarity. He ended the paper by claiming that all objections to diamagnetic polarity had now fallen away, placing it 'among the most firmly established truths of science'.

Tyndall's last and relatively short paper, the 'Sixth Memoir', addressed Faraday's statement that the magne-crystallic force is neither attractive nor repulsive, but directive. Tyndall gave a clear explanation of the complex effects of attraction and repulsion in explaining the direction of movement of spheres and bars of substances in different magnetic circumstances.[115] In his view, 'the most complicated effects of magne-crystallic action are thus reduced to mechanical problems of extreme simplicity; and inasmuch as these actions are perfectly inexplicable except on the assumption of diamagnetic polarity, they add their evidence in favour of this polarity'. The memoir ends: 'The whole domain of magne-crystallic action is thus transformed from a region of mechanical enigmas to one in which our knowledge is as clear and secure as it is regarding the most elementary phenomena of magnetic action'.[116] Airy wrote to Tyndall in March 1856, congratulating him on reducing diamagnetism to a 'mechanical and calculable' form, since, 'It has been a matter of no small grief to me to find that till a comparatively late time, a totally different theory, a theory of extreme vagueness, has been advocated by the highest authority' (by which he meant Faraday).[117] Tyndall was pleased that the Astronomer Royal should see him in this light, but stood up for Faraday, arguing: 'a hypothesis may be valuable in two ways. It may, by its truth, help us to systematic arrangement or lead us to discovery. But it may also…prove a source of energy and be a spur to action.

In this way while vague, or even erroneous, it may be, and I doubt not has been, productive of glorious results'.[118]

Tyndall's work on diamagnetism was at an end, and in February, he gave his last major Discourse on magnetism, 'On the disposition of force in paramagnetic and diamagnetic bodies'.[119] Even the dispute with Plücker eventually subsided. In September 1856 he met Tyndall in Vienna. Tyndall reported a cold encounter: 'I stretched out my hand which he accepted, but so frigidly that the value of the acceptance was negative'.[120] But finally, in 1858, as Plücker published a definitive account of his work,[121] fences were mended when Hofmann brought them together in London. Plücker 'laid aside his coldness and we talked together for a long time in a very friendly manner'.[122]

How are we to judge Faraday's and Tyndall's explanations of the phenomena of magnetism? Both are self-consistent, and both can be put into mathematical form; Faraday's initially by Thomson and then through the great work of James Clerk Maxwell. The mathematics for Tyndall's approach—a vector algebra—existed at the time, but Tyndall had no Thomson or Maxwell to develop it. Tyndall's matter-based approach to magnetism remains valid, but Faraday's field theory was deeper and more productive in the long run. At the time, many physicists struggled to understand Faraday. As Tyndall expressed in a letter to Hirst: 'It is amusing to see how many write to Faraday asking him what the lines of force are. He bewilders even men of eminence…I heard Biot once say that he could not understand Faraday, & if you look for exact knowledge in his theories you will be disappointed'.[123] Faraday's theoretical greatness was not apparent to many of his contemporaries, including Tyndall, though Thomson and Maxwell had the insight and mathematical ability to develop his ideas. Given the visibility of Faraday and Tyndall, Maxwell recognized the need to bring his own work to their attention. He made sure to publish a significant paper in the *Philosophical Magazine* 'because Faraday reads it and so does Tyndall'.[124] Nevertheless, convinced of the merits of his own views, Tyndall published his collected papers in 1870, after Faraday's death, as *Researches on Diamagnetism and Magnecrystallic Action*.[125] Throughout his life he never changed his position on Faraday's field theory.

* * *

The year 1855 drew to a close with a run of social engagements and time for reading, including George Lewes's *Life and Works of Goethe*, dedicated to Carlyle. On Christmas Eve he dined with Bence Jones and the children. Christmas Day was spent with Bevington, and the next day found him at a dinner of partridge with the Huxleys. Tyndall and Netty Huxley (Figure 13)—'a very superior woman, a fine natural intellect and well cultured'—had taken to each other, and he was invited back on New Year's Day for what became an annual event.

Fig. 13. Henrietta 'Netty' Huxley (1825–1914).

Right at the end of the year, Tyndall moved lodgings after problems with his land-lord in Mount Street, who he thought was overcharging him. He took the second floor of the house of Mr and Mrs Flores in South Audley Street for £1 2s a week, per-haps swayed by the sight of Miss Flores, who was 'more than respectable i.e. pretty'.[126] He had two sets of lectures to prepare at the start of 1856; a series of six at the London Institution on Mondays, and one of nineteen at the Royal Institution on Thursdays, on 'Light', running through to 12th June. There were many social engagements in early January: at White-Cooper's, discussing Goethe and science with Owen; at Bevington's; with the Lions; and at Miss Moore's, where Babbage ranted at the Royal Society. But none to compare with the weekend of 11th January.

Tyndall had received an invitation to Lord Ashburton's country residence in Hampshire, The Grange, discovering with anticipation that Carlyle (Plate 9) would be there. Ashburton had chaired the British jurors for the Paris Exhibition, and was a Manager of the Royal Institution. He was a moderate Whig who became a Conservative, and though he never reached high office, he was particularly commit-ted to education, advocating the teaching of 'common things' in national schools.[127]

He had invited Tyndall partly to pick his brains prior to raising educational matters in the House of Lords, giving Tyndall the opportunity to experience his first aristocratic country house party. It was a heady affair. Tyndall was struck by the total absence of formality, as guests including James Spedding, Hofmann, De la Rue, and Henry Cole—organizer of the 1851 Exhibition—had free range of the house, talking, admiring the paintings, and listening to Lady Ashburton reading Browning in the library, with Carlyle at her left and his wife, Jane, opposite. Tennyson had been there two days previously, reading *Maud*. Tyndall's first direct encounter with Carlyle was over breakfast. Imagining for some reason that Tyndall was a supporter of homoeopathy, Carlyle heaped ridicule on it while Tyndall responded, provoking him about the great effects produced by apparently trivial causes, until the whole table was engaged. He got to 'revere the old brute more than ever', talking with him about Emerson and about the incomprehensibility of life.[128] As he put it to Hirst: 'I tried to stir him by starting some theoretic views of the nature of life—mere physiological life; ascribing it to molecular force'.[129] He sat next to Jane Carlyle at dinner, talking of Carlyle rather than her, and of Carlyle's current project on Frederick the Great.

At the end of January, Tyndall examined Artillery and Engineers candidates for commissions, and finished marking the papers a fortnight later. Moseley's report of the examinations did not include commentary on the areas for which Tyndall was examining, prompting Joseph Galbraith to write to Tyndall to complain on behalf of himself and Samuel Haughton, who had prepared candidates and felt that their achievements had not been recognized.[130] Since he had not sent in any comments, Tyndall resorted instead to the pages of the *Times*, writing: 'In justice to the candidates for commissions in the Artillery and Engineers examined by me in natural philosophy and chymistry you will, perhaps, permit me to state that the general level of the answers in the last examination was much higher than that attained in the first. Many of the papers returned to me gave evidence of rare ability'.[131] He recognized that in doing so he might have overstepped the mark. Sure enough, a missive from the War Office observed that he had written without the sanction of the authorities. Having sought advice from Faraday, who told him it might result in his dismissal, he wrote to Lord Panmure, acknowledging his error but taking the opportunity to make the case for the importance for future Artillery and Engineers officers of natural philosophy and chemistry in developing intellectual and inductive powers and practical capability, beyond the deductive qualities he saw promoted by the study of mathematics. Lord Panmure showed forbearance, and he was not dismissed.[132]

Lectures were interspersed with social engagements, until Easter found Tyndall at Queenwood for ten days, staying with the Hirsts. Anna was not well, but Tyndall had 'little fear of her lungs'.[133] In this he was sadly mistaken. As ever, he relaxed in the country, riding through the grounds of Norman Court on Fanny (the Hirsts' horse), and visiting Stonehenge with Hirst and Debus. Many of the boys trusted and idolized Tyndall. Charlie Cumming, in tears, confessed to going to Romsey Fair without permission and lying about it. Tyndall was stern—especially about the lying—but supportive, because he could sense Charlie's strong morality.

A month after Tyndall left, Hirst wrote a long letter about the expulsion of two boys as a result of 'the evil habit with which we have so long been fighting'. One of them was the son of Taylor, the Mayor of Romsey, the other of Le Quesne of Jersey. Hirst described Le Quesne as a 'puny, weak-minded boy. He was only detected about a month ago, but it proved that he had been a slave to the habit for some years. I spoke to him very earnestly and he made me many solemn promises to amend. In less than a week after, he was detected again and I thrashed him, but not very severely... I thought I had frightened him, but in less than a fortnight he was detected again. On being accused he stoutly denied it and I gave him a very severe thrashing. This broke him down and he made some revelations which astounded and afflicted me greatly'.[134] Hirst reported an agonizing meeting with Taylor's father, desperate at the disgrace that would ensure for the family and the boy's future—'you have damned him for life'—who argued that some medical men even asserted the physiological necessity of masturbation. Hirst stood his ground. The boy was contaminating others and had to go for the reputation of the school, but they would expel him quietly, not publicly.[135] He could not see the end to it: 'Week by week we find more and more implicated and our preventative measures grow more severe'. Tyndall was fully supportive.[136] When Tyndall suggested that Hirst should write about education, Hirst's response was that his theory of education consisted 'almost entirely in the prevention and care of one fatal habit. Do you think the time has come to tell the world this and to assert what is taking place in every school in the kingdom?'.[137]

Tyndall spent another Sunday at the end of March with Wheatstone in Hammersmith. As he walked home, cannons fired and bells jangled. He supposed that peace had been proclaimed. Indeed, the occasion was the signing of the Treaty of Paris, ending the Crimean War. Having finished a paper on colour blindness, in which he argued for the value of physics in helping understand the causes,[138] he spent several days reading about binocular vision in the context of the dispute between Wheatstone and Brewster over the invention of the stereoscope.[139] Tyndall was becoming a good friend of Wheatstone and took his side more than Brewster's, commenting that it was 'a great pity that those personal feelings should be allowed to mix so largely with scientific questions... How simply grand Faraday stands out in comparison with these disputants... if he died tomorrow he would not leave a foe behind him'.[140]

With the end of his lectures in sight, and no immediate research programme, Tyndall threw himself into the fray; partly with scientific friends—he chaired the anniversary dinner of the Philosophical Club, and presided over the Lions—and partly with wider society. At Barlow's in April, he met the Irish politician Sir James Emerson Tennent. He talked with him about table-turning and Goethe, sitting opposite Thackeray. Tennent, who admired his success as an Irishman triumphing over prejudice, invited him to Sunday lunch. Tyndall, perhaps to avoid further interaction with table-turners, claimed that he 'rarely dined out but had no more objection to Sunday than any other day'.[141] Indeed his engagements in May alone, apart from scientific meetings, included dining with Faraday before hearing Owen's Discourse (at which he sat with the Carlyles); at Lady Ashburton's; at the Duchess of Northumberland's ('some glorious specimens of English beauty present'); at Mrs Pollock's ('a pretty pleasant little creature Miss Spedding appears to be'); and at Sir George Clerk's.[142] He found time to deposit a further £50 in the London Joint Stock Bank. Francis held the receipts in his private ledger in a fire-proof safe.[143]

Tyndall was now an established researcher and lecturer, and increasingly welcome in society. It remained for him to give the last Discourse of the season on 6th June on a 'Comparative view of the cleavage of crystals and slate rocks'. The response it stimulated gave a completely new focus to his research.

7

❖⟹ ⟸❖

GLACIAL EXPLORATIONS
1856–1857

Thomas Huxley is arguably responsible for Tyndall's great love affair with the Alps, a consequence of his presence at the Royal Institution on 6th June 1856, when Tyndall gave his Discourse on a 'Comparative view of the cleavage of crystals and slate rocks'. This Discourse, on a seemingly obscure topic, would set the direction of his work for the next three years.

Tyndall's interest in slate cleavage stemmed from his early experiments on diamagnetism, and the effects of pressure on his sample substances. He imagined that atomic and molecular arrangements within materials were critical to explaining their properties, such as magnetism and the transmission of heat. The plane of cleavage, along which a substance splits if sharply hit, seemed to him to be a strong clue to that underlying molecular arrangement.

Some eight months before the Discourse, Tyndall had noted in his journal: 'one beautiful problem I think I have solved and that is the question of slate cleavage'.[1] The answer he came to, building on the work of people including Adam Sedgwick, Henry Sorby, and Daniel Sharpe (who fell off his horse and died just a few days before the Discourse) was that the cleavage plane in slate was caused not by a force of crystallization, as Sedgwick had suggested, but by enormous pressure at right angles to it. This process was separate from the stratification, bedding or layering seen in many rocks, which arose from phases of slow deposition of different materials. Herschel picked up on this, impressed by the simplicity of Tyndall's explanation and its applicability to other situations, such as iron rails and coin manufacture.[2] As Tyndall noted of his work over the past year: 'I have never eaten a biscuit during this period in which an intellectual joy has not been superadded to the more sensual pleasure; for I have remarked in all such cases cleavage developed in the mass by the rolling-pin of the pastrycook or confectioner'.[3] He had indeed co-opted Mrs Croombe, the Queenwood cook, to make puff pastry for his experiments.

Sorby had suggested that flat plates of mica guided the creation of the cleavage plane in slate, as they would tend to line up perpendicularly to the pressure. But Tyndall showed with white wax that no such process was required. Under pressure the wax exhibited a clean cleavage surpassing even slate. Pressure alone was necessary, uniting phenomena from slate to puff pastry, and from cream cheese to iron. This led Huxley to suggest that these ideas might help explain the laminated structure of glaciers. Frankland and Tyndall were planning to go to Switzerland and the Tyrol that summer, taking in the conference of the Deutscher Naturforscher und Ärzte in Vienna. It required but a slight modification of plans to arrange an excursion with Huxley to the glaciers.

Tyndall's lecturing for the year was nearly at an end. The day after the Discourse he set off with Francis for a weekend ramble from Tring to Windsor, looking down as they passed on the lawn and woods of Chequers. He completed his final lecture of the season on 12th June, the last in his series on 'Light' stretching back to January, writing 'thank the gods for this liberation'.[4] Hirst had taken Anna to Ireland and then to the warmer climes of Bordeaux for her consumption, though Tyndall now had no hope of her recovery. He was able to relax in Hirst's rooms at Vine Cottage in Broughton, while preparing examination questions for the Royal Artillery and Engineers examinations, admiring the vegetable garden and lawn, and catching up with Debus, Wright, and Fox, a local doctor. The countryside around Queenwood, under a cloudless sun, with the song of birds, the floating of butterflies and the perfume of the flowers 'fell softly on my soul and made me content'.[5] After returning to London for the examiners' meeting, he decided on the spur of the moment on a week's break near Christchurch, in the company of the Wrights and their new baby. He walked on the beaches and cliffs, indulging in his pastime of examining inscriptions in local graveyards, flirting with a waitress, and making excursions to Lymington and the Isle of Wight—which he described at length in verse—snatching a tantalizing glimpse of Tennyson's house. Back again in London, the social rounds continued: at the Harveian Oration at the College of Surgeons; at the Taylors in Richmond with Poggendorf; with the novelist Geraldine Jewsbury; and at the Moores in Wimbledon, where Babbage treated him to an account of scandalous aristocratic liaisons.

It was an evening with the Hookers at Kew in late July which proved significant for the glacier studies. From the conversation, Tyndall was induced to read James Forbes's work on glaciers, *Travels Through the Alps*, and the papers of William Hopkins. As he wrote to Hirst: 'I would content myself with the glaciers of the Tyrol, but in order to appreciate Forbes's theory I must see what he has seen'.[6] This meant a visit to the glaciers of the Aar and Rhône. The plan was for Tyndall to meet Huxley, Hooker, and their wives in Basle, with Frankland joining them later as they headed for Vienna. Tyndall left London Bridge on 12th August (missing the British Association meeting in Cheltenham), and went on from Dieppe to Paris and a sleepless night at the Hotel Strasbourg as wagons rumbled all night, whips cracked, and the clock struck every hour. A hot day's travel took him on to Strasbourg and Basle, rattling between the vineyard-clad hills of the Vosges and the ridge of the Schwartzgebirge. Spotting Huxley in the corner of a diligence he ran to join him. They set off for Bern as thunder and lightning broke over the distant Jura mountains, to put up at Die Mohren, finding the Hookers there at breakfast in the morning.

* * *

On a good day, the distant panorama of the Alps from Bern is spectacular, and the sense of anticipation as one approaches from the plains is palpable. The mountains were looking magnificent as the party approached Thun, where they spent a night before taking the steamer to Interlaken. Here the glacier observations started in earnest, as the Hookers departed and Huxley secured two donkeys to take Tyndall, himself, and his pregnant wife to Lauterbrunnen and then up to the Wengen Alp, beneath the glaciers tumbling from the north faces of the Eiger, Mönch, and Jungfrau. Perhaps with that apprehension which all mountaineers know, of challenges to come on the morrow, Tyndall failed to get any sleep. He experienced a stunning dawn as the Sun's beams struck the Silberhorn, the exquisite satellite of the Jungfrau: 'I never saw anything so beautiful. The white cone so high up in heaven covered with snow of perfect purity'.[7] He set off alone in the dawn light, scrambling up above the deserted chalets to the Guggi glacier under the Jungfrau. Clambering onto it with a sense of timidity and unfamiliarity, startled by the ghostly rattle of debris falling from melting ice, he started to examine the fissures and the banded structure of the ice.

After breakfast he returned with Huxley, and they puzzled over the origin of the bands. Huxley imagined they had been formed in the snowfield above, a suggestion Tyndall found unconvincing. A violent thunderstorm drove them down, and the next day, dodging rain again, they climbed up to the Lower Grindelwald glacier under the Eiger. Here Tyndall examined the ice-fall where the glacier steepened, already interpreting what he saw in terms of tension and pressure. The ice, under tension, was split by great transverse chasms above the ice-fall into blocks which themselves were split at right angles, along the axis of the glacier. The pressure of the ice on more level sections seemed to close up the transverse chasms again. Further up, the banded structure was clear, showing a lenticular shape like the green flattened spots in roofing slate.

While Huxley went down to Interlaken, Tyndall took to the glacier with Hooker, observing the 'dirt bands' on the glacier below the ice-fall; these were dark lines across the glacier, widely spaced apart, first described by Forbes. He then set off alone over the pass to Meyringen, to meet the guide Kaspar Bürki at the Grimsel Hospice.

On the way, he noticed scratches on the granite slabs very like those which would be produced by glaciers. He remarked: 'if the existence of these glaciers be assumed, they must have reached nearly to the mountain tops, for the rocks all upward are similarly marked'.[8] The idea that glaciers could have once reached such a height in an 'Ice Age', propounded by people such as Louis Agassiz, was novel and disputed.

Meeting Huxley again at the Grimsel, Tyndall arranged to explore the Unteraar glacier, beside which lay the historic Dollfus hut of 1846, with the names of Agassiz, Forbes, and Ramsay scratched into it, and the altitude 2,404 metres marked in paint on the door. Higher up, on the central moraine between the Finsteraar and Lauteraar glaciers, and moving downhill at glacial pace, was a rudimentary shelter beneath a huge rock, the famous Hotel des Neuchâtelois of Louis Agassiz. This whole area was the site of Forbes's exploration. The party spent an uncomfortable night in the hut, concerned that their guides had lice, and with eyes smarting from the smoke of the fire. Rain and mist prevented exploration the following day, but Tyndall had seen further evidence of the relationship of pressure to banding. They went over to the Rhône glacier, through that high, remote mountain environment with no tree, no sound, and the Schreckhorn rising majestically above. Recharged with cold mutton, bread, and wine at the inn, they went to inspect the névé, or snow-covered part of the upper glacier. The 'dirt bands' were noticeable from a vantage point on the way, and they found pieces of ice with alternating layers of white and transparent ice which could be cleaved and appeared just like weathered slate. Higher up, overlooking the whole glacier, they stopped to admire the fabulous view: the Finsteraarhorn lording it over surrounding summits; the Weisshorn, Monte Rosa, and the Matterhorn 'like a black savage tattooed with streaks of snow'.[9]

Huxley left with his wife, while Tyndall set off on a trip of three weeks by foot, coach, and boat to Vienna. He met Frankland on the Finstermunz pass and they walked over the Stelvio pass to Bormio in Italy. Tyndall noted with wonderment the blue colour in a hole created by thrusting his stick in the snow, an observation he follower up later. Fuelled by Asti wine, they retraced their steps, and spent a week rambling through the Tyrol to the white walls of Innsbruck.

The impetus for this trip was scientific, but Tyndall delighted in the landscape and in the excitement of mountaineering. Descending the Stelvio as night advanced, a ghostly gleam of white light rested on the summit of the Ortler. Onward in the distance the summits deepened into a dusky neutral tint and were finally lost in space. Tyndall felt that he was 'part of the great whole and lost my individuality in this conception'.[10] The beauty and scale of the Alps were beginning to cast their spell. From here on, Tyndall's writing about mountains is suffused with emotional intensity and poetic imagery, even if the modern ear may find some of it overblown.

The meeting of the Deutscher Naturforscher und Ärzte in Vienna was the model for the British Association, though Tyndall discovered to his regret that there was no discussion after papers, and that the dining arrangements were 'very different from those reunions where English intellect and English beauty mingle during the evenings of the British Association'.[11] He spoke on ice and slate structure, and the presentation drew applause, the first that he had heard in the meeting.[12] The next day, he escaped from the lengthy and stifling general meeting in the gas-lit Redouten Saal into Vienna's dusty, windy streets, to visit the zoological and mineralogical museums.

* * *

Apart from the glacier studies, this Alpine trip set off a number of lines of thought for Tyndall, especially connected with colour: the blue inside shafts driven into snow, and the colours of rivers and lakes. He mentioned these to De la Rive,[13] asking if the cause was known. De la Rive responded that he took blue to be the natural colour of pure water and big blocks of ice, and that the other colours that water takes on would be due to the presence of other substances, or to the reflection of daylight from the bottom.[14] This puzzle would exercise Tyndall over many years. But it was the work on glaciers that became his primary focus and occupied him on his return to London in October. Nevertheless, the regular commitments at the Royal Institution and elsewhere had to be met, as he finished his six lectures on 'Light' at the London Institution on 19th January 1857 and prepared a course of eleven lectures on 'Sound' at the Royal Institution, starting on 22nd January. It was a new topic that required much time.

Tyndall worked up his summer explorations into a paper 'On the Structure and Motion of Glaciers'.[15] Despite his view that women were not capable of the same original scientific research as men, he did value the comments of female friends, and he asked Frances Hooker to look over the draft before he sent it to his co-author, Huxley. Huxley could find no flaw in his reasoning, and thought he had 'run an oiled rapier through James Forbes'.[16] Tyndall was aware, as he expressed in letters to several people, that Forbes might think he had some deliberate design on him. He wrote to Hirst that he had 'smashed up this theory; so utterly annihilated it, that its author will hardly acknowledge it'.[17]

The theory that Tyndall believed he had destroyed was Forbes's idea that a glacier is a viscous substance. Tyndall acknowledged that the glacier did seem to behave in many respects as if it were viscous, but thought that it was 'not viscosity; but a property known some for years, but wholly overlooked by glacial theorists' to which the phenomena were due. Tyndall called this property 'fracture and regelation', the breaking and re-freezing together of ice. To add weight to his argument, he carried out laboratory experiments with ice under pressure to demonstrate the effect.

117

It was this extensive practical work in the controlled conditions of the laboratory which formed the substance of his publication, complementing the observations made in the natural environment. The paper was read at the Royal Society on 15th January 1857, and Tyndall gave it as a Discourse at the Royal Institution a week later entitled 'Observations on Glaciers'.[18] Despite months of work, significant elements were not decided until late, not least with respect to the term 'fracture and regelation'. As late as 6th January, Tyndall was toying with Huxley's term 'frangi-flexibility', as well as Hooker's suggestion of 'regelation'. He told Huxley: 'Flexibility in a solid implies a power of regaining its shape when the strain upon it is removed. Our ice would not do so…I had set down the "break and make" theory; a better term…would be one to include the idea of breaking with that of refreezing'.[19] This is what became the theory of 'fracture and regelation', although Tyndall appreciated that the motion was complex: a glacier moved under its own weight, fracturing and 'regelating', but also by sliding along its bed.

The Discourse gave a clear account of his theory. Referring to the early ideas of Louis Rendu, Bishop of Annecy—later the focus of much discord about who had priority for the theory—and acknowledging the apparent power of Forbes's viscous theory to explain glacier motion and the lamellar structure of ice, Tyndall pointed out that it was opposed to our ordinary experience of ice. It was said, in support of the viscous theory, that the true nature of ice could be inferred from large masses, and that the apparent viscosity shown by those large masses could not be seen in small specimens. This Tyndall demonstrated was incorrect, working like an ice magician in front of the audience to transform bars of ice into rings, and blocks into cups, simply through moulding ice under pressure. He attributed this to a discovery Faraday had announced in 1850; that two pieces of ice, placed in contact, freeze together by the conversion of a thin film of water between them into ice. So ice fractured under tension could be reformed by what he called 'regelation'. This could explain how a glacier accommodates itself to its bed, while continually forming and re-healing its fractures. The process also explained the fissured nature of glaciers, which could not be accounted for if the glacier were truly viscous.

Next, with reference to his ideas on slate cleavage, and referring to his observations when two glaciers join, Tyndall attributed the laminated structure to pressure. Forbes, by contrast, had imagined that the veined or lamellar structure was due to sections of the glacier moving past each other.[20] Tyndall could not see that this differential motion would produce structures at an angle to the motion of the glacier. He backed up his argument with more laboratory experiments, by exploring the behaviour of a thick mixture of water and pipe-clay to simulate viscous motion. For Tyndall, the oblique position of the veins near the sides, the transverse lamination of the centre, the lenticular structure, and the relation of the veins to the crevasses, were all in harmony with a compression theory of the veined structure of glacier ice.

Finally, he explained the 'dirt bands', the widely spaced lines of debris that can often be seen on the surface of glaciers, by tracing their origin to the base of ice-falls. At these steep sections of the glacier, where the ice tends to break into huge blocks, the dirt seemed to collect between them. He ended with words of appreciation for Forbes's experimental work, while dissenting from his theory. Frederick Pollock, ever an admirer, wrote that the viscous theory 'may be now considered as relegated to the same lumber-room of obsolete theoretical apparatus, in which the crystalline spheres of the old astronomy, phlogiston, & various other similar things which have done their temporary work may be supposed to repose'.[21] Darwin showed both penetrating thought—inviting Tyndall to test his regelation theory by squeezing pieces of ice together 'quite dry as far as water is concerned, but wetted with something that will not freeze'—and premonition of discord ahead by writing: 'the Lord have mercy on you, when Forbes answers you is my prayer'.[22] Just as Grove had done with Tyndall's paper on Forbes's rockers, John Phillips commented on reviewing this paper: 'It appears that the references to the "viscous theory" of Professor Forbes, have something too much the air of <u>controversy</u>'.[23] Tyndall's card was marked early on, not that it had much effect.

Tyndall knew his conclusions would be challenged. The *Saturday Review* published a measured account of his Discourse,[24] arguing that while his idea might be part of the explanation for glacier motion he had not disproved Forbes, and that the term 'viscosity' was vague.[25] So Tyndall continued to experiment, applying to Sabine for £30–40 from the Government Grant for further work. He had also taken care to keep in touch with respected figures to gain support, writing to Clausius, who had reached similar conclusions but not followed them up.[26] Clausius sent Tyndall details of the work of Agassiz and others, and Tyndall read out at the Royal Society part of Clausius's letter in his support.[27] Following the Discourse, he wrote a summary to De la Rive and to Herschel. Others, such as Charles Lyell, had heard the ideas by report, or in person. Forbes, who knew from correspondents that his ideas were being challenged, got sight of Tyndall's conclusions in March when he received the *Proceedings of the Royal Society*. He asked for the paper as soon as it was published.[28] In an attempt to pour oil on the waters in advance, Tyndall replied that he hoped Forbes would find in it 'no expression at variance with the Philosophic Spirit in which a question of the kind ought to be discussed'.[29]

These questions of Philosophic Spirit became entangled at this point in matters of romantic emotion. There is part of a wistful note in the diary, the rest torn out, expressing the sentiment: 'I often look forward with tranquil eye to a ripe old bachelorhood, much opposed as this approach appears to be to my constitution. I shall always love children, and have no doubt that there will always be two or three upon the face of the earth that will love me'.[30] These are perhaps the first hints of the problems which were to ensue with the sophisticated daughters of Maria Drummond: Mary, Emily, and Fanny. He revealed not a word of this in his letters, even to Hirst.

As early as March, Tyndall was planning a more extensive visit to the Alps in the summer of 1857, starting on the Mer de Glace, the huge glacier below Mont Blanc which Forbes had explored. He wrote to De la Rive in Geneva to ask to borrow a hydraulic press, a dye for water infiltration experiments, and a theodolite and chain.[31] He also sought agreement with William Hopkins as to how they should acknowledge each other's work. Hopkins told him that he had 'felt unwilling to write again on the subject in any way which might involve me in controversy with such men as Forbes and Whewell, neither of whom I am convinced can carry on a controversy without a large spice of ill-humoured feeling'.[32]

The world of science was now open to Tyndall—he was elected to the Royal Society's Council from 1856 to 1857—and the wider world of culture and Society was starting to tempt him and welcome him, through the connections of the office-holders and Managers at the Royal Institution, and of people such as Francis and Chapman in publishing circles. Tyndall went to Carlyle's house for the first time, taking tea with him and his 'excellent wife'. He recalled earlier thoughts of 'how every word of his told upon me, how I looked forward to meeting him as the crowning point in my life's history'.[33] For an Easter break, Tyndall and Debus took Harry, Bence Jones's son, to the Isle of Wight. They walked the island, often in pelting rain, and scrambled on the cliffs. Tyndall again visited Sterling's grave at Bonchurch—primroses on the graves, old elms spreading their bare brown arms above, with the song of the lark and the sound of the waves making melody for the dead. As ever, he enjoyed the juvenile company (Harry was about thirteen): 'I almost think that all that is pure in love may be felt by a man towards a little boy'.[34]

Tyndall had come to the Royal Institution because it gave him the opportunity for research, and Faraday had supported and encouraged his appointment on that understanding. But he took his lecturing seriously, and now faced a further eight lectures on 'Sound' after Easter. He wrote to Faraday that he found that the nineteen lectures for which he was employed, delivered between January and June, took half the year to prepare and greatly reduced his opportunity for original research. He sought to lessen the load, pointing to the historical precedent of Brande's arrangement for fewer lectures, with similar remuneration.[35] Hooker noticed the strain on him. He urged him to come to Kew, 'and rusticate with books and papers and a room to yourself'.[36] Lecturing at the Royal Institution was not the only demand, and when the War Office announced a cut in pay to £25 for examining candidates for the Royal Artillery Academy he wrote to Lord Panmure's office to resign, regretting the lack of appreciation of experimental science in military matters, an issue that became a continuing campaign in parallel with his attempts to remedy the neglect of scientific education in schools. When Colonel Portlock—a member of the Council of Military Education, which was formed in 1857—discovered later in the year that Tyndall was not on the list of examiners, and Tyndall told him why, he asked him to become one of the examiners of the Commissioners, to which Tyndall agreed. This invitation stemmed from the introduction of competitive examinations for entry to the military academies at Woolwich and Sandhurst in 1857 and 1858, following the debacle of the Crimean War.[37] As he told Hirst, he was committed to '10 examinations a year & for this the sum of £150 is allowed. It is too small, but the labour I am told will not be great'.[38] He was still an editor of the *Philosophical Magazine*, a role he did not relinquish until the end of 1863.

Nevertheless, he found time for more laboratory experiments on ice, exploring the conduction of heat through ice using Harrison's ice machine (developed by the Scottish-born and Australia-domiciled inventor James Harrison, who later tried unsuccessfully to launch a refrigerated meat trade from Australia),[39] and writing a short note on hail for the *Philosophical Magazine*.[40] He was at the Royal Society on 7th May to hear James Thomson's paper 'On the plasticity of ice, as manifested in glaciers', in which Thomson suggested that local melting under pressure, followed by refreezing when the pressure was released, could account for the plasticity of ice required by Forbes's theory. Tyndall responded that he preferred resorting to the mountains rather than indulging in theoretical speculation, and reported that he was applauded for his remarks. It was a discourteous response, and reflects his growing antagonism to the Scots (James Thomson was William Thomson's brother) and a scepticism of the value of theoretical over experimental approaches.

Matters of romantic emotion now took centre-stage. Tyndall was unwell for a period with bronchitis—reading Goethe's autobiography *Dichtung und Wahrheit* as he recovered in bed—but he was soon back in the social fray from which he could never quite extract himself. In early May he visited the Drummonds. He found Mary Drummond laid up with a cold, but talked of Shakespeare, Goethe, and Tennyson with the others. They thought Tennyson unforgiving in *Locksley Hall* and absolutely revengeful in *The Epitaph*. Tyndall wrote: 'I had feared these ladies were too purely scientific but today's experience disposed of this fear ... The second eldest girl, who is called "Pussey" is rather small and her features are at first sight not strikingly beautiful; but on closer examination there is a depth of good nature, a capacity for affection, a sweetness and kindness shining in her eye when she looks at one, that make her countenance more than beautiful'.[41] A few evenings later he was at Mrs Busk's, temporarily forgetting about the Drummonds—'there were two of the most beautiful girls there that I have seen for a long time'[42]—and was invited by Mrs Crosland, the author Camilla Toulmin, for a communion with the spirits a few days later. He was not impressed, to the great relief of Sarah Faraday. Some scientific activities demanded his attention—including a visit to Blackwall to see the magneto-electric light proposed for a lighthouse, aiding the French savant Lissajous at a Discourse on his acoustic experiments,[43] and completing what seems to have been his last piece for Chapman's *Westminster Review*—but it was the Drummonds who occupied his waking thoughts.

At an evening in 18 Hyde Park Gardens in May, Tyndall took Miss Emily ('Pussey') to dinner, sitting between her and Mrs Grove, with her 'fair sister', Mary or 'May', the eldest sister, opposite. After dinner the distinguished party included the Marquis of Lansdowne, Wheatstone, Babbage, and the Marcets. Further visits followed, with a trip 'through difficult streets' in a carriage with the four women to visit the *Great Eastern*, docked at Millwall.[44] He now fell dangerously in love with one, and in due course two, of the three Drummond daughters; dangerously since, as he readily recognized, they were from a different social background and he found it hard to judge their attitude towards him. Was it just a welcoming sociability, or something more? His journal at this point reads like a teenager's, peppered with remarks such as: 'all my life I have loved the <u>unattainable</u>'.[45] He thought 'Δ' quite beautiful: 'her soft smooth white hand rested in mine for the shade of an instant longer than is due to a common shake hand'.[46] He tried to resist temptation, dining on 10th June at White-Cooper's: 'Miss Pussey was there ... She is a beautiful and gifted girl, but I kept away from her—bade her a kind good night. That was all'.[47] Emily was eighteen or nineteen at this point (about half Tyndall's age), and Mary a year older. He was with them again on 12th June at the Royal Institution, remarking: 'Whether I shall preserve a sunny memory of last evening by thinking of it alone, or make myself unhappy by setting its enjoyment side by side with a future in which as regards this maiden I have nothing to hope ... I will choose the sunshine'.[48] He dined at Bence Jones's between Gassiot and Sabine, went to the Royal Society, now in its new home of Burlington House, having just moved from Somerset House, and finished with tea at Mrs Drummond's, claiming: 'this folly is now all gone, and my enjoyment was therefore without alloy'.[49]

But the temptation was too strong. Just four days before leaving for the Alps, having finished his paper 'On the sounds produced by the combustion of gases in tubes' (stimulated by a visit by Count Schaffgotsch, who could extinguish a gas flame with his own voice),[50] he decided like Caesar to cross his Rubicon: 'Posted at half past

9 o'clock PM this day extracts of my journal to Δ ... I have not quite followed my own natural bent in this matter: that bent would undoubtedly be to make sure of the maiden herself first. I would not have one through a mother's mediation ... I am prepared for the worst. There is no unmanliness in the excerpts which I have posted ... One thing is manifest, that science would not get much out of me if my life continued in this fashion'.[51]

* * *

Hirst had written in mid-June from Paris, where he was supported by the Simpsons, to tell him of Anna's impending death.[52] One lung was completely diseased and the other failing fast. Death came on 1st July, while Tyndall was still in England. The distraught Hirst pleaded in anguish: 'Anna, hast thou not a word for me before thou goest?' She breathed 'I love you', and then, as she painfully struggled to breathe, 'Oh, Christ take me'.[53] Life, and too often young life, was frequently cut short by disease now avoidable or treatable, and Hirst was still disposed to think of it as God's will. Death was all around them, whether by tuberculosis, typhoid, cholera, or scarlet fever.

Just days after his Discourse on glaciers, Tyndall had rushed to Bossington churchyard to see the mourners for Wright's little daughter departing. Now, on 2nd July, he raced to Paris, carrying with relief in his pocket a letter from Mrs Drummond: 'I asked for nothing, I hoped for nothing, but I made a frank avowal; and surely I have been responded to with kindness and forbearance considering that my error was so great'.[54] He found Hirst in the street and accompanied him to the funeral in Montmartre, amid the tombs, cypresses, and the tender glimmer of acacias: 'As I looked down upon the coffin I could see the grain of the wood, and thought of my own enquiries upon the subject: and I thought at the same time that the human mind was the residence of powers that needed far other exercises than those which belonged to scientific culture. I thought that as long as woe and bereavement enter into the experience of man, that the essence of religion however its form might change, was out of the reach of danger'.[55] Hirst, devastated, never married again. Wishing to keep him from scenes which would incessantly revive the memory of his loss, Tyndall encouraged him to come to Switzerland. They had, after all, planned a joint trip from Marburg years before, when Hirst was called away by his mother's unexpected death.

In Paris, Tyndall took the opportunity to pay visits to many of the French savants, dining with Despretz and seeing among others Verdet, Lissajous, Wertheim, Moigno, Foucault, Becquerel 'hard and not cordial',[56] and the eighty-three-year-old Biot, who was astonished at the recent marriage of the seventy-five-year-old David Brewster to Jane Purnell, some forty-five years his junior. He wrote to both Juliet Pollock and

Grove that he hoped that Forbes, who he thought already in Chamonix, would not pitch him into a crevasse.[57]

Travel from Paris to Geneva in 1857 was still tortuous. Tyndall set off by train from the Lyon station to Macon, gliding past the apple trees and the vineyards to stay overnight at the Hotel Champs Elysées. The early steamer down the river to Seyssel followed, then 6 hours crammed into a diligence to Geneva in intense heat and dust, with France on one side, Savoy on the other. Tyndall would rather have walked. De la Rive was at his country home outside Geneva, with its grounds like an English park, where Tyndall joined him. Mont Blanc that afternoon was hidden by cloud, but the next day, as he travelled on to Chamonix, Tyndall had his first glimpse as 'he rose without a cloud with snowy brow and shoulders amid his mountain companions'.[58] He wasted no time in going straight up to Montanvert and then to the glacier above Trélaporte, with the guide Joseph Simond and the young Balmat, who told him that Forbes was on his way over—this turned out to be mistaken, as Forbes was taken ill at Folkestone and was unable to travel.

They made their base at Montanvert, with its stunning prospect down and across the Mer de Glace, and its views of the sharp aiguilles with their granite spikes piercing the heavens. There was a rudimentary hotel, but Tyndall took possession of his 'castle' close by; a largely unused building, but with a bed, long table, stone floor with goatskin rug, crackling pine fire, and the window at which Goethe sat when he feasted his eyes on the glacier. The words 'A la Nature' stood over the rough, heavy door. Sitting there at night, listening to the wind moaning over the glacier, and to the distant rumble of the stones upon the moraines, Tyndall 'never felt so like a philosopher in my life'.[59] His only regret was that he had not brought a pair of Nichol prisms with him, to measure the polarization of light.[60] Hirst arrived, and they made their first observation with the theodolite, detecting the movement of the glacier in just an hour, and estimating that in 3 hours it had moved 1 foot.

Several glaciers fed into the Mer de Glace, including the Géant, the Léchaud, and the Talèfre (Figures 14a and 14b). Exploration began in earnest, as Tyndall examined the structure of the ice at different places up the main glacier and at the join of the Léchaud and Talèfre, looking at the stratification and banding. The effect of pressure was evident to him: 'It is only when compressed that ice is enabled by its peculiar physical character to simulate plasticity. If the test of tension be applied the analogy fails utterly'.[61] Seeking a better view with Balmat, he climbed a pinnacle which Forbes had not been able to reach, while chamois observed them from a distance. He was now comfortable with exposure, with not a trace of giddiness as his feet dangled over a precipice. Across the glacier, a French Count was attempting to climb the Aiguille

Verte, on which an Englishman had failed a few days previously; it would not be climbed until Edward Whymper's stunning season of 1865. From their vantage point, Tyndall observed the dirt bands, tracing them to the ice-fall and noticing smaller bands below the ice-fall of the Talèfre but none on the Léchaud, which did not seem to have an ice-fall. The glacier appeared to him less and less like a viscous body. As its slope steepened below Trélaporte, a multitude of transverse crevasses formed. Balmat was pleased with this summit, so Tyndall took the opportunity to suggest they climb Mont Blanc, after an expedition to the Col de Géant, to which Balmat immediately agreed.

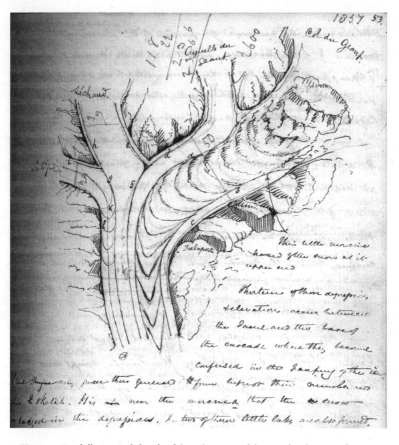

Fig. 14a. Tyndall's journal sketch of the tributaries of the Mer de Glace, 21 July 1857.

Fig. 14b. The Mer de Glace, showing the constriction at Trélaporte.

Tyndall set off for his first mountaineering trip on 24th July 1857, from a bivouac above the Talèfre cascade, armed with chicken, mutton, bread, cheese, chocolate, and two bottles of wine, and wearing a pair of dark glasses given to him by White-Cooper. After a wonderful sunset, unable to sleep, but warmed at 5am by a fire of juniper and rhododendron, he left with Balmat for the Col du Géant, the high glacier pass between Chamonix and Courmayeur. Neither of them had knowledge of the route, and a guide had told them they would either not succeed, or die trying. It was only the third ascent of the Col that year, and their way lay through dangerous crevasses and séracs, or ice-cliffs, potentially raining down ice avalanches at any moment. Tyndall girded his loins and enjoyed the challenge: 'If I fall, there is nothing to be gained by falling as a coward'.[62] Clouds on the Italian side blocked their view as they reached the Col. They descended immediately, worried that Balmat's feet were getting frostbitten, and arrived back with blood-shot eyes, burnt cheeks, and blistered lips. As he recuperated at Montanvert, Tyndall twice noticed older men with young wives. He remarked prophetically: 'another case of comfort if I like to make use of it'.[63]

Tyndall was reflexive about his way of working, as he filled pages of his journal and notebooks with observations and measurements, writing: 'My intellectual mechanism appears to be cumbrous: it gathers a quantity of collateral matter and scarcely ever solves a point, without solving others connected with it. Some men have those quick single flashes that illuminate detached problems... but I do not think I ought to envy this power, for my machinery... if confined to the solution of a single point, is valuable where a whole class of subjects is to be dealt with'.[64] By now he was convinced that the veined structure of glaciers could only be explained by pressure. But he was not sure if the stratification of the névé was preserved lower down. On one day, startled by sounds of cracking and hissing, he realized he was witnessing the formation of a crevasse. It was instructive, confirming his ideas: 'where the maximum motion is not more than a foot a day the glacier is not able to accommodate itself without cracking'.[65]

He was humbled by the mountain environment, though conscious of human power: 'And yet small and feeble as he is, trembling on the verge of fissures and precipices, he claims by innuendo his superiority to it all, when he seeks to lord it over the laws which these mighty masses obey'.[66] Huxley arrived, and sketched Tyndall as a cat while he carried out measurements which showed the gradual diminution of the speed of the glacier from top to bottom. As he fixed stakes in a dangerous position in an ice-wall to measure the speed at different heights, Tyndall said to Huxley: 'If I am killed let no-one say it was through my own foolhardiness—you see the importance of the observation'.[67] Huxley understood this manly morality.

The Chamonix regulations required four guides to accompany a client up Mont Blanc (Figure 15), which Tyndall realized could seriously hamper his movements and harm his wallet. He argued that he was no traveller but a scientific man come to make observations, and took just Simond, after an argument with the young Balmat. Simond was not keen to go alone, since he thought Hirst and Huxley might not make it and would need looking after, but reluctantly agreed. A porter took their supplies up to the hut at the Grands Mulets—one bottle of cognac, three Beaujolais, three *vin ordinaire*, three large loaves of bread, three halves of roasted leg of mutton, three chickens, raisins, and chocolate in cakes—and on 12th August the four of them set off directly across the slopes. The hut was a simple affair, of rough stones without mortar, roofed and lined with wood, and with a rickety plank for a table. Huxley, tired and only recently arrived from England, decided not to make the attempt. Tyndall did not sleep, and they were away shortly after 2am.

Fig. 15. Mont Blanc and Glacier du Géant from the Jardin.

These Alpine ascents, and this was Tyndall's first major ascent, are intense experiences. Tyndall and Hirst wrote pages of description. It was a long, trying climb, often in deep snow, and the first ascent with only one guide since the initial ascent in 1786 by Jacques Balmat and Michel Paccard. Tyndall took the lead for a while when Simond seemed almost ready to give up. They were finally on top at 3.30pm, firing a 'cannon', and drinking to Saussure, Faraday, Huxley, and to the ladies Mrs Huxley, Mrs Pollock, Mrs Drummond, Mrs Coxe, and Mrs Faraday.[68] Hirst shook Simond's hand on impulse, and assured him that although he had grave faults as a guide he was a decent fellow. Tyndall and Hirst narrowly survived a long fall on the descent, when Hirst slipped and pulled them off on a steep slope. They made it back to the hut at 7.30pm, desperately thirsty. Huxley was extremely relieved to see them, for they had been expected back by 3pm, exclaiming: 'To the end of my life, Tyndall, I can never forget the sound of your batons when you reached the rock'.[69]

They had a bad night. The hut was cold and crowded, so they lay on the hard boards. Descending the next day, Hirst was almost snowblind. Then the world reasserted itself, as the white and black of the high Alps gave way to the colour of the lower slopes. At the *auberge* by the Cascade du Dard, Tyndall noticed a beautiful maiden: 'who was more refreshing to the eye than her beverage to the physical man. A finely formed, handsome country girl; with even white teeth, brown eyes full of light and beauty, sunny cheeks...We all appeared to walk more cheerily after we had met her acquaintance. Thus it is that beauty without trouble to itself can confer benefit on men'.[70]

The weather broke, and for three days Tyndall, Hirst, and Huxley, cooped up at Montanvert, could do little, though they 'were so thoroughly comfortable, steeped through and through with satisfaction, that the presence of women would have been simply a bore'.[71] Then it brightened and they spent a final three days on the glacier, measuring the movement of the stakes on the ice-wall, showing in 48 hours that the top moved 12.5 inches, the middle 8 inches, and the lower 4 inches. Tyndall commented: 'I think it is the best and most instructive measurement of the kind that has yet been made'.[72] As they descended to Chamonix, they passed the newly married George Stokes and his young wife on the glacier. The ascent of Mont Blanc had cost them just 55 francs each, not much more than £2—a traveller who had taken the normal complement of guides had paid 640 francs—but Tyndall still did not have enough money to pay Simond. Hirst offered to go to Geneva for cash, but Simond kindly lent them 500 francs. Huxley remained for a few more days, while Tyndall and Hirst retraced their steps; Hirst to Paris, and Tyndall reaching home on 25th August.

* * *

London was hot. Tyndall went to Queenwood to write, staying until 9th September, in the company of Wright, Debus, and Fox. While he was in Chamonix, Airy had congratulated him on his paper on the sounds of tubes produced by gas flames, writing: 'You are so completely master in everything that relates to interference of undulations that I very much wish I could enlist you to thoroughly study the geometrical and algebraical theory of this phenomenon of depolarization...Our physicists in general and our optical experimenters in particular (always excepting Stokes, the prince of mathematicians) have been such wretched mathematicians that these subjects are sealed to them: I wish greatly that you would enter into them'.[73] Airy's high opinion of Tyndall's mathematical abilities is intriguing. Was Airy mistaken, or was Tyndall a better mathematician than he appears from his papers? Certainly the recently discovered notes made by Tyndall of his lectures on mathematics in Marburg demonstrate a sound facility, including with basic calculus. But they fall well short of the capabilities of Thomson, Stokes, and Maxwell.

Tyndall continued to work on his next glacier paper, interspersed with social weekends with the Moores and the Pollocks, meeting James Spedding and hearing about his monumental twenty years editing the works of Francis Bacon. He did not go to the British Association meeting in Dublin. Nor was he able to go to Bonn to counter Plücker at the meeting of the Deutscher Naturforscher und Ärzte as

Knoblauch urged.[74] He felt thoroughly well after his time in the Alps and, in the midst of the Sepoy Mutiny in India, 'could contemplate a trip to India and the shooting of half a dozen sepoys with great tranquillity'.[75]

Frankland was now back in London, having left Owens College to lecture at St Bartholomew's Hospital. Hirst stayed in Paris to pursue his mathematics, turning down, with Tyndall's support, another request to return to Queenwood. Tyndall resigned from the Royal Society Council after only a year, 'partly at my own desire, as my lectures interfered with attendance',[76] but had proposed Frankland for the Royal Medal, which he heard had been awarded. He had decided not to push the claim of Bunsen in opposition to Lyell for the Copley Medal.[77] That was a sound diplomatic move: Lyell was awarded the Copley Medal the following year, and Tyndall's mentor Bunsen received it in 1860. The London Institution lectures restarted in November, alongside a full social life. Tyndall met the Unitarian James Martineau for the first time at the Carpenters (the physiologist William Carpenter was also a Unitarian), dined at the White-Coopers and at the Moores with Babbage, and spent time with the Franklands, Hookers, and Wheatstone.

Huxley had sent Tyndall a long letter in September entitled 'Observations on the Structure of Glacier Ice', which Tyndall published in the *Philosophical Magazine*.[78] Tyndall commented: 'talent is marked in every line, but I should fear writing with my facts so meagre'.[79] It was the result of just two days work. This paper concentrated on observations of the structure of glacier ice at surface and deeper levels, prompted by Tyndall, and on the dirt bands. It was partly aimed at disproving the idea, supported by Agassiz and Forbes, that the whole glacier was threaded by fissures that could contain water. However, it also seems to have stimulated Tyndall to experiment on the effect of the Sun's rays on ice. While Tyndall followed this up in the laboratory, he made progress at learning Italian (his friend Pollock was a considerable scholar), and told Hirst that he was also considering writing a book on glaciers in 'the hours of the evening which I dare not employ at taxing work'.[80] He was finally able on 21st October to send Forbes his original paper with Huxley;[81] publication in the *Philosophical Transactions* was notoriously slow. Forbes replied with no hint of animosity, thanking him for the 'friendly opinion which you express in your accompanying letter of my observations on Glaciers'.[82]

In mid-November, Tyndall received 'a clever paper from Mr Ball on ice structure' (this was John Ball, the alpinist and shortly to be first president of the Alpine Club), which made the case for preservation of the stratification of the névé into the veined structure of the lower glacier.[83] This was still an open question for Tyndall. Meanwhile, he experimented on Norway and Wenham lake ice, noticing congeries of water cells entangled in the skeleton of Norway ice, and liquid rings encircling bubbles.

He sought Clausius's advice to confirm his idea that water drops associated with air bubbles inside the ice were formed by conduction of heat through the ice, not the heating of the air bubbles by sunlight, as Agassiz and others had proposed.[84] He also asked him if he thought the deep colour of Lake Geneva could be caused by fine suspended particles, though he stated: 'I know that pure water is blue'. Clausius agreed with Tyndall's explanation of the conduction of heat through ice, and gave insightful comments about the temperature range over which melting takes place, which he thought might explain the apparent plasticity which accompanied the fracture and regelation of pieces of ice. He said that he believed the colours of water were very dependent on circumstances, and not yet understood properly.[85] Tyndall left this question for now.

Tyndall gave Faraday a draft of his new paper, and sent it in stages to Hopkins.[86] Though Hopkins saw Tyndall's views as much nearer to his than to Forbes's, he believed Tyndall had carried his argument about the inextensibility of ice too far, as Clausius had also implied.[87] Tyndall's paper 'On some physical properties of ice', the second of his four major papers over three years on glaciers, was read at the Royal Society on 17th December 1857, and given as a Discourse a month later.[88] In it, Tyndall revealed more laboratory experiments on the properties of ice. He linked them to aspects of glacier structure, and emphasized his conclusion that the conduction of heat through ice, which might seem counterintuitive, could explain both the formation of liquid spaces in the ice in sunlight (in six-petalled flower shapes in the plane of freezing), and the 'regelation' or refreezing of pieces of moist ice in contact. He regarded this paper as a 'very heavy discharge of artillery against the viscous theory', and was concerned to hear that Forbes was 'in a very disturbed state of mind about the subject'.[89] He told Hirst he would give Forbes the 'full measure of credit which I believe to be his due'.[90] Time would tell if Forbes and his friends would agree.

8

⊰⊹⊸⊙⊷⊹⊱

STORMS OVER GLACIERS
1858–1860

The final years of the 1850s found Tyndall, now in his late thirties, extending his social circle, cementing his place at the Royal Institution, and completing his work on glaciers, interspersed with some notable feats of mountaineering. They also brought to a head the saga of the Drummond daughters.

Tyndall spent Christmas Day 1857 with the Hookers at Kew, having already started his series of lectures at the London Institution, again 'On the Nature and Phenomena of Light'. More social engagements followed in January 1858: at Sir George Clerk's; with Spedding, Babbage, and the Pollocks; with the Barlows; and with the Huxleys, who had a daughter, Jessie, in mid-February. Tyndall wondered 'when an equal blessedness is likely to fall upon me'.[1] It was a thought that frequently recurred. He had made a wistful note in his journal the previous year on hearing that Huxley 'had a young son presented to him as a New Year's gift'. He wrote: 'I wonder have the gods such a gift in store for me'.[2] A few years before he had said to Hirst: 'if I had a little son I should love him so I think I could lift him up from the ground by the pure force of will alone'.[3] He delighted in his friends' children, and they seem to have much enjoyed his company, and the little presents he gave them. His friends of both sexes regularly suggested that he should get married, though the perceptive advised him to resist 'the temptations with which a man like me, and at my "interesting age" is sure to be beset'.[4] He did not heed that advice. About the eventual childlessness of his own marriage he appears not to have written a word.

At the Royal Institution his reputation was now secure. He was a phenomenally popular lecturer, and gave the first Discourse of the year 'On some Physical properties of Ice'.[5] His lecture series in 1858 proved significant. The topic was 'Heat Considered as a Mode of Motion'. It anticipated the focus of his later researches, and when developed in 1862 became a springboard for much more. Tyndall had written to Stokes, as part of his preparation, about what caused the dark lines in the solar spectrum and the bright lines of electric spectra. Stokes explained that he thought the dark lines were due to absorption of solar light by the atmospheres of the Sun or the Earth, and that for the bright lines he thought that 'matter sets the ether in vibration in a manner analogous to a finite number of notes of particular pitch'.[6] Within a few years, Tyndall made his major, and related, findings about the phenomena of absorption and emission of radiant heat in the Earth's atmosphere.

The lecturing success spun over into more demands for his writing. Tyndall's ability to communicate attracted publishers, for scientific works and for other topics of interest. He was already writing up an account of his expedition to the Col du Géant, which was published by William Longman in the first volume of *Peaks, Passes and Glaciers*, the forerunner of the Alpine Club's *Alpine Journal*, and Longman urged him to publish his lectures and a book on Switzerland. He was still planning to write a book of general interest about natural phenomena and glaciers, an idea he had been considering since 1856. But despite Longman's approach he offered first refusal to John Murray, just before he left for the Alps.[7] The suggestion of Murray came from Hooker, who wrote: 'I cannot bear the idea of your going to that screw Longman with your wares when John Murray is in the same street'.[8] Yet Longman published almost all his subsequent books. Much later, in 1868, Tyndall explained why, when a cheque appeared from Murray for the American edition of *Glaciers of the Alps*. Murray had claimed part of the profits from American sales of an edition he had not published.[9] Tyndall told Murray that Longman offered two thirds of the profits, paid him on the day of publication, and did not seek control over books beyond the editions published in England.

Tyndall's writing took place in spare hours between lecturing and military examination duties, while responsibilities were in flux at the Royal Institution. Bence Jones approached Tyndall, on his return from a long weekend at Queenwood to try to drive away his insomnia, to tell him that Barlow was thinking of resigning as Secretary, and that he would be prepared to take it over if Tyndall took responsibility for the Discourses.[10] Tyndall was happy to do this, but sought to reduce the burden of lectures to give more time for research. The Managers, with the best of intentions, suggested that Bence Jones certify that the load be reduced on health grounds, so the salary could be maintained. Tyndall objected to this, as anyone knowing him well might have anticipated, and wrote to Faraday to remind him gently of the expectations of increasing salary originally made to him, and of his commitment to the Institution.[11] At Faraday's prompting, the Managers offered to reduce the load on scientific grounds, maintaining his salary (it was increased to £300 in 1859, and to £450 in 1868).[12] Money was in any case becoming less of a concern. He spent a whole day 'stock jobbing' with Francis, who acted as his financial mentor and agent, paying

£150 into his bank. At the beginning of February, he examined the candidates for Direct Commission and for Staff College, preceded by candidates for entry to Sandhurst, noting that 'the answers were extremely bad'.[13] He wrote a report to the generals 'which will probably make them wish they had a more pliant examiner in my place', again critical of the current system of education for officers and arguing the importance of a stronger scientific education.[14]

Towards the end of February, Tyndall gave a soirée lecture at the London Institution 'On the Analogies Existing between Light and Sound'.[15] The demands of preparation for this and for his Royal Institution series resulted in his declining many social invitations, even from the Drummonds and Pollocks. He gave priority to scientific meetings, hearing Gassiot's Bakerian Lecture at the Royal Society and spending a day at the Institution of Civil Engineers. The anniversary of his brush with death in the River Barrow on 2nd March passed with a recollection 'so vivid as to cause the intervening time to shrink to the dimensions of a dream'.[16] He was still working on his third glacier paper, but found time to dine with Carter from his railway days and with Bevington, and to read Whewell's *Of the Plurality of Worlds*, which he found 'idle and empty' with 'an association with religion which is forced and unnatural'.[17]

As in 1857, Tyndall spent Easter on the Isle of Wight in charge of Harry Jones. Again the weather was mixed. Below their hotel, where Tyndall found a goodbye note from George Wynne, off to China, the east became black and the rain came down, the sea 'boiling along the base of the cliff, tossing itself in vertical leaps and whipping itself to whiteness'.[18]

This visit was memorable for Tyndall's first meeting with Tennyson, through an introduction to Emily Tennyson from Juliet Pollock, who asked Tyndall to present her with her book *New Friends*. Never one to push himself forward too obviously, this was to Tyndall 'a great pleasure and a great privilege come without my seeking'.[19] They hit it off immediately, the start of a lifelong friendship. Tyndall described how Emily Tennyson's 'brown eyes light up with expressive beauty' and he 'a fine frank-looking fellow with large forehead and dark beard'.[20] At dinner, joined by Tennyson's two boys (they were aged five and four), 'the best poems that Tennyson ever produced', was Benjamin Jowett, the theologian and future Master of Balliol, then Regius Professor of Greek at Oxford.[21] That, too, turned into a friendship. They talked of daffodils, landscapes, and poems. Tyndall made the *faux pas* of pronouncing 'knolls' with a short 'o' and was exhorted to employ a long one. Tennyson revealed himself sensitive to criticism of *Maud*, which he thought one of the best things he had ever done, a portrait of the time, not personal, he claimed (though its autobiographical origins are strongly evident). Later, in Tennyson's study upstairs, with its apparatus on the chimney piece like a test-tube stand in which fifteen or twenty pipes were stuck, they puffed sociably side by side. There were few, Tennyson said, to whom he could say so much about his feelings and beliefs: 'we parted as if we had known each other for years'.[22] Invited for the morrow, but beaten back by wind and rain, Tyndall saw him the next day and talked over interpretations of poetry. Tennyson maintained that the talent for explaining such concentrated imaginative expressions was very different from that of writing them.

The juggling of work and social commitments continued after Easter. Faraday took him to Meyerbeer's *Les Huguenots*, one of the most popular operas of the nineteenth century. He dined at Lord Ashburton's, discovering that Carlyle was the only guest he knew and managing at dinner to sit between him and Lord Stanhope, who had recently been to Chamonix, in lively conversation about Forbes and glaciers.

Throughout this period the Drummond drama was never far away. The emotional turmoil of the summer had been eased in the New Year when Tyndall received an invitation to Hyde Park Gardens. He saw all the Drummond daughters: 'A load is thrown off my heart and lest it should again accumulate I will write no more about it. I have learned by bitter experience the result of journalising feelings. It falsifies the life and withdraws us from its real duties'.[23] This resolution was not kept, and a joyous evening followed on 9th January when he recorded that '"Pussey" is the most delicious little being that I know'.[24] So, despite the advice of his friends, Tyndall was again drawn into the intoxicating mix of the Drummond household, bowled over by their ease and culture: 'the whole family appear to be utterly lifted out of the sphere where a trace of anxiety regarding worldly matters is felt. Beauty is around them, beauty hangs upon them, and beauty is within them'. He tried to keep a level head, 'lest this should blossom again into folly'. But blossom it did. He went with them in March to Swindon, on the central line of the annular eclipse, the last major eclipse in England until 1917, though it clouded over halfway through. He thought Pussey's 'childlike audacity, and withal a light of intellect and dauntless freedom in her countenance enough to provoke one to kiss her twenty times a day'.[25] A few days later, after a Discourse by Henry Buckle, the Drummond girls surrounded him. Tyndall was close to Pussey throughout: 'I can hardly tell how I love the little girl…I think I love her better even than I used to love Alexander Cumming'.[26] As they left, Babbage helped them into their carriage. The love-struck Tyndall watched as 'a fair face looked towards us, and a white hand waved a last goodnight, after which the window was drawn up again, and we were left alone to the gaslight and the stars'.[27] Just before leaving London for Easter, he was allowed into the girls' work room, their inner sanctum, with its drawing materials, birds, baskets of primroses, and glorious prospect over Hyde Park.

There were frequent social visits to the Drummonds in early summer. Tyndall's attraction was now clearly for 'May', the eldest daughter Mary, a year older than 'Pussey' (the middle daughter Emily). He thought Mary 'exceedingly beautiful' and started wondering how she was filling her hours. His 'courage accelerated at an augmented pace...I patted her little hand today and she did not frown'.[28] But the situation was confusing, and potentially damaging to his reputation. Frankland told him that people noticed how much time he spent with them. Tyndall left for the Alps in July, without a clear resolution. He asked Mrs Drummond to ask Mary to sign a note 'which should effectively cut away every hope. This was not done'. In consequence he decided to cease visiting, and simply 'bade them a friendly goodbye'.[29] Much later, on 5th September in Chamonix, he wrote out what he called 'this folly': 'I may have been a deluded coxcomb...but the facts were these. In our little meetings that fair girl was always beside me...many, many times I have squeezed her little hand, and as often had a sudden cordial squeeze in return'.[30]

* * *

The aim of Tyndall's alpine trip in 1858, this time with Ramsay, was to resolve questions left open by the previous year's explorations. He sought advice from Airy and Clausius, among others. Airy wrote with insight: 'In the subjects of regelation and viscosity or plasticity, it seems to me that there is no essential difference between you and Prof. Forbes: for I can hardly imagine viscosity except as a destruction and restoration of organization: so that, in my view, the regelation is not an opposition to Prof. Forbes' viscosity, but is an explanation of its...modus operandi. But in the formation of the laminae of ice, your theories are entirely opposed'.[31] Clausius was particularly impressed, even though, given the delays in publication of papers in the *Philosophical Transactions*, he had not yet seen the latest work.[32] Before leaving for Switzerland, Tyndall communicated his third paper on glaciers to the Royal Society, 'On the Physical Phenomena of glaciers—Part I. Observations of the Mer de Glace', followed by a Discourse.[33] He described his measurements of the motion of different parts of the glacier: from side to side, depending on the curvature of the valley, and from the surface to its bed. Airy notwithstanding, he reiterated his conclusion that 'no quality which could with propriety be called viscosity is possessed by glacier ice'. Instead, he explained all the phenomena on the basis of the fragility of ice around its freezing point, aided to some extent by the partial liquefaction of ice by pressure as pointed out by James Thomson (despite what he had said about Thomson before), coupled with its power of regelation. This paper was envisaged as a pair to the previous one. Given the glacial speed of Royal Society publishing, it did not appear in the *Philosophical Transactions* until 1859.

Tyndall and Ramsay had a wet crossing of the Channel, enlivened by an exhibition of porpoises. Tyndall spent some time with Hirst in Paris, calling on Wertheim before setting off with Ramsay again, on a hot sunny day, to an overnight stop at the Hotel de Paris in Mulhouse. They travelled on along the Rhine, flanked by high hills and gloomy slopes of pine, watching the reapers among the yellow corn and the fruit gatherers in the orchards. Passing through Zurich, seeing Clausius and Liebig, they crossed the lake to Horgen, as Ramsay read sonnets by Keats under the stars. Two more days via Zug and a rainy Alpnacht took them to Meiringen, where they met Christian Lauener and headed to Rosenlaui. The next day, 23rd July, they went straight onto the Rosenlaui glacier, inspecting the flattened liquid discs, and observing the local development of structure caused by the pressure of the end of the glacier against the rock. On the glaciers above Grindelwald, the critical observations were at the ice-fall above the junction of the two main glaciers. Here Tyndall saw the structure formed at the base of the ice-fall, giving rise to the transverse blue veins. Ramsay agreed it was nothing to do with the stratification of snow higher up (Figure 16): 'The glacier breaks across the top of the fall. It descends subsequently in peaked ridges with spaces of debris between them. These spaces are excessively dirty, but not the slightest trace of structure is observable. This appears first towards the base of the fall, and at first very imperfectly'.[34]

After a further day on the Obergrindelwald glacier, Tyndall and Ramsay crossed the Strahlegg pass with Lauener to Grimsel, noticing evidence of ice erosion 2,000ft above the valley, and nearly encountering Airy, with his two sons, who arrived the

Fig. 16. Structure and bedding of glacier ice, sketched by Ramsay on the Aletsch glacier.

day after they left.[35] Descending past the end of the Rhône glacier, they based themselves at the Hotel Jungfrau under the Eggishorn, at the southern end of the great Aletsch glacier, the longest in the Alps (Figure 17). From the summit of the Eggishorn, which Tyndall climbed on successive days, there are glorious views up the glacier to the Eiger and Jungfrau, and across to the Aletschorn. On its lower slopes, beside the glacier, is the Märjelen See, now a small lake distant from the glacier, but then right next to it, with ice carving off and icebergs lying listlessly on its surface 'like a painted ship upon a painted ocean', as Tyndall quoted Coleridge.[36] On one occasion Tyndall used the icebergs as boats, capsizing them into the icy water.

A local guide, Johann Bennen, was attached to the hotel. Tyndall asked to be accompanied up the Finsteraarhorn, the highest mountain in the Bernese Oberland, in what would be the first ascent with a single guide. The ostensible aim was to meas-

Fig. 17. The Aletsch glacier from near the Märjelen See, looking towards Belalp.

ure solar radiation on the summit, with Ramsay taking readings on the glacier. In this way, Tyndall sought to explore the transmission of heat through the atmosphere. It was not without risk, and Tyndall gave Ramsay 'some testamentary directions to be carried out in case I should not return'.[37] This partnership of Tyndall and Bennen, a large man with round shoulders bent forward and a thick pedestal of a neck, bearing a massive head and earrings, was broken only by Bennen's untimely death in an avalanche in 1864. Blankets, wood, and hay were sent up the glacier to the Faulberg cave, where they spent the night, with Tyndall's boiling water apparatus, telescope, haversack, bottles, wine keg, and mattock strewn around in confusion. Bennen's snoring prevented Tyndall sleeping, and they left at 3am.

For a short while Tyndall considered switching his attention to the Jungfrau, 'the very fact of the conquest being difficult lent a piquancy to the idea', but changed his mind and turned up the Grünhornlücke towards the Finsteraarhorn.[38] They climbed up rocks and steep snow, unroped and cutting steps, to the ridge overlooking the glaciers on the north side. Tyndall felt increasingly calm in the face of the obvious danger of slipping: 'even the thought of death itself loses a great deal of its horror and discomfort when you are made familiar with it'. As they climbed the splintered ridge of the mountain, which protected them from a fierce north wind, the clouds rolled in. The experiment with Ramsay was thereby stymied (as were two subsequent attempts, including one on Mont Blanc). But Tyndall measured the boiling point of water (187°F) and left a minimum thermometer on the summit before a speedy descent, glissading down the snow. Back on the glacier, he saw the veined structure of the ice cutting the stratification at right angles, just as in rocks. For Ramsay it was decisive, and in *Glaciers of the Alps* Tyndall gives it as his first observation of structure and bedding seen together.[39]

The next objective was Monte Rosa, the highest mountain in Switzerland (Figure 18). But on arriving in Zermatt, Ramsay heard that his mother had died unexpectedly, and he departed immediately for England. Tyndall was not to be deterred. He left the Riffelalp at 4am on 10th August, with Christian Lauener, who had not climbed the mountain before but had the way pointed out to him, by his brother Ulrich, from the viewing point across the glacier. The ascent of Monte Rosa from Zermatt is in glorious scenery, next to the stately Lyskamm and the ridge of peaks linking it to the Breithorn and the Théodule pass, beyond which stands the Matterhorn (Ulrich thought the Matterhorn could hardly be climbed, but that he would try if he had no wife and child). After the lower snows, and a steep slope to a saddle, the final ridge is rocky, with a little chimney to ascend before the summit. It was snowing, so the spectacular view down the precipices of the east face and south across to the Italian plains was obscured. But out came the boiling point apparatus (184.92°F). After an hour on the summit, they left at 12 and were back by 3.30pm, in remarkably quick time.

Fig. 18. Monte Rosa from the Gornergrat.

More remarkably still, a week later, waking before 6am on a glorious morning, Tyndall thought the view from the summit must be unspeakably fine. He had intended to go to the Schwartzsee with a local guide, having let Lauener take an English party up Monte Rosa, but the opportunity was too good to miss. So he set off with a telescope, half a bottle of tea, and a slice of bacon and bread, to make the first solo ascent. Although the other party had 3 hours start, Tyndall met them just below the summit, borrowing a neckcloth from Lauener since he had taken no coat and it was intensely cold. He found himself completely calm and master of the situation, though he did drop his ice axe at the summit and only realized on the descent what a disaster it would have been had it slid out of reach. He later wrote that 'if climbing without guides were to become habitual, deplorable consequences would assuredly sooner or later ensure'.[40] Tyndall was ahead of his time. Indeed, he wrote before his ascent of Mont Blanc: 'Were it not for the bother of carrying clothes and provisions I should certainly climb the mountain alone'.[41] It was many years before guideless climbing, let alone solo climbing, became generally acceptable and common.

Juliet Pollock, when she heard of this exploit, gave him a sound ticking off: 'I call upon you to remember that however muscularly strong you may be, you must, in common with the rest of humanity, be subject to casualties...and to my thinking it is really not right to risk a valuable life in such a way'.[42] Tyndall's description of the feat to Sarah Faraday was sent on by Faraday to the *Times*,[43] much to Tyndall's surprise. After this fine ascent, a day spent on the Furgg glacier under the Matterhorn gave Tyndall the final confirmation of the separation of stratification from structure: 'Here upon the Furgg glacier; with the solemn Matterhorn as witness and earnest Nature looking one everywhere in the face, no man could resist the evidence that the structure and the stratification of glacier ice were things as distinct as the cleavage and bedding of rocks'.[44] The evidence had convinced him.

The weather broke. A fierce storm prevented a crossing of the Weisstor into Italy, so Tyndall socialized with passing Victorian society, including two pretty Scotch girls, before going round to Saas. Here he met the curé Imseng, a keen mountaineer, and went with a large party up to the Distel Alp by the Mattmark See. Tyndall remained overnight, alone and in wet weather which deposited about 3 feet of snow on the Monte Moro pass, which he then climbed. Breaks in the weather allowed him to explore the Allalin glacier, and his lonely hotel was invaded by Thomas Hinchliffe and others aiming to cross the Adler Pass to Zermatt, but another wild wet night and stormy morning forced them back.

The weather finally drove Tyndall down too, and he met Hirst by chance before setting off for Chamonix. Here he explored the Glacier des Bossons, making inconclusive experiments on water cells, and argued with Michel Bossonney, the chief guide, who complained that he had been molested and reprimanded for granting Tyndall so much liberty the previous year. He insisted that Tyndall take four guides up Mont Blanc to place his thermometers; his plan was to measure the maximum and minimum temperatures at the summit over the winter. Tyndall protested to the President of the Commission of Guides, and threatened to write to Faraday and to the government in Turin if necessary. He went back up to Montanvert, finding Magnus and his family there. Alfred Wills arrived over the Col du Géant with Balmat, and on 13th September, the three set off for Mont Blanc, despite Bossonney threatening Balmat with legal proceedings for going alone, and sending a spy after them. The night was clear and starry, with the bright Donati's comet behind them.[45] Tyndall, Wills, Balmat, and five porters all reached the summit, where disaster nearly struck for Balmat, who suffered incipient frostbite digging holes for the thermometers. Tyndall and Wills were desperately worried as Balmat rubbed his hands furiously in the snow, the accepted way of countering frostbite. They descended at speed, as feeling came back into Balmat's swollen hands. The maiden at the chalet was declared 'drowsy and dingy-looking' compared with the maiden of the previous year.[46]

Examination duties called Tyndall back to London on 21st September. Balmat was prosecuted for accompanying the party up Mont Blanc as a single guide, and Tyndall went to the British Association meeting in Leeds in late September—giving a paper praising the frost-bitten Balmat's selfless exertions in the cause of science—where a resolution was passed asking the Sardinian authorities for increased facilities in making observations on alpine summits.[47]

* * *

Tyndall retired to Queenwood in early October to start writing the book which was to become *Glaciers of the Alps*. Despite the earlier misunderstandings about his romantic intentions, he was again invited by the Drummonds, this time to stay in the countryside for several days. He worried in a letter to Hirst that he was compromising himself, but he still had a faint hope that marriage was possible, although he saw the disparity in wealth as a real problem: 'I would not drag her down, I am unable to maintain her at her present level, and the thought of her dragging me up is so terribly repugnant to me that even supposing what is far too daring to suppose, I could not bear it'.[48]

The complicated romance dragged on. Tyndall was nothing but honest, writing once to Mrs Drummond that he hated her opulence because it threw an impassable barrier in his way, but he found her letters perplexing. Perhaps he was simply not self-aware enough to recognize the difference between being welcome as a trophy guest and as a potential member of the family. His poor Irish origins and religious 'heterodoxy' were giant barriers.[49] Nevertheless, he was invited for Christmas Day, along with Joseph Kay, a barrister and younger brother of the educationalist and founder of the first training college for school teachers, Sir James Kay-Shuttleworth. He liked Kay, who he thought might possibly love somebody else, not Mary. Mary had just turned twenty-two, and Tyndall understood that a dozen fellows at least had proposed to her within the last four years—Tyndall had known her for three and a half—but he thought she loved him. Even a letter just after Christmas from Mrs Drummond asking him not to write to Mary did not put him off: 'For now that my sight is clear I see plainly enough that May has felt kindly towards me for a long long time—but my sceptical, proud heart, kept me from believing this'.[50]

Tyndall's third series of lectures 'On Light' at the London Institution had finished before Christmas. In the early part of 1859, he started to prepare his next lecture series at the Royal Institution on the 'Law of Gravitation', for which he sought advice on demonstration models from Airy.[51] Now reduced to just twelve weekly lectures, they ran up to Easter.

On 17th January a further move into Society beckoned as he was entered into the candidates' book for the Athenaeum Club, a few minutes' walk from the Royal Institution and the Royal Society. Roderick Murchison and Thomas Bell, the surgeon, zoologist, and vice-president of the Royal Society, were his proposer and seconder. The Athenaeum, founded in 1824 and housed in a grand purpose-built edifice on Pall Mall (Plate 10), was the non-political home of the intellectual elite, where men of science, literature, and the arts could rub shoulders with politicians and men of business (all the members were men, until 2002). Tyndall was elected on 31st January 1860.

In the glacier disputes, Forbes knew he was under a sustained attack (he had been writing worried letters to Whewell for some time about the traction that Tyndall's ideas were gaining). He wrote to Tyndall in February to ask him, as one of the editors of the *Philosophical Magazine*, to publish a letter in response to Tyndall's paper, 'Remarks on Ice and Glaciers', in the February issue.[52] In this paper, Tyndall had challenged both William Thomson's explanation of the formation of six-petalled shapes inside melting ice and Forbes's explanation of the freezing of water between two blocks of ice, as well as Forbes's account of the conversion of névé into ice in the upper part of the glacier, and the origin of the veined structure.[53] Forbes gave notice that he was publishing all his collected papers in the hope that his meaning 'would be found to be both definite and intelligible'.[54] Faraday weighed in with a paper on regelation in the same issue, also challenging Forbes's explanation of the freezing process. He expressed the view, shared by Tyndall, that regelation of a thin film of water between two pieces of ice was due to the ordered ice promoting crystallization of the water.[55] Airy and Brewster both wrote seeking clarifications. To add to the confusion, Brewster now claimed priority over Forbes in the observation of the veined structure of glacier ice.[56] Tyndall, seeking to buttress his position, wrote to Clausius asking for a paper by Albert Mousson, which he had lost, and one by Rendu.[57] Rendu was seen by many, including Tyndall, as the originator of the idea that a glacier flows. Clausius speedily responded, having raided Escher von der Linth's library without his knowledge to source Rendu's paper in a volume of the *Memoirs of the Savoyen Academy*, 'the only one to be found in Zurich'.[58] It was the start of a dispute that would extend well beyond Forbes's death in 1868.

Tyndall presented his work on the veined structure of glaciers to the Royal Society in February, and revealed it to the Royal Institution audience in a Discourse on 4th March.[59] The main argument, contradicting Forbes's conclusions, was that the veined structure was produced by intense pressure, for example at the bottom of ice-falls or where two glaciers merge, and was independent of the stratification of the névé. Liquefaction under pressure expelled air and left the clear blue veined ice between the white ice (the white bands he thought were formed from snow filling crevasses over the winter). He differentiated between the 'marginal structure' caused by the faster motion of the centre compared to the sides, the 'longitudinal structure' caused by two tributary glaciers clashing, and the 'transverse structure' produced by changes of inclination. The paper also dealt with the shapes of bubbles in the ice, the overall motion of glaciers—owing partly to the sliding of the glacier over its bed and partly to its yielding under pressure—and the formation of the 'dirt bands'. The Discourse stimulated Ball to publish a paper largely retracting his earlier contradictory remarks on the veined structure of glaciers.[60]

Tyndall's journal is silent about how relationships with the Drummonds ended, but as they passed him on the evening of the Discourse, he turned his head away: 'I thought in the glance I had that there was something "forward" in the aspect of the eldest. Probably the clever woman thinks that she will again succeed in twisting me round her finger'.[61] Finally, the affair had come to a head. As Tyndall described to Hirst: 'Everything went as merry as a marriage bell...strange to say the lady so behaved as to leave the general impression that they made advances to me and not I to them, which indeed was substantially the correct impression. At length things went so far that some of the lady's friends warned her regarding her conduct. Her replies to them were circulated and were highly unsatisfactory to me'.[62]

Tyndall broke off contact, to the relief of his friends, feeling he had been badly deceived by both mother and daughter. Indeed, after his death, Julia Moore (Figure 19)

Fig. 19. Tyndall with Faraday, Mrs Faraday, and the Misses Moore in 1858.

wrote to Tyndall's widow that they looked on Mary as a 'heartless flirt'. She added that Mary 'entirely lost her looks many years ago and is singularly plain'.[63] As it transpired, Kay did marry Mary in 1863 (she refused him twice and then changed her mind). But as late as July, Tyndall remained concerned that Mrs Drummond was the source of false rumours about his behaviour. He confided in Hirst over several evenings, and eventually in Faraday, who knew Mrs Drummond well.[64] Faraday interceded, and a line was drawn under the protracted episode.

* * *

With lectures finished for Easter, Tyndall set off with Frankland for the Lake District.[65] Reaching Windermere on 15th April, they walked to Orrest Head before dinner to view Wordsworth's 'beautiful romance of Nature'.[66] Next morning they rowed over to Ambleside in a snow storm, the lake black and rough, and walked to Rydal Mount to look round Wordsworth's 'nest' (he had died in 1850); Frederick Pollock had once spent two pleasant days with Wordsworth at Rydal, having met him as a guest at Lowther Castle.[67] They stood for some minutes beside his grave, as Tyndall remarked prophetically: 'A clear stream rushes near it, a few trees are at hand; one black cypress gives character to the scene, and there the poet sleeps…If I can manage it I will be buried in a country churchyard'. From Grasmere they climbed Helvellyn, blasted by a cold north wind, snow showers, and hail, finding the summit ridge edged with a beautiful cornice. Tyndall left Frankland on the summit to his sandwiches, and dropped down Striding Edge to Red Tarn and back again. They descended together and 'marched along the high-way towards Keswick' through further snow storms, the hedgerows laden and the whole landscape white. According to Frankland, who drew a dramatic picture of the scene, they also climbed Scafell.[68]

Frankland was Tyndall's companion again on the regular summer alpine trip. Tyndall's main objective was to place thermometers at several places on Mont Blanc, including the summit. There he proposed spending two nights, to measure atmospheric temperature and the depth to which the cold of winter penetrates. Snow conditions were good and the ascent straightforward. In worsening weather they spent one night on the summit, the first by anyone. After that, as promised, he took up an invitation from De la Rive to attend a special meeting of the Societé Helvétique in Geneva to meet Agassiz.[69] Their conversation must have touched on the dispute with Forbes.

This trip coincided with the major upheaval of the second war of Italian independence, which took place from 1859 to 1861 and resulted in the establishment of a kingdom of Italy, forerunner of the modern Italian state. In the process, Savoy became part of France. Hirst noted in his journal on 14th June the 10,000 Austrians killed or wounded in the Franco-Austrian phase of the war, probably a reference to the battle of Magenta on 4th June, in which the Franco-Sardinian and Piedmontese coalition defeated the Austrian force.[70] The bloodier Solferino, which led indirectly to the Geneva Convention and the founding of the Red Cross, followed on 24th June, when some 250,000 troops clashed in the last major battle in which all the armies were under the personal control of their monarchs: Napoleon III of France, Victor Emmanuel II of Sardinia, and the young Emperor Franz Josef of Austria. The famous Alpine guide Jean-Antoine Carrel, who later climbed with Tyndall, fought in this battle which brought to an end this phase of the war.

Tyndall was back in London on 29th August, now in lodgings at 31 Aberdeen Place, Maida Vale. He decided not to go to the British Association in Aberdeen, in order to finish his book, and based himself at Queenwood during September, finding it hard work. He wrote to Hirst on his return to London to start his experiments that he would 'never undertake to write a book again'.[71] Many authors have likewise failed to keep such resolutions. Hirst moved into lodgings at 27 Aberdeen Place, close to Tyndall, and they spent time together in the evenings, or with friends like the Huxleys and Franklands, with long walks on Sundays to Hammersmith, Highgate, or Kew. Hirst found Huxley 'the king of them all', though he remarked that 'Huxley dislikes Jews and said some clever things at their expense', and observed that his mathematical colleague Sylvester went under an assumed name to hide 'a somewhat too Jewish one'.[72]

Stokes resigned his Professorship of Physical Science at the School of Mines and Tyndall told Hirst that Murchison in the handsomest manner had offered it to him.[73] Urged on by Huxley, he checked that Faraday was comfortable with the arrangement and accepted, writing to Barlow to explain to the Managers that the lecturing was routine and would not affect his activities at the Royal Institution.[74] It might have been routine, but it was to a very different audience of students, and Tyndall took his duties seriously as ever.

Engagements started to crowd upon Tyndall: dinner with the Duke of Argyll on 3rd November, and the next day with Gassiot. The Royal Institution, the London Institution—he gave six lectures 'On the Radiation and Absorption of Heat' on Mondays in November and December, his last series there[75]—and now the School of Mines had their demands, as did his examination work, his book, and a request to examine the Great Bell of Westminster, known as 'Big Ben'.

The saga of the Great Bell involved the Whitechapel Bell Foundry under George Mears, who cast the bell, and Edmund Beckett Denison, who designed it, despite being a barrister with no experience of manufacturing bells. The bell first rang in July 1859 but it cracked in September, and was out of service for the next three years. Having gone with Percy to the Office of Works to arrange to inspect the bell,

Tyndall examined it on 5th November. He identified a crack that might be superficial, though needed its depth establishing. But the Great Bell proved more problematic than first thought. Tyndall discovered more cracks a week later, and spent several Saturdays up the tower in freezing cold as it snowed outside. The report was finished on 18th December, by which time Denison had written to the *Times* to withdraw his legal case against Mears. Tyndall thought he was wise to do so.

At the School of Mines in Jermyn Street, Tyndall's commitments were substantial. He gave thirty-six lectures on 'Physics', covering statics, dynamics, heat, electricity, magnetism, and optics, starting on 3rd October, lecturing from 2 to 3pm, generally on Tuesdays to Thursdays, until February 1860. This general course in physics remained substantially unchanged during his time at the School of Mines, where he taught until 1868, making preparation easier. But he put real originality into his examination questions; one can sense his objective to engender understanding of physical principles. The first question in his first examination for the School of Mines, taken on 15th December 1859, was 'What is the meaning of magnetic polarity?'. A question on the same paper illustrates the type of reasoning he was after: 'Two bomb shells are accurately filled with water and securely stopped; one of them is exposed to strong heat, the other to intense cold. Both of them burst. Why?'. His students needed to be able to argue like physicists; rote learning was not enough.

Following his dispute with Thomson, and now with Forbes, Tyndall had plenty of enemies in Scotland. But he had friends too, and when Forbes was appointed Principal of St Andrew's College in succession to Brewster, Playfair wrote to see if Tyndall would be a candidate for the Chair of Physics at Edinburgh, which he estimated would offer £650 per annum.[76] Tyndall could see the advantage of founding a true school of experimental physics, which the country lacked, but had to judge it against the advantage of London and his proximity to Faraday; he claimed that he had declined an earlier offer of St Andrew's from Brewster for this reason. He added up his current earnings: £300 for twelve lectures annually at the Royal Institution; £200–300 from the School of Mines; examinerships of at least £100; and £100 'which I cannot help making from other sources, even if I tried to do so, and you have a total of between seven & eight hundred a year'.[77] Playfair suggested he could make much more, as he had built his own income from £450 to £1,250.[78] Letters went to and fro, with Playfair suggesting 'a nearly unanimous election'. Tyndall had representations from many friends, including a letter from Willie Cumming, brother of Mary Coxe, who intimated strong support from James Simpson (discoverer of the anaesthetic properties of chloroform for medical purposes).[79] Juliet Pollock advised strongly against, unlike her husband. As ever, Tyndall consulted Faraday, who while professing not to advise him said that he had turned down the Chair of Chemistry, as 'if I had a sufficient moderate income in London, nothing would make me change London for Edinburgh'.[80] Although he discovered that Brewster would support him, and that Forbes was not expected to oppose him, he decided against proceeding, writing that he could not leave London and his friends, and hoping that men in power would 'grant to London the school of physics which the country so much needs'.[81] Murchison wrote that he would never forget his magnanimous gesture in staying in London 'to support the advancement of physical science in this great Metropolis'.[82]

Forbes was keen to advance his claims against Tyndall with respect to his theory of glacier motion. At Barlow's, Tyndall heard from Wheatstone that Forbes was canvassing the Royal Society for a medal for his glacier work. He had been proposed for the Copley Medal by Whewell and W. H. Miller, while Wheatstone and Gassiot had proposed Weber. Tyndall commented: 'I don't think he quite acts with the dignity of a philosopher in sending scraps of reviews to members of Council of the RS'.[83] Huxley, without Tyndall's knowledge, made representations about this to the Council through Frankland. Dining with Ball, Tyndall discovered that Forbes had been sending letters to him too. He remarked: 'Forbes is not a great man'.[84] Thomson, supporting Forbes, made private representations to Stokes, as he felt that Tyndall lacked the 'common candour, honesty and good feeling' to properly acknowledge the work of Forbes and his own brother.[85] Further questions arose in Tyndall's mind about Forbes's claims to priority and he wrote to Clausius, 'anxious to put Agassiz's labours on the glaciers in a fair light before the English public as I think he has been neither justly nor generously treated'.[86] He suggested that Agassiz had published his measurements of glacier motion first, and asked Clausius and others for information. Escher von der Linth wrote to give him his account of the work with Agassiz, but asked him not to use it in any case of priority warfare: 'wars that always seem to me as unfortunate and not serving the advancement of science'.[87] In the event, Weber was awarded the medal. The differences between Forbes and Tyndall remained unresolved.

To add to his evidence, Tyndall now planned a rare winter visit to Chamonix to observe glacier motion, and received advice from Auguste Balmat about likely conditions on the Mer de Glace in December.[88] The rest of the year, up to his departure, was occupied with examining duties—for the Council of Military Education and at the Kensington Museum, examining schoolmasters—and lecturing in several places. Attendance at his last Jermyn Street lecture on 8th December was double those of previous years, though he felt this was not enough to make the place a success. He was at the Royal Society for Sir Benjamin Brodie's Presidential address on 30th November, pleased with the award of the Copley Medal to Weber. Hirst gave an insight into Tyndall's impact at Royal Society meetings: 'Tyndall is evidently looked upon as their orator. He not only speaks fluently and well but beneath his words there is always solid thought'.[89] But Tyndall readily acknowledged scientific expertise in others. Much later, at the Society of Telegraph Engineers in 1880, when William Preece had spoken on the photophone, he prefaced his remarks by saying: 'My scientific seniors are in the room, and I would rather hear their observations than listen to my own'.[90]

Balmat had written from Chamonix to inform Tyndall that the weather would make work on the glacier impossible. Nothing daunted, he left for France on 22nd December.[91] He stayed overnight in Geneva on 24th December, spending the evening with De la Rive. On Christmas Day he set off for Chamonix, and bought a ticket for the diligence to Sallanches, despite being informed that there was no communication with the village.[92] The winter ski-season was not yet invented. Arriving eventually in Sallanches he did find a carriage. Deep snow flanked them, there was 'no moon, no lamps, and a dense gray cloud canopy overhead cut away from us the feeble light of the stars'.[93] Having passed the Pont Pelissier, they alighted to lessen the strain on the horses. Tyndall went ahead: 'Bare, brown, and motionless the trees stood right and left, while the cliffs and precipices, mottled with the snow which clung to their ledges took any form which the imagination choose to give them…It was the silence of a churchyard; and the huge black pines which threw their gloom upon the road seemed like the hearse-plumes of a dead world'.[94] As they approached Chamonix, late at night, the drifting snow obscured the road, and one of the horses sank in to the shoulder. Chamonix was like a city of the dead: 'no sound, no light; the thawing snow splashed as it fell from the loaded eaves, the fountain made a melancholy gurgle; here and there a loosened window shutter, swung creaking in the wind, and banged against the object which limited its oscillations. All was desolation…We rang the bell at the Royal Hotel, but the deep bark of a watch dog…was long our only answer…so I tried the energy of my boot heel against the door, and by perseverance roused the sleepers'.[95]

It snowed heavily on 26th December, but the next day Tyndall set off for Montanvert with four companions, in snow up to five feet deep: 'The Mer de Glace was quite glorious, not as in summer wasted and dirty, but pure and white with its frozen billows steep, high, and sharply crested'.[96] He tied the men together for security on the glacier, since the crevasses were covered by snow, and they placed two lines of stakes, measuring them with difficulty in a snowstorm. The beauty of the winter snow captivated Tyndall, yet he thought: 'some flatter themselves with the idea that this world was planned with strict reference to the human use; that the lilies of the field exist simply to appeal to the sense of the beautiful in man. Whence these frozen blossoms? Why for aeons wasted?'.[97] 'Prodigal nature rained down beauty, and had done so for ages before a human eye existed to enjoy it'.[98] On 30th December, Simond took him to Sallanches by sledge, and he just caught the diligence, to stay in Geneva overnight and reach London via Paris at 4am on 1st January 1860.

Lectures started again at the Royal Institution on 12th January, a series of twelve on 'Light' ending before Easter on 29th March. The opening lecture was attended by the largest audience he had ever seen, larger again the following week, with a Discourse on 20th January 'On the influence of magnetic force on the electric discharge'.[99] Tyndall was still unable to escape the Drummonds, suffering an edgy dinner sitting next to Mary at Mrs Buxton's, who was unaware of the delicacy of his position. He did not exchange a word with her.[100] He was too busy for the next few months to maintain his journal, but Hirst documented his movements as they spent much time together in February and March. At St James's Theatre one evening, he watched Tyndall laugh at a piece of 'nonsense' before them, thinking to himself: 'I sometimes wonder that Tyndall has preserved his liking for the theatre amidst his high occupations'.[101] Tyndall was a regular theatre-goer, liking the low-brow entertainment as well as Shakespeare. He seems to have attended the opera and musical concerts less often. In contrast to Hirst, who was an accomplished cellist and pianist, Tyndall was no musician.

A scientific highlight, on 10th February, was Huxley's Discourse 'On Species and Races, and their Origin'. Huxley expressed the view that the production of new species by selection had not been observed, and that in that sense Darwin's theory, published in November in *On the Origin of Species*, fell short of being satisfactory. He argued that nevertheless all should be free to discuss it and explore its implications, including for human origins.[102] Tyndall's immediate reactions to Darwin's book are hidden from us. There are no mentions in his extant correspondence, or later explicit recollections. He was present a few months later at the British Association meeting in Oxford, where he experienced the now famous confrontation between Huxley and Bishop Wilberforce over Darwin's ideas. Some twenty-five years later, as the mythology of this event was created, Tyndall described pushing through the crowd to the front when he imagined his friend might suffer a 'physical mauling', but he did not comment on the substance of the debate.[103] The death of Wilberforce in 1873, from a fall from his horse, led to one of Huxley's memorable *bon mots*: 'for once reality and his brains came into contact and the result was fatal'.[104]

Work on *Glaciers of the Alps* continued while Hirst, now a mathematics master at University College School after some help from Tyndall, tried to 'modify John's severity' with respect to Forbes.[105] The book was in final form at the end of April, and Tyndall lined up supporters to review it. Hopkins said that he would be happy to do so for a magazine such as the *Quarterly* or *Fraser's*, and told him: 'Yours will be the first work in which anything like justice has been done to Rendu. If Forbes knew his writings, (as I doubt not he did), he has acted most unworthily towards him. But it is the same want of candour which he showed towards Agassiz'.[106] Hopkins had also reviewed Darwin's *On the Origin of Species* for *Fraser's*, writing to Tyndall: 'Tell your friend Huxley he mustn't quarrel with me for dealing severely with his friend's theory. I like scientific freedom, with personal courtesy, but I hate scientific quarrelling'. In a long letter, Hopkins gave Tyndall editorial advice, suggesting that he acknowledge William Thomson's explanation of the influence of pressure in forming strata of liquefied ice, since Thomson was likely to be pitched into a camp against him.[107] The alpine researches were the subject of Tyndall's Discourse on 1st June, 'Remarks on some Alpine Phenomena', in which he recounted the winter visit to the Mer de Glace and some observations on light, describing the blue colour of the sky as being due to reflected light and red to transmitted, with their possible origin in particles suspended in the atmosphere.[108] Later in the decade he would take up this study systematically.

Tyndall was on holiday in Ireland for the last two weeks of June, visiting the Lake of Killarney for the first time: 'The weather was on the whole atrociously bad, but it could not quite disguise the extreme loveliness of the region'.[109] It was an experience that would have a far-reaching impact on his Alpine explorations.

Glaciers of the Alps was published on 6th July 1860, and Tyndall gave the first copy to Hirst. The first review, by Huxley, appeared in the *Saturday Review*, in which he called the book the 'mightiest evangel' of the muscular philosophers, with the Alpine Club their best-known church.[110] To counter Tyndall's threat, Forbes had published his collected *Occasional Papers on the Theory of Glaciers*, with a prefatory note explaining his position. But when he saw Tyndall's book, with its section entitled 'Rendu's Theory', he promptly issued a separate pamphlet entitled 'Reply to Professor Tyndall's Remarks'.[111] He had sought advice on his strategy from Whewell, telling Whewell later that he thought Tyndall duplicitous in the way he described the contributions of different people.[112] Reviews of *Glaciers of the Alps* appeared in rapid succession in July. Hirst noted that they were all extremely favourable. The supportive review from Hopkins came in December.[113] But the dispute with Forbes was far from settled, and such unanimity would not last.

PART III

1860–1870
THE PEAK YEARS

9

❖⇒◉⇐❖

RADIANT HEAT
1859–1862

Tyndall's scientific reputation rests firmly on his demonstration and measurement of the absorption of 'radiant heat'—thermal radiation—by gases. This was no esoteric discovery. Tyndall explained the ability of the Earth's atmosphere to trap heat. Now known as the 'greenhouse effect', it results in our planet having a temperature suitable for life. Tyndall's work has had a major impact on our understanding of the atmosphere, weather, and climate.

Tyndall started experiments on the absorption of heat by gases in the spring of 1859,[1] while he completed his glacier studies. His interest had a long gestation. In his youth he had read about Humphry Davy's experiments on radiant heat, and remembered the longing which they incited in him to do something of the kind.[2] He had considered the topic for several years; he read Macedonio Melloni's work on the absorption of heat by liquids and solids around 1850, and frequently discussed the issue with friends. His work on glaciers rekindled that interest. He had explored the existence of air bubbles in ice, the conduction of heat through ice, and the formation of flower-shaped structures in ice by a focused beam of light (Figure 20). Now his attention turned to the atmosphere, to examine its interaction with solar and terrestrial radiation, and to investigate the remarkable conditions of temperature in mountain regions. His aim was to do for gases what Melloni had done for liquids and solids. There was a further motivation. He was convinced that not only the physical but also the chemical composition of substances—and specifically their molecules—played a part previously unrecognized in radiation and absorption.[3] He would be probing the nature of molecules themselves using radiation.

Samples of air in the laboratory absorb very small quantities of heat. It was difficult to measure the absorption, and even more difficult to measure the small differences in absorption between different gases. Tyndall's critical experimental breakthrough was his differential spectrometer. Heat from a cube containing boiling

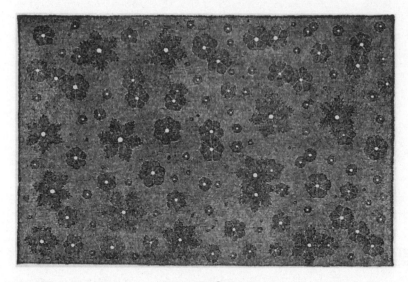

Fig. 20. Ice flowers.

water was passed through the tube that would contain his sample gas, and detected on one face of a thermoelectric pile. He used a compensating cube, which radiated against another face of the thermoelectric pile, to balance the heat from the cube radiating through his sample tube (Figures 21a and 21b). This enabled very small amounts of absorption, previously undetectable, to be measured by difference. Initial success came quickly. On 9th May he tried various experiments without effect, but on 18th May, he could write 'the subject is completely in my hands' (Figure 22). A week later he announced his results to the Royal Society, followed on 10th June by a Discourse at the Royal Institution chaired by Prince Albert, 'On the transmission of heat of different qualities through gases of different kinds'.[4]

Tyndall took as his starting point the idea from Joseph Fourier, Claude Pouillet, and William Hopkins that the atmosphere could allow heat from the Sun to pass through it more easily than heat emanating back from the Earth. Melloni had not been able to show the absorption of heat by air. Tyndall, with his sensitive apparatus, showed it clearly. He also demonstrated the ability of many individual gases to absorb heat, some of them powerfully, though the only one he specified in the report and lecture summary was coal gas. But he concluded: 'Thus the atmosphere admits of the entrance of the solar heat; but checks its exit, and the result is a tendency to accumulate heat at the surface of the planet'.[5] Keen to establish his priority, he

Fig. 21a. Tyndall's radiant heat tube.

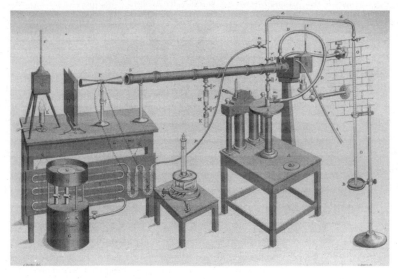

Fig. 21b. Tyndall's radiant heat apparatus.

157

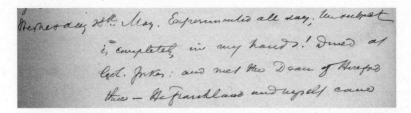

Fig. 22. Tyndall's journal entry noting his quick discovery of the absorption of radiant heat by gases, May 1859.

ensured that reports were also published in continental journals, including *Cosmos*, *Il Nuovo Cimento*, and the *Bibliothèque Universelle*.[6] It was only in the last of those that he specifically mentioned his discovery of the absorption of heat by water vapour and carbon dioxide. He may not have realized their full significance. And at this early stage, he was perhaps running the fine line between revealing too little and revealing too much. Though there is no evidence that he knew it, the American Eunice Foote had discovered the absorption of heat by water vapour and carbon dioxide before him in 1856. Her discovery was announced in a presentation to the American Association for the Advancement of Science in Albany, New York, that has only recently come to light again.[7]

There is a year's gap between this 1859 work and the next phase, caused by a combination of ill-health, the demands of finishing *Glaciers of the Alps*, and his lecture commitments in the early part of 1860.[8] He was committed to thirty-six lectures on 'Physics' at the School of Mines in January and February, having recently taken up the professorship, and twelve weekly lectures at the Royal Institution on 'Light including its Higher Phenomena' from January to the end of March. In addition, he felt that he had made his priority clear, and wanted time to improve his apparatus.[9] But he made detailed measurements in June and July, as he explored the absorption of radiant heat by gases such as air, nitrogen, oxygen, hydrogen, and carbon dioxide, and others including coal gas and chloroform. He published these results later, in April 1862, during a dispute with Magnus.[10]

Tyndall left on 2nd August for Switzerland, where he remained until early September, but not before he had written two articles for the *Saturday Review*. One is a remarkable piece entitled 'Physics and Metaphysics'.[11] It is the first substantial statement of Tyndall's 'materialist' philosophy, ascribing sensations such as colour and sound to the purely mechanical effects of motion impinging on eye or ear, interpreted by consciousness. It appeared anonymously, as was the practice in the *Saturday*

Review. That was advantageous, given its religious sensitivity; Tyndall had intended to publish it in *Glaciers of the Alps*, until discouraged by friends.[12] In the article, Tyndall imagined that the same molecular motion induced by internal rather than external causes would produce the same sensation. In other words, thought and feeling depend on mechanical processes. Byron and Tennyson, he claimed, were poets who could express by intuition this truth that a man of science deduced from physical principles. He thought it would be possible in time 'to infer from the molecular state of the brain the character of the thought acting on it, and conversely to infer from the thought the exact molecular condition of the brain'. He went on: 'Casting the term "vital force" from our vocabulary, let us reduce, if we can, the visible phenomena of life to mechanical attractions and repulsions. But even then, a "mighty Mystery" still looms beyond us. Whence come we—whither we go?'. There were some questions that even science could not answer.

Tyndall's experimental notebook restarts on 21st September 1860. It was a break-through day: 'After a fortnight of constant effort, being…baffled daily…I seem to have so arranged my apparatus that I get constant results throughout the entire day'. The autumn was a key period, as he established that there is 'no sensible thickness of an olefiant gas [ethene] that does not absorb an appreciable quantity of heat', and that 'gases not only radiate, but their radiation is proportional to their absorption'.[13] His experiments were extremely sensitive. As he told Debus, while seeking advice on the chemistry of nitrogen and ozone, he could measure the effect of 0.008 per cent of certain gases mixed with air. He added, still wary about his priority: 'Do not mention these things to anybody. Faraday has warned me against speaking of the subject at all'.[14] The volatile liquids such as ether that he was using for his experiments were making his head ache. This reinforced his mechanical view of the functioning of the brain: 'Dreams are the consequence of certain mechanical actions set up within the brain itself, which actions are the same as those which would be excited by the objects seen in a dream if seen objectively'.[15] But he found time for the social rounds: at the Pollocks for a game of squails after dinner; at Admiral Fitzroy's; and at Barlow's, with Geraldine Jewsbury, Grove, and his son Crauford Grove (the moun-taineer and later President of the Alpine Club). By December, Tyndall had a wealth of quantitative results. He wrote them up for his first major paper on radiant heat. It was immediately selected by the Royal Society as the Bakerian Lecture, Tyndall's second.

It was a social period up to Christmas, while Tyndall continued his examination work at Sandhurst and Kensington. He had breakfast at Sir Henry Holland's, tea at the Pollocks with Spedding there, and dinner at Busk's and at Buxton's. Further din-ners followed with the Lubbocks (John Lubbock was a naturalist, banker, and friend of Darwin), Hopkins, Busks, Barlows, and White-Coopers (after a Turkish bath). At a country house weekend at Bedgebury Park in Kent, home of the society hostess Lady Mildred Beresford-Hope and her husband, a Conservative MP and High Churchman, who co-founded the *Saturday Review*, Tyndall was the only one of the party who did not go to church. He stayed over Christmas at Glympton House outside Oxford with the Barnetts, whom he had probably met at the Riffel,[16] then took the train to Chester. There he joined Busk and Huxley, and they all repaired to the Penrhyn Arms in Bangor. This was a supportive male friendship outing, after the recent death of Huxley's four-year-old son Noel. They walked in biting wind and snow to Capel Curig, buying rake handles for batons on the way. Picking up a porter, Robert Hughes, at Pen-y-Pass, they climbed Snowdon in deep snow, breaking through a large cornice on the summit ridge to find the huts at the top encased in ice. After descending to Llanberis and walking over to the Pen-y-Gwryd—now the famous mountaineering hotel—they went towards Beddgelert to take a carriage to the railway at Carnarvon. But deep snowdrifts soon brought the carriage to a halt, so they walked the remaining 10 miles to Carnarvon with their luggage tied to their backs. On New Year's Eve they were back in London for the traditional Huxley dinner.[17]

In January 1861, between the starts of two lecture series, Tyndall gave a Discourse 'On the action of gases and vapours on radiant heat'. It was a preview of the Bakerian Lecture for February, which was published as the First Memoir in his collected papers on radiant heat (Plate 12).[18] Hirst commented that the Discourse was not as popular as usual, owing to the nature of the subject, though it was treated 'in his usual masterly style'. But at the Royal Society, according to Tyndall, 'they heard me with breathless attention from beginning to end'.[19] The audience included Faraday, Hirst, Frankland, and Tennyson—who would be elected a Fellow in 1865, even if he then rarely attended (Tyndall signed his certificate, and Tennyson came to visit him shortly after his election, to discuss matter and mind).[20] Their attention was fully justified. Tyndall presented his findings, arguably the most significant of his career, about the absorption and emission of radiant heat, and the remarkable experimental design and skill he had brought to bear to reveal them.

On 20th November he had determined for the first time the substantial absorption of radiant heat by water vapour compared to dry air. It was immediately apparent to him not only that this absorption could explain differences between the temperatures at midday and evening, or the temperature at the top of a mountain, but also that changes in the amount of water vapour, carbon dioxide, or hydrocarbons, all of which he found absorbed radiant heat, could have climatic effects. He wrote: 'if, as the above experiments indicate, the chief influence be exercised by aqueous vapour, every variation of this constituent must produce a change of climate. Similar remarks would apply to the carbonic acid [carbon dioxide] diffused through the air'.[21] He saw this in geological terms, especially with respect to water vapour, arguing: 'Such changes may in fact have produced all the mutations of climate which the researches of geologists reveal'.[22] The importance of carbon dioxide became apparent later, despite the evidence to Victorians of the belching of the gas into the atmosphere from factory chimneys, and Tyndall drew no explicit connection between climate change and human activity. It was not until 1896 that the Swedish chemist Svante Arrhenius estimated the actual warming effect of carbon dioxide. Though Arrhenius was aware of the potential of anthropogenic emissions, he also saw them as a potential benefit in delaying a possible ice age. It was Guy Callendar, in 1938, who made the quantitative connection between global warming and emission of the gas through human activity. But Tyndall's vision of geophysical changes was not limited to geological timescales. He once predicted that the temperature of Ireland could be raised by four degrees if it was drained.[23]

Tyndall showed that heat absorption was proportional to density for small amounts of a gas. He concluded, because he was working on gases, in which the molecules were perfectly free from each other unlike in liquids and solids, that these experiments probed the absorption of radiation by the molecules themselves. This molecular mechanism was fundamental to his thinking. When he started his work, the 'caloric' theory of heat, which imagined heat as a fluid, was being abandoned in favour of the idea of heat as matter in motion. As with magnetism, and the motion of glaciers, Tyndall imagined that the transmission of heat depended on the structure of matter. Most importantly, and stimulated by the work of Balfour Stewart in 1858 and 1859 on radiation and absorption by solids, he showed that those gases that absorbed well also radiated well. He thought this 'a simple mechanical consequence of the theory of an ether', since whatever quality 'enables any atom to accept motion from the agitated ether, the same quality must enable it to impart motion to still ether when it is plunged in and agitated'. He was giving a molecular explanation that underpinned the big idea of the 'mechanical equivalent of heat', or the conservation of energy.

But why were oxygen and nitrogen (the main components of ordinary air) such poor absorbers of radiant heat? Tyndall thought that this might be due to their existence as single atoms—though we now know them to be diatomic—and that the far stronger power of other substances, such as water, carbon dioxide, and coal gas, was due to their molecular structure as oscillating systems of atoms. These compound molecules, Tyndall imagined, 'present broad sides to the ether', unlike the simple individual spherical atoms. They have more sluggish motions, so tend to bring the period of oscillation into synchrony with the slower undulations of radiant heat compared to those of visible light.

For Tyndall, these results were explicable in terms of mechanical principles. They required atoms and molecules to be physically real, unlike the theory of equivalent proportions still held by many chemists. Mary Somerville, the well-known scientific writer, was delighted by this evidence, writing: 'although the quantities of vapour are minute beyond imagination, the experiments give the conclusion that they and their molecules have as true an existence as the objects around us'.[24] Finally, Tyndall made the link between conduction and radiation, looking back to his 1853 paper on the transmission of heat through cubes of wood and other substances, and finding that good conductors were generally poor radiators. His explanation was that 'motion, instead of being expended on the ether…is in great part transferred directly from particle to particle, or in other words, is freely conducted'. As Tyndall sought to cement his priority, the work was also published in *Poggendorf's Annalen* and summarized,[25] with the help of Frances Hooker, for the *Comptes Rendus*.[26]

Thomson and Stokes, who refereed the paper for the *Philosophical Transactions*, recognized its significance immediately. Both had detailed comments to make, but Thomson expressed the view that Tyndall's results, as far as he could judge, were 'perfectly novel, and constitute a most important contribution to science'.[27]

* * *

The rhythm of Tyndall's year meant that research was restricted to the latter half. He might have found more time by reducing his socializing—a temptation he managed to resist only intermittently, or when unwell—but the demands of lecturing and examining were burdensome. The work on radiant heat was suspended until mid-July, and did not start in earnest until he had returned from his annual Alps trip in September. Lectures started on 8th January 1861 with a Tuesday evening course of ten lectures for Working Men at the School of Mines on 'Magnetic and Electrical Phenomena'. The series cost just 5s and was aimed, as Hirst put it, at 'a class only at liberty in the evenings'. There was a thirst for this knowledge, and seats rapidly sold. Royal Institution lectures started on 24th January, a series of twelve weekly on 'Electricity', running up to Easter. As Tyndall's reputation spread, Anderson commented that he had never seen so many new members at the Royal Institution. Tyndall was also an examiner for several classes of military candidates, and for the University of London. The pressure resulted in his late marking of Woolwich papers, despite the help of Frankland, and on receipt of a tart letter from Captain Greentree he sent in his resignation. Greentree tactfully ignored it, and when Tyndall received a pleasant letter from him in March he assumed that the resignation had not been accepted.

February and March found Tyndall frequently at social evenings—with the Hookers, Moores, Barlows, Bence Joneses, Pollocks, Colonel Yorke, and at Sir

Charles Hamilton's with his 'fair niece', Mary Adair—and temporarily suspended from the Athenaeum for forgetting to pay his subscription. Though he was 'in good company; two Bishops being also upon the list'.[28] He was soon back, chatting there with James Froude and impressed by the vigour of Brewster, in his eightieth year. He dined regularly at the Royal Society and Philosophical Clubs, contrasting them to the dullness of an Alpine Club dinner. He nevertheless dined individually with mountaineers like Francis Vaughan Hawkins, and paid his Alpine Club subscription of a guinea in April, the day before the first shots of the American Civil War were fired at Fort Sumter. Tyndall was continually torn between enjoying social events and feeling that he ought to be working. He had time for the theatre, seeing *The Rivals* at the Haymarket and going to the gymnastics at the Alhambra with a party including Helmholtz, who was over in London to lecture. It is a charitable interpretation of Tyndall's work pressures that he failed to sign the certificate for Fellowship of the Royal Society of his closest friend Hirst, whose candidature Frankland had championed. He likewise omitted to sign the certificate for Debus, also engineered by Frankland, but as a member of Council (he had been elected for his second spell from 1860–1862) he did work behind the scenes to ensure that both went through.

In the period after Easter up to his summer break in the Alps, Tyndall's extensive lecturing commitments continued. There were thirty-six lectures on 'Physics' at the School of Mines from 15th April every Monday to Thursday, lectures on 'Elementary magnetism' to primary school teachers at the South Kensington Museum, and the final three weekly lectures of his course on 'Electricity' at the Royal Institution, where he also he spent time helping Helmholtz with preparations for his invitation lectures. Examining added its load, but social life was a whirl. April found him dining at Gassiot's, Wheatstone's, Busk's, Major Danach's (meeting Cardwell again), at Barlow's, Pollock's, and Sir George Everest's, where he met Everest's successor, Sir Andrew Waugh, who surveyed Mount Everest itself. He was also at the house of William Spottiswoode, mathematician and later a close friend. He read Spencer's *First Principles*, Eliot's *Silas Marner*, and Jowett's *Essay on the Interpretation of Scripture*. This was Jowett's contribution to the notorious *Essays and Reviews*, a volume by liberal Anglicans arguing that the Bible should be read like any other ancient book. Two of the essayists were found guilty of heresy by the Church's Court of Arches, though the verdict was overturned by the Judicial Committee of the Privy Council.[29] Tyndall declared it 'very mild to a heretic like me'.[30] In May the social round encompassed a country house weekend at Broome Hall with Sir Benjamin and Lady Brodie, and evenings with friends such as Huxley and Ramsay, at the Royal Society and Philosophical Clubs, at Ball's, and at Miss Coutts's. A couple of weekends were spent walking with Hirst, as they planned to rent Huxley's house at 14 Waverley Place together. Hirst took possession on 8th June, having been elected a Fellow of the Royal Society two days earlier, and Tyndall joined him ten days later.

Tyndall realized that his discovery that good absorbers of radiant heat are also good emitters was closely related to Kirchhoff's recent revelation that metal vapours absorbed and emitted light of the same frequency. Indeed he claimed that he would have found the law himself in 1859, 'had not an accident withdrawn me from the investigation'.[31] He linked his and Kirchhoff's work in a Discourse on 7th June, 'On the Physical Basis of Solar Chemistry', turning down a further country house weekend with the Brodies to work on it.[32]

He started this lecture by demonstrating, for the first time in public, the relative abilities of oxygen and ethene to absorb and emit radiant heat, and the relationship of radiation to absorption: good radiators being good absorbers, and bad ones likewise bad. Then he showed the emission spectra of various metals, stunning his audience with the colours achieved as the light was projected through carbon disulphide prisms. He demonstrated that each metal has its own light signature, which he argued it should show anywhere in the universe. Dark lines in the solar spectrum had been noticed by Wollaston and examined by Fraunhofer. But Tyndall said that it was Kirchhoff who had fully explained the connection between the bright emission lines and the dark bands in the solar spectrum. The dark bands were due to absorption, by substances in the Sun's atmosphere, of parts of the continuous spectrum of light emitted by the Sun. The bands corresponded to the emission lines of heated metals on Earth, enabling us to identify those substances in the solar atmosphere. Tyndall acknowledged that others had drawn similar conclusions, including Thomson, Balfour Stewart, and Foucault, but declared that it was Kirchhoff who had stated the law clearly.

Herschel, who had not been present, sent a piqued letter to say that in the 1820s he had published similar observations of the possibility of detecting substances by their flame colours, and that everyone else bar him had been mentioned in this 'admirable lecture which must have created quite a sensation and will be accepted as a résumé of the history of the subject'.[33] Tyndall apologized profusely and managed to add a reference to Herschel in proof for the report in the August edition of the *Philosophical Magazine*.[34] He made further amends by publishing an open letter 'Remarks on Radiation and Absorption' to Herschel in November.[35] Then, with examination commitments over, his thoughts turned to the Alps. He was planning an attack on the Matterhorn, as he told Clausius: 'In company with a friend I attempted it last year, and had we time I think we should have surmounted it. This year it will probably be ascended before I arrive, as several of the best climbers in England start in July to make the attempt'.[36] He had time for a few experiments on ozone and chlorine before leaving for the Alps on 31st July.

Back in England in early September, having tried but failed to climb the Matterhorn, he started his second major period of work on radiant heat. It was all-consuming. Between 12th September and 13th December he recorded experiments on most available days, including Saturdays from mid-October, though he did find time to write up his Alpine explorations for possible publication. Towards the end of September, he visited Bence Jones in Folkestone and then the Lubbocks in Kent. After that only Sundays were rest days, typically consisting of long walks with Hirst in the countryside around Guildford, to Kew, Richmond, or Harrow, or in Hyde Park. For once, his health was good. He told Debus: 'My hairs are beginning to turn gray, but my body is usually full of vigour. Indeed, it is many years since I have been so strong'.[37]

Tyndall's initial focus in his experiments was on water vapour in the air. As early as 13th September he had recorded that 'aqueous vapours in the atmosphere this day exerted an absorption of at least 25 times the absorption of the atmosphere itself'.[38] Over the next month he explored—in addition to air—bromine, chlorine, hydrogen chloride, and ozone. By 10th October he could write: 'it is certain that the foreign contaminants in the air of the laboratory exert at least 67 times the absorptive power of the air itself, and the aqueous vapour alone exerts about 40 times the absorption of the air'.

Radiation from the Moon, as well as the Sun, intrigued him. In mid-October he carried out several night experiments from the roof of the Royal Institution, noting with astonishment that his thermopile 'when turned on the moon lost more heat than when turned on any other part of the heavens of the same altitude. It was practically a radiation of cold'.[39] He gave his explanation in an open letter to Herschel, suggesting that the radiation from the moon evaporated water droplets in the atmosphere, enabling his thermopile to radiate more in this than in any other direction. Tyndall concluded that 'the cutting off of the moon's heat must be entirely due to aqueous vapour, but if this can cut off all the heat of the moon it is evident that the heat of the earth must be absorbed by its own atmosphere'.[40] Herschel replied pointing out a possible flaw in Tyndall's argument, and Tyndall worried that he had made a bit of a fool of himself.[41] In a follow-up note, he reiterated his main finding but recognized that the irregularities of the London atmosphere might be affecting his results, so it was not conclusive.[42]

But by mid-November Tyndall was struggling with his experiments. He wrote pleadingly to Nevil Maskelyne at the British Museum for a large diamond: 'Never was mortal baffled & perplexed as I have been. My work is tremendously difficult and if I am not helped it will conquer me'.[43] He wrote for plates of rock-salt in December, and to Stokes with a question on the mechanism of convection cooling.[44] He sought assistance from others, asking Debus if he could supply volatile liquids and help with calculations.[45] Nevertheless he was soon back on track, working every weekday and Saturday for the rest of November. He may have taken time out for the Royal Society Anniversary meeting on 30th November, where his patron, Sabine, succeeded Brodie as President.[46] The Copley Medal went to Agassiz and the Royal Medals to William Carpenter and Sylvester, the latter principally thanks to Tyndall. He continued to record experiments until 13th December, after which no more entries were made in his notebook until 26th August 1862. But he now had sufficient experimental results to write up a further substantial paper, based on improved apparatus that took much time to master.

He was unwell again, and perhaps upset by some humorous remarks made by Leslie Stephen at an Alpine Club dinner about scientific measurements on mountains, to the extent that he spent a few days at the oasis of Queenwood before returning for the start of his lectures. He decided not to publish the accounts of his alpine exploits in the next edition of the Alpine Club's *Peaks, Passes and Glaciers*, and they appeared instead in his book *Mountaineering in 1861*. In late December and early January, he gave his first Christmas Lectures, a series of six on the topic of 'Light'. Faraday had resigned the lectures, and when Tyndall discovered this, he had written saying that he would cheerfully undertake them.[47]

* * *

Hirst returned to London on 12th January 1862 to find Tyndall preparing both for his Discourse 'On the Absorption and Radiation of Heat by Gaseous Matter' on 17th January, and a weekly series of twelve lectures on 'Heat' at the Royal Institution—the lectures that became his major popular book *Heat as a Mode of Motion*.[48] To complicate matters, Tyndall was now engaged in a delicate dispute with Magnus. In March 1861, just as the Government Grant Committee voted him £200 for his experiments on radiant heat, Tyndall had received a letter from Magnus which caused several years of arguments, both about priority and the validity of Tyndall's results and conclusions. Unlike Tyndall's disputes with Forbes, and later with Thomson and Peter Tait, this one was conducted in the best spirit. It was only ended by Magnus's death in 1870, by which time there was little doubt that Tyndall was correct.

Magnus had seen the summary of Tyndall's recent work in the *Comptes Rendus* and was surprised by his results on the absorption of radiant heat by water vapour, which were 'not intelligible to me, except on condition that your air was cloudy or perhaps that a layer of water had settled on your rock-salt plates'.[49] Tyndall replied immediately, having discussed the issue with Faraday. He pointed out that notes of his earlier work, which Magnus had apparently not seen, had been published in foreign journals, and he was keen to assert his priority.[50] He made sure that Clausius knew his position too, writing: 'The fact is that all my results were obtained in 1859, a year before Magnus touched the subject: I took possession of the field by the note I published and circulated in France, Italy, Switzerland and Germany; and having done so waited until my apparatus had reached a high degree of perfection before I published in full. So that as regards both my published and my unpublished work I was first in the field'.[51] Magnus, who bore him no ill will, sent a pleasant letter back.[52] But he did not change his mind on the experimental facts. He offered to hold back his own results until the position could be clarified, though he was concerned that a third party might intervene.[53] As Tyndall finalized his paper, and prepared to give it both as a Discourse and to the Royal Society, he wrote to Stokes for advice on convection and the effusion of gases, as he sought insights for the behaviours of his gases in radiation and absorption. Stokes's explanations and ideas were penetrating and detailed, comparing theory and experiment, but taking it for granted that Tyndall did not want to go into the mathematics.[54]

Tyndall's new paper, the 'Second Memoir', extended and deepened the findings of the First Memoir.[55] It included comparison of the relatively weak absorption and radiation of the 'elementary' gases chlorine and bromine with that of compound gases such as hydrochloric acid, ozone, water vapour, scents, and the products of Tyndall's breath and those of his assistants, Anderson and Cottrell, as they exhaled after drinking Trinity Audit Ale and brandy.[56] Tyndall interpreted his results with ozone, giving a far larger effect than oxygen, as evidence that ozone was molecular, as opposed to elementary oxygen.[57]

The sensitivity Tyndall achieved was remarkable, as he measured the radiation from a billionth of an atmosphere of boracic ether. He also discovered that he could do away with the complexity of an external heat source, through using the idea of what he called 'dynamic' radiation and absorption. A gas cools when it expands against a piston, and heats up when it is compressed. Tyndall used this to make the gas itself a source or sink of heat, achieving the same results as with his external heat sources. He also imagined that it might be possible to measure the temperature of space directly, by taking a thermopile high enough in the Earth's atmosphere. Herschel was impressed by his results, writing: 'you have made a grand step in meteorology in showing that the dry air is perfectly transcalescent & that the invisible moisture is what stops the sun's heat'.[58] Tyndall replied: 'Indeed you are one of the chosen few whose approval I think it well worth the labour of a life to obtain'.[59] Poggendorff, who was arranging for the paper to be published in the *Annalen*, wrote to say that he need not worry about Magnus: 'he is far from allowing the dispute with you to become personal'.[60]

Having once said while writing *Glaciers of the Alps* that he would never write another book, Tyndall found himself not only publishing a book on his Alpine exploits in 1861, working on the proofs in February, but also setting out his lectures on 'Heat' for a general readership. He argued: 'nothing like the written lectures has yet appeared and I think the publication of them will do good'.[61] In the run-up to Easter he found time to write a note for the April *Philosophical Magazine*, 'On the regelation of snow-granules', in which, having observed the way in which a layer of snow slid down over a warmed greenhouse roof, he subtly reinforced his view of regelation enabling the snow layer to bend 'as if it were viscous'.[62] Weekends were spent walking and visiting with Hirst: to Ramsay, Frankland, and the chemist Alexander Williamson. April also saw the publication of his paper 'Remarks on recent researches in radiant heat'.[63] This dealt in detail with the priority question and the disputed experimental results between him and Magnus. Tyndall explained what effort and time it had taken to remove sources of error, 'especially when one of the most conscientious experimenters of modern times is found falling into error on some of the points which most perplexed me'.

Tyndall now found the absorption of radiant heat by water vapour to be at least sixty times that of dry air, while Magnus could detect no effect. Magnus considered that Tyndall's results were caused by condensation of water on the plates, but Tyndall showed not only that there was no evidence of this, but that he could obtain the same result without using rock-salt plates at all. He suggested that Magnus erred in allowing his gases to come into direct contact with his source of heat, chilling it. He argued that he had found that the vapours of strongly absorbing liquids were strong absorbers themselves. The well-known strong absorption of heat by liquid water therefore implied that its vapour would strongly absorb too, as he claimed to confirm. Tyndall may have hoped that this paper would be the end of the argument, but Magnus was not persuaded. Much later, writing up the history of 'calorescence', which he had defined as the transmutation of radiation from a longer to a shorter wavelength, Tyndall remarked that he was forced to suspend his work on obscure heat (infrared) and calorescence for about two years, from August 1862, while he addressed Magnus's objections.[64] Such are life's vicissitudes at the frontiers of science.

With all these pressures, it was time to recharge batteries. Tyndall spent about ten days alone over Easter, starting in Weymouth and then walking along the coast to Torquay. He wrote to Juliet Pollock: 'I saw a ruined castle, gray & mournful as the evening itself: thither I wandered, and sheltered by one of its broken walls listened for a long time to the sound of the wind and of the sea. The part of human nature which comes into play under such circumstances is that which puzzles me most. That solemn unison which the soul experiences with nature, and which is a thing essentially different from the intellectual appreciation of her operations'.[65] After Easter, and Sabine's Presidential Soirée at the Royal Society, at which Tyndall showed experiments on metallic spectra and singing gas jets, lecture commitments started at the School of Mines. Tyndall was billed for forty lectures on 'Physics', at 1pm every weekday. The audience was almost entirely male, unlike the preponderance of women at the Royal Institution, but it did include Geraldine Jewsbury. She thanked him for letting her attend, writing: 'I only hope that your pupils are not disgusted at seeing one solitary female amongst them—for to tell the truth I feel something like an intruder'.[66] In parallel with the lectures Tyndall researched details for the publication of his lectures on 'Heat', writing to Helmholtz and Clausius to ask if they knew the extent of Julius Robert Mayer's contribution to determining the mechanical equivalent of heat and whether he, or Joule, was the first to do it.[67]

This simple question would lead inexorably to a further public row over priority with Thomson and his Scottish colleague Peter Tait in particular, although in this case it was nothing to do with Tyndall's own discoveries. Helmholtz responded that Mayer had calculated it, but not in a manner safe from objections.[68] Clausius's initial response, without having been able to get hold of all Mayer's papers, was that 'you will not find very much in them, since Mayer does not have a sufficiently clear understanding of the basic concepts of mechanics'.[69] However, he wrote again a few days later, having read more of Mayer's works and completely changed his mind: 'I must retract what I wrote…I am amazed at the number of fine and correct ideas that they contain'.[70] Both Helmholtz and Clausius were impressed that Tyndall was aiming to publish a book on heat for a general audience, which they saw as a valuable activity. Tyndall was still not well, despite a brief trip to the Isle of Wight to recover from a bad head, which had interrupted his lecturing at the School of Mines. He turned down invitations from the Duchess of Northumberland, Lady Holland, Mrs Pollock, and Lady Millicent (Bence Jones),[71] though he was at the Royal Society on 22nd May to hear Hopkins's contribution to the glacier disputes.[72] It supported his views.

A fortnight later, it was Tyndall's Discourse 'On Force' that launched an acrimonious battle with the 'North British' physicists.

10

✦═◎◎═✦

HEATED EXCHANGES
1862–1865

The mid-nineteenth century witnessed the culmination of several decades of effort to understand energy conversions and their relationship to heat. Force, as energy was then broadly termed, could neither be created nor destroyed but could change its form, for example in a steam engine or an electric motor. The idea of the 'conservation of force' could be expressed as the mechanical equivalent of heat, as people made quantitative calculations and measurements of the amount of heat that could be obtained from mechanical work of any type. This 'dynamical theory of heat' as William Thomson termed it, became the science of thermodynamics, and the idea of the conservation of force became the principle of the conservation of energy. As the field developed, claims of priority and scientific reputations were at stake across Europe.

Tyndall entered the lists on 6th June 1862 with his Discourse 'On Force'.[1] He started by describing what we now term potential and kinetic energy and their conversion into heat: 'the motion of the mass is converted into the motion of the atoms in the mass'. This heat could be used to do work, such that the annual burning of coal could enable work equivalent to 108 million horses working continuously all year. But, he said, it could be calculated that if the Sun were made of coal it would burn for just 4,600 years, so whence its power? Tyndall pointed to meteorites falling with huge velocity through gravitation as a possible source of continuing energy (this, many decades before the discovery of nuclear reactions). He explained the Sun as the ultimate source of wind and water power, and of the energy necessary to build plants by incorporating carbon dioxide. The argument extended to animals, which eat plants and so also derive their being from the Sun.

Tyndall then revealed that these generalizations were the work of Julius Mayer, a German, of whom the audience would not have heard. Mayer, he asserted, had calculated the mechanical equivalent of heat earlier than Joule, and had suggested the meteoric origin of the Sun's energy before John Waterston and William Thomson. Tyndall claimed that he did not wish to overstate Mayer's claims, but 'to place him in that honourable position, which I believe to be his due'. He might have anticipated that others would rush to defend their priority.

The first shots in the controversy over Tyndall's support for Mayer were fired, most politely, by Joule himself. Joule expressed his admiration for Tyndall's second paper on radiant heat but claimed his priority over Mayer: 'I think in a case like that of the Equivalent of Heat the experimental worker rather than the mere logical reasoner (however valuable the latter) must be held as the establisher of a theory'.[2] He followed this up with an open letter in the *Philosophical Magazine*, in which he provided examples of others before Mayer, particularly Séguin, who had also developed a dynamical theory of heat. But he argued that to demonstrate it required experiment: 'I therefore fearlessly assert my right to the position that has been generally accorded to me by my fellow physicists as having been the first to give a decisive proof of the correctness of the theory'. Even then, Joule acknowledged that apart from Rumford, Mayer, and Séguin, merit should be shared with Thomson, Rankine, Helmholtz, Holtzman, Clausius, and others.[3] Tyndall replied to this open letter with one of his own, written from Switzerland. He quoted from his Discourse that he had given every acknowledgement of Joule's work in establishing that 'under all circumstances the absolute amount of heat produced by the expenditure of a definite amount of mechanical force is fixed and invariable'. He trusted that Joule would find nothing to question his claim of being the experimental demonstrator of the equivalence of heat and work. Nevertheless, he argued that Mayer had calculated the mechanical equivalent from reasonable assumptions of the properties of expanding gases and deserved more recognition, which did not in the least diminish his respect for Joule's work.[4]

Tyndall's first recorded experiments after his return from Switzerland in 1862 were on 26th August, giving results obtained during the previous days with aqueous vapour.[5] Many of the continental physicists were in London for the International Exhibition, and Tyndall hosted Poggendorff and his daughter at Waverley Place.[6] He entertained Bunsen, Kirchhoff, Wartmann, Clausius, Henry Roscoe, and others to dinner. Wertheim was sadly absent. He had committed suicide by throwing himself from the tower of the cathedral at Tours.[7]

In the presence of Faraday and Poggendorff, Tyndall reproduced twenty times his experiment to show the power of absorption of ordinary air compared to dry air (Plate 13). Later, at the Anniversary Meeting of the Royal Society, and at Kirchhoff's request, he accepted Kirchhoff's Rumford Medal for research on the solar spectrum and on the inversion of the light lines in the spectra of flames. Kirchhoff imagined that the award was partly due to Tyndall's influence.[8] In this he was correct. At the suggestion of William Sharpey, Tyndall had withdrawn Kirchhoff's name for consideration for the Copley Medal the year before, to make it possible for him to receive the Rumford Medal.[9]

Hirst rejoined Tyndall at Waverley Place on 23rd September after a long trip away, and there is a gap in Tyndall's recording of experiments between 26th September and 6th October, including the period at the beginning of October when he was at the British Association meeting in Cambridge. He gave a lecture there on the 'Action of Water' to a large audience. It impressed Edward Youmans, the visiting American scientific lecturer and writer who would later be important to Tyndall. Youmans wrote that it was 'altogether the most brilliant affair of the kind I have ever seen'.[10] He was equally struck by Tyndall's appearance: 'He is a single man of forty with a scanty strip of forehead, and big, straight, prominent nose—the most restless, nervous creature I ever set eyes on'.[11] It was the energy and evident sincerity and commitment of Tyndall that engaged his audiences, coupled with the skill, clarity, and often beauty of his experimental demonstrations, and with his ability to conjure up compelling verbal images.

After that digression, measurements were made almost every day except Sundays until 19th December, even when Tyndall was away for a week on the Isle of Wight, having gone there from Cambridge after straining his back trying to lift a hundredweight sack of ice just before his lecture.[12] Some of the work on radiant heat was delegated to his assistant, Chapman, and Tyndall's laboratory support was increased the following year when he and Frankland (who had just joined as Fullerian Professor of Chemistry, a post he held until 1868) were given paid assistants for the first time. Tyndall was also able to engage a second assistant, whom he paid out of his own pocket until 1875. With Chapman at work in the laboratory for him, Tyndall had time between giving directions to finish his book. Longman called on 15th November to agree the price and profit. Two days later the final manuscript of *Heat as a Mode of Motion* went to the printer. As he had with *Glaciers of the Alps*, he declared: 'I shall never write another—it is awfully heavy work if done well. To forget oneself and be simple is the highest quality of a scientific writer'.[13]

The autumn's work resulted in Tyndall's 'Third Memoir', 'On the Relation of Radiant Heat to Aqueous Vapour'.[14] Here Tyndall set out his firm reiteration of his view that meteorologists might without misgiving apply his finding that aqueous vapour is a powerful absorber of radiant heat, with its important implications for understanding weather and climate. As he wrote to Hooker, whom he had asked to look over the paper: 'Herschel and Airy have both said to me that it is a grand step in meteorology! Magnus denies the action; but he has not a foot to stand upon'.[15] Thomson, in his review for the *Philosophical Transactions*, was thoroughly supportive, considering the whole subject 'of extreme importance'.[16]

In this paper, Tyndall sought to meet all Magnus's objections, and to emphasize the significance of water vapour in the atmosphere. He first countered the charge that condensation of water on the rock-salt plates might be responsible for his findings (plates of rock-salt were used at the ends of the tubes because they enclosed the gas in the tubes but were transparent to radiant heat). He disproved Magnus's charge, both by moving the heated pile very close to the outer surface, and by packing the drying-agent calcium chloride close to the inner. He had even shown the plates to Magnus, who could see no trace of moisture on them. But he went further by cunningly designing a set-up which required no plates at all. Now he could examine the difference between dry and humid air in open tubes. The results were the same. He made the remarkable calculation that 10 per cent of the entire terrestrial radiation would be absorbed by the water vapour within 10 feet of the Earth's surface on an average day. Rising air containing water vapour radiates its heat, thereby cooling and condensing the water to fall as precipitation. Tyndall described, in a beautiful image, how a cloud therefore 'constitutes the visible capital of an invisible pillar of saturated air'.[17] He went on to discuss the role of mountains as condensers, and why in desert regions where daytime temperatures are high, the nights can nonetheless be very cold, for there is no water vapour in the dry air to radiate back the heat.[18]

Meteorology was a subject of great interest to Tyndall, further stimulated by his later work on lighthouses. Given its capacity to absorb and radiate heat, Tyndall could see the importance of the movement of water vapour in the atmosphere, and its complexity. He wrote to Lyell in 1865, in response to a question from him on the possible differential heating and cooling of the Earth's hemispheres: 'the existence of our atmosphere & the transport of water in the shape of snow from the equatorial regions to the polar ones, render the actual problem a complicated one'.[19]

Just before Christmas Tyndall paid a further visit to the Isle of Wight, this time with Huxley, Busk, and John Lubbock. The afternoon of Christmas Day was spent with Lubbock and his wife Nelly, as they became increasingly close friends (Figures 23a and 23b). Tyndall went with Hirst the next day to the artillery shooting ground for some rifle practice. Hirst remarked that Tyndall 'who has won many shooting prizes could scarcely "get on the target" at all'.[20]

Tyndall's 1863 lectures at the School of Mines started on 16th February, a series of thirty on 'Physics', on Mondays to Saturdays, excluding Tuesdays. Geraldine Jewsbury again attended, writing with thanks and a recommendation of novels to read, based on her experience as a reviewer.[21] The Royal Institution lectures, a series of seven on 'Sound', did not start until 24th March, but on 23rd January Tyndall gave a Discourse 'On radiation through the earth's atmosphere' which Hirst described as 'a capital lecture' to open the campaign.[22]

Fig. 23a. John Lubbock (1834–1913) in 1856.

Fig. 23b. Nelly (Ellen) Lubbock (1835–1879) in 1856.

Explaining his findings on the absorption and radiation of heat by water vapour, Tyndall maintained: 'This aqueous vapour is more necessary to the vegetable life of England than clothing is to man. Remove for a single summer-night the aqueous vapour from the air which overspreads this country and you would assuredly destroy every plant capable of being destroyed by a freezing temperature'. Tyndall even considered that this property might explain the blue colour of the sky. He wrote to Herschel: 'Assuming that the action of aqueous vapour upon radiant heat is established beyond doubt (which it is) we have in the vapour a body which absorbs the same rays as the water which produces it—in other words it shares the colour of water. Through the operation of this cause our atmosphere is certainly a blue medium. The quantity of vapour might not be sufficient to produce a sensible blue, if the rays went straight through it; but the reflections within the body of the atmosphere must be innumerable, and thus its virtual depth increased. I wonder is this likely to throw any light on the blue of the sky!'.[23] Herschel replied that he liked the idea, but could not see that there was enough water to show it.[24] It was not until 1868 that Tyndall was able to explain the actual cause of the sky's blueness.

February 1863 found Tyndall in shaky health. Though he attended various social events, including dinners at the Sharpeys, Busks, and Franklands, he fled to Queenwood for a few days in the middle of the month. He was again unwell in March. Hirst reported that he was in a most abnormal condition for two weeks with dyspepsia, captious in the extreme, difficult to talk to, taciturn and irritable, and fickle in his desires for food. The wisest plan was to leave him alone.[25] Tyndall spent the first part of April trying to recover his health on the Isle of Wight, probably alone, as was often his practice. At the end of the month he was with Hirst at a Royal Society Soirée, showing the Prince of Wales the experiments of Jules Lissajous. Then the lectures started: seven weekly lectures on 'Sound' at the Royal Institution from 28th April, and eight weekly evening lectures on 'Heat considered as a Mode of Motion' at the School of Mines from 29th April.

Heat as a Mode of Motion was published by Longman on 4th March. It became one of Tyndall's most successful publications, bringing him an initial cheque for £186, and eventually £2,300 from the English editions alone over his lifetime. In publishing it, Tyndall joined the ranks of the legion of writers of popular Victorian science books. The majority of authors were not practising men of science. Indeed many were women, including Mary Somerville (Figure 24) and others such as Rosina Zornlin, Agnes Clerke, and Arabella Buckley (who had worked with Lyell). Zornlin and Clerke were relatively rare examples of women, like Mary Somerville, who wrote on the physical sciences. Several of Tyndall's books ran to ten or more editions, in both England and America. Their sales were robust, putting them among the better performers of the time, even if they did not reach the levels of blockbusters like Brewer's 1847 *Guide to the Scientific Knowledge of Things Familiar* (195,000 copies by 1892) and Chambers's 1844 *Vestiges of the Natural History of Creation* (39,000 copies by 1890).[26] His best books probably sold in similar numbers to Somerville's 1834 volume *On the*

Fig. 24. Mary Somerville (1780–1872).

Connexion of the Physical Sciences (17,500 copies), and his library contained a copy of her *On Molecular and Microscopic Solids*, which she gave him in 1869.

* * *

Unfortunately, just as *Glaciers of the Alps* had brought to a head the dispute with Forbes, *Heat as a Mode of Motion* highlighted the priority dispute over the mechanical equivalent of heat. Tyndall had sent Clausius part of the manuscript, asking him if he thought he had been fair to Mayer and Joule, and Clausius had replied with detailed and supportive comments in several letters, saying that he had been perfectly fair to them both.[27] Some of the 'North British', a group of mostly Scots Presbyterian physicists and engineers, with which Joule was associated, had other ideas. It was an opportunity for them to fight their corner against one of the 'Metropolitans', the London-based 'scientific naturalists', and particularly Tyndall, their opponent in the glacier controversy.

In October 1862, Thomson and Tait had written an article entitled 'Energy', in the Evangelical magazine *Good Words*. They explained for a popular audience (the magazine had a monthly readership of 120,000) the concept of the conservation of energy, claimed Joule as the founder of the modern dynamical theory of heat (the equivalence of heat and work) and criticized Tyndall, without directly naming him, for championing Mayer. They also explained the fact that since heat cannot be completely converted into the equivalent mechanical work, the energy in the universe gradually becomes uniformly diffused heat, from which it can never afterwards be changed. Their religious beliefs were evident as they commented: 'dark indeed would be the prospects of the human race, if unillumined by that light which reveals, "new heavens and a new earth"'.[28]

Tyndall was not aware of this article until February. He saw Thomson as his real antagonist, who he understood had been unsparing in his censures over Forbes, but never openly. He felt he had to respond, and he immediately took issue with the insinuation from Thomson and Tait that he had suppressed the claims of Joule, his own countryman, in favour of Mayer.[29] Joule, however, did not appear to take any offence after staking his own claim following Tyndall's Discourse in 1862. Tait, at the request of Thomson, sent a robust rejoinder which gave no ground, and brought up the old arguments about Rendu and glacier motion.[30] Tyndall's reply was swift and addressed in an open letter to Thomson, whom he saw as the senior party.[31] The electrical researcher Sir William Snow Harris, who was having his own battle with Thomson, much approved, as did Pollock and Airy.[32] Tyndall suggested that Thomson was not in full possession of the facts about Mayer's discoveries. In response to the charge that Mayer's paper had 'no claims to novelty or correctness at all' he quoted Helmholtz's and Verdet's views of the significance of his work. He also emphasized Mayer's thinking about the relationship between food and energy, and the ultimate importance of the Sun. He ended in barely suppressed anger that his honour had been called into question, let alone his understanding of the facts. Thomson, himself doubtless angered by being addressed in such a manner in an open letter that he had not seen before publication, refused to engage further.[33] Tait could not resist, still asserting that Mayer's work rested on a false analogy.[34] Tyndall was goaded again, pointing out that Tait had, in his view, quoted in a selective manner from Joule's early paper, though he regretted appearing to cast a slight on Tait in his first response.[35]

Tyndall proceeded to have three of Mayer's papers translated and published in the *Philosophical Magazine* between November 1862 and June 1863, to put the evidence on the record.[36] Clausius told Tyndall of his own difficulties with Thomson, writing: 'I must guard my priority, in particular from Thomson and Rankine, who are supported by Tait, and constantly praise one another and claim for themselves the achievement, and do their best not to mention my work, or, when that is not possible, only briefly touch upon it'.[37] He thought Tyndall's response to Thomson 'somewhat harsh', but considered Thomson's behaviour 'unmanly'.[38] Bunsen also wrote in support: 'Your response to Thomson's attacks is pretty devastating for him. May he take a lesson from it and stop arousing suspicion about the works of others, in fully unfounded ways and making sideswipes, as he has done so frequently in recent times'.[39]

Later, in June 1864 just before he left for the Alps, Tyndall set out his case for Mayer in the *Philosophical Magazine*.[40] It included a résumé of Mayer's second paper of 1845, which had not yet been translated into English, emphasizing his priority in calculating the mechanical equivalent of heat, and his recognition of the significance of the Sun. He showed that Mayer had ideas about the meteoric origin of the Sun's heat in 1848, six years before Thomson. And he expressed in forthright terms his opinion of Thomson's continued silence on Mayer, that it was '*not* dignity, nor even manliness as defined in England'. He ended by reiterating his respect for Joule as 'the Experimental Philosopher' beside Mayer, 'the Thinker and Generalizer'. Mayer was effusive in his thanks.[41] Tyndall hoped that Mayer would be at the British Association meeting in Newcastle. In the event, worried about his command of English, he went instead to a congress in Szczecin.[42]

Thomson and Tait's article in *Good Words* had highlighted the fact that some heat was always dissipated in energy conversions, and no longer available for use. Clausius coined the term 'entropy' for this dissipation of energy in 1865, as the basis of the science of thermodynamics was established. So, while energy overall was always conserved in transformations, useful energy was not. The universe, by extrapolation, would eventually run down to a thermal equilibrium, effectively death. It had a beginning and would have an end, consistent with theologies of creation and divine directionality. Tyndall wrote extensively about the conservation of energy, which he regarded as one of the great ideas of the nineteenth century, but rarely about the consequence of entropy. He saw the universe as an organic whole, rather than as a machine, and his view may have been closer to those of the German cyclical cosmologists, who imagined an endless universe transcending the finite limits of its individual parts.[43] One person who was stunned by the problem that equilibrium meant death was Spencer. His philosophy of development envisaged equilibration as 'the ultimate and highest state of society' (and of the universe), in which all men would be organically just, and 'none have any desires at variance with the welfare of others'.[44] Spencer, like Tyndall, seems to have disregarded a too literal and mechanistic interpretation of the concept of entropy.

Against this backdrop, Tyndall gave the next instalment of his work on radiant heat in a Discourse 'An Account of Researches on Radiant Heat' at the Royal Institution on 12th June 1863, before presenting it at the Royal Society on 18th June as the 'Fourth Memoir'.[45] Here he described his new and accurate measurements of the absorption of different thicknesses of gases, showing that a shell of ethene two inches thick would absorb 30 per cent of the reflected solar radiation, raising the surface of the Earth to 'a stifling temperature'. Comparing the properties of liquids and their vapours, Tyndall showed that strong absorbers and emitters in one state were also strong in the other, indicating that absorption and emission were molecular phenomena: 'the molecule carries its power, or want of power, through all states of aggregation'. This was one of his crucial arguments against Magnus with respect to water, and the real focus of his work on heat, as he sought to explore the conservation of energy in molecular terms. He found it necessary to state his argument in a detailed paper in the *Philosophical Magazine* in July, responding to Magnus's paper defending his position in the same issue.[46] Relations between them remained cordial. Tyndall helped Magnus to election as a Foreign Member of the Royal Society, for which he was grateful.[47]

Meanwhile, Irish relatives in droves were discovering Tyndall. As Hirst drily observed: 'Tyndall's kinsmen are increasing remarkably in number, as he himself begins to acquire a European reputation'.[48] It is a reasonable assumption that Tyndall had provided financial support to his family at least since the death of his father. He now gave his mother £100 a year, as well as £20 to Dr Phillips to look after Emma and nearly £30 to John McAssey, his mother's brother.[49] The family of Caleb Tyndall, his uncle, was a continual drain, and the money was not always easy for him to find. In 1889 he described to Elizabeth Stuart, who often acted as his agent and intelligence-gatherer in Leighlinbridge, 'the hundreds of pounds which I have already expended in aid of the family of Caleb Tyndall', for a marriage settlement, a cow, or even the purchase of a farm.[50] He sent gifts for the poor at Christmas, or contributed to the Leighlinbridge Clothing Fund, and gave specific help to many individuals beyond his immediate family.

Throughout his life Tyndall was generous with his money, even if at times he worried about his future solvency. He frequently contributed to memorial appeals, but also to the living. Two examples stand out. Shortly after his return from America in 1873, when he had already loaned Huxley £1,000 to help him buy a house, Tyndall was instrumental with Spencer in raising nearly £2,000 from eighteen subscribers to give the ailing and over-worked Huxley some breathing space. Nelly Lubbock donated, to the perturbation of the men (including her husband, who gave £200). Their friend Thomas Farrer agreed to transfer the money from his bank so it would be untraceable to the obvious suspects, and Darwin wrote on everyone's behalf to Huxley, though he rather wished that Tyndall had done so. Huxley, to their relief, accepted with good grace. Darwin commented that 'it made him think better of human nature than he had previously done'.[51] In another venture, in 1878, Tyndall donated the substantial sum of £100 to a fund managed by Pollock to enable the mathematician and philosopher William Clifford to go to the Mediterranean, Switzerland, and perhaps South America in the winter, to seek respite from his tuberculosis.[52] Darwin sought Tyndall's advice on how much people were donating, conscious of the need to provide for his many children, and sent £50, a short while before Clifford died in Madeira, aged thirty-three, on 3rd March 1879.[53] Tyndall contributed about £300 to such causes in 1878 alone, even though his aim, which he had not yet then reached, was to ensure he was entirely independent of government or other aid. He even offered to help Hooker, then President of the Royal Society, with his plan to reduce fees to Royal Society Fellows by a subscription from the wealthy. Hooker did not feel that the 'worker bees' like Tyndall should be expected to contribute but Tyndall was nevertheless willing to contribute £100 in instalments if Hooker really needed it. Hooker would have none of it, though he had felt obliged to subscribe himself to set an example, even though he had just had to write to his bankers for advice.[54] Life was more expensive for the family men like Hooker and Huxley.

Hirst found Tyndall 'in a grim state' in his work in early November 1863, and took him to the Haymarket for distraction. The grim state may have been caused by tension over laboratory arrangements and some uncharacteristically unkind words from Faraday, perhaps over the appointment of laboratory assistants to Tyndall and Frankland. Tyndall had asked Faraday if William Barrett, whom he had been employing at his own expense, could be appointed assistant at £50 or £60 per annum, with a second assistant for a similar sum. He also wrote, rather pointedly: 'The expenses of the original investigations which I have presented during the last ten years have been defrayed in small parts by myself, but mainly by funds voted to me, on my personal application, by a committee of the Royal Society. In framing their new decisions the Managers I suppose will not calculate upon this foreign aid, but proceed on the assumption that the Researches conducted at the Institution shall be carried out at the expense of the Institution'.[55] Bence Jones got wind of Tyndall's stress from Faraday and urged Tyndall to talk to him about it.[56] In the event, Barrett was employed at the Institution's expense, as Tyndall had requested, though Tyndall then paid for a second assistant himself. He may have been put out that both Frankland's assistants were paid for by the Institution. Faraday would have supported the Managers' decisions.

There had earlier been some friction or misunderstanding over moves by Bence Jones to develop the Holland Fund which supported research. Bence Jones wrote to say that his only objective was to help Tyndall and Frankland, but that he would withdraw the plans if they were in any way problematic.[57] In a major report in 1862, he had expressed the view that the original educational function of the Royal

Institution had been taken over by others, with public scientific positions now paying £1,000 and more a year. Instead, the Institution should capitalize on its reputation for original research. He argued that the professors only stayed for the research, and that they should be properly supported and their salaries raised, so they were not tempted elsewhere. Believing it inappropriate to seek government aid, he appealed for men to endow funds (and to women to encourage them to do so, even though, he maintained 'It is not our intention to invite them to assist in the laboratories').[58] The consequence was the establishment of the 'Fund for Promoting Experimental Researches' (the 'Holland Fund' or 'Donation Fund'). It built on Sir Henry Holland's donation of £40 each year from 1859, but did not succeed in attracting major sums. About £4,000 had been donated by the time it was closed in 1872, including one legacy of £2,000. Tyndall even donated £70 himself in three tranches.[59] When Tyndall's salary was raised to £450 in 1868, it was still well short of prime professorial positions elsewhere.

Just before Christmas, Tyndall spent two days walking with Busk and Lubbock, and the three friends Hirst, Debus, and Tyndall dined together on 23rd December. Debus and Wright were considering leaving Queenwood following its sale after Edmondson's death earlier in the year, effectively ending Tyndall's long connection with the college. Soon Tyndall was immersed, for the second time, in the Christmas Lectures. He had hardly finished them—on 'Electricity at Rest and Electricity in Motion'—when his main lecture series on 'Experimental Optics' started at the Royal Institution, a series of twelve on Tuesdays and Thursdays, finishing on 25th February 1864. This course was interrupted by a Royal Command to spend a weekend at Osborne House in late January to lecture to the Queen and Princesses, a singular honour. As those Royal Institution lectures ended, another series started at the School of Mines, the general course of thirty-two on 'Physics' each weekday. For relaxation, Tyndall was often out with Hirst over this period, dining with Ramsay at the Geological Society, and entertaining or being entertained by the Lubbocks, where Tyndall retreated for a weekend in February when he was not well.

The significance of Tyndall's work on radiant heat was such that his Fifth Memoir, 'Contributions to Molecular Physics', was selected as the Bakerian Lecture, his third. He gave it on 17th March, and as a Discourse the following evening with the Prince of Wales in the chair.[60] The Discourse, according to Hirst, was highly successful, but ended with heterodoxy not mentioned in the published summaries, 'on the danger of ascribing motives to Providence—apropos of the beneficial effects of the vaporous envelope of the earth—and of the groundlessness of the fears that the investigation of truth in science could be injurious to religion'.[61] This paper, the experimental work for which was done wholly by Tyndall's assistants for the first time, established clearly that the relative power of a liquid to other liquids as an absorber is that same as that of its vapour.[62] Again, this reinforced the strength of the absorptive power of water vapour, which Magnus disputed. Tyndall predicted that radiant heat, if rendered sufficiently intense, would be able to heat substances to incandescence, and argued that radiation and conduction were reciprocal phenomena: 'molecules which transfer the greatest amount of motion to the ether, or, in other words, radiate most powerfully, are the least competent to communicate motion to each other, or, in other words, to conduct with facility'. Both Thomson and Stokes took issue with the solidity of the latter point.[63]

After the Bakerian lecture and the Discourse, Tyndall went again to the Lubbocks for the weekend of 19th and 20th March and to the Isle of Wight at Easter, reading Esmond and Vanity Fair: 'What a noble writer poor Thackeray was!' (he had died in December).[64] By early April, he had started a course of six Working Men's lectures at the School of Mines,[65] but in mid-April, he told Hirst of a 'sad calamity' which had befallen him and might necessitate his going to Ireland.[66] This was an illness of his mother: 'Years ago I thought if she lived long enough she must finally give way. Her mind was so feeble, so rambling, so unoccupied, and so saturated with the enthusiasm of what people call religion'.[67] Hirst stood ready to go instead, but it proved unnecessary, so he left for France and Italy; he would be away until late August. Tyndall finished his Jermyn Street lectures in early May and planned to go to Ireland on a ramble through Wicklow to Carlow, leaving Debus to look after his examination papers. While he was there he arranged for Miss Roche, a local friend, to help with managing money, arranging accommodation, and choosing suitable clothes for his mother and sister, 'dark, sober and decent'.[68] Elizabeth Steuart later seemed rather put out by this arrangement, which she thought usurped her relationship with him.[69] Tyndall was giving his mother a substantial allowance and other expenses but was unhappy that his mother and sister seemed confined to a bedroom, and about how the money was being spent. He wrote later to Miss Roche asking for help in arranging the tomb over his father's grave, which she kindly did.[70] In a letter to Hector Tyndale, he told of his visits to the Gorey Tyndalls and to a Robert Tyndale, proprietor of a large estate in County Kilkenny. He looked forward to Hector's news on the American Civil War, revealing his Confederate sympathies: 'I would sooner be fighting at your side, or rather on the other side than living a life of inaction on the fairest portion of the earth's surface...The exponent of Federal news in the House of Commons is a personal friend of mine. We shall probably make a tour together this year in the Caucasus—I mean Forster the member for Bradford' (this was William Edward Forster, the Liberal MP).[71]

Tyndall had been unwell in April and was still not well by the end of May, rarely dining out and exercised by the preparation for his forthcoming Discourse 'On a Magnetic Experiment' on 10th June.[72] The task had bothered him, although it was mostly a description of the work of others, and a straightforward description of magnetism. He needn't have worried: 'There was a Queen's ball, and a meeting of the Astronomical Society—still the house was full'.[73] Juliet Pollock wrote him a lovely letter afterwards, expressing her gratitude: 'I must really thank you for the quantity of interest you have added to our life in so many ways'.[74]

After the Alpine season of 1864, Tyndall stayed in London in September, apart from a trip to Folkestone early in the month. He was 'working incessantly' at the second edition of *Heat*,[75] and discussing French editions of several of his books.[76] To free up time he subcontracted his examining at South Kensington to Debus; about 100 hours at 3/6d per hour.[77] He could not be inveigled out, neither by Nelly Lubbock to the British Association Meeting in Bath, nor by Juliet Pollock to the Isle of Wight, despite her entreaties: 'What! not a Saturday and Sunday to spare?...Tennyson is come back, and Mrs Cameron is in...and I should like a moonlight walk with you if only to show off my courage'.[78] He was finally able to travel down on 19th October, staying for a couple of days. Tennyson climbed with him on the chalk cliffs, as the poet and the physicist risked themselves on the crumbling precipices.[79] Weekends found him mostly with friends, visiting the site of a recent powder explosion at Erith, walking round Hampstead or Hyde Park with Hirst, and at the Lubbocks.

* * *

The next phase of Tyndall's research on radiant heat—from the end of September until mid-December 1864—brought both discoveries and controversy.[80] His interest now was in 'black' or 'obscure' heat; infrared radiation beyond the visible range. He revealed his discoveries to Stokes: 'I think you will be interested to learn that I have succeeded in raising platinum to vivid redness by black heat. The same black heat sets paper instantly ablaze'.[81] These were significant findings: that 'black heat', or invisible radiation, could not only set substances ablaze, but make a metal glow with visible light. It was an example of 'ray-transmutation', the emission of visible light following absorption of 'black heat' of longer wavelength (infrared). This was the opposite of Stokes's discovery of fluorescence, the emission of visible light following absorption of invisible radiation of shorter wavelength (ultraviolet).

But a few days later he asked Stokes 'to consider that note unwritten'.[82] The reason was a most unpleasant dispute with Dr C. K. Akin. Akin was a Hungarian physicist who had first approached Tyndall in 1862, asking him as editor of the *Philosophical Magazine* if he would help with the English of the paper he wanted to submit.[83] Akin considered that the idea of the possibility of ray-transmutation was his, and demanded that he be allowed the opportunity to demonstrate it first. In response to what must have been a most confrontational letter, now lost, Tyndall replied with great generosity: 'I cannot help expressing my deep regret that you did not write to me in a different manner. I should have gone almost any length to gratify you, had you adopted another tone; and I shall prove my sincerity in writing thus, by doing that which still remains in my power to do. I have to say, then, that from this 3rd of November 1864 to the 3rd of November 1865 I shall not make known, publically or privately, any experiments on "ray-transmutation". Nearly seven months ago I was, as you know, ready to take this subject up. Out of deference to you I did not do so; so that at the termination of the period referred to I shall have held back for nearly nineteen months. That you will be able to release me long before, by the actual performance of this experiment, is almost beyond a doubt; and in this event I hope to muster sufficient greatness of heart not to envy you'.[84]

In his paper 'On luminous and obscure radiation', Tyndall described a means of examining 'obscure' heat by separating it from heat in the visible spectrum.[85] He had found that iodine dissolved in carbon disulphide completely absorbed visible radiation but let obscure radiation through. By this method he showed that the amount of obscure heat increased with increasing temperature of flames, with the visible radiation remaining the same. The heating power of the obscure radiation often exceeded the visible by 25 to 1. He noted that a significant proportion of radiation reaching the retina was incapable of stimulating vision. Taking his sight in his hands he directed into his eye a converging beam of obscure rays, filtered from an electric lamp, and detected no ill effect. The retina and optic nerve seemed impervious to the obscure heat. These experiments showed to him that light and heat are essentially the same, and also that radiant heat could not be used for fog signalling since it is so powerfully absorbed by water vapour.

One aspect of his work, given his dispute with Akin, he decided he could not pursue, writing: 'The rendering of metals incandescent by obscure rays has not yet been accomplished. This is a question on which Dr Akin has been engaged for some years, and it is not my intention to publish anything relating to it until the very promising arrangements which he has devised have had a sufficient trial'.[86] By the time this paper was published in November, Tyndall had accomplished it, and asked Stokes to ignore the fact that he had.

At the end of November, a letter from Akin innocuously asked Tyndall to present two portraits to the Athenaeum and a book to the Royal Institution. A couple of days later, Tyndall saw Akin's response in the *Philosophical Magazine* to his paper 'On luminous and obscure radiation'. He wrote: 'I have read Akin's article—it is vile'.[87] In his own paper, Tyndall had referenced Akin's discoveries, which had been presented at the British Association meeting in Newcastle, and described how he had independently come to the conclusion that obscure rays could be 'transmuted' into visible rays of shorter wavelength. Akin accused Tyndall of appropriating his ideas, perhaps by reading accounts of his presentation at the British Association in the *Athenaeum* or *Reader* and 'forgetting' that he had.[88] He also charged him with carrying on experiments in private when he had undertaken not to, publishing them, and failing to reference John Draper's earlier work on the link between obscure and visible heat as the temperature increased. He reproduced the text of a letter he had just sent to the President of the British Association, in which he lamented that fact that he was unable to carry out experiments he had planned for two years, with the help of George Griffith (a lecturer in science at Oxford) and the encouragement of the Association, such that Tyndall would now doubtless take the credit. Tyndall was shocked, as were his friends. Pollock declared: 'I hold him to be an unmitigated s—nd—l. When he wrote it he must have known from yourself that you had actually made the discovery in question'.[89] Tyndall prepared his defence, asking Debus to write up his recollection of a conversation they had had much earlier on the possible conversion of long waves into short ones, which Debus confirmed.[90] Tyndall spent some days drafting his reply to Akin, which he sent off on 17th December. Huxley, Pollock, and Hirst all read it, and considered it a settler.

Tyndall's reply to Akin, 'On the history of negative fluorescence', appeared in the *Philosophical Magazine* for January 1865.[91] He described how his own thoughts had developed in parallel with Akin's, and that he had already publically acknowledged Akin's priority of explaining that the heating of platinum in a hydrogen flame showed a change from a longer to a shorter wavelength. Tyndall initially called this 'negative fluorescence', a term coined in 1861 to describe this opposite of fluorescence. Akin, in 1863, had termed it 'calcescence', and his earlier work was republished in the *Philosophical Magazine* in the same issue as Tyndall's paper.[92] Tyndall criticized Akin's dearth of experiments. He stressed that his own conclusions were founded on experimental data, and excused missing the report of Akin's presentation to the British Association in the *Athenaeum* by claiming that he did not generally pay attention to newspaper reports of the meeting. Tyndall saw the problem being 'to raise the refrangibility of invisible rays of long period, so as to convert them into visible rays'. He claimed this could only be done with the invisible rays from strong sources, such as the Sun and electric light. This required separation of the luminous from the invisible rays, and concentration of the invisible rays themselves. He described how he had done this, and the information he had shared throughout, which should have allowed Akin to make the experiments that he had failed to do.

Holding back no longer, and deeming Akin's action in publishing his attack an unwarranted breach of trust, Tyndall sent in a short paper to the Royal Society on 13th January 'On the Invisible Radiation of the Electric Light', promising a fuller account in due course.[93] It was in this paper that he first used the term 'calorescence' (coined by James Challis) as a counterpart to 'fluorescence', and he reinforced it in a letter to the *Philosophical Magazine* published in February, effectively shutting out Akin, who was using the term 'calcescence'.[94] He started the demonstration of his results at the Royal Society on 19th January by showing that the point of maximum heat in the spectrum for the electric lamp was beyond the visible red. In a vivid metaphor he termed this peak 'the Matterhorn of heat' (Figure 25). It was much more prominent than the corresponding peak in the solar spectrum, which Tyndall put down to absorption of solar radiation by water vapour in the atmosphere. Filtering out the visible radiation by a solution of iodine in carbon disulphide, Tyndall showed that the invisible rays beyond the red could be focused to burn charcoal, wood, and metals, and even to light a cigar. But these were chemical actions, and Tyndall went on to demonstrate that platinum could be raised to incandescence by the invisible rays. For Tyndall this was true calorescence, the transmutation of rays from a longer to a shorter (visible) wavelength, the opposite of fluorescence. The following evening Tyndall demonstrated the same ideas to his Royal Institution audience, in a Discourse 'On Combustion by Invisible Rays'.[95] Hirst was there, sitting near the Busks, Lubbocks, and Ramsay. He commented that Tyndall gave Akin every credit he could with justice claim, and more. At the Discourse the Count de Paris, in the chair, lit a cigar at the invisible focus of the radiation. It became a party piece.

Akin responded in the next issue of the *Philosophical Magazine*. He accused Tyndall of breaking their agreement behind his back, carrying out experiments when he and Griffith had been given a grant by the British Association to prosecute them, and

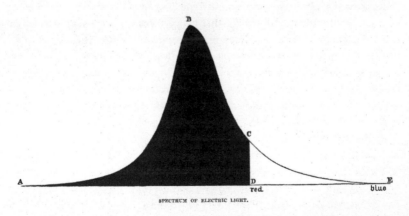

SPECTRUM OF ELECTRIC LIGHT.

Fig. 25. The 'Matterhorn' of heat: the spectrum of radiant heat, showing its maximum beyond the red.

suggested that Tyndall was driven by the prestige he would gain by being the first to show the direct transmutation of rays from a longer to a shorter wavelength.[96] Francis, as editor, regretted that these personal matters were being aired, though readers doubtless enjoyed them. He gave Tyndall the right to reply, then declared the subject closed.

Tyndall's reply came in the March issue.[97] He traced the whole history, and especially the separation of invisible from visible rays, to the time before Akin appeared. He wrote to Sabine and to William Armstrong,[98] President of the British Association that year, and received statements from both that the grant to Akin and Griffith could not be seen as preventing work by anyone else. Privately, Griffith told him that Akin was disagreeable to everyone to whom he was introduced and was a man 'morally incapable of working in concert with anybody'.[99] Tyndall's response effectively put an end to this episode. Francis reiterated that no further correspondence would be published, since Akin had started it. This is one dispute for which Tyndall does appear to be completely blameless.

11

❖⇒◉⇐❖

THE X-CLUB
1864–1866

W hile the Akin dispute rumbled on, a significant event in scientific politics took place. So significant that Tyndall made his first diary entry for two years to note it. On 3rd November 1864, the X-Club met for the first time, at St George's Hotel in Albemarle Street. Hooker and Huxley were the prime movers, as eight scientific friends, anxious to keep social contact in a busy world, attended their first dinner: Huxley, Tyndall, Hooker, Hirst, Frankland, Lubbock, Busk, and Spencer. At the second meeting in December, they were joined by the mathematician William Spottiswoode.

This informal gathering of rising stars became a powerful group over the next two decades. It met between October and May on the first Thursday of the month at 6pm, to enable members to go on to the Royal Society, and in June with wives in the country. All nine members of the X-Club were Fellows of the Royal Society except Spencer. He refused the honour on principle when it was offered, arguing that it should have been awarded to him earlier in life when it would have made a difference. All were, or became, members of the Athenaeum. Five became President of the British Association (Hirst and Frankland declined the honour), and three President of the Royal Society. No other members were ever appointed, through two were invited. First William Carpenter, physician and zoologist, and close friend of Huxley, and when he turned them down, James Fergusson, an architectural historian and friend of Hooker and Busk. Busk was deputed to invite him, but he too declined.[1] The X-Club was important to Tyndall. Despite his commitments and regular bouts of ill-health, he missed only sixteen of the first hundred meetings, and remained a steadfast member thereafter, for as long as he was well enough to attend.

Public life and club life were distinctly male affairs in Victorian society, but women were significant behind the scenes. The wives of the married X-Club members were a knowledgeable and characterful bunch, with whom Tyndall enjoyed lively relationships: Netty Huxley, who called him 'brother John', and relished arguing with him; Nelly Lubbock, with whom he seems to have fallen in love, so spirited and unconventional that the straitlaced Hirst destroyed her letters to him after her death; Frances Hooker; Ellen Busk; Lise Spottiswoode; and Sophie Frankland.

It was at the first meeting of the X-Club that Tyndall heard that he had been awarded the Royal Society's biennial Rumford Medal for his researches on the absorption and radiation of heat by gases and vapours. (Darwin was awarded the Copley Medal in a curious symmetry with the award of Royal Medals in 1853.)[2] Tyndall told Miss Roche: 'It consists of a lump of gold worth sixty pounds; and some separate and additional lumps of the same total value—So that if I should fall into poverty I can pledge the medal, and thus keep my head above water for a time'.[3] He received his Rumford Medal from Sabine at the Anniversary Meeting in November. On presenting the Copley Medal to Darwin, Sabine announced that the award specifically excluded *On the Origin of Species*. Huxley strongly objected to this omission, a consequence of the sensitivities engendered by Darwin's theory of evolution by natural selection. Curiously, Sabine made Hirst reply, though what he said is unknown.[4]

One of the first questions to occupy the X-Club was the future of the *Reader*, a weekly magazine on contemporary topics in science, religion, and the arts. It was founded in 1863 by Thomas Hughes—author of *Tom Brown's Schooldays*—and John Ludlow, both Christian Socialists, with the support of the publisher Macmillan. By 1864 the magazine was foundering. Huxley, through his contact with Norman Lockyer, who was providing much of the scientific content, was keen to see a successful vehicle for the wider dissemination of scientific ideas. He had written to Tyndall in April 1858, with the proposal, initiated by Maskelyne, to generate more scientific coverage in the *Saturday Review* following the difficulties with a previous venture of the *Scientific Review*. The editor had offered space and payment for a scientific article once a fortnight and for reviews. Huxley, Sylvester, Maskelyne, Tyndall, Frankland, Hooker, Ramsay, and Smyth were to form the 'corps d'elite', contributing at least one article in three months.[5] Tyndall was enthusiastic—'I think it may become a very important agency and exercise a salutary influence in this quackridden country'—but the idea did not last.[6]

Now Spencer took a hand. He raised the question of the *Reader*'s future, and the X-Club undertook to support it if there was a liberal editor. At a meeting in Lincoln's Inn Fields, it was agreed that the new regime would have the paper for £2,250 and issue thirty-five £100 shares. Tyndall invested £100—at this time he was reasonably flush with funds, despite just buying £100 of Natal Lands shares and £200 of London and South Western stock—and sent circulars to Herschel, Holland, and Bence Jones.[7] Holland (who became President of the Royal Institution early the next year when the

Duke of Northumberland died) and Bence Jones both invested, as did Huxley, Spencer, Lubbock, Spottiswoode, Darwin, Francis Galton, and John Stuart Mill. Herschel declined, declaring that he had no available funds.[8] Tyndall joined the editorial committee, on the understanding that Huxley would do all the work, and was soon aware that 'Spencer and Lockyer in their haste have committed us; but we must now face the problem'—of making it pay. Tyndall contributed some articles. In 'Vitality' he argued that all vital energy is ultimately derived from the Sun, and that all organic forces are the same as those of inorganic nature: there is no 'vital force'. He expressed here for the first time his belief that matter is 'essentially mystical and transcendental'.[9] In December he added an article on 'Science and the Spirits', rubbishing spiritual phenomena.[10] But by March 1865 the *Reader* was in difficulties. John Stuart Mill, 'a thin man, with clear decisive face' was with them for a meeting on 21st March, as Pollock, who had taken an editorial role, announced that he was retiring from it. Tyndall reflected: 'Altogether the affair has been a muddled one. No man knew what he was doing when he embarked in it'.[11] The paper was sold in August to Thomas Bendyshe, who removed the science section edited by Lockyer.[12] A useful outlet for the men of science disappeared.

Tyndall had continued his experiments on obscure heat before Christmas, but lecturing commitments now took priority. His series of eight evening Working Men's lectures on 'Electricity' started at the School of Mines on 28th December; he dined with Pollock and Lord Houghton, and they all walked to Jermyn Street to hear the first lecture. As these finished—to a very attentive and appreciative audience, according to Hirst—his Royal Institution lectures started, a series of twelve also on 'Electricity', scheduled on Tuesdays and Thursdays in January and February. Almost every weekend in January was spent at Lamas, the Lubbocks' house near Chislehurst, with them and their children Amy and Johnny. Tyndall's friendship with the Lubbocks was close at this time. Nelly Lubbock helped him with the proofs of *Heat*, and Amy Lubbock was one of his favourite children. He spent several days just before Christmas walking with Lubbock and Huxley in Derbyshire, from Matlock through Chatsworth to Baslow, over the moors to Castleford and the druid's circle, on to Buxton and back to Matlock. On New Year's Eve he had been at Lamas, finding the children joyous and beautiful. He returned to London on New Year's Day 1865 and joined the Huxleys as usual for dinner. They finishing by reading Shelley and Tennyson.

Tyndall was sleeping better, which caused him to observe: 'This dependence on sleep and its dependence on food tend to make me a materialist'.[13] His frequent insomnia may indeed have resulted partly from poor eating habits when he was under stressful work pressures. February found him visiting Carlyle with Pollock, at Lamas with Hirst and the Busks for Nelly Lubbock's birthday, and dining with Murchison, William Siemens, and the Spottiswoodes. For the first three weekends in March he was again at Lamas, though his joys in the place were not to last. On 14th June he went with Lubbock, who was seeking a political career, to a meeting at Bromley. A week later Lubbock's father was dead, and Lubbock became the 4th Baronet. Tyndall realized the days at Lamas might be over: 'they will probably quit Lamas & and go to High Elms. Change! Change! & Change!—The world is so ephemeral'.[14]

During the first part of March, Tyndall dined out repeatedly, but remarked in the middle of the month: 'This dining out does not suit me—I have had here a glut of it'. He suffered the continual tension between the demands of work and the enticements of society. Tyndall and Hirst planned a long walking trip in Ireland over Easter, but spent a few days first on the Isle of Wight, meeting at Southsea to give moral support to Wright, whose school was failing. Back in London, there was just time for dinner at Lubbock's, with Hirst, Lord Houghton, Huxley, and Spencer, before Hirst took Tyndall to Ireland. Tyndall was in very low spirits over the Akin affair, as Hirst recorded: 'It pains him to think that a stain will cling to his chivalrous action, and he writhes under the thought of a blot on his character, on the ground that he has come out of the conflict a gainer...His work has been utterly spoiled this year by this miserable episode'.[15] They travelled across to Dublin and on to Belfast.

A three-week trip followed, much of it on foot up to 20 miles a day, across the coast and countryside of the north of Ireland. Highlights included the Giant's Causeway, climbing Errigal—which Hirst described as 'the Matterhorn of Ireland'—and Slieve League. Hirst found this an exposed climb, commenting: 'This necessity for a strong stimulus in the shape of personal danger is quite characteristic in Tyndall. There is no doubt of it that this day has been of more service to him than any other during our journey'.[16] Tyndall's climbing, which included sea cliffs, was the subject of talk in the whole neighbourhood. The wild life was not safe from them. From a boat they blasted divers, gulls, rock pigeons, and other birds with guns. Hirst fired unsuccessfully at a seal. This is a curious episode. Twenty years earlier, after an evening shooting water hen, Tyndall had written: 'good sport tho' my conscience does not clear me for indulging in it'.[17] Later he argued strongly against cruelty to animals, in the context of support for the use of animals for medical research. Tyndall found himself refreshed by the trip on his return, though he had not felt it at the time: 'the food was bad and the drink far worse'.[18] He was back in London by 1st May, when he restarted his lectures at Jermyn Street after the Easter break, and made preparations to give the Rede Lecture in Cambridge.

Tyndall's invitation to give the Rede Lecture on 13th May signals his growing reputation.[19] He preceded all his contemporaries, such as Thomson, Ruskin, Tait, Maxwell,

Huxley, Galton, Lubbock, and Stokes. The previous arrangement of Rede lecture-ships had fallen into abeyance in the mid-1850s. Now most famous for C. P. Snow's centennial lecture in 1959 on the 'Two Cultures', the lecture had been re-established in 1859, as a one-off annual event, when Richard Owen gave it. Fred Pollock, the Pollocks' eldest son, was up at Trinity College. Tyndall dined with him and his parents in his room after a walk in the college gardens, 'the foliage massive and beautifully coloured, and the noble chestnuts crowded with their milk white cones'.[20] The following day, missing a 'very painful sermon' as Whewell buried his own wife, Tyndall dined in Trinity Hall as the guest of the Greek scholar Thomson, seeing the kindly geologist Adam Sedgwick, now over eighty. He was amused by the sudden disappearance of the 400 undergraduates 'like that of an Alpine mist' when they had finished their meal. On the Monday he dined with the Vice-Chancellor, Henry Cookson, opposite Airy, and woke on Tuesday with a 'clouded head'. A ramble into the countryside refreshed him, after which he strolled to the Senate House and stood outside until 2.30 in the afternoon before walking in. The hall was packed. Concerned about the acoustics and the potential liveliness of the undergraduates if he bored them, he spoke to the most distant person in the room. He need not have worried, as they heard him for an hour and a half, motionless, and silent apart from occasional bursts of applause, and cheers when he mentioned the parts played by the Cambridge figures of Stokes and Hopkins.

The lecture was a summary of all his work on radiant heat, including his striking analogy of the 'Matterhorn of heat', the peak of radiation beyond the visible red. For the students he emphasized the role of imagination in picturing atoms, molecules, vibrations, and waves, which eye has never seen nor ear heard, a theme he developed five years later in 'Scientific Use of the Imagination'. Sedgwick gave the vote of thanks, and the room rang with the clapping of hands and the thunder of boot heels and walking sticks. One listener wrote to him after the lecture to say: 'Your future biographer will epitomise you as the one man who ascended the Matterhorn, and smoked a cigar in the Senate House',[21] thereby anticipating Tyndall's ascent of the as yet unclimbed peak by three years. A few days after the lecture, the Vice-Chancellor wrote to say that the Senate had decided to offer him an Honorary LLD, to be awarded at the same time to the headmasters of Eton and Harrow. Tyndall went up on 30th May, walking down the street in a blaze of scarlet to the Senate House where, following a colonial bishop, he was 'doctored'. After a couple of games of croquet with the Vice-Chancellor and his wife, and an encounter with Carlyle, he returned to London.

Tyndall's experimental work continued until he left for the Alps in mid-July. He narrowly escaped serious injury on one occasion, upsetting burning alcohol on himself and leaving his leg and hand scorched. It was not enough to prevent his attending the first X-Club summer outing with wives. Only the Hookers were absent on this excursion from Paddington to Maidenhead with a trip on the river and to Burnham Beeches. Tyndall, Busk, and Lubbock climbed trees, while Huxley read from Tennyson's Œnone. Max Müller, the Oxford philologist who became a good friend of Tyndall, was invited to join them for dinner at the Orkney Arms.[22] The following year, with their growing wealth and sense of social position, they took a special carriage in the train from London to Maidenhead and back. Hirst revealed their nicknames, giving an insight into the characters of the group: the Xquisite Lubbock, Xcellent Spottiswoode, Xperienced Hooker, Xalted Huxley, Xcentric Tyndall, Xemplary Busk, Xpert Frankland, Xtravagant Hirst, and Xhaustive Spencer, the only one missing from the second trip.

* * *

Back in England after the 1865 Alpine season, Tyndall went to Birmingham for the last part of the British Association meeting. He nearly encountered Forbes: 'I had one long look at Principal Forbes there, and could have shaken hands with him had the thing been proposed. I strike when struck, but I cannot hate'.[23] Perhaps not, but he could certainly be self-righteous. He was invited to the Martineaus, and went on to Lord Wrottesley's, where he met Armstrong and Grove—the incoming President—and enjoyed some croquet. Forbes was not his only concern with respect to glaciers. He found himself defending his previous work on glacier theory against Helmholtz in the pages of the *Philosophical Magazine*.[24] Helmholtz supported Tyndall's view of the importance of pressure and regelation in forming glaciers, and of fracture and regelation in explaining their motion, rather than liquefaction by pressure and refreezing as James Thomson claimed. But he differed from Tyndall in the cause of regelation, stating that it was due to pressure only. Tyndall countered that pieces of ice even just touching in warm water will freeze together. He argued that the true cause was the promotion of crystallization of water on the surface of the ice, just as a salt crystal promotes the crystallization of salt from its solution. He was glad to find that Clausius agreed.[25]

The honour of a Cambridge LLD was followed in the autumn by an attractive invitation from Oxford. On the death of the incumbent, Robert Walker, and urged by colleagues including Sylvester,[26] Tyndall was approached to become Professor of Experimental Philosophy. It was a singular offer for someone with his known religious outlook. He went to Oxford to discuss it. But as Hirst observed the post required six-months residence and Tyndall would not do that. He was tempted nevertheless, writing: 'It might have been well had I done so. Still in declining it I was guided by the best light of my mind and the best feeling in my heart'.[27] The pull of the Royal Institution and of London friends was too strong, as he revealed to Juliet Pollock: 'I came to the

Rubicon and found its waters cold; I looked back and saw the warm homesteads among which I have spent the last dozen years of my life. And at the door of one of them I saw a face which said silently "don't cross" '.[28]

Tyndall's work on infrared radiation ('obscure heat') was now coming to fruition. After dinner at the Philosophical Club on 23rd November, he performed experiments at the Royal Society, showing the 'transmutation' of the invisible heat rays from the electric light, lime light, and the Sun into the shorter wavelengths of visible light, while the secretary read his paper 'On Calorescence'.[29] This was one of two new papers on calorescence that Tyndall prepared in the autumn. It was his definitive statement about the invisible radiation from the electric light, its relative intensity compared to visible radiation, the means of separating it from visible rays, and its power to cause substances to glow and burn.

His second paper, 'On the Influence of Colour and Mechanical Condition on Radiant Heat', which became the Sixth Memoir, was read at the Royal Society on 18th January. Tyndall had submitted it a few days before Christmas, clearing space to give the Christmas Lectures on 'Sound' (on which he was also planning a book).[30] Here he rebutted two misconceptions, 'correcting a great number of prevalent errors': that dark colours were the best absorbers of heat and light colours the worst, and that the physical condition of a body was more important than its chemical constitution in determining its ability to absorb heat. He found, contrary to the belief at the time, that the radiative power of powders is affected by their chemical constitution, and he showed that others had been deceived by the absorptive power of the varnish present in all of them as a fixative. As with gases and liquids, he demonstrated again the reciprocity of radiation and absorption. He wrote later in the introduction to this Memoir in *Molecular Physics in the Domain of Radiant Heat*: 'throughout these memoirs radiation and absorption have been looked upon as the action of atoms and molecules; chemical constitution rather than physical condition, being regarded as the really potent agency'. The realization that colour alone was not a predictor of the absorption of heat led Tyndall to question the accuracy of the black bulb thermometer, because the black glass might be transparent to invisible radiation, especially at high altitudes— for example in balloons—where it had not been intercepted by water vapour.[31]

This reference to balloons may explain an approach Tyndall made to the aeronaut Henry Coxwell to take him up in one. At the British Association meeting in Leeds in 1858, Tyndall had been appointed to a Balloon Committee, with a sum of £200 at its disposal, to consider ascents for scientific purposes. He had volunteered to make a balloon ascent in 1859 but was unable to attend on the day, although bad weather anyway prevented the flight. Coxwell was employed to make the ascents, and the proceedings of the 1862 British Association meeting carry an extraordinary description of a flight by him and James Glaisher, supposedly to a height of 36,000 or 37,000ft. At various points on the ascent, from Wolverhampton, they released pigeons and observed their behaviour. Some flew off in surprise; others apparently died and plummeted to Earth. Glaisher passed out at an estimated height of 29,000ft, the height of Mount Everest, but recovered and reported that Coxwell had lost the use of his hands, which had turned black. He poured brandy over them, and then had the presence of mind to pull the valve-cord with his teeth. The balloon dropped 19,000ft in fifteen minutes, and landed safely near Ludlow. The Balloon Committee continued in place until 1867 and Tyndall still wished to make an ascent, probably for a high-altitude experiment on radiation. He wrote to Coxwell, who proposed a fee of 50 guineas for a flight from Munich or Geneva, but it does not appear that the ascent was ever made.[32]

From the latter half of January until the end of June, in the typical pattern, Tyndall's life was driven by lecturing commitments. The day after his Royal Society paper he gave the opening Discourse of 1866 at the Royal Institution on the same topic.[33] A stellar audience was present, including the Prime Minister Earl Russell (previously Lord John Russell), William Gladstone (then Chancellor of the Exchequer), and Carlyle. The assembly was treated at the end of the Discourse to a plea about the influence of science as a means of intellectual culture, and the defects of any system of education in which the study of nature is neglected or ignored. Tyndall dined beforehand with Sir Henry Holland, with guests including Russell and Gladstone. Four days later his series of ten lectures on 'Heat' started at the Royal Institution, two each week in January and February. Once those had finished, he restarted his experiments, while giving evening Working Men's lectures on 'Sound' at the School of Mines until mid-March.

A break in lecturing followed until May, which he filled with a combination of socializing, laboratory work, official business with the Commission on Heating, Lighting and Ventilation of the South Kensington Museum, and public arguments about the power of prayer. The socializing included seeing Carlyle at Countess Russell's, where he also found Emily Drummond and her mother. He longed to talk to the 'little girl', who he believed 'once liked me'.[34] A few weeks later, Tyndall took it upon himself to accompany Carlyle to Edinburgh.[35] He had come to see himself as Carlyle's 'lieutenant', calling Carlyle 'General', in an allusion straight out of *On Heroes*. It was a very different relationship from that with his other hero, Faraday, whom he saw as a father-figure. When they were in the laboratory together, Faraday had once said: "Tyndall, you and I are brothers and will talk together as such'. Tyndall had replied: 'Let me be your son'.[36]

In 1865 Carlyle had been appointed Rector of Edinburgh University, his alma mater, elected by the students in a contest with Disraeli, after Gladstone had retired from the office. His installation was to take place on 2nd April 1866. Tyndall escorted him to Kings Cross for the train to Yorkshire, where they were met by Lord Houghton and taken to Fryston Hall to break the journey. Carlyle had a bad night, disturbed by the noise of distant trains, and wished to leave. So Tyndall took him riding as a distraction, which invigorated him. He repeated the medicine the following day. The refreshed Carlyle, with Tyndall, Ramsay, Huxley, and Hirst, set off by train from York to Edinburgh, where Carlyle was to stay with his brother, and Tyndall with Sir James and Lady Coxe. James Coxe had been appointed 'Lunacy Commissioner' in 1857 (Tyndall noted the salary of £1,000–£1,200 per annum), and knighted in 1863; he was a phrenologist who campaigned against restraint in asylums and advocated the training of women in medicine.

Easter Sunday found Tyndall at lunch with Playfair, Ramsay, Hirst, and Miss Chambers, daughter of Robert Chambers, 'the reputed Author of the *Vestiges of Creation*', as Hirst (correctly) described him. On the Monday, Huxley, Ramsay, and Tyndall were 'doctored' in alphabetical order. Then Carlyle gave his address, a stirring oration 'On the Choice of Books' for an hour and a half without notes, to huge applause. According to Moncure Conway, Tyndall, whose emotional responses were never far from the surface, had tears in his eyes.[37] The students mobbed Carlyle in the street afterwards. Tyndall telegraphed to an anxious Jane Carlyle: 'A perfect triumph'.[38] She was thrilled, and took the news to John Forster's birthday dinner, sharing it with Charles Dickens and Wilkie Collins. Carlyle was appreciative. He wrote to Jane: 'Tyndall's conduct to me has been Loyalty's own self; no loving son could have more faithfully watched a decrepit father'.[39]

At the Royal Society of Edinburgh later, Tyndall saw Tait for the first time: 'he appeared much older than I had imagined him to be; but though bald he is I am told young...He would not join the dinner because I was there'.[40] He learnt that Forbes had said 'a rigid no thank you' when Playfair asked if he might introduce him. It was the last chance for them to meet. On 5th January 1869 came news via the *Pall Mall Gazette* of the death of Forbes on 31st December. Tyndall confided to his journal: 'I have often wished to and thought of writing to Forbes; but I was told he was so cold. And now he is beyond the reach of my writing'.[41] As events were to prove, his death did not bring to an end the disputes over glaciers.

Tyndall stayed in Edinburgh until the end of the week, based with the Coxes at Kinellan. He called on Sir David and Lady Brewster, and strolled round Edinburgh with Ramsay and Hirst. He returned south for a weekend at High Elms, the house (near Darwin's Down House) to which the Lubbocks had moved, where he stood godfather to little Rolfe Lubbock alongside Lord Cranworth, the Lord Chancellor. Hirst was in the church 'for the first time in I will not say how long',[42] after rescuing

Tyndall the previous day from Nevile Lubbock's beautiful mare Kate, a fiery coquet-tish little animal who 'teased' Tyndall so much that he swapped her for the steady old hunter that Hirst was riding.

Du Bois-Reymond was over in London for a long visit, which included giving a Discourse in April on the speed of nerve transmission, based on the work of Helmholtz and others. Tyndall invited him to dinner with Hirst at the Athenaeum and went with him for several days to the Isle of Wight. They dropped in on Tennyson and climbed the cliffs of Scratchell's Bay. At Southampton, as he returned, he was stunned to pick up a copy of the *Times* and learn that Jane Carlyle had died suddenly in her carriage two days earlier. Carlyle was still in Edinburgh, and rushed south to take her body back up to Haddington, breaking down repeatedly as he described to Tyndall her goodness, nobleness, and loyalty to him. Their relationship throughout had been tempestuous, and it was only on her death that he realized how badly he had treated her. Geraldine Jewsbury, her friend of many years in a further tempestu-ous relationship, sent a letter to Tyndall the day after her death to say how delighted she had been to receive his telegram of Carlyle's Edinburgh success.[43]

Tyndall's duties at the Royal School of Mines started on 1st May: thirty-two lec-tures on 'Physics' to a cohort of sixteen students every weekday at 2pm, finishing on 21st June. Once the lectures were underway, and while he continued to work on his book on *Sound*,[44] the social events took a back seat. One reason was a further episode in the long-running dispute with Magnus, who wrote warmly to warn Tyndall of his forthcoming paper on dew, not wanting publication of conclusions at variance with Tyndall to affect their friendship.[45] In this paper Magnus reiterated his view that transparent aqueous vapour was not a good absorber of heat. He claimed that absorption was due to the presence of 'nebulous vapours'.[46] Tyndall, unhappy about the implications for understanding meteorology if the conclusion went unchal-lenged, responded just before he left for the Alps, pointing out the likely sources of error in Magnus's experimental arrangement.[47] He also responded to a question from Lyell as to whether there might be a difference in radiation from the oceans and dry land that might help explain glacial epochs.[48] Tyndall thought any difference would be insignificant.[49]

As his Royal School of Mines lectures began, Tyndall found time for experi-ments on vibrating strings.[50] They may have been stimulated by ideas emerging as he wrote his book on sound, and from Helmholtz's recent work. Bence Jones had also asked if Tyndall would offer an extra Discourse, and he gave it on 15th June as 'Experiments on the vibration of strings'.[51] Hirst was impressed by the experi-ments, but considered the lecture itself less successful. It was simply a demonstra-tion of the factors influencing the rate of vibration of strings: length, diameter, tension, and density of material. He also thought Tyndall was playing to the gallery. Tyndall had been 'disturbed in the day by aristocratic friends desirous of being present at a rehearsal and to please the ladies experiments were repeated too much in showy ways, strings vibrating in blue, red, green, purple light etc'.[52] But the aristocracy were impressed. Tyndall gave a private showing later to a distinguished small audience including the Duke and Duchess of Argyll, Lord and Lady Belper, and Lady Lubbock, after which Tyndall and Hirst went to Lords to watch Alfred Lubbock playing cricket.[53]

Tyndall left for the Alps on 5th July 1866, without Tennyson whom he had hoped to inveigle out.[54] He planned to make a detour to Le Havre to inspect the lighthouse, so he took the train from Fenchurch Street to Blackwall to board the Trinity yacht, but a gale forced them to take shelter with other vessels in the Downs and forestalled the visit. Tyndall caught the boat from Dover instead and travelled on to Paris, where he bought some instruments and looked up friends.[55] In the Alps the weather was dreadful, and he climbed little.[56] Bence Jones gave him news from England: Faraday was weak but slightly better (he was now nearly seventy-five years old); there had been rioting in London as the Reform League campaigned for male suffrage.[57] Bence Jones was looking forward to getting down to Folkestone where he hoped Tyndall would join him, away from the cholera in the capital.

Tyndall returned to the news of the unexpected death in Brazil of Ginty, the great friend of his surveying days.[58] Margaret Ginty, with two daughters and three sons to look after, took the death badly. Tyndall wrote in uncompromising fashion to exhort her to pull herself together.[59] For the next twenty years he acted as her Trustee, and looked out for the education and employment of her children. He also found not only that Faraday's health was failing, but that Bence Jones felt that he too might no longer be able to carry out his duties. Bence Jones asked Tyndall to discuss plans with Sir Henry Holland, and suggested that John Gladstone might take on the lecture arrangements, or at least someone to whom Tyndall could speak plainly when he became the head of the house, as was anticipated on Faraday's eventual death.[60] As Faraday declined, William Thomson, who had brought the initial mathematical expression to Faraday's field theory, grew in recognition. His contribution to the success of the Atlantic cable engendered a knighthood that Tyndall did not begrudge. He told Clausius: 'He well deserves it, and I have no doubt he likes it'.[61] The old order was changing.

12

EYRE AFFAIR AND DEATH
OF FARADAY

1866–1868

The autumn of 1866 was an unsettling time for Tyndall. One event above all reveals his moral and political sympathy with Carlyle, against that of liberals such as John Stuart Mill. A rebellion in Jamaica in 1865 had divided public opinion in Britain when Governor Edward Eyre, fearing a revolt on the scale of the revolution in Haiti that had been led by Toussaint Louverture, declared martial law, and sent in the army.[1] Several hundred black Jamaicans were killed directly by soldiers, or by execution. Most prominent among those executed was George Gordon, a former slave and Jamaican politician who had played no direct part in the insurrection. His summary trial and execution under martial law, and the extended and savage nature of the response of the authorities, generated fierce debate in Britain. Prominent MPs and others active in the abolitionist movement, including Henry Fawcett, raised their concerns about the legality of the execution and the harshness of the reprisals. Eyre was suspended in December, and a Royal Commission of Enquiry established to investigate his actions. At the same time, a Jamaica Committee was set up to maintain the pressure on the government, initially under the chairmanship of Charles Buxton MP and then of John Stuart Mill.

The report of the Enquiry, completed in April 1866 and presented to Parliament in June, praised Eyre for the 'skill, promptitude and vigour which he manifested in the early stages of the insurrection', but declared the punishments excessive. The Jamaica Assembly passed an Act indemnifying him, but Eyre was removed from his post and retired without a pension, which the government hoped would quell criticism. But the Jamaica Committee had other ideas, and sought to bring a private prosecution for murder against him. The ultimate purpose, from Mill's perspective, was to uphold the due process of law for all citizens. Opinion split on this constitutional and moral point: did black Jamaican citizens have equal rights to whites under the law, or did the threat of rebellion justify violent suppression by a strong ruler?

The figurehead for the justification of Eyre's behaviour and of the Eyre Defence Committee was Thomas Carlyle, proselytizer for authoritarian figures and author in 1849 of *The Nigger Question*, a racist characterization of former Jamaican slaves as inherently inferior, lazy, pumpkin-eating 'Quashees' who needed to be whipped to work. Tyndall had read it in 1853, before he knew Carlyle. He wrote then in his journal: 'Of course it is all right. This is because it expresses the facts of nature'.[2] From a scientific perspective, black Jamaicans were seen as inferior to whites, effectively 'savages'. Nevertheless, this strong strand of racism did not determine the side on which individuals came down. Huxley shared Tyndall's 'scientific' belief in the superiority of European whites over the 'savages', and of men over women, based particularly on a consideration of their relative brain sizes.[3] Yet he took the side of the liberals—as did Frankland, Spencer, Lubbock, Darwin, and Lyell—the only significant issue apart from the American Civil War on which he and Tyndall ever differed (Tyndall had Confederate sympathies). Tyndall's position was shared by Hirst, De la Rue, Tennyson, Ruskin, Dickens, and Nelly Lubbock (in opposition to her husband).[4] Hooker did not take a partisan position in public but wrote privately. Tyndall 'rejoiced to find that you are not against me'.[5] Hooker's view was a mixture of moral principle and belief in the inferiority of the black population: 'I am as averse as anyone to bloodshed...I regard the negro's life as sacred as a pheasant's, or yours, or mine. I know no distinction but in degree, but when you or I, or the negro, or pheasant, become dangerous to law, order, and the general weal they all are subject to the same nemesis; and that the Negro in Jamaica and even in the free towns of W. Africa is pestilential I have no hesitation in declaring, nor that he is a most dangerous savage at the best'.[6]

Tyndall sent two guineas to Hamilton Hume, the honorary secretary of the Eyre Defence Fund, on 20th September, and made public his support.[7] The Jamaica Committee wrote an open letter to remonstrate with him, but Tyndall argued that one should not judge Eyre with hindsight.[8] According to Tyndall, Eyre faced this challenge: 'How, with the force at his disposal, to preserve the lives of 7,000 British men, and the honour of 7,000 British women from the murder and the lust of black savages capable of the deeds which history shows to be theirs in St. Domingo; and a re-enactment of which had, to all appearance, begun under the said Governor's eyes'.[9] In the Carlylean sense, Might was here allied to Right, and the decision taken by the hero was a moral one. Tyndall had chosen his position, as ever with a clear moral sense. But that did not make it comfortable. Hirst noted his long walks to recover from the 'rupture of equilibrium' as he took sides, and found himself in opposition to some close friends.

In the midst of the furore, Tyndall made a brief visit to Ireland, and he attended the X-Club dinner at the beginning of November, at which the opposed positions on Eyre would have been evident. The most difficult rupture was with Huxley, and Tyndall wrote to express his pain at finding himself in opposition.[10] Huxley replied, observing that their fundamentally different political principles had become obvious during the progress of the American War. He wrote that 'there is nothing for it but for us to agree to differ, each supporting his own side to the best of his ability, and respecting his friend's freedom as he would his own and doing his best to remove all petty bitterness from that which is at bottom one of the most important constitutional battles in which Englishmen have for many years been engaged'.[11]

The first prosecution of Eyre took place in March 1867 at Market Drayton, Shropshire, but he was discharged by a local grand jury. Tyndall played a visible role in the weeks before the trial. A meeting of the Eyre Defence and Aid Fund took place in Willis's Rooms on 18th February. Tyndall proposed a resolution for the application of the fund in Eyre's defence, prefaced by an address in which he quoted verbatim Hooker's description of the characteristics of negroes. According to Tyndall this was not a question of colour, but of the attributes of those with colour: 'I do not object to black. I rather like it; but I accept black as indicative of other associated qualities of infinitely greater importance than colour'. Tyndall and others did not see this as prejudice; it was simply the way things were. He also dismissed the argument, as the Jamaica Committee was claiming, that a precedent had been set for the treatment of Englishmen: 'We do not hold an Englishman and a Jamaican negro to be convertible terms, nor do we think the cause of human liberty will be promoted by any attempt to make them so'.[12] He bolstered his argument by referring to Carlyle's earlier call for the anti-slavery movement to treat with moral urgency the dreadful circumstances of starving workers in England, who in his view suffered worse conditions than slaves. It was a stirring speech in which Tyndall, calling on the necessary defence of the virtues of English wives and daughters, reiterated his view that Eyre's actions had prevented far worse, even if he might have made legal errors and his subordinates committed excesses. Tyndall argued that this was not a party political issue since Liberals as well as Tories supported Eyre. He made his own politics clear, having heard himself called a Tory for refusing to denounce the conduct of Eyre: 'In fact my leanings are quite the other way. When asked for example, prior to the late general election if I had a vote for Westminster my reply was "No! but if I had a hundred votes I would give them all to John Stuart Mill"'.

Through the efforts of the Jamaica Committee, Eyre was brought to trial again in May 1868. The presiding judge dismissed the case on the precedent of Nelson and Brand, Eyre's army officers, whose prosecution had also failed despite a ringing address from Sir Alexander Cockburn, Lord Chief Justice of England. He had argued that they should be indicted for Gordon's murder and that martial law was inimical to English constitutional principles. The Committee tried for a third and final time under a different Act, on the basis of the unconstitutionality of martial law, but failed again when the case came to a grand jury in June. Mill lost his Parliamentary seat in the election of 1868, owing in part to opposition by supporters of Eyre. In 1872 the Gladstone government compensated Eyre for his legal expenses, fulfilling a promise made unofficially by the previous Conservative administration. Finally, in 1874, the Conservatives reinstated Eyre's pension, at a reduced figure.

* * *

In parallel with the first stage of the Eyre affair, after his return from the Continent and then from the British Association in Nottingham, Tyndall worked hard on a lecture to the Edinburgh Philosophical Institution on miracles. Hirst thought the pressure was damaging Tyndall's health, so before the autumn season started—with the X-Club meeting at the beginning of October and a weekend at Spottiswoode's 'lordly country house'—Tyndall and Hirst spent time in the familiar landscape of the Isle of Wight. They walked the 24 miles from Ventnor to Freshwater Gate, explored the Needles, Scratchell's, and Alum Bays, and returned via Cowes and a walk to Carisbrooke and Brading before taking the train to Ryde and home. In the event, the lecture on miracles was deemed too controversial, and Tyndall lectured instead on 'Sonorous Vibrations'.[13]

Just before Christmas the 'impetuous' Tyndall, as Carlyle fondly described him to Ruskin,[14] took Carlyle away to Lady Ashburton's house in Mentone. Ever solicitous, he persuaded Murchison to write to the ambassador so that he could bring Carlyle's tobacco and brandy out without being interfered with by Customs.[15] They spent a night in Paris, joined next morning by the kindly Jamin, and arrived in Mentone at 2am on 23rd December. He wrote to Ellen Busk: 'Mentone is charming and Lady Ashburton is charming and her little daughter is charming and her brandy was charming and the mountains and sea are charming and the orange trees loaded with fruit are charming. On Xmas day Carlyle astride a donkey reached a height of 2,000 feet accompanied by Lady A and her daughter... They returned and I went on to the top of the highest Aiguille whence I had a most glorious view of Alps and sea'.[16] Tyndall was driven to Monaco to catch the steamer for Nice and the long journey to Paris, with a horrible night passage from Calais to Dover. He was back by the end of December, walking with Hirst and Debus round Chelsea before the year was out. Lady Ashburton wrote: 'Mentone is very dull without you'.

Sonorous vibrations became the theme of early 1867. They underpinned, in time, a whole programme of research on coastal sound signalling and discoveries in meteorology. The lecture round commenced at the Royal Institution, starting with a Discourse on 18th January 'On sounding and sensitive flames'.[17] Tyndall demonstrated the notes which could be obtained by flames in open tubes of different lengths, which rise in pitch as the tube diminishes in length. Flames, silent initially, could be caused to 'burst into song' by a siren (a mechanical device driving air under pressure through holes), a human voice, a clap of the hand, or another flame. In a bravura performance, Tyndall used different burners and tubes, whistles, and anvils to cause flames to jump and emit sounds, transforming them from short, forked, and brilliant to long and smoky, and vice-versa. The *Philosophical Magazine* published a beautifully illustrated version of Tyndall's sixth lecture on Sound, soon also to appear in his book, as 'On the action of sonorous vibrations on gaseous and liquid jets', thus extending his Discourse to include water jets.[18] The book *Sound*, for which Clausius in particular had given helpful comments, appeared at the beginning of July just before Tyndall left for the Alps.[19] He dedicated it to Richard Dawes (who had died in March), presented a signed copy to the Athenaeum, as he did with many other books, and sent copies to friends. Magnus much appreciated the gift, and Weber particularly valued Tyndall's recognition of the work of Chladni, whom he knew.[20] The publisher Gauthier-Villars wrote to finalize a deal for publishing the work in France, with a translation by Abbé Moigno.[21] Sound was also the theme of Tyndall's Royal Institution series of twelve lectures 'On Vibratory Motion with Special Reference to Sound', which started on 22nd January and continued every Tuesday and Thursday until the end of February.

The X-Club meeting on 7th February was an unusual one, as the forceful Nelly Lubbock held court at Price's Hotel in Dover Street to the X-Club members and guests, Armstrong and Spedding. Hirst was scandalized when she ordered dinner for them in a spirited manner: 'It was an unauthorised act of interference which was as successful in its results as it was audacious in its character. None of my acquaintances except Lady Lubbock could have done it'.[22] Tyndall's speech to the Eyre Defence and Aid Fund on 18th February had no discernible impact on his relationship with Lubbock, a firm supporter of the opposing Jamaica Committee, who also had the opposition of his wife to deal with. Tyndall was at High Elms for both the following weekends. The Lubbocks' country house, like the Spottiswoodes' at Combe Bank near Sevenoaks, was a haven for the X-Club and their friends.

The terrier-like Magnus intruded again in March, writing to Tyndall to give friendly notice of a paper in which he claimed to have shown that adhesion of water to the walls of the tube was responsible for the difference in their results.[23] He had been working away for nearly a year to counter Tyndall's further objections to his assertion that water vapour was not a strong absorber or emitter of radiant heat. Tyndall wrote to ask for details so he could repeat the experiments. Magnus responded, hoping they might meet shortly in Paris during the Second Paris International Exposition.[24] Tyndall used his relationship with Francis to have a letter inserted with Magnus's paper in the *Philosophical Magazine* expressing his commitment to continuing to meet Magnus's objections with 'clear and irrefragable facts', as he had done by going to the Isle of Wight in a south-west wind to get pure air when Magnus suggested that London air was causing his results.[25] The respect in which Tyndall nevertheless held Magnus is evident in the pride of place he gave to a photograph of Magnus in his room. The pictures of Magnus and Faraday stood over his chimney piece with Hector Tyndale's between. Immediately below was 'a crayon drawing of one of the most beautiful children in the world: a little maiden whom I know and love [it was probably Amy Harriet Lubbock].[26] The mother of the child—a glorious woman—knowing my feeling drew the portrait herself for me'.[27] The Second Paris International Exposition, which had been initiated by Napoleon III to rival the Second London International Exhibition of 1862, opened on 1st April 1867. Tyndall and Hirst went out over the Easter period, taking a long walk on Easter Sunday through the Forest of St Germain en Laye, lunching at Maison Lafitte, and dining at Champean's. Tyndall was the juror for Class 24, Apparatus and Processes for Heating and Lighting,[28] and was only able to see Magnus briefly before he returned to London, not long enough to resolve anything.

Back in London, Tyndall was due to give a course of thirty-two lectures on 'Physics' at the Royal School of Mines every weekday, in addition to the Working Men's evening lectures. But the start of the course was interrupted as Tyndall learnt of his mother's death. Hirst recorded that the unexpected news 'broke him down utterly…abandoning his habitual reticence on such subjects he spoke with earnestness and deep affection for her rare qualities'.[29] He left the next day for Ireland, returning a few days later in low spirits. He appointed Elizabeth Steuart as his liquidator to settle the estate, and lodged Emma under the care of Dr John Tyndall, a relative he had discovered in Gorey. Faraday's health was worsening, and at the request of Holland, Tyndall gave a Discourse on 21st June 'On some experiments of Faraday, Biot, and Savart'.[30] He concentrated on Faraday's demonstration of the rotation of polarized light by a magnet, of Biot's demonstration of the effect on light of the vibration of a piece of glass through which it passed, and of Savart's on the effect of sound on a water jet.

* * *

Tyndall returned early from his Alpine break in 1867, after encountering poor weather. He started planning for the British Association meeting in Dundee, which included a special inaugural lecture to the working men. The lecture illustrates Tyndall's reputation and commitment to widening education, a commitment that extended beyond his lectures at the School of Mines to supporting the growing movement for Working Men's and Women's Colleges. Early the following year, Tyndall and Lubbock were on the Council when the South London Working Men's College was established with Huxley as the Principal.[31] Tyndall used his relationship with Hirst, now General Secretary of the British Association, to optimize his visit to Dundee, writing: 'I hope the lecture will be got over early...I love the Highlands better than the Sections. It would also suit me if the Report of Schools were introduced to the General Committee at an early sitting. Three days are quite enough to demonstrate my loyalty to the Association'.[32]

A few days after he wrote this, Faraday died, sitting in his chair on an August Sunday morning 'as peacefully as if he had merely fallen asleep'.[33] As a Sandemanian, the funeral was private to the sect and none of his scientific friends was able to accompany him to the grave. Tyndall had been Faraday's designated successor as Superintendent of the House for many years, and had attended Managers' meetings for a couple of years. He now took up the mantle. In addition, he effectively inherited Faraday's responsibilities as scientific adviser to the Board of Trade and Trinity House on lighthouse matters; responsibilities that would later loom large.

Tyndall arrived in Dundee on 4th September for the opening of the British Association meeting, and stayed at the Royal Hotel with a large party including Lady Colville, Jane Strachey, Spottiswoode, Lubbock, Busk, and Hirst. He lectured to more than 2,000 of the working men of Dundee on 'Matter and Force', a great success for this initiative. Nevertheless, he stoked early controversy by avowing that the physical philosopher must be a 'pure materialist', arguing in vivid phrases: 'Depend upon it, if a chemist, by bringing the proper materials together in a retort or crucible, could make a baby, he would do it. There is no law, moral or physical, forbidding him to do it'.[34] It was a remarkable public declaration, given the dangerous associations of materialism with atheism, though he qualified it by saying that there were questions of existence beyond the powers of science to elucidate. Despite this startling statement, which elicited muted responses compared to his Belfast Address seven years later, hatchets were buried with the religious Scots, temporarily at least. Thomson met him in the Kinnaird Hall, 'blocked my passage, smiled and stretched out his hand'.[35] He shook hands with Tait afterwards at St Andrews and smoked and drank beer with them both. He was impressed by Tait's capacity for alcohol (and he was partial to drink himself). As he later wrote to him: 'I would give a thousand pounds tomorrow for your capacity to smoke & drink beer. It is simply the want of that power which prevents me being a real worker'.[36] Huxley was not present at this meeting but was glad at the reconciliation: 'It might be better for me if the like could happen with Owen, but it can't happen, for the man is base, and I could not trust him if I would'.[37]

Afterwards Tyndall roamed through the Highlands for several days, up the Pass of Killiecrankie, visiting the Parallel Roads of Glen Roy, to Fort William, past Glencoe and on by Oban, Arran, and the Kyle of Bute to Glasgow and Edinburgh. He finished with several glorious days in Northumberland at Cragside, the seat of Sir William, 'the great gun-man', and Lady Armstrong. In his element, Tyndall was 'every day upon the moors wandering knee deep in heather & indulging at times in a little climbing; for the moorland is bastioned by fine crags'.[38] He travelled back to London with Armstrong on 25th September.

One consequence of Faraday's death was Tyndall's refusal of an offer to go to India for two years to investigate aspects of its meteorology, which in other circumstances it seems he would have been glad to accept. He had been approached at the instigation of Sir Stafford Northcote, and he recommended to the India Office that Colonel Richard Strachey, already in India, be asked instead to lead the work.[39] By chance he met Northcote in a railway carriage, both bound for the same social event, and was able to express his views personally. Strachey's wife Jane, the authoress and later supporter of women's suffrage, had been with him and her friend Lady Colville in Dundee. He saw her just before she left for India at a dinner at Sir James Colville's, and gave her a letter for her husband in which he expressed his regret.[40] Holland was particularly grateful that he had declined the invitation: 'My ambition for you as well as for the R. Institution has been that you should be considered as the direct successor of Faraday, taking his place in all ways. It is no mean succession, that of Davy and Faraday. But it was a succession which would have been broken even by a year's residence in India'.[41]

Faraday's legacy now demanded Tyndall's attention. He had visited the ailing Bence Jones in Folkestone and been persuaded to give two special Discourses in Faraday's memory. From the beginning of November he vowed not to go out except to fulfil existing commitments, because of the demands of the Faraday Discourses and preparation for the Christmas Lectures, which it was his turn to give this year on 'Heat and Cold'.[42] He did attend the X-Club as usual on 7th November, when Hirst resigned the Treasuryship to Spottiswoode, and later in November spoke at Sion College with Huxley, at the inaugural lecture for the use of this hall for the benefit of the clergy of London.[43] For much of November Tyndall worked on his lectures and a book, *Faraday as a Discoverer*, writing to people such as Herschel, De la Rive, Clausius, Magnus, and even Tait for information about Faraday's life. He asked Helmholtz and Clausius for advice on Tait's comments, to ensure that he gave Thomson the proper credit.[44] The task prevented him from accepting social invitations, including from Lady Emily Peel, wife of Sir Robert Peel (son of the Prime Minister he had 'memorialised'), whom he was getting to know.[45]

The third edition of *Heat* was published during the month, and he was delighted with the German edition: 'It looks far better than the English edition and leaves the French nowhere'. He sent Anna Helmholtz a brooch in thanks, which much pleased her.[46] While at the Paris Exposition, he had been told by the Duke of Leuchlenberg, the nephew of the Emperor of Russia, that the book had been translated more than once into Russian. Indeed, Tyndall's *Heat* makes an appearance in Tolstoy's *Anna Karenina*.[47] To Juliet Pollock he wrote that it 'ran like wildfire through America, and both in Italy and France its sale has been very great. These books reach as I have said tens of thousands: is this a thing to be despised?'.[48] He wrote to Helmholtz to ask if a German translation of *Sound* would interfere with Helmholtz's *Tonempfindungen* and sought his advice on interest in Germany in a translation of the book on Faraday on which he was working.[49] Vieweg wrote to offer to publish *Sound* in German sight unseen, and suggested that Helmholtz and Wiedemann again look after the translation.[50] Wiedemann was keen that his wife and Anna Helmholtz should undertake it, which was agreed before Christmas.[51] Gauthier-Villars was unhappy, since he had intended to sell French copies into Germany and urged Tyndall to reply so that the French edition would appear first.[52] Jamin later brokered a deal with Gauthier-Villars for the publication of *Faraday as a Discoverer*,[53] though Moigno had assumed that he would be asked and had already completed it when he heard of the deal through Jamin. He wrote in some anguish to be allowed to issue it for a different French edition.[54] Tyndall must have written some adverse comments on his proof since Moigno wrote again to protest the criticism of his work.[55]

As Christmas approached, Tyndall was still working hard on his lectures, which included on 4th December a special lecture at the Whittington Club to the printers employed by Spottiswoode. It was effectively a repeat of the Dundee lecture. Lady Colville was expected to be there, but as he wrote to her: 'I fear you will find me dull; for…Spottiswoode has forbidden all heresy which might have added spice to the occasion'.[56] The Christmas Lectures began on 26th December, forcing Tyndall to decline an invitation from Lord Stanhope to spend Christmas at Chevening. He saw in the New Year with Hirst; they shook hands on the occasion, then dined and spent the evening as usual with the Huxleys. At the weekend a cold and wet excursion to High Elms followed; then dinner at home with Hirst, Wright, and Debus on the evening of Tyndall's last Christmas Lecture. Debus, for whom Tyndall had secured a teaching job at Clifton College, took offence at a remark of Tyndall's about his accent and refused to speak to him for the rest of his visit. Hirst saw this response as a weakness of Debus, but observed that Tyndall's mode of correcting him varied with his own moods and thought allowances should be made on both sides.

Tyndall's ordinary lectures for the New Year at the Royal Institution focused on Faraday, a course of nine in January and February on 'The Discoveries of Faraday'. But first, following an evening with Hirst at the Prince of Wales's Theatre to hear Boucicault's new play *How she loves him*, he gave the two Discourses 'On Faraday as a Discoverer'.[57] Tyndall felt pressurized: 'had my own wishes been consulted I should have remained at least a year silent', and he visited Faraday's grave for the first time on the day of the second lecture, which threatened to disturb his equilibrium. But the house was crowded and Hirst called the lectures real masterpieces of deep thought clothed in appropriate expressions. Carlyle attended the second one, having failed to get to the first (at which he had asked for seats for himself and his sister) because it was raining in torrents and he was unable to get a cab.[58] The course of lectures on 'Faraday' had no sooner finished than his series at the School of Mines began, with thirty-two lectures on 'Physics' to a group of thirty-six students.

* * *

The Royal Institution was now Tyndall's domain. Towards the end of January 1868, he left Waverley Place for good to move into the Royal Institution, which would be his home until he built his house at Hindhead. He had lived with Hirst for nearly seven years, and Hirst was sad to see him go. Not that their interactions ceased significantly. Hirst was in the chair at the Philosophical Club on 30th January with Tyndall, Huxley, Frankland, Gassiot, Grove, Carpenter, and others, and both were at the X-Club on 7th February, where Francis Galton was the guest. Galton was the joint General Secretary of the British Association with Hirst, and they discussed Section Presidents for the upcoming Norwich meeting. Tyndall was at High Elms on successive weekends in February with Hirst, including celebrating Nelly Lubbock's birthday. In between he dined at the Spottiswoodes, with the Lubbocks, Colvilles, and Hirst. The following week brought Spencer's election to the Athenaeum at the head of the poll, proposed by Grove and seconded by Hirst. He was the final member of the X-Club to join, and it was his bachelor's haunt for many years. For the X-Club, the Athenaeum was a place in which they could interact with other powerful intellectuals in the Church and the worlds of politics, literature, journalism, and business. It was the venue of many X-Club meetings from the 1870s.

Faraday as a Discoverer was published on 5th March, the day of an X-Club meeting with Darwin as their guest. Tyndall had been worried, after a hint in a letter from Jane Barnard, that Sarah Faraday and she were not happy with some of the references to Faraday's personal characteristics in the draft. He wrote an anxious letter, making clear that he would not publish if they had any concern whatsoever.[59] It seems that he was being over-sensitive, as Barnard wrote to give their full approval.[60] She was also thoroughly pleased with the full series of lectures, of which she attended every one.[61]

Tyndall received thanks for his book on Faraday from many friends including Lady Ashburton, Lady Peel, and Emily Tennyson, with a note from the Prince of Wales, lamenting 'not only the great man Science has lost, but…the good and upright man England has lost'.[62] Tyndall missed the X-Club meeting—he was feeling low, and spent a couple of days on the Isle of Wight staying with Tennyson—which included a discussion about a memorial to Faraday in Westminster Abbey.[63] Hooker wondered if the X-Club, or a select group of themselves, Huxley, Lubbock, and Spottiswoode, should create itself into a Committee to take the idea forward.[64] Hooker, Huxley, and Tyndall wrote to the Managers of the Royal Institution, with the support of the Dean of Westminster, but Bence Jones was unhappy with the plan.[65] He established from the Dean that his fee would be 200 guineas and was unimpressed by the possibility of a tablet and bust, 'near Sir Humphry Davy's tablet which I find is a poor thing; put up by Lady Davy, I should think at a cost of £50'.[66] He refused to put it to the Managers, saying that to be worthy of Faraday, and the Dean's expense, they should commission a statue like Newton's costing some thousands of pounds. He argued that the initiative ought to come from the Presidents of the various societies, who would carry more weight even than the Council of the Royal Society. Tyndall wrote to Hooker to suggest they carry on themselves, but learned from Bence Jones that Sarah Faraday was unhappy about the idea of a public subscription for a memorial.[67] Perhaps she changed her mind, as some years later, John Foley was commissioned to produce a sculpture. Tyndall visited the studio and found the countenance 'full of truth and strength, and dignity'.[68] He wrote to Pollock and to Spottiswoode, who was to give the final word over the choice of electro-magnet or induction ring to be held in Faraday's hand. Tyndall preferred 'the less obvious, and I would add, less commonplace ring'.[69] His view prevailed and the statue now stands at the foot of the main staircase in the Royal Institution.[70] A memorial floorstone was placed in the Abbey in 1931, at the same time as one to Maxwell.

March was a social month. Weekends were spent with close friends, with dinners at the Athenaeum and at the Dean of Westminster's and Lord Tankerville's. At Dean Stanley's the guests were the Prince and Princess Christian, who remembered him from Osborne, Carlyle, Froude, Robert Browning, and Arthur Russell. He tried to persuade Carlyle to come to the Isle of Wight but without success, as Carlyle feared for his sleep in the hotels. But by the end of March, Tyndall was feeling very low, confiding to his journal: 'My life is very much of a blank…I should rather have preferred the continuance of my old life to coming to the Institution'.[71] A few days later, after hearing a very good lecture from Frankland on London's water supply, he wrote: 'I am lower even than I used to be—not caring much for the world or the things of the world. I wonder has my bachelor life anything to do with this? Of late I have seen but little of children, and they have often saved me from depression'.[72] Despite this depression he arranged his apparatus for future work,[73] corrected some of his older papers for publication, and went out socially, dining with a large party at the Spottiswoodes where he sat with Nelly Lubbock almost the whole evening. At both Carpenter's—unimpressed by Carpenter's willingness to accept spirit-rapping—and Mrs Douglas's he saw Mary Adair (a relative of Sir Charles Hamilton, a Manager of the Royal Institution), who had piqued his interest: 'A fine head and a fine heart that girl assuredly possesses'.

At Spottiswoode's Tyndall had talked to Nelly Lubbock about going to see Vesuvius. Though he had thought of spending some time over Easter on the Welsh hills with Huxley, he bowed to Hirst's judgement and changed his plans.[74] He left England on 9th April, having arranged to meet John Lubbock in Amiens to visit the archaeological site at St Acheul. Tyndall was the only one on board who avoided being sick on the Channel crossing, and on arrival in Amiens he went sightseeing, finding himself blocked in at the cathedral as a service took place. The master of communication was impressed by the priest and the singing. He remarked: 'I suppose the Church of Rome has studied all such effects'.[75] On Good Friday he went to St Acheul with Lubbock, who bought flints, and on via Paris to Marseilles, travelling all night.

It was desperately cold and windy in Marseilles, and boats would not venture out until the following day, Easter Sunday, on which they left for Naples and Vesuvius. He described his pleasure at the trip to Juliet Pollock, revealing at the same time his depression: 'I have rarely passed so delightful a day. The blue of the sea; the blue of the air, the noble line of coast, and the perfect absence of all internal disturbance rendered the voyage exceedingly enjoyable…I intended to call to see you before I came away, but I was most horribly low & very occupied. I hope the journey will diminish the lowness, for really one feels at times as if the boon of life were scarcely worth acceptance'.[76] Having taken the boat from Leghorn to Naples, they put up at the Hotel Brittanique, where Tyndall's sleep was interrupted by braying donkeys. Sightseeing started with the Solfaterra crater, surrounded by sulphur deposits and sulphurous fumes. Temples followed, and then a grotto in which carbon dioxide was welling up. A woman carried a little dog with beautiful eyes into the grotto and Tyndall watched while it spasmed and twitched as it was asphyxiated. It recovered out of the grotto, but Tyndall considered the action scandalous: 'The brutal experiment which is in no way instructive ought to be put an end to'.[77]

The next day they took the train to Pompeii and explored the ruins, where they were impressed by the security measures to prevent tourists stealing artefacts and the refusal by the guide of any recompense. Tyndall climbed Vesuvius twice, and visited Rome before returning at the end of April as the lecture course at the School of Mines restarted.[78] Lady Peel had discovered that Tyndall had given Lady Tankerville tickets to attend and asked if she could join.[79] He was delighted to have two such pupils, who must have stood out like Geraldine Jewsbury against the regular students, and suggested she start with his new topic on electricity.[80]

Tyndall's desire to reduce his commitments was not just due to concerns about his health. He found he could not undertake research and lecturing simultaneously. The Managers of the Royal Institution had offered him the option of giving just six lectures, so that if he dropped the thirty or more he was giving at the School of Mines he could devote the autumn and the spring to one continuous stretch of original investigation.[81] He planned to do that for the next two or three years, for which one impetus was his need to deal with the continuing objections of Magnus, to whom he wrote for details of apparatus.[82] He tendered his resignation to Murchison, and Frederick Guthrie succeeded him at the School of Mines, though Tyndall had tried to have Alexander Herschel, Sir John's son, appointed.[83] The examination for the School of Mines was on 18th June, and Tyndall aimed to leave for Switzerland in early July once he had finished marking the papers. He broke off marking for the second day of the X-Club's summer excursion, this time to Oxford, where he joined them briefly at the Randolph Hotel.

Before he left for the Alps, he brought Clausius up to date with his plans to reduce lecturing and devote more time to research. He had found producing the books on *Heat* and *Sound* hard work, writing: 'I do not intend to publish anything of the kind for a long time to come—perhaps never again, because such work takes my thoughts so much away from original investigation. Still I think those two books on Heat & Sound have done great good, by giving numbers of people an interest in scientific matters who had previously no interest in them'.[84] He would not write a further book of that type, unless one counts the publication of the *Six Lectures on Light Delivered in the United States in 1872–1873*. Nevertheless, he left a legacy of popular science books and lecture notes that had a major influence throughout the world.

13

---✦≈◎≈✦---

PRAYER, MIRACLES, METAPHYSICS, AND SPIRITS
1865–1880

In the mid-nineteenth century, one issue became a lightning rod in the debate between scientific and theological claims to truth: the power of prayer. Tyndall had revealed his position in *Mountaineering in 1861*. In a chapter entitled 'Reflections', an unusual interpolation in a mountaineering book, he described how the savagery and apparent unpredictability of the mountain world might have led people to believe in supernatural agents and to pray to them for protection. For Tyndall, physical processes were explicable solely on the basis of natural laws, which could be established through scientific investigation alone. There was no evidence of miracles, which many people prayed for, nor any need for them. As he put it: 'The region in which prayer is opportune is the human heart: retire to that citadel and you are safe from the attacks of science, but come out from it into that system of things which is in the vocation of science to express, and confusion will infallibly attend your operations'.[1] Pollock, for one, was concerned at the way he had thought to express what were widely seen as heterodox beliefs.[2] But Alfred Wills, the mountaineer, bought the book and approved of the chapter: 'It is bold, manly & suggestive & will do good ... The old forms of faith cannot co-exist with the ascertained results of modern investigation, whether in philosophy, science, criticism or history—and those who help to prepare the human mind for ... the great change in our modes of thought & of belief ... do a great service'.[3]

The issue was sharply posed in October 1865 when the *Pall Mall Gazette* ran an article on 'Prayers against the Cholera',[4] after the government had proposed that the country should unite in a special religious act to counter the disease. Sabine drew the piece to Tyndall's attention, considering that it met the difficulty he had set out in *Mountaineering in 1861*.[5] Tyndall had argued that 'the age of miracles is past', and that people no longer sought to pray for things 'in *manifest* contradiction to natural laws'. The case in point was the instruction by Bishop Wilberforce in 1860 for prayers to be said for fine weather. The newspaper argued that though the physical laws of the universe were constant, their effects could be directed by the application of human will and activity. It was not, therefore, unreasonable to think that prayer to God could achieve specific ends, consistent with natural law. Tyndall was having none of this. He pointed out that only where, as in the case of cholera, we did not yet understand the physical antecedents, did some people think of resorting to prayer. He gave the example that vaccination had proved far more effective than prayer against smallpox.[6] Tyndall noted that Frances Power Cobbe, a leading women's rights campaigner and later founder of the National Anti-Vivisection Society and the British Union for the Abolition of Vivisection, whom he had met in Birmingham, shared his view.[7] The *Pall Mall Gazette* held its position, on the grounds that our knowledge of nature and of the freedom and power of the human will was incomplete.[8] We could not prove a negative, that prayer could not be effective. Tyndall argued that the scale of the effect was irrelevant, and that it was 'as great a miracle to suspend the gravity of a straw as to extinguish the force which holds the solar system together'. The question of free will remained. Tyndall had manifest difficulty in holding simultaneously the idea of free will and of physical law and necessity. Imagining the act of reaching for an inkstand he asked: 'Is the cause physical or super physical? Is it a sound, or a gleam, or an external prick or pressure, or some internal uneasiness that stimulates the nerves to unlock the muscular force, or is it free will?'.[9] He gave no answer.

The question of miracles and natural law stimulated Tyndall's first significant foray into the world of intellectual periodicals. The mid-1800s saw an explosive growth of publications, as new technologies and reduced taxation increased the speed and reduced the cost of printing. Every taste and commitment was catered for, from the radical socialists to the staunch Tories, and from the atheists to the Anglican divines. Tyndall had already contributed to the science reviews in the *Westminster Review* and written pieces for publications like the *Reader*. He now started publishing major articles, eventually producing more than thirty in his career, often on contentious topics and predominantly in Liberal-leaning periodicals. The first was a piece in George Lewes's new *Fortnightly Review*, entitled 'The Constitution of the Universe', which appeared in December 1865.[10] Tyndall gave a firm statement of the importance of the principle of conservation of energy and linked it to the question of miracles. He wrote: 'A miracle is strictly defined as an invasion of the law of the conservation of energy... Hence arises the scepticism of scientific men when called upon to join in national prayer for changes in the economy of nature', such as making water flow uphill. He saw that prayer produced its effect on those who prayed, but suggested that 'if our spiritual authorities could only devise a form in which the heart might express itself without putting the intellect to shame, they might utilise a power which they now waste, and make prayer, instead of a butt to the scorner, the potent inner supplement of noble outward life'.[11]

The issue became pertinent when the Archbishop of York instigated a 'day of humiliation' in his diocese on account of the cattle plague. The *Times* reported the Dean of York's sermon, in which he attributed the plague to God's work and declared that therefore He alone could avert it. The North Staffordshire Church Protestant Association blamed it on the government's failure to stand up to Romanism by increasing the grant to the Maynooth seminary—in 1845—which it argued had caused first the famine in Ireland and now the cattle plague. It railed against the government for not appointing a day for fasting and humiliation, 'being more afraid of the sarcasms of Professors Huxley and Tyndall and the Pall Mall Gazette than of the warnings of the Word of God'.[12] In a letter to the *Spectator*, Huxley denied that he had claimed that miracles were impossible, but asserted that they required evidence of the same kind and quality as scientific propositions.[13] Tyndall backed him up in his own letter to the magazine, writing: 'The evidence of the permanence of natural laws is not that of a generation, but of many generations... The real strength of science is not, I think, manifested either by logic or by authority... It gradually attunes the mind to the methods of nature, and superstition dies, not through any formal demonstration of its error, but by its simple incapacity to live in the presence of scientific knowledge'. He added that the knowledge gained had 'deepened indefinitely the impression which the mystery of this Universe has produced... on every profoundly religious mind'.[14]

Unlike the Dean of York, Dean Stanley of Westminster, who had read Tyndall's piece in the *Fortnightly Review*, wrote asking for his advice when informed by the Bishop of London that 23rd March 1866 was the day set for the cattle plague service.[15] Tyndall responded: 'With regard to humbling ourselves in the expectation that a single beast the less will die I would say in all frankness that I consider it to be mere paganism... I hope and think... that you will not pray as others will for the staying of this plague, but will ask on the contrary for strength of heart and clearness of mind to meet it manfully and fight against it intelligently'.[16] Stanley used just these words—in stark contrast to most churchmen—sending Tyndall a copy of the prayer with an apologetic note: 'The best that I could do'. Tyndall replied: 'your "best" is thoroughly good'.[17]

Two years later, Tyndall's 'Miracles and Special Providences', published in the *Fortnightly Review* for June 1867, reopened the debate.[18] It was still a highly sensitive topic. Tyndall had intended to deliver it as a speech in Edinburgh, but the organizers got cold feet and asked him to give an experimental lecture instead. Bence Jones urged him not to publish the piece, given its sensitivity, which led to hot words between them.[19] In the article, Tyndall gave his response to the Reverend James Mozley, whose Bampton Lecture on miracles he had read in the Alps in the bad weather of 1866. Mozley's argument was that the evidential value of miracles accompanying a revelation lay entirely in their deviation from the order of nature. They showed the existence of a higher power. Special providences, by contrast, were not so clear cut. Tyndall wryly suggested that they be called 'doubtful miracles'. In the case of miracles, according to Mozley, the 'affection' must urge reason to accept the apparently impossible events. While recognizing the role of the affections and emotions for religious belief and moral elevation, Tyndall held firm to the importance of reason alone in matters of natural philosophy. He also challenged Mozley on his own ground: surely the truly religious soul needs no proof through miracles of the goodness of Christ? Then he outlined his reasons for believing in the validity of the methods of science and the constancy of natural laws, the importance of 'pondering' on the observed phenomena, and the induction from these, or inspiration, that led to generalizations that could be subjected to deductions and testing. At the root of our confidence in these natural laws was our understanding of the forces underlying them, and the necessity that the phenomena would follow in a predictable and explicable manner. For example, we calculate eclipses before they have occurred and find them true to the second. The principle of miracles would have free scope without this 'rock-barrier' of natural knowledge which we now possessed. Built on reason, he declared, such knowledge 'breaks to pieces the logical pick and shovel of the theologian' which, left unchallenged, leads cultivated men to be 'harried on to deeds, the bare recital of which makes the blood run cold'.

This article was later published in *Fragments of Science*. That version contains an intriguing addition, written at the time but not included in the original article, in which Tyndall imagined 'transferring our thoughts from this little sand-grain of an earth to the immeasurable heavens, where countless worlds with freights of life probably revolve unseen'.[20] It is a vision consonant with Tyndall's view of life as inherent in matter. Mozley took the article in good part, though he did not agree that a detailed knowledge of nature affected his argument in any way.[21] The Duke of Argyll was equally unperturbed, inviting Tyndall to dinner more than once shortly after publication. Hector Tyndale appreciated the piece, though with his religious outlook he accepted the possibility of miracles. He said it had been widely read in America.[22]

* * *

Spiritualism was hugely popular in Victorian times, challenging the boundaries of the natural and the supernatural. Tyndall, like Faraday and many other scientific men, had spoken out about spiritualism in the past. In 1860, at the Lubbocks', Tyndall had an intense exchange with the author William Thackeray over spirit rapping, urging him not to 'damage the respect which scientific men feel for you, by lending the authority of your name and of your magazine to this atrocious stuff'. When Thackeray replied that he must bear witness if he saw grapes jump up to the ceiling, Tyndall replied: 'They won't do it. God and the Universe is against any such feat'.[23] Alfred Russel Wallace, Darwin's co-discoverer of evolution by natural selection, wrote to Tyndall in 1868 with a typical view, convinced that a Miss Nicholl had been lifted bodily in her chair to the centre of the table by spirits: 'You know Miss. N.'s size and probably weight, and can judge of the force and exertion required to lift her and her chair, on to the exact centre of a pillar table; and the immense surplus of force required to do it almost instantaneously and noiselessly, and without pressure on the sides of the table, for that would lift it up. Will any of the known laws of nature account for this?'.[24]

Tyndall was soon engaged in a dispute in the pages of the *Pall Mall Gazette* with the famous medium Daniel Home. He offered to meet him at any time, provided only that he was allowed to use his scientific methods as he saw fit. Home had claimed that Faraday had declined an earlier offer to examine him, at the request of Sir Emerson Tennent. Tyndall, who said he had been due to accompany Faraday on that occasion, disputed that interpretation and offered to explore any phenomena with just the experimental approach he would apply in other areas. He wrote in response to criticism from the poet Francis Palgrave of his publication of a letter from Faraday to Tennent: 'the investigation can have but one of two results: either his [Mr Home's] phenomena will be proved delusive, or I shall be converted to the ranks of spiritualism'.[25] The *Pall Mall Gazette* invited Home to accept the challenge, but he evaded it. Instead, they republished an article 'Science and the Spirits', which Tyndall had originally written for the *Reader* in 1864, an amusing tale of a séance during which Tyndall was dubbed by the spirits 'the Poet of Science'.[26] Though many considered Home a fraud, he was never decisively unmasked.

A year later Tyndall received a letter from Wheatley Kenneth, Honorary Secretary of the London Dialectical Society, which gave itself the mission of investigating 'phenomena alleged to be spiritual manifestations'. Members included Alfred Russel Wallace and William Crookes, respected men of science. According to Kenneth, while sceptical of the spiritual phenomena, and making every effort to avoid trickery, they believed there might be 'some power (magnetic or otherwise) which has not yet been generally recognized by men of science'.[27] They asked Tyndall if he would attend two or three sessions to assist them. Given the reputations of Wallace and Crookes, Tyndall accepted in principle.[28] He did mention the opinion of Cromwell Varley, Fellow of the Royal Society and enthusiast for spiritualism, who commented that the presence of Tyndall at a séance resembled the introduction of a big magnet among smaller ones, throwing all into confusion, which he thought might disqualify him from participation. But Crookes wrote to say that Kenneth should not have quoted his name; his relationship with them was informal. He hoped Tyndall would come to his own laboratory as a friend and in a private capacity if he thought he had phenomena to show him.[29] It is not clear that Tyndall ever did.

The tension between scientific and religious perspectives stimulated a number of attempts to flush out people's positions through public declarations, by demonstrating the numbers who held religious beliefs. The 'Oxford Declaration' of 1864, just before the prayers for cholera episode, was an attempt by the clergy to condemn the liberal Anglicans who had contributed to *Essays and Reviews*, and to maintain the Divine Authority of the Scriptures. Though it garnered much clerical support, it failed to engender any prosecutions for blasphemy.

Later that year, in a complementary move, Tyndall and many other men of science were approached to sign the 'Declaration of Students of the Natural and Physical Sciences'.[30] This declaration, drawn up by religiously minded science students, invited signatories to deplore the perversions of research into scientific truth that were used to cast doubt upon the Truth and Authenticity of Holy Scriptures. The declarationists argued that Science and the Scriptures would always in due course be reconciled, and sought a 'harmonious alliance between Physical Science and Revealed Religion'.[31] While Brewster and Joule signed it, many leading men of science, religious and non-religious, refused; among them Faraday, Owen, Huxley, Herschel, Stokes, Airy, and Sabine. The Duke of Argyll also refused, as did the vast majority of Fellows of the Royal Society. Their grounds for refusal varied, but were largely the desire not to cause discord in the scientific community, and a view that the clergy should not have any right to interfere with science through religious tests. They argued for free enquiry. Tyndall, like Herschel, gave it short but polite shrift, writing: 'If it be of God—that is if it be true—I cannot overthrow it—if it be of men—that is if it be untrue—it will come to naught'.[32]

* * *

The continuing tensions between theologians and men of science about the scope of 'science', and who could speak with scientific authority, gave birth in 1869 to the Metaphysical Society.[33] Proposed at a meeting in 1868 between James Knowles, Alfred Tennyson, and the astronomer Reverend Charles Pritchard, this semi-secret group was initially envisaged as a body of religious believers, formed to discuss how to deal with the challenge from science. The membership was quickly broadened, at the insistence of the Unitarian James Martineau and others that a wider range of views be accommodated. It was a stellar group, involving sixty-two members over its lifetime. As at the Athenaeum Club and the Royal Society they were all men: the men of science—Huxley, Tyndall, Lubbock, Sylvester (elected 1873), and William Clifford (1874); literary figures—Tennyson and Ruskin (1870); religious leaders—Stanley, Martineau, and Manning (Catholic Archbishop of Westminster); journal editors— Knowles, Hutton, Froude, Morley (1876), and Leslie Stephen (1877); and politicians— Gladstone, Arthur Balfour, and Lord Arthur Russell. Spencer was invited by Lubbock to join but declined as 'too much nervous expenditure would have resulted'.[34]

At the planning meeting Huxley coined the term 'agnostic', though he did not publically define it until 1889. Those who were invited to join the Metaphysical Society showed themselves largely disposed to an open debate and to an attitude of enquiry toward the limits of knowledge, as they sought to persuade others of their beliefs.[35] Members brought and discussed papers at the monthly meetings. Many were published in the leading periodicals, particularly in the *Contemporary Review*, of which Knowles became the editor in 1872. They dealt particularly with the differences between intuitive and empirical grounds for belief. That became an almost unbridgeable dividing line, as the intuitionist theory of knowledge, that truth is anterior to sense, justifies Christian theism. Related topics included idealism and materialism, the significance of emotional and psychological influences, and the existence and importance of free will.

Tyndall was one of the original twenty-three members. He attended twenty-four of the ninety-four meetings, but was not the most enthusiastic of participants and never presented a paper. After hearing Froude speak on 'Are Numbers and Geometrical Figures Real Things?' he commented: 'It is a curious…waste of thought:—nothing firm or coherent in the discussions'.[36] He did not attend for Huxley's paper 'Has the Frog a Soul?', nor Lubbock's on 'The Moral Condition of Savages'. But he was eclectic in his choice of subjects, which ranged from 'On the Doctrine of Human Automatism' to 'The Personality of God'. At the first meeting, Tennyson read 'The Higher Pantheism', one of his perhaps less impressive metaphysical poems. Henry Sidgwick commented: 'After he had done reading there was a pause, and then Tyndall (who is entirely devoid of shyness) said "I suppose this is not offered as a subject for discussion"'.[37]

The Society cemented some important connections for Tyndall. He heard two papers from Lord Arthur Russell, who served for several years as a Manager of the Royal Institution, and they were often at meetings together. Tyndall's friendship with Russell developed despite the fact that Russell tended to side with the Christian metaphysicians. It extended to Russell's first cousins, Agatha and Rollo (who also became a Manager), and their parents the Earl and Countess Russell, regulars at the Royal Institution. The Earl had been Prime Minister from 1865–1866, and from 1846–1852 as Lord John Russell. Arthur Russell argued a strongly pro-vivisection position in 1875 in his paper 'The Right of Man over the Lower Animals'. It generated furious argument, but would have appealed to Tyndall, Huxley, and Lubbock. The antivivisection campaign at the time was highly vocal. The Cruelty to Animals Act, passed the following year, allowed vivisection but imposed the use of anaesthetics unless the experiment required the animal to be alert in order to yield results. (Busk became the first National Inspector under the Act.)

The Metaphysical Society ran until 1880, and has been viewed as a forum for established elites to manage their differences against the greater danger and challenge of popular belief and resistance to authority from others such as feminists and socialists.[38] Whatever their religious persuasions—and some, like Carpenter, were both men of science and religious—all sought to appropriate science to their own beliefs and ends. Tyndall's Belfast Address of 1874 can be seen as his response to many issues raised in the Society's discussions, as well as a culmination of his earlier writing. It is perhaps no coincidence that he attended most regularly in the period leading up to and following the Address.

In 1872, when the Metaphysical Society was in full swing, Tyndall threw in his own contribution to further public debate on prayer just before he left for the Alps. A letter to him, probably stimulated by claims for the power of prayer over medical treatment for the recent recovery of the Prince of Wales from typhoid, proposed a means of testing the efficacy of prayer. Tyndall had it published in July's *Contemporary Review* entitled 'Prayer for the Sick', with a supportive introductory note, thereby giving the visiting clergy on Belalp plenty to talk about.[39] The anonymous writer—in fact the surgeon Sir Henry Thompson—pointed out the various ends to which prayer to the Deity was believed to be effective, including spiritual and moral improvement, national supremacy, suitable weather, and preservation from pestilence, famine, and battles. Most were difficult to test experimentally. But praying for the sick ought to be testable, he argued, if the focus were specific enough. A prayer for general health would not be sufficient, as its impact would affect all. So the suggestion was for prayer to be directed, for three to five years, to one hospital ward containing patients with diseases of known mortality rates, to see if it made any difference.

This became known as the Prayer-Gauge Debate. Tyndall fronted up in public behind the author of the proposed test, in his first significant discussion of the relationship between science and religion.[40] He pointed out that the claim for the physical value of prayer was a legitimate subject for scientific enquiry, dealing as it did with effects caused by physical energy in the ordinary course of events. This, he claimed, was not an attack on religion, which was now 'freeing itself slowly' from physical errors such as the fixity of the Earth, the period of Creation and the age of the Earth. Even Darwin's recent *Descent of Man*, from the old point of view 'more impious' than *On the Origin of Species*, had been received more calmly than its predecessor. James M'Cosh, Presbyterian Scot and President of Princeton University, objected that the project was not consistent with the methods and laws of God's kingdom (that it was always his Will whether to grant the petition or not, and that in any case it required faith), nor of the spirit in which Christians pray. M'Cosh argued that people 'who have excelled in physical experiments are not *therefore* fitted to discuss philosophical or religious questions'. Their arguments proved merely their ignorance of the 'kind of evidence by which moral and religious truths are sustained'.[41]

The debate raged in the columns of the *Spectator* and other newspapers over subsequent months.[42] Francis Galton added his piece in the August *Fortnightly Review* on 'Statistical Enquiries into the Efficacy of Prayer', using statistics to address the simple question: are prayers answered or not? His evidence relied on a multitude of individual examples, including the observation that members of Royal houses, frequently prayed for, nevertheless lived shorter lives on average than other classes of aristocracy or gentry. The *Spectator* recognized that this question of the efficacy of prayer went to the root of the dispute between the physicists and supernaturalism: 'If prayer is a delusion, so are the creeds, churches are organised superfluities, priesthoods are impostures'.[43] The Duke of Argyll joined in, with a lengthy critique of the argument that spiritual and physical domains are separate, and prayer only valid in the first of these.[44]

* * *

One specific focus for disputes over theology, science, and the place of religion in society was the Lord's Day, Sunday. While the Metaphysical Society was being formed in 1869, Huxley and Carpenter had been instrumental in launching the Sunday Lecture Society. Whereas they might have seemed to be on opposite sides of debates in the Metaphysical Society, here they made common cause with the aim of prising Sunday from the grasp of the established Anglican church, and of the fundamentalist dissenters who wished to restrict what people could do on a Sunday.

In 1855, the secularist National Sunday League had been established to campaign for opening museums, art galleries, libraries, and gardens on Sunday afternoons. The X-Club became involved in 1866, when two meetings were taken up with discussing the League's plans for Sunday Evenings for the People, offering 'discourses on science and the wonders of the universe, thus producing ... a reverence and love of the Deity, and raising up an opposing principle to intemperance and immorality'.

The many supporters listed included Tyndall, Huxley, Spencer, Lubbock, Frankland, Carpenter, Martineau, Owen, Lyell, Mill, and Dickens. Huxley gave the first lecture, on 7th January 1866 in St Martin's Hall, 'On the Desirableness of Improving Natural Knowledge'.

The organizers had to walk the tightrope of the 1781 Lord's Day Act, which prohibited charging for 'entertainment'. They closed down after four weeks because of the legal threat, and reformed as the Recreative Religionists with a more accommodating style that Huxley and Carpenter considered an evasion. In competition, the Sunday Lecture Society was then created to deliver talks on science, history, literature, and art, aiming them at moral and social improvement. The supporters were more diverse and radical this time, including women such as Frances Power Cobbe and Sara Hennell. Hennell was a friend of Marian Evans (George Eliot), and had given Tyndall a copy of her *Comparative Metaphysics*, though his response to it is not known. Many women lectured, consistent with the Unitarian and Christian socialist interest in Sunday lectures, and the involvement of supporters of women's higher education and political participation. The X-Club were no devotees of radicalism or the emancipation of women (only Spottiswoode and Busk were particularly supportive of women's educational aspirations).[45] Just Huxley lectured—and agreed to be President after Carpenter's death at the end of 1885, having previously refused when he was President of the Royal Society. Tyndall was publically encouraging, for example appearing with Huxley and Hirst at a meeting of Sunday Evenings for the People at St James's Hall in April 1867, but he was not a leading activist.[46]

Tyndall's most extensive statement of his views on Sunday observance was in St Andrew's Hall in Glasgow in October 1880. This lecture, 'The Sabbath', was delivered to a packed hall as the Presidential Address to the inaugural meeting of the Glasgow Sunday Society.[47] As in the Belfast Address, he started with a historical perspective, based on his summer alpine reading of Renan's *The Sources of Christianity* and Caird's *Introduction to the Philosophy of Religion*. He drew a parallel between the development of organisms, set out by Darwin, with that of the mind, described by Spencer, leading to the comparative work by Max Müller and others in founding a 'science of religion'. Over time, he said, the 'materialistic and figurative' conceptions of God become purer and more distilled, helping distinguish myth from underlying truth through discussion and reason.

Tyndall argued that Sabbath observance was one of these mistaken beliefs, as he countered at length the various arguments made from Scripture or tradition. He poked fun at the knots in which some tied themselves—'Bishop Blomfield, for example, seriously injures his case when he places drinking in gin-shops and sailing in steamboats in the same category'—and he countered the idea that immoral behaviour might result from Sunday activities. He recollected seeing Prussian soldiers on a Sunday in Dresden shortly after the suppression of an insurrection. They were impeccably behaved as the girls of the city passed through them. Motions were carried to send delegates to the second National Conference of the Sunday Society, in the hope that just as the public libraries and galleries of Manchester had been opened on Sunday afternoons and evenings after the first meeting, so might those in Glasgow following the second. Tyndall's lecture was published the following month in the *Nineteenth Century*, a new magazine founded in 1877 by Knowles (who had left the *Contemporary Review*).[48] Knowles wanted the sort of papers which his Metaphysical Society friends wrote, and many, like Tyndall, followed him from one journal to the other.

Tyndall believed that there was a strong moral imperative to offer working people opportunities for educational and spiritual improvement, to take them out of an environment of enforced idleness and often ensuing drunkenness: 'The physics of the drunkards' brain are incompatible with moral strength'. Tyndall believed that the true spiritual nature of man was bound up with his material condition. Wholesome food, pure air and cleanliness, with hard work, fair rest, and recreation, were all necessary to physical and spiritual wellbeing. It was a belief that carried through to his work on germ theory, and led to his support of initiatives in health education and his contribution to public debate.[49] Aware that he was addressing an audience of 'Liberals, perhaps Radicals—perhaps even Democrats or Republicans', Tyndall declared himself a Conservative in the true sense of the word, with the foresight to anticipate potential revolution and smooth the path of inevitable change. Brought up in a Conservative Protestant household and radicalized by his earlier experiences, Tyndall was instinctively Liberal in his politics, with a Whig's belief in the importance of an enlightened governing and aristocratic class. The battles with Gladstone and his own brief flirtation with the idea of becoming a Conservative MP lay ahead.

Alongside Huxley and others, including open-minded clergymen, Tyndall played a visible role in promoting the opening of places such as libraries and museums on Sundays. While Sunday lectures could be perceived to be in opposition to sermons, the opening of cultural venues was not such a challenge to Anglican orthodoxy. In 1875 a new Sunday Society took up the aims of the National Sunday League to campaign for opening museums, art galleries, libraries, and gardens on Sundays. Dean Stanley, liberal Anglican but not a listed supporter of the Sunday Lecture Society, chaired a public meeting on the issue, with Huxley and Tyndall on the platform.

Two years later, in May 1877, Tyndall attended a meeting of the Sunday Society at the Freemason's Tavern for its second annual report, supporting the Dean—the President—with Huxley, Morley, and others.[50] He attended again in 1884, after pressure from the Dean and from Mark Judge,[51] the founder and Secretary,[52] armed with Hooker's evidence of the beneficial effect of opening Kew Gardens and Museum on Sundays. Hooker believed it had resulted in a 'moral gain' in the better behaviour of 'certain classes of visitors', and pointed out that about half of all visits were on Sunday afternoons.[53] Even so, it was not until 1896, after Tyndall's death, that the great national museums and art galleries in London were opened to the public on Sunday afternoons.

Tyndall is often portrayed as a standard-bearer for the defence of the scientific worldview against the religious. In some respects this is true, but Tyndall's understanding of religion was subtle. He was never one to belittle the religious sense. Late in life, following an article by Huxley on miracles, he revealed to Hirst his dislike of the ribald secularists and his fondness for the moderate Christians: 'there is one thing that I feel probably more than either you or Huxley; and that is the goodness, tenderness, even loftiness of heart, that have got mixed up with these beggarly elements of Christian doctrine. I have been many times forced to put before myself the question:—Given the intellectual accuracy of many of our Freethinkers, plus the ribaldry to which they give expression in Secular journals…on the one side, and others known to me as believers in the creation of Adam and the resurrection of Christ on the other, I confess that I can take the latter to my heart, while I push the former far away from it'.[54] Hirst agreed: 'You and I were agnostics long before our friend Huxley invented the word, and we arrived at our views of Christianity, and of the Miracles which still hang on to it, in spite of Paley's Evidences and all such Treatises'.[55] Tyndall had no time either for internecine tussles, writing in the same period from Belalp: 'The Bel Alp hotel—about 300 feet below me—is a great gathering ground for the parsons. They have built a church here, and mounted a bell, while within the church some of them play extraordinary ritualistic antics. The ritualists and evangelicals swear at each other, and dislike each other; and thus into this grand region we have introduced the follies and stupidities of sacerdotalism'.[56] Yet he could appreciate and respect high-minded Christians: 'I often think that Christianity, to show itself in its best colours, requires a naturally high and serious and truthful character through which to act. Where it finds such a vehicle, it is beautiful, where it fails to find it, it is often the reverse'.[57]

For Tyndall, as for Spencer, the ultimate mystery was unknowable.[58]

14

MOUNTAINEERING IN THE 1860s
1860–1868

Tyndall went to the Alps every summer from 1856 until the end of his life, except in 1891 when he was too ill to travel. What started as a scientific exploration of glaciers on the way to a conference in Vienna evolved into a deep bond with the mountains, cemented when he built his chalet on Alp Lusgen in 1876. Almost by accident, Tyndall became a central figure in the 'Golden Age of Alpinism', traditionally set between Alfred Wills's ascent of the Wetterhorn in 1854 and the triumph and tragedy of Edward Whymper's ascent of the Matterhorn in 1865, when four of the party of seven died on the descent. This period of alpine exploration was dominated by the British and Irish. Tyndall's public position and visibility helped give mountaineering respectability, not least when many were questioning its morality after the Matterhorn disaster.

Tyndall was one of the most notable mountaineers of his generation. The Alpine Club for mountaineers was formed on 22nd December 1857, at a dinner in Ashley's Hotel chaired by Edward Kennedy and attended by ten others. John Ball became the first President. The idea had been suggested by William Mathews in February, who took it further with Kennedy during their ascent of the Finsteraarhorn—the first British ascent—on 13th August, while Tyndall was on Mont Blanc. Tyndall was not at the dinner. He had been approached in November by Mathews, who had written to Kennedy to say that he thought Tyndall would make an excellent President.[1] Tyndall declined to join, while making some suggestions about the proposed rules.[2] He asked to be excused 'simply for the sake of my pursuits', and hoped 'that the scientific side of the Alpine question will not suffer by this arrangement'.[3]

He accepted election a year later on 27th November 1858 (the same day as Joseph Chamberlain), and might have succeeded Ball as President had the controversy with

Forbes not caused offence to some leading alpinists.[4] Instead, he was elected Vice-President in 1861, but resigned shortly afterwards. This was ostensibly due to pressures of work, but he may have been provoked by sarcastic comments from Leslie Stephen about the carrying out of scientific experiments on alpine summits.[5] If so, it did not damage their personal relationship; Stephen was not one of the close friendship group of Tyndall and Huxley, but he later described Tyndall as 'among the most chivalrous and warm-hearted of men'.[6] He said that Tyndall never mentioned the incident to him, and that Longman had told him that the reason was that Tyndall did not want the Club to have any claim on his writings.[7] Indeed, after publishing 'A Day Among the Séracs of the Glacier du Géant' in the first volume of Peaks, Passes, and Glaciers, forerunner to the Club's Alpine Journal, Tyndall did not publish in the later volumes, nor in the Alpine Journal itself.[8] Though the Alpine Club was a gentlemen's club like so many others, he never found dining there congenial compared with the Philosophical or Royal Society Clubs. Of the X-Club, only Spottiswoode was also a member. He had travelled widely in mountainous regions, as had Hooker, who reached a height of at least 18,000 feet in the Himalayas on a botanical expedition. All the other X-Club members went to the Alps at some point, as did so much of Victorian society, but they did not join the Alpine Club.

Tyndall planned his Alpine season of 1860 to include scientific observations on glaciers. He had thought in 1859 that he would not return, but on the trip to Killarney he found the air too moist and warm for his temperament. In his weariness 'the mere thought of snow peaks and glaciers was an exhilaration'.[9] The lure of the Alps, and its ability to refresh him, was too much to resist; he had perhaps not realized the profound relationship he was developing with the landscape. By accident he met Vaughan Hawkins on the way, which precipitated some additional mountaineering exploits. After a few days measuring the motion of the Grindelwald glacier, and prevented by bad weather from climbing the Wetterhorn, they set off over the Lawinentor—a difficult pass from Lauterbrunnen across the mountain wall to the Aletsch glacier—and down to the Eggishorn, saved from a benighting by the yodelling of the guides for help.[10] There they remained for a week, making further glacier measurements, before heading for the still-unclimbed Matterhorn with Bennen, who had prospected it with Hawkins the previous year. The weather was bad in 1860, but Tyndall, Hawkins, Bennen, and Jean-Antoine Carrel reached the highest point yet attained.[11]

Then Tyndall left for Chamonix over the Col du Géant, to inspect the site of an accident on 15th August in which three Englishmen and one of their three guides, Frédéric Tairraz, had been killed. He described the scene in gory detail in a letter to the Times, having found fragments of Tairraz's skull—'an extremely thick one'—at the bottom of the rocks. His ire was reserved for the Chamonix guides since on such an easy slope, which had not avalanched, he thought a slip ought easily to have been arrested. At this time many guides tied their clients to each other round the waist but held the rope in their hands, as Lauener had done when Tyndall climbed Monte Rosa with him. On failing to stop the slide, as Tyndall wrote, 'the two men who led and followed the party let go the rope, and escaped, while the three Englishmen and Tairraz went to destruction. Tairraz screamed, but, like Englishmen, the others met their doom without a word of exclamation'.[12]

Tyndall went on to climb Mont Blanc for the third time, for an unsuccessful look for the thermometers left the previous year.[13] When he returned to England he found that Ball had wanted him to succeed as President of the Alpine Club. But with the glacier controversy in the background he hoped that Tyndall would understand that the choice of the next President—Edward Kennedy—should be completely neutral. Sensing Tyndall's uncertainty about the Club he entreated him not to leave it, arguing that many men were starting to take an interest in scientific questions, not just in climbing,[14] and because 'others look on you simply as the boldest and most successful mountaineer'.[15] He urged Tyndall to come to the dinner on 13th December to drink Kennedy's health even if he did resign. Tyndall did attend the dinner, where Owen made kind remarks about *Glaciers of the Alps*, and did not resign—for now.

* * *

'Hail to the Alps! thine they are, pyramid beyond pyramid, crest above crest'.[16] Tyndall's heart sang as the Bernese Oberland came into view the following year, on 2nd August 1861. The seasickness of a windy Channel crossing and the stifling heat of a train journey from Paris were banished. Tyndall had travelled out with William Forster, MP for Bradford and later architect of the 1870 Education Act. Meeting Bennen, Ramsay and his wife, and 'a fair maiden' on the quay by Lake Thun and arranging to meet Ramsay again at the Grimsel in two or three days, he set out with Forster, Bennen, and Wenger, another guide, up the Urbachtal to the Gauli and Lauteraar glaciers, reaching the Grimsel Hospice in the late evening of 4th August. They climbed the Siedelhorn, taking in its panoramic view of the Weisshorn, Matterhorn, and Finsteraarhorn. Leslie Stephen, 'a thin bony man, capable no doubt of great endurance', arrived from climbing the Finsteraarhorn.[17] Ten days later he would climb the Schreckhorn, perhaps his finest achievement, and one of his many first ascents. After a day ill in bed at the Grimsel, Tyndall went down to Fiesch with Ramsay and Bennen. Encountering Hawkins there, Tyndall prospected up the Aletsch glacier with him for an ascent of the Jungfrau. But the weather closed in and he decided instead to head down the glacier to Belalp with Bennen and Hirst, who had just arrived.

Tyndall revealed his mountaineering objectives to Faraday, who sent the letter straight to the *Illustrated London News*: 'There are two grand mountains, which above all others have of late years excited the attention of our best climbers...but which have hitherto successfully resisted all attempts to reach their crests; these are the Matterhorn or Mont Cervin, and the Weisshorn (Plate 14). I have always regarded the latter as the noblest mountain in the Alps and from its position and magnitude it forms a grand and striking object from every point of importance in Switzerland'.[18] Tyndall climbed the Sparrenhorn above Belalp to inspect the white pyramid of the Weisshorn, standing across the Rhône valley, both for a practicable route and a possible bivouac site. He travelled the following day to Randa, the village which sits under the eastern flank of the Weisshorn. Illness prevented activity the next day, but on Sunday 18th August, he climbed with Bennen and Wenger to a bivouac under an overhanging rock. Typically he had no sleep, looking at his watch at 12 and at 2am before breakfasting too early at 2.30. Needing the initial dawn light to start, they set off at 3.45.

All the ridges of the Weisshorn are long. The climbing was tiring, mostly unroped, with the added incentive of two people spotted on the glacier below, following them up. They quenched their thirst with draughts of champagne though Bennen refused it, fearing inflammation of the eyes. 'Onward we went; we scaled the last turret of rough rock, backing it was an arête of snow which ran up to a pyramidal point—on this point the fraction of a minute placed Bennen. He went beyond it a little along the arête on the other side and thus allowed me room to stand upon the topmost point of the Weisshorn, two men could not do so'.[19] The two followers below saw them and turned back, but they were glad to have witnesses since no-one would believe that a man who 'for two days previously was the object of a female waiter's constant sympathy on account of the incompetence of his stomach to accept what she offered' could possibly climb it.[20]

On the summit, as Tyndall described to Faraday, 'I opened my note-book to write a few words concerning the view, but I was absolutely unable to do so. There was something incongruous, if not profane, in allowing the descriptive faculty to meddle with that which belonged to the soul alone, so I resigned myself up to the silent contemplation of the scene, completely overpowered and subdued by its unspeakable magnificence'. At a time when so many people regarded the mountains as objects to be conquered, emblems of masculine prowess, Tyndall saw it differently. Though he was competitive, he did not describe his success as a conquest: 'the delight and exultation experienced were not those of Reason or of Knowledge, but of BEING:—I was part of it and it of me, and in the transcendent glory of Nature I entirely forgot myself as a man'.[21] This is Tyndall's authentic response to mountains and landscape. Great mountains look different when one has climbed them, and ever afterwards, Tyndall viewed the Weisshorn with a proprietary yet generous affection. He was

grateful to Bennen, and wrote in Bennen's 'Führerbuch' that he bore 'the same relation to the common run of guides as Wellington to an ordinary subaltern'.[22]

In the long descent, despite his exhaustion, Tyndall observed by chance the under-
side of a glacier exposed above a hollow. He noticed that just as the rocks over which
a glacier passes are smoothed and fluted, the same was true of the ice. He took this as
direct evidence of the sliding of a glacier, since if it were plastic it would simply
accommodate itself to the shape of the bed.[23] The party reached the chalets at 9pm
and went on down to the hotel. In a coincidental postscript, Ramsay made the first
ascent of the Lyskamm on the same day. The major alpine peaks were falling to the
British and the Irish.

The Matterhorn, that other great prize, remained to be climbed (Plate 15). Tyndall
went up the valley to Zermatt, where he found men gathered in the Monte Rosa
Hotel with thermometers under their tongues; they had been persuaded by Francis
Galton to participate in an experiment. On the way up to the Riffel, having sent
Bennen over the Théodule Pass to prospect the Matterhorn, Tyndall looked back to
see that his 'own fair pinnacle looked most majestic'.[24] The servant at the Riffel
exhorted him not to attempt the climb, asking: 'what will be the value of your last
achievement if you are smashed on the Matterhorn?'.[25] Nevertheless Tyndall went
over the glacier to Breuil, the base for climbing the Matterhorn from the Italian side.
He enjoyed being alone on a glacier, though many people disapproved: with reason,
since the danger of hidden crevasses is always present. He found in solitary travel in
high mountains that 'you contract a closer friendship with the universe than when
you trust to the eye and arm of your guide'. But when he had crossed the Théodule,
Bennen told him that the Matterhorn couldn't be done from Breuil in a day, and that
there was no sleeping place on the mountain for two people. Tyndall 'saw the point
that we had attained last year, and arrived at the all but certain conclusion that the
top, under the circumstances was unattainable'.[26]

It was a surprisingly cautious response, as they simply gave up the idea. The waiter
and the proprietor at the inn in Breuil, doubtless anticipating the fame and custom
should an ascent be made, were unimpressed by their backsliding. Tyndall went back
to Zermatt and walked round to Saas, finding Galton and Hooker there. After a rest
day he decided to cross the Monte Moro pass to Macugnaga, to see the eastern side of
Monte Rosa. He returned to Zermatt by forcing a route over the mountain wall above
Macugnaga with Bennen and Andermatten—up a dangerous couloir in which
Bennen deflected with his axe a stone that came whistling down at them—and then
headed for home.

By the time the next alpine season came round, in 1862, Tyndall was quite run
down. To the stress over the Mayer dispute was added the criticism by Magnus of his

results on the absorption of radiant heat by aqueous vapour. Three days after he left for the Alps, the Managers of the Royal Institution resolved: 'that Professor Tyndall be relieved from Lecturing during the continuance of his present investigations'.[27] This appears to have been a generous and open-ended resolution, since Tyndall did not have lecturing commitments in the autumn. However, he did not give the Christmas Lectures, and did not lecture again at the Royal Institution until after Easter 1863, when he gave just seven lectures on 'Sound'. He had been planning for some time to meet Huxley in Grindelwald, and suggested that Lubbock might join them too. Just as he left for Switzerland at the beginning of July, he wrote to Hector Tyndale in America, who he knew was caught up at Harpers Ferry, where there was soon to be a major battle in the American Civil War: 'Well my dear Hector however my general sympathies may tend in this quarrel be assured of it that there is at least one man in England who would glory in your success, and sympathise with you in reverse … Were I a praying man I would say may God guard you'.[28] Tyndall's sympathies, like those of many in England at this early stage of hostilities, lay clearly with the Confederates.

As planned, Tyndall joined Huxley in Grindelwald, where they spent a quiet week before walking to the Rhône glacier to be joined by Lubbock. He was on his first visit to Switzerland, and Tyndall climbed the Galenstock with him. They went on to the Aletsch glacier but were stymied on an attempt to climb the Jungfrau when they came across an accident. Bielander, one of their porters, had fallen about 40 feet into a crevasse. Bennen, in a 'frenzy of lamentation', declared that he was dead. Lubbock demanded evidence, as a moan issued from below. With their coats knotted together, Lubbock let Bennen and Tyndall down into the crevasse, where they scraped away the ice and snow burying Bielander, extracted him, carried him to the bivouac site, and piled him under warm clothes all night. As Tyndall wrote: 'There was certainly no more real danger of falling into a crevasse on the Aletsch Glacier, than there is of being run over by a cab in crossing from Albemarle Street to James's Street. Recklessness however makes both positions dangerous'.[29]

Tyndall reached Zermatt on 24th July, leaving Huxley at Visp to return to England. He was feeling recovered in health, and aimed to go the next day to Breuil.[30] He wrote to Juliet Pollock: 'The Matterhorn is at hand; but I shrink from augmenting the excitement that already exists regarding it'.[31] But he did not shrink. Finding at Breuil that 'the enemy' (Whymper) had made an attempt and decided it could not be climbed without ladders, Tyndall set off with Bennen and Walter on Sunday 27th July for an attempt which took him beyond Whymper's highpoint. Whymper, who lent Tyndall his tent, must have waited anxiously. The negotiation about the ascent beforehand, and whether Whymper would join the party, led to acrimonious and conflicting accounts in their later publications about the episode. The climbing was desperate, though for an hour, Tyndall was sure of the summit.[32]

But they were stopped at the final steepening: 'Had the crags been planted on a plain we should no doubt have attempted them, though the chance of slipping approached almost to a certainty, but we looked around and saw the fate which awaited us should the footing of any one give way. A fall of at least 5000 feet, tumbling from precipice to precipice would be the consequence. The thing looked horrid'. Tyndall was devastated: 'This defeat has fallen upon me like the chill of age, and I must "mix myself with action" to shake it off. On quitting the mountain it poured down hail like grapeshot upon us…Well goodbye to him and goodbye to you and goodbye to my climbing. For there is nothing else in the Alps that I should care to do'.[33] Tyndall had reached the point now known as the Pic Tyndall, but it was not the summit. A small consolation came the following year, when Clarence King, the American geologist and mountaineer who became the first Director of the United States Geological Survey, wrote to say that he had climbed and named Mount Tyndall in the Sierra Nevada, a positively Alpine peak at 4,275m (14,025ft).[34] The Pic Tyndall lies at 4,241m.[35]

After this deep disappointment Tyndall set off with Bennen on the 'Tour of Monte Rosa', crossing the Col Betta Furka to Gressoney, in the Val de Lys, from where he climbed the Grauhaupt. He wrote to Juliet Pollock: 'As far as I am concerned therefore the Matterhorn is inaccessible and may raise its head defiant as it has hitherto done—the only unconquered & unconquerable peak in the whole Alps'.[36] He walked on over the Col de Val d'Obbia to Alagna, over to Macugnaga and eventually to Saas, subsequently crossing the Alphubeljoch. He had been aiming for the Italian lakes but, driven off by the heat, went instead to the Jura and Neuchâtel, looking for signs of glaciation which he had also traced in the Italian valleys running down from Lyskamm and Monte Rosa.

Tyndall had long had an interest in geological matters. He had attended occasional meetings of the Geological Society, usually at the invitation of Lyell, and doubtless enjoyed its reputation for heated discussions.[37] He was more than prepared to challenge papers presented.[38] Nevertheless, though he became a Fellow in 1868 he did not publish in geological journals. On this trip, Tyndall found evidence of glaciers extending from Mont Blanc all the way to the Jura, and sent a paper 'On the Conformation of the Alps' to the *Philosophical Magazine* on his return in mid-August.[39] He argued that the Alps had been formed by uplifting, a theory championed by Lyell and Darwin. Applying his understanding of forces of tension from his study of glaciers, he thought that the valleys could not have been formed by fracturing, as some believed, since if so they would be at right angles to the present valleys, as the valleys of Lys, Ayas, and Tournanche were to the main ridge stretching from Monte Rosa to the Matterhorn. Water he also considered insufficient to explain the formation of the valleys by erosion, since towards the end of old glaciers the impact of water had caused relatively little more erosion. He pointed to the widespread signs of glaciers, now much melted. These glaciers, evidenced by the scarring and fluting of rocks on the valley sides high above current rivers, had operated on a stupendous scale, competent to plough out Alpine valleys. He imagined the glaciers forming after the uplifting of mountains, creating a glacial epoch, yet sowing the seeds of their own destruction as they sank down and the heat generated in the new valleys rose and melted them: 'To account for a glacial epoch, then, we need not resort to the hard hypothesis of a change in the amount of solar emission, or a change in the temperature of space traversed by our system. Elevations of the land which would naturally accompany the gradual cooling of the earth, are quite competent to account for such an epoch; and the ice itself, in the absence of any other agency, would be competent to destroy the conditions which gave it birth'. That may be an incomplete analysis, but Tyndall's understanding of the ability of water vapour to absorb and radiate heat led him to think on global scales.

* * *

The weather in the Alps the next year, 1863, was atrocious. Tyndall left for Switzerland on 17th July—Hirst planned to follow in ten days or so—accompanied by Sclater, the Secretary of the Zoological Society, who remained with him until the end of the month. He intended to be away for about six weeks, but the weather drove him home after four, though they did climb the Jungfrau. Tyndall wrote to Hirst on 29th July from the Eggishorn, having had continuously bad weather and nearly having killed himself descending to the Grimsel, to say that he had given up plans to go to the Engadine and to Marburg with Hirst, and would remain in the Oberland before returning to England and going on to Ireland to meet some newly discovered relatives.[40] Hirst did not receive this letter before setting out. He wrote to Tyndall from the Engadine to say that the weather had been fine and that the Swiss Naturforscher were meeting there in Samaden, where Tyndall's absence was much regretted by Clausius, Wartmann and others.[41] Tyndall's theory of the formation of glaciers was attacked, and Hirst wished he had been there to defend himself.

On his return from Ireland, Tyndall made some amends for the poor Alpine season by spending several days with Huxley in the Lake District *en route* to the British Association meeting in Newcastle, over which Armstrong presided. Starting at Windermere they crossed Helvellyn to Ullswater (Tyndall again descended the famous Striding Edge but sent Huxley down the easier Swirral Edge), and the following day over to Keswick in Borrowdale.[42] He noticed there, after his Alpine experience, that until this visit he had not had 'any adequate idea of the magnitude of the ancient glacier action of the region'.[43] Crossing into Buttermere, they traversed Red Pike into Ennerdale and walked over Black Sail Pass to Wasdale. They stayed there for two days while Tyndall made an attempt on Pillar Rock, the huge formation on the side of Pillar mountain, which had probably been ascended fewer than thirty times by then, and remains a remote crag for connoisseurs of rock-climbing.[44] The sport of rock-climbing was then barely known in Britain. He wrote: 'I climbed the Pillar rock, but the driving rain and cloud and the wet slippy crags were too much for me. You might break your neck very respectably on the Pillar Rock'.[45] They went on over Bowfell to Langdale, where Huxley left for London, much refreshed, and Tyndall headed for Newcastle. He was Armstrong's guest for the week, after which he spent some days at Alnwick Castle, residence of the Duke of Northumberland, and then at Chillingham Castle, seat of the Earl of Tankerville, where he stalked the famous wild cattle with Richard Owen.[46]

After the disappointment of 1863, Tyndall initially proposed going to Norway in the summer of 1864, but changed his mind in favour of the Engadine and set out at the beginning of July.[47] He explained the plan of campaign to Hirst: 'I go straight to the Engadin, and make Pontresina my headquarters. Ball joins me there…We are afterwards to be found by Forster at Santa Caterina—not far from Bormio, where we all there hope to attack the Ortler Group, and spend a week together. Afterwards Ball purposes that we should visit the Dolomite Mountains, but this is a part of the expedition to which I have not yet committed myself'.[48]

After two days in Paris in early July, meeting scientific friends, Tyndall went to Basle and on to Chur, with a particular scientific objective in mind. He had already entered the debate about whether valleys were fissures created by upheaval of the land, or exclusively the work of erosion, since for him the formation of mountains was 'one of the great questions in geology today'.[49] The large width of the valley and the detritus washed down and accumulated suggested to him that, though fissures might start a process, it was erosion by water and ice which had scooped out the valleys, or dissolved rocks in the case of limestone, creating deep gorges. He had heard that the Via Mala was held up as an example of the fracture theory, so he took a diligence there from Chur. He found clear signs that the river had once been hundreds of feet higher and wrote up his observations in September, as a follow-up to his paper two years earlier stressing the glacial origin of valleys and cols.[50] Ramsay was delighted when Tyndall wrote in support of his response to Murchison's Anniversary Address to the Geographical Society, in which he challenged Ramsay's—and Tyndall's—views on the glacial origins of valleys and lakes, sticking instead to the fracture theory of their formation.[51]

Meeting Hirst in Pontresina on 9th July—by chance they arrived within fifteen minutes of each other—Tyndall settled down to glacier measurements and mountain climbing. He had also taken some work with him, revising *Heat* for the second edition, which was published in 1865. He found the landscape beguiling, writing to Juliet Pollock: 'This is a most lovely country... The mountain forms are very fine, and the valleys crowded with pines very beautiful'. He went on: 'The main problem which has occupied me is the formation of these mountains and valleys, and on this subject Nature has taught her pupil a great deal. In fact I do not think I should enjoy Switzerland were it not for this teaching. Mere beauty, mere grandeur, the enchantment of emotion—are not sufficient to fill a human life. Intellectual occupation must be added... the growth of knowledge, the acquisition of new truth... These are the things which shaken by the emotions make the music of the Alps to your friend'.[52] But the weather was capricious. Tyndall caught a chill on the glacier which put him out of action for four days, but he spent several days making measurements on the Morteratsch and Rosegg glaciers, climbed the Piz Languard, and was prevented by bad weather from attempting the lofty Piz Bernina. He was focused on measuring the speed of the glacier where it slowed down towards its end, to help explain the widening of the moraine there, findings that were relevant to his views on the formation of valleys and lakes by glaciers.

This trip was most memorable for Tyndall's accident on 30th July while descending the Piz Morteratsch, which he climbed with the Pontresina guides Jenni and Walter and two Englishmen, Thomas Hutchinson and Henry Lee-Warner, masters at Rugby School. Jenni led on the descent, with Tyndall next. All went well until, against Tyndall's instinct to keep to the rocks, they started to cross a couloir of steep ice in which Jenni cut steps.[53] Jenni got across, followed by Tyndall, and turned to warn the party to keep carefully in the tracks, adding that a false step might detach an avalanche. Tyndall recalled: 'The word was scarcely uttered when I heard the sound of a fall behind me, then a rush, and in the twinkling of an eye my two friends and their guide, all apparently entangled together, whirred past me. I suddenly planted myself to resist their shock, but in an instant I was in their wake, for their impetus was irresistible. A moment afterwards Jenni was whirled away, and thus all five of us found ourselves riding downwards with uncontrollable speed on the back of an avalanche which a single slip had originated'. Tyndall and Jenni both tried to break the descent with their batons, which were ripped from their grasp. Jenni deliberately plunged into a crevasse but was catapulted out by the force on the rope. As the avalanche swept them further, Jenni 'rose incessantly and with desperate energy drove his feet into the firmer substance underneath. His voice shouting "Halt! Herr Jesus, halt!" was the only one heard during the descent'. After an immense struggle they came to a halt at the edge of an ice cliff, within a short distance of complete disaster. They had fallen about 1,000 feet. None had suffered serious damage, but Tyndall found a portion of his watch-chain hanging round his neck. The watch was gone.

On the day after the Morteratsch accident, Tyndall and Hirst set off over the Bernina Pass on foot to Poschiavo and then to Bormio. They climbed Monte Confinali together and eventually met up with John Ball, with whom Tyndall spent several days before returning to Pontresina. From his knowledge of physics, Tyndall deduced that his watch might be findable: 'Both the guides and myself thought the sun's heat might melt the snow above it, and I inferred that if its back should happen to be uppermost the slight absorbent power of gold for the solar rays would prevent the watch from sinking as a stone sinks under like circumstances'. With five friends, including the MP Frederick North, Tyndall set out to search for the missing watch. They located the track of the original avalanche, and had not spent twenty minutes when a cheer from one of the guides, Christian Michel of Grindelwald, announced its discovery and 'the application of its key at once restored it to life'.[54] Tyndall returned via Zurich for the meeting of the Swiss Naturforscher as he had promised Clausius.[55] He was back in England on Saturday 27th August and dined with the Lubbocks on the Sunday with Hirst and the Hookers.

* * *

It was the following season of 1865 in the Alps that was the most memorable and the most notorious. Tyndall may not have had immediate designs on the Matterhorn, but Whymper did. The simultaneous triumph and tragedy of his party of seven on the first ascent saw the end of the innocence of the Golden Age. Descending after their success, the inexperienced Douglas Hadow slipped and knocked the guide Michel Croz off his feet, dragging Charles Hudson and Lord Francis Douglas after them. The rope snapped between Douglas and the guide Peter Taugwalder, leaving Taugwalder, his son, and Whymper on the mountain while the other four fell over the north face to their deaths thousands of feet below. The body of Lord Francis Douglas was not found. It remains undiscovered to this day.

Tyndall and Hirst left London on 10th July, four days before the accident. They started their Alpine wanderings in Stachelberg, meeting there Sigismund Porges, who had made the first ascent of the Mönch in 1857 and the second ascent of the Eiger in 1861. Prevented by extortionate guides from climbing the Tödi, they had reached Gadmen on 19th July when they heard rumours of a disaster. Indeed Tyndall heard it as his own death: 'He is killed, sir... killed upon the Matterhorn'.[56] But they continued their planned tour, because the accounts were vague, taking in Grindelwald and Mürren—visiting the Schilthorn to see where the newly-wed Alice Arbuthnot had

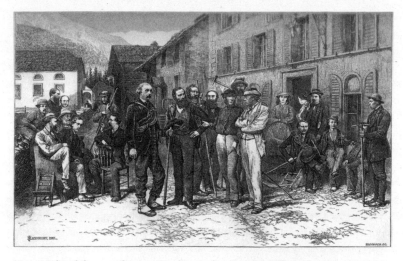

Fig. 26. The Clubroom of Zermatt in 1864. In the centre are Ball, Mathews, E. S. Kennedy, Bonney, Ulrich Lauener, Tyndall, and Wills. Lucy Walker stands by the doorway of the Monta Rosa Hotel.

been struck dead by lightning—on to Kandersteg and over the Gemmi Pass to Brig. After a detour to visit Bennen's grave at Ernen (following his death in an avalanche in 1864) they walked on to Zermatt, arriving on Monday 31st July. Lubbock arrived on 3rd August and, with Hirst, witnessed Tyndall's extraordinary plan to find the body of Lord Francis Douglas.

As Tyndall wrote to Faraday: 'I had been informed that Lord Francis Douglas's Mother suffered much from the idea of her son not having been found. I therefore resolved to make an effort to regain the body. I proposed to climb to the point from which they fell, fix irons there in the rocks, attach ropes to the irons, and descend by the ropes along the precipices, examining the mountain right and left of the line of ascent'.[57] It was a mad idea, inspired by Tyndall seeing the road-menders at work boring holes in cliffs to fix stanchions. He engaged one of them, a mountaineer called Lochmatter, to help him, and sent a man to Geneva to purchase 3,000 feet of rope, which were brought to Zermatt and taken up to the chapel at Schwartzsee. Perhaps fortunately the weather was abominable; heavy snow made the rocks of the Matterhorn thoroughly dangerous. Tyndall waited eighteen days, enjoying excursions with Lubbock, Hirst, and others, before giving up to attend the Jubilee of the Helvetic Society of Natural Sciences. Although he was not well in Geneva, he enjoyed De la Rive's hospitality and met friends such as Clausius, Magnus, Dove, and

Wiedemann. Tyndall, Hirst, and Lubbock left Geneva on 24th August and were back in England two days later.

Alpine climbing techniques, including safe use of the rope, developed slowly in this period. More than ten years after Tyndall's accident, another Alpine catastrophe was brought to his attention by Du Bois-Reymond. On 18th August 1878, four climbers, including Du Bois-Reymond's assistant Carl Sachs, were killed on Monte Cevedale. According to Du Bois-Reymond, the three travellers with two guides were crossing an ice slope near the top when Sachs missed his footing, dragging the others down over a 200ft ice-wall and a further 1,500–2,000ft until one of them, Salomon, was caught by the leg in a small crevasse. His leg was broken twice, but at this juncture the rope snapped as it had on the Matterhorn. Salomon was saved from death and the four others sped to their doom below. Du Bois-Reymond asked Tyndall to use his influence with the Alpine Club to instruct the guides to cease this practice of roping all together. He suggested it would be safer for the two guides to take each traveller in turn over such dangerous places.[58] Tyndall sent the letter to the *Times* with his response. He told Du Bois-Reymond that he thought his solution would not be practicable in all situations. It would be better to split large parties up, but part of the problem was the expense of guides which led inexperienced parties to take too few, and the guides to take the risk.

It was another dreadful summer of weather in the Alps in 1866. Tyndall went to Engsteln and climbed the Titlis, then on to Rosenlaui hoping to climb the Wetterhorn. Driven from it by bad weather he was further driven off the lowly Faulhorn by calf-deep wintry snow. Aiming for the familiar comforts of the hotel at the Eggishorn he crossed the Petersgrat to Platten in the Lötschental and over the Lötschen-Lücke to the Aletsch glacier and the trek down to the Eggishorn. There he encountered the Reverend Arthur Girdlestone, 'a tall fine fellow' with whom he planned to try the Aletschorn (Figure 27). Girdlestone, whom he must have narrowly missed the previous year in Zermatt, became famous, or infamous depending on the point of view, for publishing *The High Alps Without Guides* in 1870.[59] Guideless climbing was still frowned upon by the majority of climbers, and by the Alpine Club itself. To seek to obviate the danger, Girdlestone would gather his porters around him in prayer before his ascents. That would not have impressed Tyndall, who was nevertheless a natural partner after his solo ascent of Monte Rosa. But the weather remained fickle for their attempt: 'the heavens blackened utterly. The summit which first rose clear above us became enveloped: we worked up to a bunch of stones where we proposed having a bit of food. But before we had finished a gust of storm so violent struck us that we instantly backed up and commenced the descent. We paused again—exceedingly unwilling to yield as long as there was any hope. But hope was more and more shut

Fig. 27. Aletschorn from the Eggishorn.

out by the utter blackness of the air...And now the snow began to fall and my companion who had quite set his heart upon the ascent saw that all possibility of effecting it was cut off'.[60] The weather remained poor, so they explored the Italian lakes, going together as far as Milan.[61]

By comparison, the Alpine season of 1867 was notable for an ascent of the Eiger. Tyndall went straight from England to Grindelwald. He had spent two training days on the glaciers, 'when the clouds dissolved and the mountains rose so gloriously in the sunlight that a longing to climb one of them beset me. I chose the most difficult—a splendid pyramid, which as point of steepness and real mountain character beats all others in Oberland'.[62] The Eiger, despite coming in at just under the magic 4,000m (3,970m, or 13,020ft), is still a prized peak today. The profile of its majestic Mittellegi ridge looms above Grindelwald. The only route up the Eiger at that time was by the west flank from the Wengen Alp, skirting the fearsome north face, which remained unclimbed until 1938. Leaving at 1.30am, Tyndall was on the summit by 9, finding mostly ice on the exposed upper slopes above the north face, which in those days before modern crampons required deep steps to be cut to aid the descent. There is an old mountaineering saying that one has not climbed a mountain until it has been safely descended, and with tiredness and thawing conditions the descent is often more difficult and dangerous than the ascent. That was the case for Tyndall on the Eiger: 'For the first time in my life I leaned my face to the ice slopes and went backwards down them' (the author confesses that he did likewise on the same slopes, despite the aid of modern crampons). It was a long descent and they were not down until 6pm.

After an enforced rest day at Lauterbrunnen in bad weather, Tyndall went over the Petersgrat again and then to Belalp for a couple of days before heading for Zermatt and the Riffelberg: 'There was the grim Matterhorn in front of me. There was my own beloved mountain, the ever beautiful Weisshorn. Nothing in the whole alps approaches it in beauty and nobleness. But my thoughts were now turned toward the other monster'. He crossed to Breuil and tried to make arrangements with the guides there if the weather played fair. But this year it was not to be. He stayed for three weeks, with 'the Matterhorn encased in ice, and swathed in cloud'.[63] 'No sooner however did I leave Switzerland than the clouds vanished, the rain ceased, the weather brightened, and a gentleman in whose company I intended to ascend the Matterhorn succeeded in getting to the top. Altogether the mountain has treated me very badly'.[64]

But the 1868 season changed all that, despite an inauspicious start. Hirst noted that Tyndall arrived in Paris in July 1868 depressed and unable to sleep. They travelled slowly through Switzerland together as Tyndall regained his spirits and his strength, staying at the Giessbach Hotel for several days before exploring the Simmental. After nearly a fortnight, Tyndall was fit enough to climb the Diablerets, from where they went on to Martigny in the Rhône valley, to 'stop in that wretched oven of a place all night'. But he was still not well, which makes what happened next all the more remarkable.

Tyndall's three most notable climbs are the first solo ascent of Monte Rosa in 1858, the first ascent of the Weisshorn in 1861, and the climb he was now about to attempt, the first traverse of the Matterhorn, from the Italian side into Switzerland. He took an Einspanner[65] with Hirst to St Bernard and they went on to Chatillon, with Tyndall still feeling very ill. But the next day, 22nd July, he left with Georges Carrel on mules to Breuil. The Matterhorn was unfinished business for Tyndall but it remained a major challenge. When Tyndall set out, it had only been climbed four times since the first ascent three years before, all from the Italian side. The only descent on the Swiss side, after its initial ascent, had resulted in four deaths. But Tyndall reasoned that since three had succeeded in getting down alive it must be possible to repeat the feat. He was nearly prevented from starting by a combination of doubtful weather and the need to leave early on a Sunday, 26th July. That could not be done until after Mass. A little persuasion resulted in a service at 2am, after which they set off.[66]

The weather was kind. As Tyndall described it: 'after 2 or 3 hours of climbing the fog was folded up like a scroll by the south wind and rolled grandly over the Théodule into Switzerland'. Avoiding stone bombardments by keeping to arêtes, they gained the rudimentary hut where they made their coffee and soup, and where they planned to stay for the night. The piercing cold kept Tyndall awake; he was so stiff he felt obliged to stand up. His guides 'wrapped me up for a time in their sheepskin and by degrees restored the proper temperature...I should have lain down with them, and perhaps accepted a proffered night cap, had I not seen the neck of one of the porters covered with flea bites'.

Day eventually dawned. The guides 'fed on beefsteak réchauffés, I did not touch food, purposing to reach the top of the Matterhorn in a spirit of prayer and fasting'. From the top of the Pic Tyndall, 'a very formidable piece of work lay before us...The slope to the left was terribly steep, and the precipices to the right abysmal. Herein consists the unparalleled grandeur of the Matterhorn. For savagery in its crags and precipices on the southern side it has no competitor...From the arête we passed to the final cap of the mountain. About 700 feet of exceedingly difficult rockwork now lay before us. And it was trebly difficult today through the crust of ice which covered the rocks...At length we came to the end of another rope stretching down over the iced rocks. The rope had been frayed by the wind and the guides hesitated to trust it. It was in part covered with ice which rendered any firm grasp of it impossible...This obstacle once surmounted an easy climb of a few minutes placed us upon the summit of the Matterhorn'.

They had arrived on the Italian summit. It is a few feet lower than the Swiss summit which is reached along a narrow arête. The cloud swirled around them, then cleared, allowing them to see straight down to Zermatt. At both ends of the arête were planted pieces of the old ladder which Tyndall had taken with him in 1862. At the Swiss end, to their surprise, they found foot prints. An ascent had been made the day before by Julius Elliott (who was to die the following year on the Schreckhorn): 'This abolished all wavering as to descending on the Zermatt side, for though none of us knew anything about this side of the mountain we trusted to the traces to help us. I here took about an ounce of food:—for the first five hours I had eaten nothing, and this ounce of food was all that I consumed in what turned out to be a day of nineteen hours. The descent at first was in no way formidable, but it soon became so. The snow was "like flour" and rested so loosely upon the steep rocks that the most tender treatment was necessary to keep it attached to the surface on which it rested'.

Taking enormous care, and by doubling the rope at various points to lower themselves down off projecting rocks, they reached safer ground after three hours. Tyndall wrote to Clausius: 'I found the slope on which those poor fellows slipped in 1865, exceedingly dangerous'.[67] They now had the added danger of falling rocks, and kept to the arête as far as possible to avoid them. 'Down we went; but at length we went astray, and it was dark before we could free ourselves'. Losing their way in the valley, amid trees and brushwood, they finally reached Zermatt at 1am. Felice Giordano, the Italian alpinist and Inspector of Mines, who had desperately wanted to make the first ascent of the Matterhorn, repeated the feat on 4th September and wrote to Tyndall, sending a sketch of the geological section of the mountain.[68] Tyndall rested at the Riffel, where he stayed for a week working on his address for the British Association meeting in Norwich. He had succeeded on the Matterhorn, and accomplished his last significant climb.

15

CLOUDS OF
IMAGINATION
1868–1870

Fresh from his triumph on the Matterhorn, Tyndall barely had time to pause before setting off for Norwich and the 1868 Annual Meeting of the British Association. Hooker, proposed by Tyndall and seconded by Lubbock, had eventually been persuaded to preside, having initially refused an approach by Tyndall and Hirst.[1] A large group gathered on 19th August for dinner at the Maids Head Hotel before going on to hear his Address. Most of the X-Club were present: Tyndall, Huxley, Frankland, Hirst; and the Lubbocks, Spottiswoodes, and Busks. Tyndall was President of Section A, the Mathematical and Physical Section, and gave his Address the following day.[2]

The Norwich Address is one of Tyndall's most significant statements about the nature and scope of science, its power, and limits. On the one hand, with an allusion to other debates, he observed that 'miracles are wrought by science in the physical world'. On the other, he acknowledged the existence of 'that region to which the questionings and yearnings of the scientific intellect are directed in vain'. Tyndall rooted his scientific philosophy in the self-organizing power of matter, illustrated by the formation of salt into crystal structures, guided by molecular forces. He argued that these forces necessarily led to the structures, in consistent and predictable ways. But then the ear of corn, the plant, and even the animal are logically no different; they are just more complex structures. Life is self-organizing. He declared: 'Incipient life, as it were, manifests itself throughout the whole of what we call inorganic nature'. This is Tyndall's materialism, the opposite of a vitalism that imagines a force directing and animating matter. He imagined the formation of a crystal, a plant or an animal as a purely mechanical problem of self-assembly. The challenge of understanding was of complexity, not quality. Tyndall acquired around this time a translation of *De rerum natura* (On the Nature of Things) by the Roman poet Lucretius, and was immediately interested by the man who 'deduced the world and all that therein is from the "fortuitous concourse of atoms" '.[3]

But he thought there were limits to understanding. He asked: 'how does consciousness infuse itself into the problem?'. He could 'hardly imagine there exists a profound scientific thinker...unwilling to accept the extreme probability of the hypothesis, that for every fact of consciousness...a certain definite molecular condition is set up in the brain'. So, given the state of the brain—Tyndall would have been fascinated by modern brain imaging—the corresponding thought or feeling might be inferred. But here is the problem: 'even if we could associate a definite thought and a definite molecular arrangement in the brain...the "WHY?" would remain as unanswerable as before'. The chasm between the two classes of phenomena, say of molecular structure and love, would remain intellectually impassable. Tyndall could envisage a psycho-physical parallelism, a naturalistic dualism, but could go no further. Meanwhile, he thought, 'the mystery is not without its uses. It certainly may be made a power in the human soul; but it is a power which has feeling, not knowledge, for its base'.

This is Tyndall's account of the grounds of difference between religion and science, to which he held in all his further writings. In a remarkable postscript, *Musings on the Matterhorn*, dated to his ascent just three weeks before, Tyndall mused: 'Did that formless fog contain potentially the *sadness* with which I viewed the Matterhorn? Did the *thought* which now ran back to it simply return to its primeval home?'.[4]

The Address made an immediate impact. Armstrong, having seen a report in the *Times*, wrote from Cragside: 'What a contrast you have provided to the narrow dogmatism of materialists and spiritualists!'.[5] To understand Tyndall, one needs to appreciate his transcendental view of matter. He wrote earlier: 'Matter...is, at bottom, essentially mystical and transcendental'.[6] Perhaps he moves towards what would today be called panpsychism, an intermediate position between physicalism and dualism that imagines consciousness as a fundamental feature of the universe.[7] Matter is not just 'dead' or 'brute'.

Judging from his laboratory notebooks,[8] Tyndall started experimental work again in earnest on 11th September, though he dashed to Folkestone a couple of days later to see Bence Jones, who was unwell. Experiments, some carried out and recorded by his assistants, took place throughout most of the period to mid-December. Almost immediately he made two significant discoveries that set the direction of his work for many years. The first, on 5th October, was the observation that dust in the air, which he referred to as 'floating matter', could be burnt up by a flame. Tyndall claimed that it was therefore organic, a fact that was unexpected. This was a shaky conclusion, but one to which he stuck; it gave him a powerful rhetorical weapon. He later observed that for many years he had been using dust to reveal the paths of luminous beams through the air, but it was not until 1868 that he reversed the process and used a beam to explore the nature of the dust.[9] Burning up this dust was essential, as it allowed him to experiment on vapours uncontaminated by other matter that might scatter light. A few days later he made the second discovery; that solar light could decompose the vapour of amyl nitrite. It seemed to cause a chemical reaction.

Tyndall was working hard, but without conflicting demands, knowing that he could escape to the Isle of Wight if he felt any symptoms of breakdown. He made time for a visit with Hirst to Sir Benjamin and Lady Brodie at Box Hill, who were keen for him to take a house close by, 'an indescribably beautiful place in Surrey. But it is an hour and a quarter from town, and the rent is terribly high'.[10] The X-Club reconvened for the first meeting of the season in October with Asa Gray as the guest, over from America. This was Gray whom the Scottish-American naturalist and environmentalist John Muir considered to be 'a great, progressive, unlimited man like Darwin and Huxley and Tyndall'.[11] Gray and his wife were staying with the Hookers, where Tyndall and Hirst dined with them. Tyndall told Darwin, who was often not in the best of health, that he had tried to make it clear to Mrs Gray that his ill-health was a benefit, 'inasmuch as it compelled you to ponder a great deal, and this accounted for the extraordinary proportion of thought which your works display'.[12] In congratulating Tyndall on his Norwich Address, Darwin had written: 'how entirely I agreed with you in a paper published a long time ago...in which you enlarged on the wonderful power of pondering; I believe you have hit on the whole secret of scientific discovery'.[13] Later in the month, Darwin invited Tyndall to stay for the weekend at Down House and to dine with the Hookers and Grays.[14]

* * *

The work in the laboratory bore speedy fruit, and on 24th October Tyndall submitted his paper 'On a New Series of Chemical Reactions Produced by Light' to the Royal Society.[15] The speed doubtless reflected his desire to establish priority for his development of the new technique of using a concentrated beam of sunlight or electric light to probe the vapours of volatile liquids. He realized that he had a means of exploring molecular structure, by examining whether the absorption of light was a function of the molecules themselves or of their constituent atoms, a question to which he had been directed some years before through a conversation with Clausius. His conclusion was that the absorption occurred within the molecule as a whole, and that it was probably the synchronism of the vibrations of one portion of the molecule with the incident light waves that led to the shaking asunder of the molecule and its decomposition into other chemicals. The nature of these chemicals he left to the chemists to establish, urging Frankland to take an interest.[16] Tyndall recognized that similar experiments with chlorine might be evidence that it was molecular (containing more than one atom), in which he was right. This was a change from his previous (incorrect) conclusion from experiments with radiant heat that the elementary gases such as hydrogen, oxygen, and chlorine were atomic. He went on to describe the beautiful blue clouds formed by the decomposition of vapours, and asked: 'May not the aqueous vapour of our atmosphere act in a similar manner? and may we not fairly refer to liquid particles of infinitesimal size the hues observed by Principal Forbes over the safety-valve of a locomotive, and so skilfully connected by him with the colour of the sky?'. Tyndall was on the way to explaining why the sky is blue. He updated Clausius with his findings about the decomposition of chemicals by light, and the cloud forms produced, which he thought might be characteristic of the substances.[17] Clausius thought he would probably open up a new field of research.[18]

Having staked his claim he spent time at Spencer's, but turned back on the way to High Elms with Hirst the following day, unwilling to face the company. A break in the Isle of Wight beckoned, where he spent nearly a week, based at the Royal Hotel in Ventor and missing the meeting of the X-Club. There may well have been further amatory disappointments, hinted at in letters to Juliet Pollock and Hirst. Outside the hotel 'the sea moaned, and there seemed a responsive moan at the bottom of my life'.[19] He paid homage again at the grave of John Sterling, telling Hirst: 'You remember writing to me about John Sterling—indeed it was your letter that caused me to buy his life when I could not well afford the outlay. I lay hold of you my boy by many memories...It gives me pleasure to think that I have a strong hold of Huxley in the same way. I was the only friend behind him as he walked to the grave of his first-born'.[20]

Though Tyndall's mind may not have been in perfect shape his physical stamina was impressive: 'I find I am still good for 20 miles in four hours. I can walk and trot 5 miles an hour without much labour to myself...but it is my head that wearies me'.[21] Either Tyndall's spirits rapidly improved, or Hirst felt he should go out more, for in November they went to the theatre three times in the space of a week, both greatly pleased by Miss Bateman at the Haymarket. At the end of the month they were at the committee rooms in Cannon Street for Lubbock's vote in the 1868 General Election, in the constituency of West Kent, where he stood as a Liberal. It was the first election since the Reform Act of 1867, which greatly enlarged the franchise. More than two million votes were cast, necessarily all by men. The Liberals, under Gladstone for the first time, increased their majority over Disraeli's Conservatives. This was the election in which Mill lost his seat as a consequence of his opposition to Eyre. Tyndall commented to Huxley: 'Mill deserves what he got—but if I had a pocket borough at my disposal he is the first man that I should choose to represent it'.[22] Lubbock lost by a small number of votes as two Conservatives won in West Kent, but gained a seat for Maidstone at a by-election in 1870. He was an important ally in Parliament. One of his earliest initiatives, in 1871, was the introduction of Bank Holidays.

Herschel was intrigued by Tyndall's experiments on the effect of light on vapours and its connection to the blue colour of the sky and polarization of light, and a flurry of letters followed.[23] After a foggy weekend walk with Hirst, Tyndall sent Herschel detailed accounts of his observation of the polarization of light from faint clouds formed from carbon dioxide, noting that 'as the cloud in the carbon dioxide tube became coarser and whiter the polarization diminished materially'. He also observed that 'dust particles of our laboratory, strongly illuminated by the electric light show the same effects with great splendour...The track of the beam is distinctly blue'.[24] The next day he described further experiments, using dust in air and mastic in water, which also showed polarization effects. He wrote: 'I think we shall explode utterly the idea that the polarization of the atmosphere is due to reflection by the particles of air'.[25] Two days later he sent a further batch of observations and conclusions on the behaviour of several vapours: 'They increase for a time in intensity and then begin to grow feebler; after a time the blue cloud has changed to a white one and all traces of polarization disappear. This blue cloud is undoubtedly the first step towards the formation of the real cloud. It is a case of true condensation, and the blue colour is due to the smallness of the particles...may not the continuous accession of new small particles, as the old ones grow bigger be the cause of the whitening of the light?'.[26]

Tyndall knew he was on to something. He sent a letter speedily to Stokes, explaining the cause of polarization in the atmosphere: 'It is the matter thus diffused which produced the effect; for when pure clear air is admitted into the experimental tube it shows no signs of polarization: the common air admitted into the tube shows vividly the colours of polarized light'.[27] In a follow-up letter the same day, he asked Stokes to give his opinion on his explanation for the blue colour of the sky and the cause of polarization.

Tyndall's new suggestion for why the sky is blue went right back to his understanding of twenty years before. Based on his amateur reading, and perhaps conversations with Frankland, Tyndall had written a simple treatise 'On Water' for the *Carlow Sentinel* in 1848. He stated: 'Clear water is of a blue colour...the blueness of the sky depends upon the quantity of watery vapour in the atmosphere'.[28] Five years later, in a piece for the *Queenwood Observer*, he expanded on the role of water vapour: 'On raising our eyes gradually from the horizon to the zenith, we encounter less and less vapour and consequently observe a purer and purer blue'.[29] Now he described the cause of the blue as follows, effectively as the colour of water: 'May not the colour of the sky be due to light transmitted through it? Were the vapour of the air condensed to a liquid shell it would not be competent to colour sensibly the light passing through it; but you perhaps remember those observations of mine on the colour of newly fallen snow; a little crack or hole in such snow is filled with pure blue light. This is no doubt caused by the innumerable reflections that occur at the limiting planes of the granules...Similar reflections must occur in the atmosphere; you cannot whip air & aqueous vapour to a homogeneous whole; the atmosphere is an aggregate of air parcels in different states of saturation. Reflection must take place at their limiting surfaces and in this way the vertical depth of vapour is augmented & the light finally reaches us from the atmosphere as blue'.[30] Stokes thought the question still an open one.[31] Tyndall had not yet reached a convincing explanation of why the sky is blue.

Stokes agreed that a paper from Tyndall on blue sky and polarization was worthy of being communicated to the Royal Society, not least because the idea was new to so distinguished a physicist as Herschel. Though acknowledging the difficulty of showing that the blue was not caused by reflection within homogeneous air itself, he suggested that 'if you can show by experiment that you can by a sufficiently fine precipitate get a blue as deep as that of an Italian sky, and that air can be so purged as to show no sensible beam, that goes far to render improbable reflexion otherwise than by motes [i.e. small particles]'.[32] Stokes recalled that Roscoe had made a similar suggestion about the blue colour of the sky, and that he had talked about it at a Discourse.[33] Roscoe, in this Discourse in 1866, had discussed the views of Leonardo and Goethe, who believed the blue colour was due to white light passing through an atmosphere of finely divided particles, and of Newton, who thought they were due to the effects of minute vesicles of water with walls of different thickness. Some physicists assumed that air itself was blue.[34] Tyndall asked Herschel to comment on his draft paper, and Herschel saw that Tyndall's artificial blue sky had led to an understanding of why the polarizing angle changed as the cloud developed, due to a gradual increase in the size of particles.[35] This effect of the scattering of light from small particles in suspension (for example of water vapour in air, or mastic in water) is now known as the 'Tyndall Effect'. As a parting shot Herschel asked: 'How about the blue colour of water? Does that too arise from disseminated particles? or is the blue colour absorption?'.[36] That question Tyndall explored later. For now, he gave the first full explanation for the blue colour of the sky (Figures 28a and 28b).

His initial paper, 'On a New Series of Chemical Reactions Produced by Light', had been submitted to the Royal Society and published in the *Proceedings* simply to establish his priority. The second paper 'On the Blue Colour of the Sky, and the Polarization of Light' was read at the Royal Society on 14th January 1869.[37] Tyndall acknowledged his predecessors, referring to Roscoe's findings, to the work of others including

Fig. 28a. Tyndall's 'blue sky' tube.

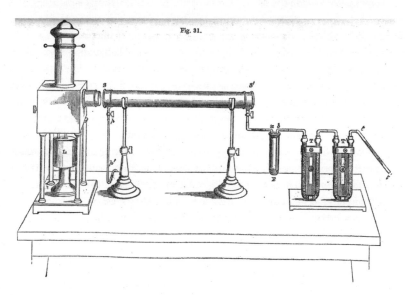

Fig. 28b. Tyndall's 'blue skies' apparatus.

Clausius, who had tried to connect the colour of the sky with suspended water-vesicles, and to Helmholtz, who had ascribed the blueness of the eyes to the action of suspended particles. But whereas many physicists had written and speculated about the cause of the blue colour of the sky, and of polarization, what Tyndall had characteristically done was to subject the phenomena, with the use of model 'clouds' in the laboratory, to a critical analysis. He showed that finely divided matter could produce a blue 'as deep as the purest Italian sky', when the light was polarized perfectly at right angles to the illuminating beam. As the particles of the cloud grew larger, the polarization became less perfect, disappearing eventually at right angles but appearing in oblique directions, while the scattered light became white. Pressed by Stokes's comments to think deeply, Tyndall now realized that he had 'no reason to doubt' that he was exploring particles with diameters 'but a very small fraction of the length of a wave of violet light'. These were not just particles of water. Tyndall stated explicitly that 'any particles, if small enough, will produce both the colour and the polarization of the sky'.

The modern understanding of why the sky is blue attributes it to scattering by molecules in the atmosphere. Tyndall attributed it to very fine particles, without specifically identifying molecules themselves as the cause. It was Einstein, building on Rayleigh's work in addition, who made the detailed mathematical calculations on a molecular basis in 1911. The evening after Tyndall's presentation to the Royal Society he gave his findings as the first Discourse of the season, 'On chemical rays and the light of the sky'.[38] He sent a copy of this to Airy.[39] It resulted in another flurry of letters, after which Airy was forced to admit that he could not provide a satisfactory explanation of sky light and polarization based on his own theories.[40] Tyndall's work intrigued both the old guard of Herschel and Airy, and the central figure of Stokes. It also amazed his audiences. The coloured clouds and artificial blue skies that he created in his laboratory and the lecture room were sensational.

Tyndall was fascinated by his artificial clouds. He called them 'actinic clouds' because they were formed by the short wavelength 'actinic' radiation—essentially ultra-violet—from the Sun or electric light. He experimented with forming clouds from different substances, generating them by the cooling effect of a sudden expansion of the vapour in a tube. In a note to the Royal Society he described the startling diffraction phenomena he saw, and the relative sizes of cloud particles, which were proportional to the specific gravity of the vapour and inversely proportional to the specific gravity of the liquid. Since aqueous vapour is very light, he argued: 'to this circumstance the soft and tender beauty of the clouds of our atmosphere is mainly to be ascribed'.[41] He imagined that cloud particles might be either spherical, reflecting light continuously, or plate-like and sparkling, like the effect of mica in Lake Geneva.

Perhaps this work reminded him of the Alps; his mountaineering writings time and again describe cloud effects, such as the continuous appearance and disappearance of a cloud blowing over a ridge. Indeed, he published two articles on 'Odds and Ends of Alpine Life' in *Macmillan's Magazine* in March and April.[42]

*   *   *

Tyndall's lecture commitments at the Royal Institution did not start until after Easter. He had resigned from the School of Mines because of pressures of work. This created time for more experiments capitalizing on his discovery of artificial clouds, which he recorded from mid-January to mid-March.[43] At the end of February, Tyndall and Hirst dined with the Hookers. A Miss Symonds was there, and it was apparent that the Hookers were angling for her to become Mrs Tyndall. As fate would have it, this was Hyacinth Symonds, who married the elderly naturalist Sir William Jardine in 1871. He died in 1874, conveniently enabling Hooker himself to marry her in 1876 after the death of his first wife. At the X-Club a few days later (where Tyndall was chafed immensely about Miss Symonds), he announced that he had a new theory of comets that he would expound to the Cambridge Philosophical Society on 8th March.[44]

Tyndall told Clausius that it was impossible to examine his artificial clouds without being reminded of the texture of a comet. He then reported the effect of his presentation in Cambridge: 'Adams the astronomer, and Stokes, Miller, and others were present, but did not overthrow the theory. In fact Miller told me afterwards that he found it impossible to pick a hole in it'.[45] Hirst considered the argument 'very ingenious', and Tyndall published it in the *Philosophical Magazine*.[46] It was Tyndall's discovery of the unimaginably small amount of matter that could create his 'actinic clouds' that led him to think about cometary structure. His ideas generated interest from Airy and Herschel. As Tyndall put it, unctuous as ever: 'Nothing could more perfectly illustrate that "spiritual texture" that Sir John Herschel ascribes to a comet than these actinic clouds. Indeed, the experiments prove that matter of almost infinite tenuity is competent to shed forth light far more intense than that of the tails of comets'. The challenges were to explain the enormous tail that could appear so quickly—in the case of Newton's comet of 1680 a tail of 60 million miles was formed in just two days—and its direction, in apparent contradiction to the law of gravitation. Tyndall's theory was that the comet was made of vapour decomposable by solar light, the visible head and tail being an actinic cloud so created: 'the texture of actinic clouds is demonstrably that of a comet'. The tail was not projected matter, but matter precipitated as the solar beam passed through the cometary atmosphere. This formation could be slow or quick, as he had shown with different vapours, explaining the possibility of near instantaneous production of a visible cloud. This theory also explained the shape of the visible tail. It was not one body, but was made up of those parts of the cloud created and illuminated as the cloud moved relative to the Sun. The tail was, therefore, the result of two forces, both originating in the Sun: the 'actinic' tending to precipitate the cloud, and the 'calorific' tending to vaporize it.

Airy thought the idea excellent.[47] Herschel too was impressed, but could not see where Tyndall supposed the real material comet to terminate, or where the head—which he thought might be solid—terminated and the tail began.[48] Tyndall confided to a friend that Herschel had taken his cometary theory to pieces, 'but in such a manner as to enable me to strengthen it'.[49] He had not provided a complete explanation of cometary structure, but he had made a substantial contribution.[50]

The vapour experiments continued until close to Easter, when Tyndall left town for Exeter and the West Country with the young Fred Pollock. Tyndall's lecture load at the Royal Institution was now much reduced: nine weekly lectures on 'Light' each Thursday at 3pm, ending on 3rd June. A few days after this last lecture, he went to Ireland for a week on Board of Trade business. Debus, in London to act as an examiner, was disappointed to find him gone and wrote a heartfelt letter about his lack of intellectual company at Clifton College (where he taught from 1868–1870 after leaving Queenwood).[51] On returning to London, Tyndall had three weeks before setting off on a continental visit with Hirst. He ducked out of a meeting of the Eyre Defence Committee, suggesting to Murchison that any remaining funds should simply be handed over to Eyre.[52] Before leaving for the continent he managed to fit in a visit to Covent Garden for *Hamlet*, in a box in the grand tier with Hirst and Ellen Busk, to present prizes at University College, and to attend the X-Club summer excursion, this time to the Oakland's Park Hotel in Surrey, with an expedition to St George's Hill.

Tyndall confided to Debus that his health was very low, 'but not at all so low as it was last year. Perhaps it may improve as I grow older, and enable me to fulfil to some small extent the promise of my youth'.[53] Tyndall and Hirst left for Antwerp in the *Baron Osy* from Catherine's Wharf on 4th July, enjoying a smooth passage. They had scheduled a month travelling round Germany, sightseeing and visiting friends, before Tyndall's habitual Alpine campaign. In Antwerp they had time to visit a gallery and the cathedral, where Tyndall paid his franc and bolted up the 600 steps of the spire to view the surrounding flat country. After a poor night in a sweltering Brussels hotel, they travelled out to Waterloo. Tyndall noticed a great change since his visit in 1848. Then all was as fresh as if the battle had occurred a year ago. Now the marks, though plain, were much less distinct. They looked into the old museum of relics and round the battlefield: Hougoument, La Haye Sainte, and La Belle Alliance. Tyndall imagined the scene: 'I peopled the ground today with fighting columns, where are they? Where the glory of Wellington as far as Wellington himself is concerned; where the agony of Napoleon as far as Napoleon is concerned?'.[54]

They travelled on to Bonn, calling on Clausius, who had recently moved from Würzburg. With Clausius the next day they climbed the Drachenfels, with its glorious view over the Rhine and the Siebengebirge, crossed the river and were taken up to Rolandseck and the castle ruins, returning from the magnificent Rolandseck station by train to dine with Clausius and his wife and children. From Bonn they took the steamer to Koblenz, sharing the passage with Sir Robert Kane and his family. Heading for the old haunts and friends of Marburg, they walked over to Ems and Nassau, taking the train for Limburg before reaching Marburg on 11th July.

Marburg, still lovely, seemed largely empty of friends. For the following week they made an extensive walk in the Harz mountains, 'through the most glorious rock and woodland scenery',[55] heading for Berlin, where friends abounded. They spent two days there; an evening with Magnus and his family, in the company of Dove, Du Bois-Reymond, Hofmann, and Poggendorff, and visits the next day to Du Bois-Reymond at the University, to Hofmann's 'magnificent laboratory', and to Poggendorf and Dove at their houses.

Bence Jones had suggested that Tyndall visit Liebig's laboratory in Leipzig and Hofmann's in Berlin to see the advantage it gave the continentals, writing: 'Our institution laboratory ought to be the type for this country instead of being an adapted back kitchen'.[56] Youmans, on an earlier visit, had been less guarded in his language, describing the place as 'a dingy hole *down cellar* which Tyndall denominated "the den"'.[57] In time, the laboratories were updated. The chemistry laboratory had been renovated in 1863 when Frankland was appointed, and Tyndall eventually agreed to the upgrade of the physics facilities for the future of the Institution, even though he was loath to see the laboratory he had inherited from Davy and Faraday disappear.[58] Plans were laid in 1871 for a new laboratory, work that Tyndall estimated at £3,000–£5,000, and which eventually took place when he was in America.

Next came a pilgrimage to Weimar and Eisenach, with a halt in Halle to call on Knoblauch. They went first to the duplex statue of Goethe and Schiller in front of the theatre in Weimar, on to Schiller's house, where Tyndall sat quietly at his table for an hour, then on to the graveyard where Goethe and Schiller rest. It was evening, the sun low and sprinkling gold over the acacias and the grass. Descending into the vault, as Tyndall described the scene, 'a series of huge coffins were ranged like packed trunks around it. Near the door were two Sarcophagi . . . I waited with expectancy to hear the names of those to whom they belonged; for by a kind of instinct I felt that they were those of Goethe and Schiller. Well the names were uttered and my instinct proved correct. I cannot tell you how the place moved me'.[59] He felt his health improving. Perhaps it was the emotional impact of the visit, or perhaps the consequence of the regular swims they took, in the Rhine, Lahn, Spree, and other rivers. Tyndall wrote to the *Times* to recommend the habit, after reading an article on the establishment of swimming baths in the Thames.[60]

The final social halts on their itinerary were Heidelberg and Karlsruhe; to Heidelberg to see Bunsen, Kirchhoff—on crutches after an accident—and the Helmholtzes, and to Karlsruhe to see Wiedemann. Wiedemann was not at home but he saw Clara Wiedemann the ('very intelligent') co-translator of his books with Anna Helmholtz ('handsome, & full of spirit and intelligence'), before going on to Strasbourg. Hirst left for home while Tyndall went to Switzerland, aiming to meet up with Fred Pollock in Grindelwald. Writing from there to Bence Jones, Tyndall compared the order in Knoblauch's laboratory to the confusion of Bunsen's, declaring his own books and papers 'symmetry itself' compared to Bunsen's.[61] If that is true Bunsen's must have been highly disordered, as Tyndall was not known as a person of method at work or at home. Tyndall regarded chemical laboratories as places where one would expect to find order, since the techniques and apparatus were well-established. In physics, by contrast, each new investigation needed new apparatus and modes of proceeding.

Tyndall arrived in Grindelwald just as the guides were retrieving the body of Julius Elliott—whose tracks he had followed down the Matterhorn the previous year—from the glacier below the Schreckhorn. Elliott had been killed by falling several thousand feet after unwisely unroping from his guides: 'after three days' search his body was found: no bone broken—not even a little bottle of glycerine in his pocket. He was 27, a member of the Alpine Club'.[62] Tyndall would have gone up to the place where he slipped had the weather not been so abominable. When it relented, he made an attempt on the Wetterhorn with a single guide, taking a direct route which had only been achieved once from the valley in a single day. They were driven back from the final ridge by a violent storm just below the summit.[63] He was evidently in good physical shape, but told Hirst that his life needed a revolution to set it right.[64] Work piled up back in England and he asked Hirst to look out for a heavy letter from the Board of Trade and a second one from Dublin in case they needed attention. He planned to stay in Grindelwald for nearly two weeks, and then to go to the Hotel Jungfrau under the Eggishorn and on to Zermatt. He met Lady Mary Egerton and her husband, the MP Edward Egerton, by accident at St Niklaus, and went with them to the Riffel above Zermatt.[65] He had come to know them a couple of years before, and had recommended she go out to Switzerland, about which she was enthusiastic.[66] But he was still not well in his mind, and went over to Belalp where he found Nevile Lubbock and his wife, with whom he climbed the Sparrenhorn. He was joined by the Egertons again, who had discovered where he had gone. He took them on an expedition to the Oberaletsch glacier, hearing only later that 'poor Egerton, then full of life and joy died suddenly in the Val Ansasca a few days afterwards'.[67] His friendship with Mary Egerton lasted many years. She quickly became one of his substantial coterie of female admirers. Like several of those admirers, she declared her religious beliefs to him, writing after reading Emerson about 'that curious Pantheism, so beautiful, so true, up to a certain point, but then suddenly leaving one afloat on a chilling sea of mist, instead of in the loving presence of the Living God'.[68]

Up at Belalp Tyndall soon 'began to feel a health and lightness altogether new to me. The food here is good, and the air delicious...Exercise taken in such air and sustained by such diet would tell upon almost any malady, and it has told most beneficially on me'.[69] On 23rd and 24th August, he observed the sky from Belalp through a Nicol prism, exploring its polarization.[70] The next day—again to observe the polarization of sky light—he climbed the Aletschorn, that cold and remote peak of the Bernese Oberland, second only in height to the Finsteraarhorn, which rises at the end of the glacier across from the Sparrenhorn. He made the ascent with a single guide, which had not been done before, making frequent observations of the polarization of the sky light. It was in the context of these that Tyndall penned the memorable phrase: 'We live in the sky, not under it'.[71]

A few days later, returning from a bathe in a stream above the hotel, he slipped and fell on a block of granite, bruising his shin so painfully that he had to take to his bed. He thought he would need to rest for two or three days to let the wound heal. But it did not, and after a fortnight he was seriously worried, despite the ministrations of an American Federal army officer who had experienced gunshot wounds to the leg. He telegraphed to Hirst, who undertook to deal with his letters, and wrote to De la Rive for a recommendation for a surgeon. De la Rive responded speedily, warning his surgeon, Gauthier, to be prepared.[72] Tyndall was taken down from Belalp on an improvised stretcher, spent a night at Sierre (where he encountered Poggendorf and his daughter as he was being carried down the hotel steps), and took up residence in the Hotel de l'Ecu in Geneva, where Gauthier drained and dressed his wound.[73] It was not until 6th October, after moving from the hotel to stay for a few days with Lady Peel, who furnished him with a bed downstairs in the library, that Tyndall was fit enough to travel back to England.[74] To his relief he heard from Hirst that the Board of Trade and Trinity House had not demanded work from him. Bence Jones kindly offered him use of his house at Folkestone to recuperate further, even offering to vacate his own room, the largest in the house and commanding a full view of the sea. Tyndall was touched, writing to Hirst: 'You know what rows he & I have had; but here in my hour of need he wishes to place his whole house at my disposal'.[75] This setback could well have damaged Tyndall's mental state, but he felt that the memory of his improvement at Belalp before the accident would carry him through, despite constipation making him somnolent and apathetic.

The Exeter meeting of the British Association, with Stokes as President and Hirst as General Secretary, took place while Tyndall was at Belalp, ending on the day he climbed the Aletschorn. He missed a paper by Tait on comets, in which Tait described modelling the motion of the particles in the tail mathematically, and two official reports with which he was involved.[76] The first was from a Committee consisting of Tait, Tyndall, and Balfour Stewart to repeat Forbes's experiments on the thermal conductivity of iron, and the second by a Committee including Tyndall on provision for Physical Research.[77] This Committee concluded that provision was far from sufficient, but that it did not have the power to recommend any scheme or extension. It proposed that the British Association call for a Royal Commission to be established, a recommendation that the General Committee presented to Council, and was a stimulus for setting up the Devonshire Commission. Huxley wrote to say that he had been elected President for 1870, with Liverpool fighting off Edinburgh as the host. He understood Tyndall had 'pledged to give a lecture even if you come with your leg in a sling'.[78]

Just before he left Geneva, Tyndall wrote at length to Huxley, telling him that for the last four years and upwards he had never known the blessing of a day's really good health, but that when he finally reached Belalp he found his strength growing, until the accident.[79] Spencer expressed his shock, saying that he had several times considered coming out to Geneva to keep him company, but pressure of work and the knowledge that he had friends in Geneva had prevented him. Tantalizingly, he wrote 'What a pity you did not, before going, take that long-contemplated step which we have discussed so often. I suspect that, on your return, recent experiences will very decidedly accelerate matters'.[80] Perhaps a reference to a forthcoming marriage proposal?

Tyndall reached Bence Jones's house at Folkestone on 8th October. Hirst visited him the next day, finding that he looked weak but that his old habit of sleeplessness had left him. On the Sunday they lunched with Lady Millicent and the children and took an excursion to Dover. Spencer visited too, but the sensitive Tyndall took himself off to the upmarket West Cliff Hotel after a few days, and Bence Jones was sorry he had not been there to keep him in the house.[81] He returned to town after a fortnight, much recovered and still sleeping well, called on Lady Lyell with Hirst, and dined with Bence Jones at the Athenaeum. He recounted the accident to Carlyle, whom he had frequently wanted to entice out to the Alps. Knowing Carlyle's antipathy to noisy hotels, he wrote: 'I schemed in thought about building a little wooden chalet apart from the hotel where you & I might sleep, and in fact the scheme seemed to me there so reasonable & so laudable that I am not sure even now whether it will not be carried out'.[82] Even Ruskin wrote to commiserate with him, though his motive was caused more by guilt: 'I ventured to send you that last book of mine because I had been impertinent enough to use your name in it but I hope not offensively to you'.[83] He had written in

The Queen of the Air a thinly veiled attack on Tyndall. Referring to Tyndall's Discourse 'On Chemical Rays and the Light of the Sky'—at which he had not been present—he noted sarcastically: 'So that the bright blue eyes of Athena, and the deep blue of her aegis, prove to be accurate mythic expression of natural phenomena which it is the uttermost triumph of recent science to have revealed'.[84] In the book he had thanked Tyndall ironically for the 'true wonder of this piece of work' and asked pardon of all masters of physical science if his words seemed to fail in the respect owing to their greater power of thought, or in the admiration owing to the far scope of their discovery. The letter was as close to an apology as one can imagine for Ruskin.

It is sad that Ruskin and Tyndall did not get on. Ruskin's vision of industrial activity potentially damaging the planet, and his observation that glaciers might be in retreat, which Tyndall had also observed, could have benefited from Tyndall's rigorous science and challenged Tyndall to explore the consequences of the burning of fossil fuels. Tyndall was more concerned that as supplies of coal were exhausted the country would no longer be able to compete with the US. He wrote to the economist William Jevons, who had recently published his book *The Coal Question*: 'I see no prospect of any substitute being found for coal as a source of motive power. We have, it is true, our winds & streams and tides; and we have the beams of the sun as a source of the power derivable from trees and animals. But these are common to all the world. We cannot make head against a nation which, in addition to those sources of power, possesses the power of coal'.[85]

Tyndall's accident severely set back his work. In a normal year, he would have started research many weeks before. He told Clausius: 'this accident cost me three months of work; and altogether fully five months were taken out of my time last year'.[86] Social commitments and government demands intervened as well. In the first week of November, Tyndall dined at Forster's, with Carlyle and Fitzjames Stephen, and then at the X-Club, going afterwards with Huxley to the Chemical Society to hear a discussion about atoms, still a disputed concept among the chemists. He also received a kind invitation to Foxwarren, the home of Charles Buxton, with whom relations had been ruptured by the Eyre affair. A few experiments are recorded on 29th October and the first week of November, and while they progressed, he prepared some of his 'memoirs' for wider publication, writing: 'I have already gone through those on diamagnetism. For I wish to bring out a volume of them'.[87] They appeared in 1870 as *Researches on Diamagnetism and Magnecrystallic Action*. In parallel he spent time 'fishing up old papers out of the Journals and Reviews, as I intend to publish a collection of them some day'. This collection became *Fragments of Science*, published in 1871 and revised and reissued in many further editions. In the evenings Tyndall relaxed by watching the billiards players at the Athenaeum—Spencer was a particularly active participant—while he carried out a running battle with Cassell, the publisher. Cassell had reprinted a series of articles Tyndall had written some fifteen years previously for Hughes's *Reading Lessons*, and published them without his knowledge or permission as *Tyndall's Natural Philosophy*. The battle took place in the correspondence columns of the *Times*, as Cassell initially rebutted Tyndall's objections but eventually apologized for not making clear the date and circumstances of the original publication.[88]

On 10th November, ostensibly because he did not feel strong enough for London, Tyndall took himself off to Clifton, staying at the Clifton Down Hotel, where Debus joined him in the evening.[89] He took with him Herschel's book *Familiar Lectures on Scientific Subjects*, telling Herschel that he particularly appreciated the sections on light.[90] Just as Airy had done with respect to magnetism, Herschel encouraged Tyndall to give a physical rather than mathematical explanation of the phenomena, writing: 'I wish you or anyone could give a satisfactory explanation of the actual mechanical process by which, on the undulatory theory, the dispersion of light is effected'.[91] A year later this grand old man of science died in his eightieth year and was buried in Westminster Abbey in a ceremony that Tyndall attended. Herschel was an admirer and constructive critic of Tyndall's work, and Tyndall regularly sought his advice and support—he was ever good at cultivating the experienced older men. He defended Herschel's reputation immediately after his death, writing to the *Daily News* to express his offence at an anonymous obituary attacking Herschel's character.[92]

The calm of Clifton enabled Tyndall to put his mind to his paper for the *Philosophical Transactions*, setting out his important work on the chemical action of light. In between writing the paper, and reading Herschel's book, he visited Hirst's sister-in-law and children and took walks with Debus across the countryside south of Bristol to Failand and Dundry. He dined with Debus, meeting William Budd, the physician who greatly reduced the impact of cholera in Bristol, and the alpinist Francis Fox Tuckett. One reason for visiting the area was Tyndall's desire to see the monument to the martyr William Tyndale, his supposed ancestor, which had been erected at Nibley Knoll, north of Bristol, in 1866. He took the train with Debus to Charfield, and they walked up to the monument, a 100ft tower, still standing today overlooking the valley across to the River Severn. The atmosphere and the romantic history of his possible ancestry were a potent mixture, making it 'one of the most interesting excursions I ever made in my life'.[93] Tyndall thought Nibley 'so charming indeed as to excite in me the desire to build a little house there where I might end my work and life'.

* * *

But there was a second reason for coming to the West Country. For about six years, Tyndall had been attracted to Mary Adair, a family friend of Sir Charles Hamilton, who had been a Manager at the Royal Institution.[94] He recalled seeing her in the library of the Royal Institution in the spring of 1863, looking with her friend at a picture of the Matterhorn. It was not just her dark hair and dark eyes that attracted him, though he had 'found those eyes fixed upon myself with a depth of expression which only in one other case have I seen equalled'.[95] As he revealed in his journal: 'Her intellect was manifestly of the richest kind but it was the perfect womanliness of her nature that most interested me'.[96] He had drafted a letter to her, never sent, while recuperating in Geneva.[97] Screwing up his courage, he went on his own initiative to Taunton, where he was met by Mrs Adair's carriage and taken with Mary Adair and her married sister to Heatherton Park.

Mary Adair, born in 1843, was the youngest of seven children of Captain Alexander Adair, who had died in 1863, and Heatherton Park was her country home. Seeing her, 'it was as if love were potential and required but the pulling of a trigger to render it active. In one evening I saw that I could love her, work for her, die for her. This was no mad feeling but a quiet deliberate knowledge founded upon my conviction of her excellence of head and her double excellence of heart'.[98] Yet through lack of initiative, or the tightness of the chaperoning arrangements, he was unable to find time alone with her. So he drafted a note, aiming to hand it to her before he left. But courage failed and he left without doing the deed. Still he felt that he must bring the affair to a decision, and posted the note in Taunton on his way back to London. Two days later, he received two notes in reply, in which she 'expressed grief and pain that she had caused me pain', and turned him down.[99] We shall never know exactly what she said, as Tyndall threw the notes into the fire. He did write back, expressing sorrow that he had been forced to write instead of speaking to her, but this marriage was not to be. Mary Adair never married—she died in 1921—and it is possible that her strong religious sense, or family duties, prevented acceptance of his proposal. Tyndall kept this blow to himself: 'no man would infer from my demeanour during this last week that I have borne this wrench'.[100] He went to the Philosophical Club on the day he received the rejection, but his soul was 'filled with a high sadness'.[101] He may not even have told Hirst. They had lunch a few days later, but Hirst recorded nothing in his journal.

As he started to put this intense disappointment behind him, there was plenty to occupy his time. He was elected onto the Council of the Royal Society for the third time. On his own initiative, he approached Airy to consider the Presidency, writing that it appeared to him that 'the society is in a difficulty as to finding a man of real eminence to preside over it'.[102] At this point Sabine was still in post, having been so since 1861. Airy replied that it would be difficult financially and practically for him to accept, but after Tyndall had pressed the case he relented and answered that he could take up the post, although he would prefer not to. He also pointed out his creeping deafness, 'which is a sad disablement for a Chairman'.[103] Two years later, in 1871, Airy succeeded Sabine, giving rise to the quip that 'our new President was deaf on both sides, our Senior Secretary deaf on one side, and blind on the other (this was Sharpey), and our Junior Secretary generally dumb' (this was Stokes, known for his taciturnity in meetings).[104]

It was a sociable run-up to Christmas. Tyndall played billiards at the Athenaeum, saw *School for Scandal* with Spencer, and met Agassiz's son, a marine biologist and 'a very nice fellow'. John Morley, now editor of the *Fortnightly Review*, sent him the proofs of his article 'Climbing in search of the sky' which he corrected and returned.[105] He felt he could not decline an invitation to Drayton Manor by Lady Peel, who had been so kind to him in Geneva, so he went on 7th December, finding her in the hall.[106] Lord Chief Justice Cockburn was there, treating Tyndall coldly at first but then thawing; Tyndall supposed that he knew his sentiments about General Eyre.

At the shooting party the next day Tyndall was initially shocked, then found himself excited by the 'animated slaughter; pheasants, hares, partridge or two, a water hen or two and one or two teal'. He thought the birds must die in some way, and perhaps their suffering would be greater if they died a natural death. But he thought: 'the whole subject needs looking into, not by a bigot, nor by a milk and water philanthropist; but a man of sound heart who can really understand the joy of the sport such as it is, and enter into the pain inflicted'. He went on to an agreeable party at Lord Wrottesley's before returning to London on 11th December and an invitation from Lady Ashburton to join her and Lady Marian Alford at dinner and the theatre. On Christmas Day, Tyndall had a quiet happy dinner with Bence Jones and a game of whist afterwards. It was his turn to give the Christmas Lectures, on the subject of 'Light', and as he developed them he started writing a book on light 'for little boys'.[107] The lectures began on 28th December, with huge audiences of 700–800, ending on 8th January 1870.

Tyndall's paper 'On the Action of Rays of High Refrangibility upon Gaseous Matter' was read to the Royal Society a few weeks later, on 27th January.[108] At a time when many chemists did not accept the idea of the physical reality of atoms, Tyndall stated his firm belief that his observations were only explicable on the basis of physical objects—atoms and molecules—that could interact with waves of radiation through the medium of the ether. Some of these molecules could be 'shaken asunder' by the waves, as Tyndall's paper then described, extending the work first reported in June 1868 and setting the basis for the field of chemical physics. Herschel, for one, could see the potential. When the paper appeared in the *Philosophical Transactions* in October, he wrote that he thought it opened out 'a new field of enquiry'.[109] Tyndall had envisaged this paper as the first part of his researches on the action of short-wave radiation on gaseous matter.[110] But in the event, it was his Discourse on a different topic that same month that redirected his mind, and stimulated another new area of research.

PART IV

1870–1880
ESTABLISHMENT FIGURE

16

⋆⇒◉ ◉⇐⋆

DUST AND DISEASE
1870–1872

With hindsight, Tyndall's Discourse 'On Dust and Disease', delivered at the instigation of Bence Jones on 21st January 1870, was a blessing in disguise, though it did not seem so at the time. It would lead eventually to his work on germ theory and the spontaneous generation of life. He had just ten days to prepare it, having tried unsuccessfully to persuade Bence Jones, as a medical man, to speak instead. And it immediately landed him in trouble. He was criticized for neither properly acknowledging the work of others nor describing the history of the subject, and by Louis Pasteur's opponents for the recognition he gave to Pasteur's experiments to counter Félix-Archimède Pouchet, who advocated the possibility of spontaneous generation.

In the Discourse he described his astonishment that air passed through concentrated sulphuric acid and caustic potash—to dry it and remove carbon dioxide—still retained dust, or 'floating matter', that was visible in a beam of light.[1] His astonishment had been all the greater when, in October 1868, he had found that passing air over a spirit lamp removed all the floating matter, creating a blackness where previously the light had been reflected and scattered. The beam became invisible. The dust had been burnt up. He concluded that it was organic in origin, not inorganic as he had supposed. The human lungs seemed to perform a similar function to the spirit lamp. Tyndall described his observation that on blowing across a light beam the space went completely dark. There was no floating matter in the exhaled air to cause the light to be scattered and visible, even though such matter was manifestly being taken in with every breath. Tyndall argued that this floating matter contained the germs that could cause disease and decay.

Following up his Discourse in a letter to the *Times*, Tyndall highlighted the surgeon Joseph Lister's observation that air from the lungs does not cause the decay of tissues, from which Lister had deduced that the airways could filter out germs.[2] What Tyndall had done for the first time was to demonstrate the absence of any matter that

Fig. 29. Professor Tyndall lecturing at the Royal Institution, *Illustrated London News*.

could contain germs. He went further, suggesting that the floating matter could be removed by breathing air through cotton wool, and that cotton wool respirators might help to protect against disease and other situations in which dust was inhaled. Convinced that his views were important for medicine he had the lecture published in the newly launched *Nature* magazine, and as a leaflet by Longmans.[3] Part of it also appeared in the *British Medical Journal*, and he expanded on the history of the subject in *Fraser's Magazine* in March.[4]

The extent to which 'germs' (definite minute organisms, or the 'seeds' of them) as opposed to chemical poisons or 'miasmas' ('bad air', emanating from decaying matter), were responsible for disease and decay, or 'putrefaction', was strongly contested. The question was linked to the idea of spontaneous generation; that life could arise *de novo* from decaying matter. Tyndall suggested that experiments by Pouchet, which he claimed demonstrated spontaneous generation, had not excluded the floating matter which he had now shown was present.

Henry Charlton Bastian, a physician at University College London, who was the major British protagonist of spontaneous generation (and a firm believer in evolution) proceeded to challenge Tyndall's conclusions. Tyndall countered him, in the start of a debate that would run for years, in another letter to the *Times*. He backed up the demonstration by Schroeder and Pasteur, that air filtered through cotton wool is deprived of its power to produce life, by his own observation that such air is visibly pure and therefore, he argued, impotent to generate life.[5] Tyndall was the first person to pull together all the complex questions of contagious disease and sepsis under a single germ theory.[6] But these forthright ideas did not endear him to the medical community, many of whom saw him as engaging arrogantly in an area of which he knew little: aetiologies of disease and decay were complex and disputed. Nevertheless, he continued his championship of Pasteur (Plate 16) with an article in *Nature* on 'Pasteur's Researches on the Diseases of Silkworms'.[7]

Tyndall was here using a beam of light as a more sensitive means even than a microscope for detecting 'floating matter', whatever that might turn out to be. He had thus developed a new and powerful research tool which was available to investigate both spontaneous generation and germ theory, as he understood it. Letting air stand for several days in a chamber, he found that it became optically inactive, as the floating matter settled to the bottom. This gave him a chamber which he could use for experiments, clear of suspended organic matter in the air. He claimed to hold himself initially agnostic with respect to germ theory, writing to De la Rive: 'It is perfectly manifest from my words that I make not the slightest claim to the germ theory. Indeed my mind is in suspense about it. That germs are in the atmosphere has been known for centuries, but what was new to me, and what surprised me was that all the floating matter of our London rooms is of organic origin'.[8] He did not seem so agnostic on spontaneous generation, writing to Josiah Whitney: 'I have been deflected for the moment from my usual course of investigation to consider the question of "Spontaneous generation". As regards the evidence my mind is already made up. So I shall make a few experiments & then have done with the subject'.[9] It would not prove quite so simple. In any case, other matters soon intervened, and it was not until 1875 that he fully picked up the topic again.

Gustav Magnus, whose private laboratory had been so important both to Tyndall and to a generation of German physicists, died on 4th April, effectively bringing to an end the long-running challenge to Tyndall's conclusions about the absorption of radiant heat by water vapour. This challenge had considerably diverted Tyndall's energies, even if posterity declares him the clear victor. Tyndall's obituary of his friend in *Nature* recollected their meeting on Magnus's doorstep in 1851, and his early work on diamagnetism in Magnus's laboratory.[10] As Magnus died, Tyndall had just completed the preface to a collection of his work on diamagnetism, in addition to sending off his paper 'On the Polarisation of Heat' to the *Philosophical Magazine*. This is a neat piece of work, in which he was easily able to demonstrate, using radiant heat filtered by his iodine and carbon disulphide cells, the polarization of heat first observed by Forbes and Melloni.[11]

After Easter, which he spent on the Dorset coast with Debus—still languishing unfulfilled at Clifton College—Tyndall gave a series of seven weekly lectures at the Royal Institution on 'Electrical Phenomena and Theories'. These were interspersed with visits to Hirst, who had been quite ill, and Carlyle, from whom he was thrilled to come away with some of Emerson's work containing a handwritten dedication to the 'old warrior'.

Tyndall was also in receipt of possible honours. He heard in late May that he had been nominated as the next President of the British Association, but told Hirst that he would turn the honour down because of his existing commitments. Tyndall never took the British Association to heart, as he did with the Royal Institution and the Royal Society. He was appointed to the Council over many years, but rarely attended. His priority was metropolitan science, and the inner circle of the Royal Society. Indeed, one of his few appearances at Council, in March 1868, was probably to support Hooker and Huxley over the transfer of the British Museum's Natural History Collections to 'a single officer of Government amenable to Parliament', as part of the long-running battle with Richard Owen.[12]

By the middle of June Tyndall was again feeling the pressure of London. He spent a week in lodgings in Folkestone, from where he wrote to Juliet Pollock: 'By quitting London I have saved myself from a breakdown—I know not how reasonable people with any intellectual work to perform live through the life of London at this season of the year'.[13] Just before leaving for Folkestone he received a letter from Huxley which was phrased as a dedication to him and published in the form of a 'Prefatory Letter' at the start of Huxley's *Lay Sermons, Addresses and Reviews*, the collection of essays that set out Huxley's philosophy of science and of education. Tyndall replied 'I hope some day to see you develop...the last paragraph of the "Physical Basis of Life". I probably am the only living man who sees the meaning shadowed forth in that page and paragraph!!'.[14] In this final paragraph, Huxley acknowledged the limits of science and materialism, a view that Tyndall shared.[15]

Tyndall left for the customary trip to the Alps on 1st July, having tried but failed to persuade Tennyson to join him. He was heading for Belalp, and travelled via Boulogne to Paris for a couple of days. The train by this time went as far as Sierre in the Rhône valley, from where he caught a diligence to Brig. The next morning he took an unfrequented path through the grey and mossy cliffs of the Blindtal, up to Belalp and his room at the hotel. Gradually his sleep improved and his strength increased. He took regular swims in the nearby lake, and walks from the hotel, writing to Hirst that he was safer in company; a solo excursion over unfrequented cols and slopes had put him in dangerous situations. Bence Jones told him that his piece on Pasteur in *Nature* was out—Tyndall was reading Bastian's papers in *Nature* while at Belalp, along with Goethe's *Farbenlehre* and Bain's *Logic*—and gave him news about the 'glorious work' of the new Thames Embankment and of physical improvements at the Royal Institution.[16] Tyndall had also sought some new management arrangements, wanting William Odling, the Professor of Chemistry, to attend Managers' meetings and to be able to act effectively as a deputy when he was away.

While Tyndall was at Belalp, events of momentous significance were taking place in Europe. Provoked by Otto von Bismarck, Chancellor of Prussia, the French Emperor Napoleon III declared war on 19th July. So began the Franco-Prussian War. Napoleon's move, a consequence of Prussia's defeat of Austria in the Seven Weeks' War of 1866, and its impact on France's position as a dominant power, had the effect, as Bismarck intended, of bringing southern German states into alliance with the Prussian-led North German Federation. The consequence, after the occupation of Paris and the defeat of the French, was the creation of a unified German state on 18th January 1871, a few days before the surrender of Paris itself. Tyndall was a strong supporter of the Prussian position, although he claimed to Hirst: 'I do not say this in consequence of any particular sympathy with Prussia; if the case were reversed, & if she behaved toward France as France is behaving towards her, I would in the same way join France'.[17]

Many of Tyndall's German and French friends were caught up in the conflict. He was fully aware of what was happening—receiving three telegrams daily, and reading newspapers from England, Germany, and Switzerland—and he pledged assistance to several friends should they need it, including an offer to take care of Jamin's wife and children in England. Jamin had in fact sent them to Brittany for safety while he remained in Paris, but communication was all but impossible. Jamin's wife even tried to contact him by carrier pigeon, without success. Clausius was caught up directly in the fighting. He went as part of a battalion that consisted mostly of Bonn University students and a few professors who volunteered as nurses and emergency staff on the battlefield, and he was lucky to survive.[18] His Third Prussian battalion, outnumbered five to one, stood at Mars-la-Tour near Metz against the whole French army from 9.30am to 3pm on 16th August, while some 16,000 German and 14,000 French casualties were sustained. Hearing that there were many wounded soldiers in

Heidelberg, Tyndall sent £75 in August to Bunsen to help care for wounded Germans there, and a sum to Jamin for wounded French. He sent a further £50 in October to Clausius, on hearing of the number of wounded in Bonn. His return journey from Switzerland, via the railway station in Paris, was obstructed by the furniture of people fleeing the war. But he got through ahead of the Prussian forces—the siege of Paris began on 19th September—and was back in England at the beginning of September.

* * *

The weather in Switzerland that summer had been poor. It made climbing exped-itions unattractive, and decided Tyndall against a planned visit to Pontresina in the Engadine, since his forthcoming lecture at the British Association in Liverpool, at which Huxley was to be President, loomed over him. After several weeks, he decided that he had no new experiments sufficiently complete to present. With few books either to refer to while abroad, he had told De la Rive that 'the consequence is that I have to develop one out of the depths of my own consciousness'.[19] The result was a major statement of his philosophy of matter and life, the 'Scientific Use of the Imagination'.[20]

Tyndall was not present for his great friend Huxley's Presidential Address in Liverpool, which tackled head on the idea that spontaneous generation, or abiogen-esis as he defined it, had been demonstrated by anyone. Yet Huxley stated his belief—and acknowledged it as no more than a belief—that at some distant time in the geological past life had emerged from non-living matter. It was a view Tyndall shared.[21] Tyndall arrived two days after Huxley's Address, on the day of his own lec-ture. It was regarded by all, as Hirst put it 'as a *chef d'oeuvre* of philosophical eloquence'. Tyndall's iconoclastic philosophical set-pieces, including his address at Liverpool, are highly personal and individual statements. Yet he took care to try them out in advance on his friends. Darwin was particularly supportive, writing: 'Your whole discourse strikes me as grand...What you say about Pangenesis is quite cor-rect...What you say about me has pleased me extremely...I feel a deep conviction that Pangenesis will some day be generally accepted'.[22]

In front of an audience of 3,000, Tyndall set out his argument that imagination, 'bounded and conditioned by cooperant Reason', enables us to step beyond a 'mere tabulation of coexistences and sequences' to envisage their causal relations and their physical bases. In other words, to describe models by which sense may be made of appearances, as the idea of the ether makes sense of the phenomena of light. Tyndall was a believer in the ether throughout his life. Yet he could acknowledge that 'although the phenomena occur as if the medium existed, the absolute demonstra-tion of its existence is still wanting', while still taking a swipe at the chemists who

1. Tyndall in the 1850s.

2. Thomas Hirst (1830–1892).

3. The Royal Institution of Great Britain in the mid-nineteenth century.

4. William Thomson, later Lord Kelvin (1824–1907) in 1852.

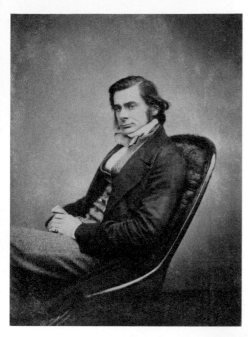

5. Thomas Huxley (1825–1895).

6. From left, Faraday, Huxley,
Wheatstone, Brewster, and Tyndall,
*c.*1865.

300

Certificate of a Candidate for Election.

(N.B. Directions for filling up Certificates are given on the other side.)

1852/13

(Name) *John Tyndall*

(Title or Designation) *Doctor of Philosophy*

(Profession or Trade*)

(Usual place of Residence) *Queenwood College, Stockbridge*

The Discoverer of {

The Author of — *A Mathematical Dissertation on a curved surface; — and of Memoirs on, the Magneto-optic Properties of Crystals, and the Relation of magnetic & Diamagnetism to Molecular Arrangement: — on the phenomena of a water-jet: — on the Laws of Electro-magnetic attraction: — on diamagnetic and magnecrystallic action: — on the plant's Bismuth: — and of several experts in the progress of the Physical Sciences —*

The Inventor or Improver of

Distinguished for his acquaintance with the science of *Natural Philosophy.*

Eminent as a

being desirous of admission into the ROYAL SOCIETY OF LONDON, we, the undersigned, propose and recommend him as deserving that honour, and as likely to become a useful and valuable Member. Dated this *15* day of *February* 18*52*.

From General Knowledge.	From Personal Knowledge.
C. *Wheatstone*	*M Faraday*
	W R Grove
	Thomas H. Huxley
Lyon Playfair	*J J Sylvester*
Edward Forbes.	*John Phillips*
T H Henry	
G Kising	

Elected
C.R.W.

Read to the Society on the *26th* day of *Feb 1852*
To be Balloted for on the *3 W* day of *June 1852*

* If of no Profession or Trade, this should be stated by filling up the Blank with the word None.

7. Tyndall's FRS certificate.

8. Henry Bence Jones (1813–1873).

9. Thomas Carlyle (1795–1881).

10. The Athenaeum *c.*1850.

11. Tyndall in 1864, drawn by George Richmond, RA.

Received as far as p. 36 inclusive
Jan 10 - 1861 C.N.W.

On the Absorption and Radiation of Heat
by Gases and Vapours
and on the physical connexion of
Radiation, Absorption and Conduction.
by John Tyndall Esqr F.R.S.

(with 2 engravings).

The researches upon glaciers which I have

had the honour of submitting from time to

time to the notice of the Royal Society, directed

my attention in a special manner to the

observations and speculations of De fruquire,

Fourier, M. Pouillet, & Mr. Hopkins, on the

transmission of Solar and terrestrial heat

through the earths atmosphere; and thence

arose on my part the desire to make the

mutual action ~~interaction~~ of radiant heat and gases of all

kinds the subject of an experimental enquiry.

acquaintance with
Our ~~knowledge of~~ this department of

Physics is exceedingly limited ; ~~so limited~~
So far as my
~~... is due~~

to an inadvertence in his mode of experiment. These

are the only observations ~~on this subject~~ of this nature

with which I am acquainted ; and ~~I believe~~ they leave

the field of enquiry now before us perfectly unbroken

ground.

12. The manuscript in which Tyndall set out his detailed work on the absorption of heat by gases.

September 16 th

Brass tube with front, 2nd & 3rd chambers.
and stopped
Divided by 3 plates of rock salt

Front chamber exhausted and common
air in 3rd chamber, the second chamber
being in 3 different conditions being

exhausted	filled with dry air		also	common air
33°.5 + 4°, 37° - 3° 35°	31.°5 + 7°, 38° 35°	with	2 nd 41°.7 38°.5 40°.5	2nd & 3 ed 44°.3 45° 45°

2nd chamber filled with common air gave a
deflection of 39° the needle was then brought to 0° and
common air was let in to the 3rd chamber and &
deflections of 17° and 16°.5   were obtained.
on pumping out and cleansing with dry air the needle
stood 38°.2 on the side of heat

Carbonic Acid gas.

The drying tubes were first exhausted, then filled with
gas, the 2nd chamber was then filled with gas, which
gave 47° the 3rd chamber was now filled and the
deflection increased to 51°, this last was pumped out
and it returned to 48°.7, and gas was again let in to 3rd
chamber which gave 51°, on again exhausting in returned
to 48°7, and gas was let in again and 51° was the result

Olefiant Gas was admitted into the 2nd
chamber 74°8 was indicated, and on filling the 3rd
chamber it was increased to 76°.3. After cleansing
several times with dry air the needle returned to 6° on
the side of cold. Total heat 77°.8 on the side of cold
80° on the side of heat,

13. Tyndall's experimental notebook, recording experiments on the absorption of heat by gases, 16th September 1862.

14. Weisshorn from the Riffel.

15. Matterhorn.

16. Louis Pasteur (1822–1895).

17. Tyndall in 1873.

JT/1/T/89

1875

11th Jan'y

My dear Miss Adair

Amid the shower of abuse which has descended on me during the last few months I have sometimes asked myself how my sins have affected you. Will you accept a copy of the iniquitous Address from me, in the Preface to which I have sought to clear up myself. If it seem to you proper fate for it pray put the book in the fire.

Most faithfully Yours

John Tyndall

18. Letter to Mary Adair after the Belfast Address, 11th January 1875.

CERTIFIED COPY OF AN   ENTRY OF MARRIAGE
Pursuant to the   Marriage Act 1949

RTA 054306

M. Cert.
S.R., R.B.D.& M

CAUTION: There are offences relating to falsifying or altering a certificate
and using or possessing a false certificate. ©Crown copyright.

WARNING A CERTIFICATE IS
NOT EVIDENCE OF IDENTITY

Registration District Westminster

18_76_. Marriage solemnized at _Westminster Abbey_ in the _Close_ of _Central &the Westminster_ in the County of _Middlesex_

No.	When Married	Name and Surnames	Age	Condition	Rank or Profession	Residence at the time of Marriage	Father's Name and Surname	Rank or Profession of Father
35	29th February 1876	John Tyndall	full age	Bachelor	Esquire F.R.S	21 Albemarle Street St George Hanover Square Middlesex	John Tyndall	Gentleman
		Louisa Charlotte Hamilton	full age	Spinster		4 Belton Square St George Hanover Square Middlesex	Claud Hamilton formerly called Lord Claud Hamilton	Privy Councillor

Married in _Westminster Abbey_ according to the Rites and Ceremonies of the Established Church, by _Licence_ by me.

This Marriage was solemnized between us, _John Tyndall_ _Louisa Charlotte Hamilton_

in the Presence of us, _Claud Hamilton_ _T. Carlyle_ _Wm N Pollock_

Certified to be a true copy of an entry in a register in my custody,

Deputy Registrar/Superintendent Registrar

Date 14th October 2016

19. Marriage certificate of John Tyndall and Louisa Hamilton.

20. Louisa Tyndall reading.

21. John Tyndall outside Chalet Lusgen (probably Louisa Tyndall in the doorway).

22. Burlington House in 1866, home of the Royal Society (1857–1968).

23. Hindhead House.

The following labels appear within the engraving:

MR. FRANCIS GALTON
President of the Anthropological...

PROFESSOR G. G. STOKES, D.C.L.
Secretary of the Royal Society; Lucasian...

MR. W. H. L. RUSSELL, F.R.S.
...

DR. WILLIAM HUGGINS
...

MR. W. H. PERKIN, F.R.S.
Chemist

LORD RAYLEIGH
Raleigh

DR. JOHN EVANS
Treasurer and Vice-President...

PROFESSOR T. RAY LANKESTER
Professor of Zoology, University College, London

MR. NORMAN LOCKYER
Editor of "Nature"

PROFESSOR A. W. WILLIAMSON
Foreign Secretary

MR. W. E. B. GLADSTONE
...

MR. FRANCIS GALTON
President of the Anthropological...

PROFESSOR G. D. LIVEING
...

MR. W. T. THISELTON DYER
Director of the Royal Gardens, Kew

MR. W. SPOTTISWOODE
...

SIR JOSEPH HOOKER
Past President

PROFESSOR J. J. SYLVESTER
Savilian Professor of Geometry, Oxford

PROFESSOR J. S. BURDON SANDERSON
...

SIR WILLIAM THOMSON
Professor of Natural Philosophy, Glasgow

PROFESSOR JOHN TYNDALL
...

DR. ARCHIBALD GEIKIE
Director-General of the Geological Survey

PROFESSOR T. H. HUXLEY
President of the Royal Society

MR. W. H. FLOWER, C.B.
Director of the Natural History Museum, South Kensington

MR. WILLIAM CROOKES
...

THE ROYAL SOCIETY
A PORTRAIT GROUP OF SOME OF THE MOST DISTINGUISHED FELLOWS

24. Some Fellows of the Royal Society: a key to the identities of the sitters. Wood engraving, 1885. In the front row are Stokes, Hooker, Sylvester, Huxley, Geikie, Tyndall, Cayley, Owen, Flower, and Crookes. Frankland stands third from right and Francis Galton second from left.

25. Tyndall and Louisa in the study at Hindhead c.1887.

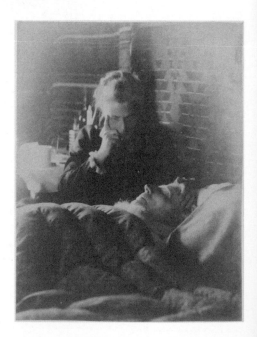

26. Louisa at Tyndall's deathbed.

'refuse to speak of atoms or molecules as real things'. By analogy with the vibrating string producing waves of sound, Tyndall's scientific imagination saw vibrating atoms and molecules, real tangible entities, producing waves of light and transmitting them to the ether.

Looking further, Tyndall pictured the atmosphere containing exceedingly small particles, preferentially reflecting and scattering the shorter blue light waves while transmitting the longer red. The sky at noon therefore appeared blue, but red at dawn and dusk as the sunlight travelled through a greater distance in the atmosphere where the particles scattered away the blue, leaving the red behind. Here the imagination could go where no microscope could penetrate and no structure could be seen, though it was surely there. Indeed, the particles were so uncountable, yet so small, that perhaps those making up the sky might 'be held in the palm of the hand'.

Considering life itself, Tyndall recognized Darwin's theory of pangenesis, the idea that the whole of an organism is involved in heredity through tiny particles called 'gemmules', as a justifiable imaginative leap. The work of Mendel, on the 'factors' involved in inheritance, was not widely known or appreciated at this time. Darwin had placed at the root of life a 'primordial germ' from which all subsequent life had evolved. But what was the genesis of this germ? Either life was present potentially in matter in its earliest and nebulous form, or it was a principle inserted into matter at a later date. For Tyndall, 'all our philosophy, all our poetry, all our science, and all our art—Plato, Shakespeare, Newton, and Raphael—are potential in the fires of the sun'. Tyndall argued for the 'Evolution hypothesis' throughout and for the right to hold and discuss it, untrammelled by dogmatism, arguing: 'If it be of God, ye cannot overthrow it; if it be of man it will come to naught'. But even this hypothesis, he declared, does not solve the mystery of the universe itself, it merely transports 'the conception of life's origin to an indefinitely distant past', as natural philosophers assert the uniformity of nature from the present into the past.

This assertion of the uniformity of nature was a bedrock of the beliefs of the 'scientific naturalists' like Tyndall, and an argument against the possibility of miracles. It is intimately tied to the idea of cause and effect; that events follow from their causes in a regular and predictable manner. Tyndall claimed that causation was 'the masterkey of modern scientific investigation', without which it would be 'impossible to keep firm hold on the methods of scientific enquiry'. Yet he admitted that it could not be directly observed. What was empirically observed was simply one event following another. The idea of causation resulted 'from our *interpretation* of observation and experience'.[23] It was, therefore, imported by the mind, a chink the Christian metaphysicians gleefully exploited. Indeed, Christian theists, like the Duke of Argyll, believed equally in the uniformity of natural law. For them it was an argument for the existence of a divine creator, who established and maintained it, in ways humans did not necessarily yet comprehend. The scientific naturalists, who argued that there was no need for concepts of the divine in scientific explanations, could not prove a negative.

Tyndall left for the Lake District with Spencer the day after his Liverpool lecture, making a ramble from Windermere to Ambleside and visiting Rydal Mount, where the proprietor of Wordsworth's property showed them over the house and gardens. After lunch at Grasmere they took a boat across the lake and climbed Loughrigg Fell. Spencer went home, and Tyndall had a scamper to Coniston and back, before returning via Manchester to stay with the industrialist Sir Joseph Whitworth and see the first stone laid of the new building for Owens College, forerunner of the University of Manchester. A few days later he was in Dublin for a week on business with the Board of Irish Lights. He managed a quick trip to Gorey to see his sister Emma, who he found in good health.

* * *

The scattering of light by small particles had set off all manner of thoughts for Tyndall. Always one to make connections, he had the idea that the colour of the Sun's corona might be caused by a kind of solar dust. A total solar eclipse was predicted for 22nd December 1870, tracking across southern Portugal and Spain, northern Algeria, and the eastern Mediterranean. Tyndall planned to join the expedition going to Oran in Algeria, largely to explore his idea, and he visited De la Rue to familiarize himself with the telescope he would be taking.

So Tyndall found himself at Portsmouth on 5th December, where the *Urgent*, an elderly steamship once used as a troop-carrier, sat at anchor in the fog, which delayed the departure of the expedition party. The weather lifted enough the following day for the compasses to be calibrated before the ship glided into the Channel. Out in the Bay of Biscay, under a grey sky, the swell was full but not violent. But off Ushant three days after leaving, the wind strengthened to a gale and the ship was tossed about. The tiller ropes came loose and the helm would not answer the wheel, as the ship wallowed side-on to the huge waves. Tyndall, alone of the passengers, fought his way to a capstan by the wheel, to watch the sailors struggling for control amid the groaning of the screw and the whistle and boom of the storm. After a full day the storm abated and they steered for Cadiz to land the first party of observers. From there they steamed past Trafalgar in calm seas, accompanied by porpoises leaving trails of phosphorescence. The African hills became visible, and finally Gibraltar with its luxuriant palm trees, cactuses, and aloes, ablaze with scarlet flowers in the British winter. They were received by Sir Fenwick Williams, the Governor and 'Hero of Kars', who had led the Turkish garrison against the siege by the Russian General Mouravieff in 1855, the last major engagement of the Crimean War. Leaving Gibraltar for the Mediterranean they sighted the white walls of Oran on 16th December and put in to the port.

The French astronomer Janssen, who had managed to escape in a balloon from besieged Paris, was already in place in open country about 8 miles from Oran. Janssen had discovered during the 1868 eclipse in India that the Sun's chromosphere is gaseous and, with Norman Lockyer, can be credited with discovering the element helium in the solar spectrum, an element later named by Lockyer and Frankland. The British party decided that the site was too far from the ship, and settled on a garden by the railway station close to a bastionet on a wall erected to defend Oran from the Arabs, where Tyndall placed his instruments. He was fascinated by the local population: 'the Jews, rich and poor; the Arabs more picturesque still, and all shades of complexion—the negroes, Spaniards and French all grouped together, each race preserving its own originality'.

Tyndall repeatedly practised the sequence of observations he planned to make during the two minutes of totality. But though he caught glimpses of the encroachment of the Moon, when the eclipse began, and the darkness rushed towards them, the clouds drew across. The total eclipse was hidden. They were back in Gibraltar on the evening of 26th December after spending Christmas night peacefully off the Bay of Almeria. Tyndall explored the caves and geology of the Rock, guiltily knocking off some stalactites as souvenirs, before they weighed anchor. The return journey gave him the opportunity to carry out experiments on the colour of the sea. He took samples at nineteen points between Gibraltar and Spithead, the colour of which varied from 'black-indigo' to 'yellow-green'. Back in London, in a blinded experiment in which he opened bottles without knowing their place of origin, he determined that the yellow-green colour was associated with water containing thick suspended matter, while the purer water had an indigo tinge. His explanation was that the red rays of the spectrum were absorbed by water in preference to shorter rays such as green and blue, and the remaining rays were reflected back from suspended particles. This gave the greenish or indigo colours, depending on the thickness of suspended matter; without this the sea would appear as black as ink. Friends were a useful source of experimental samples. Emily Peel sent a package containing Mediterranean water collected by Baroness Rothschild, which enabled Tyndall to publish an article in *Nature* in October 'On the Colour of The Lake of Geneva and The Mediterranean Sea'.[24]

The *Urgent* arrived back safely in Portsmouth on 5th January 1871, unlike the *Psyche*, the ship supporting the Sicilian eclipse expedition which included Lockyer, Roscoe, and George Darwin, astronomer and son of Charles. The *Psyche* struck a rock and sank off Catania, prompting young Hugh Spottiswoode to ask his mother: 'is jolly old Tyndall lying at the bottom of the sea?'.[25] Fortunately there was no loss life, nor of instruments. Tyndall reached London in time for the X-Club meeting in the evening, reportedly in good health and spirits despite the failure of the expedition from a scientific perspective.

The first result of Tyndall's trip was his Discourse on 20th January 1871, 'On the colour of water, and the scattering of light in water and in air'.[26] He showed, through the use of light beams, that drinking-water from different places, even when filtered, contained particles too small to be seen by the microscope yet revealed by the track of a beam of light. Water obtained from ice, which he believed to be as pure as he could find, still showed a faint blue colour. He contended that if all impurities could be removed, the colour would disappear too. In this he was mistaken. Pure water is blue, though the colour is only visible through a substantial thickness.[27] He showed that water filtered through chalk was particularly pure, but it gave the problem of hardness, with its enormous waste of soap for washing, unless it could be softened. He urged the trial of chalk in London, having seen the softening process used to supply water to Tring and Aylesbury, and supported Frankland in his responses on this issue to the Royal Commission on Water Supply. His suggestion led to correspondence in the *Times*, in which he argued that experiments should be made to supply softened chalk water, even if it could only satisfy part of London's demand, since it would be much safer.[28] He had read Budd's work on typhoid during the eclipse expedition, and later lobbied Busk to help ensure Budd's election as a Fellow of the Royal Society in June. This work reinforced his view that the 'defilements in water which are most dangerous to man are those which emanate from man himself'. Chalk water would reduce this risk.

Paris capitulated at the end of January, and Tyndall's French friends started to emerge. Moigno wrote that he had been saved by a miracle from a shell that had smashed everything in his small bedroom, and Jamin that he was now with his family in Brittany.[29] The war was not so kind to Regnault, to whom Tyndall had tried to send 1,000 francs to distribute as relief. Unable to contact him, he had instead donated it via Huxley to an appeal made by the Bishop of Versailles. Regnault eventually reached Geneva, where he had sent his children just before the siege, having been evicted from his laboratory, which was ransacked. His instruments, experimental records, and the furniture in his house were destroyed, despite the formal promises he had been given by the Royal Prince and the General Prussian Headquarters— Regnault was an associate of the Academy of Berlin, a Commander of the Red Eagle of Prussia, and a Knight of the Order of Civil Merit of Prussia. Then his second son, a leading painter, was killed in the French sortie from Paris on 19th January.[30]

Tyndall wrote to Du Bois-Reymond, with a Carlylean voice in his ear, suggesting the Count de Paris or the Duc d'Aumale as ruler in France: 'for it would be an indication that their eyes were open to the noble quality of those truly excellent men'.[31] Earlier, just before leaving for Algeria, Tyndall had been to see Carlyle. In robust form, and confessing that he had just voted for the first time—in a ballot for the Chelsea School Board[32]—Carlyle argued that the French should be thoroughly chastened into good behaviour for a century: 'if you want to rid yourself of rats you must not hesitate to stop up their holes and hunt them out of your premises utterly'.[33] Tyndall said he would prefer greater magnanimity on the part of Germany. In this he showed better judgement, since the treaty ceding Alsace-Lorraine to Germany was a factor in the subsequent Great War.

Perhaps the ultimate recognition of Tyndall's social standing came on 15th March, when he received a letter from the Lord Chancellor, Lord Hatherley, informing him of his election to 'The Club', an exclusive dining club of thirty-four men, including the Archbishop of Canterbury, Gladstone, the Duke of Argyll, and the Earl of Derby.[34] He joined with them and others such as George Richmond, Froude, and Tennyson, as only the second man of science, after Richard Owen, to be elected. Apart from his growing eminence, it was doubtless also his conversational ability that gave him entry; he was one of the great conversationalists of his age. A friend told him of his reputation of putting everyone at their ease, adding: 'You actually possess a peculiar talent to charm even those, who would like to hate your opinions'.[35] He could be a sensitive and careful listener. William Dibdin, a chemist who saw him at close hand in his later professional role as a Gas Referee, and regarded him as one of the most remarkable men he had known, described his style: 'He was generally a silent man who would listen attentively and then come out with a shrewd question direct to the point. He always gave one the impression that it would dangerous to try to "bluff" him'.[36]

After Easter, Tyndall gave a series of eight weekly lectures at the Royal Institution on 'Sound'. Sundays as usual were spent on walks with Hirst and often Debus, who had moved from Clifton to teach at Guy's Hospital. The first edition of *Fragments of Science* was published in April, in haste to try to pre-empt the fact that some of his articles were already being published in America, where his reputation was already substantial.[37] The book proved a success, as the first run of 1,000 copies quickly sold out. Vieweg wrote to arrange a German translation through Helmholtz, as he had done with *Sound* and *Faraday as a Discoverer*, work which Anna Helmholtz undertook.[38] Eight editions were published in England before Tyndall's death, and many in America.[39]

Like all authors, Tyndall had to deal with the vagaries of copyright law and the lack of international agreements, which had prompted his hurry to publish. When the Royal Commission on Copyright was set up in 1875 and 1876—with members including Sir Henry Holland, Fitzjames Stephen, Anthony Trollope, and James

Froude—Tyndall, Huxley, and Spencer all gave evidence.[40] Tyndall's view was that he obtained a fair deal in the US on a voluntary basis, because William Appleton, his American publisher, was reputable. But he thought that a formal copyright agreement would be desirable. Suggestions were made to the Commissioners that a royalty system should be introduced instead of copyright to the author. Tyndall strongly objected, since he considered that it would remove his freedom to negotiate with publishers and optimize his remuneration. The Commissioners agreed with him. Nevertheless, their report was scathing about the state of the law, which they wanted reduced to an 'intelligible and systematic form'.[41] But subsequent changes to copyright law, nationally and internationally, were slow. The Berne Convention of 1886 gave foreign authors equivalent treatment to domestic authors, and the UK signed it in 1887. The US did not do so until 1989.

Tyndall's second book of 1871, *Hours of Exercise in the Alps*, appeared in May.[42] It is a charming set of recollections of many of Tyndall's major expeditions in the Alps, including his struggles with the Matterhorn. It sold well initially in England, with a print run of 1,000 copies, and smaller reprints in July and December, netting Tyndall £200. Editions were published in America by Appleton in 1871, with many subsequent editions, and by Vieweg in Germany in 1872, translated by Clara Wiedemann under the title *In den Alpen*. But the book had to compete with Leslie Stephen's *The Playground of Europe*, and Whymper's *Scrambles in the Alps*, centred on the arresting story of the first ascent of the Matterhorn and the tragedy of the descent. Emily Peel admired the dignified way in which Tyndall had responded to Whymper's 'false attacks' on Tyndall's dealings with him and the Matterhorn, to which Tyndall responded: 'I am very glad that my way of meeting Mr Whymper meets your approval. He has introduced a new phase into Alpine literature, and in my opinion a very contemptible one. But he is eaten up by conceit, which in his case is not toned down by the culture which renders it tolerable in some men'.[43] Tyndall held no grudges; he supported Whymper's application to the Royal Society for a grant of £50 for observations of the veined structure of glaciers, even though he did not think they were likely to yield significant insights.[44] He sent many of his friends both this book and *Fragments of Science*. Clausius, who regretted that an injury sustained at the Battle of Gravelotte would prevent his coming to the British Association in Edinburgh, admired 'how you can produce rigorous scientific works without much effort, opening new fields in any branch of physics, and how you then contribute with the diffusion of science through beautiful popular presentations'.[45] Unlike Tait, Clausius was no scoffer at popular communication.

Tyndall gave his final lecture on 'Sound' on 8th June, and a Discourse the following day 'On Dust and Smoke'.[46] This time he came out firmly in favour of the germ theory of disease, regretting that he had not done so publicly before. He argued: 'as surely as the thistle arises from the thistle seed…so surely does the typhoid virus increase and multiply into typhoid fever'. In saying this, he stood clearly for Cohn's theory of fixed bacterial species, a view that was not then held by Pasteur or Lister.[47] He reinforced his view of the usefulness of cotton wool respirators to prevent infection by airborne germs.

He also recognized the opposition to this theory, especially from some medical opinion. Not wishing any dispute to prevent other possible health improvements, he described the potential use of respirators by stonemasons, colliers, and enamellers, where harmful dust was evident, to transform working conditions. He told of a Lancashire seedsman who had written to him describing the incapacitation of workers suffering from irritation and fever. On Tyndall's recommendation he had introduced a mask of a piece cotton wool wrapped in muslin and suffered no complaint at all from the men. Tyndall's thoughts led further to the invention of a respirator for firemen which, unlike the smoke jacket which required a continuous supply of air to be pumped in, could be worn easily by one person, enhancing the fireman's ability to save life, as well as extinguish fires. Finding that cotton wool, even moistened with glycerine to remove airborne material, was not enough, he recalled the effect of charcoal in absorbing noxious chemicals. His masterstroke was to combine them, removing the dust and the chemical vapours together. Captain Shaw, chief officer of the Metropolitan Fire Brigade, was impressed and helped develop the respirator. In time, it gathered interest beyond the fire service, when he was approached on the recommendation of the Commandant of the School of Military Engineering to help design one for use in military mining.[48]

The invention of the respirator prompted Tyndall to make an intriguing evolutionary speculation about the reason for the existence of nasal hair. He wrote to Darwin: 'Supposing a savage tribe to be afflicted with epidemic disease which is undoubtedly propagated by floating particles, and that some of them had by a good fortune hairs within their noses: suppose them to breathe through the nose, as many savages do. The chance of survival would certainly be with those possessing the hairs within the noses…I believe this question of hairs within the nose has been a difficulty: but it certainly falls under your general principle'.[49]

But the real message of this Discourse on dust and smoke was a different one. Tyndall had led his listeners from the actinic decomposition of vapours through the tails of comets and the blue of the sky to the dust of London, from the germ theory to the fireman's respirator, illustrating the tendency of pure science to lead to practical application. The 'wanderings of the scientific intellect which at first sight appear utterly unpractical, become in the end the wellsprings of practice'. To 'abolish these wanderings', and seek only to concentrate on practical ends, would be a 'fatal mistake'.

* * *

Tyndall left for Switzerland on 12th July, delayed by lighthouse duties. As usual he passed through Paris, this time witnessing the destructive effects of the war: 'A great change had manifestly come over Paris; moral depression on the part of the people appeared to go hand in hand with the physical twilight of the dimly illuminated streets... The great block of houses between the Rue Castiglione & the Rue Royale were gutted & tottering to their fall... It made an exceedingly odd impression upon me to see Prussian soldiers occupying the platform at Amiens, and all the principal stations along the eastern railway... Could they have felt strong enough to deal magnanimously with France at the end of the war, humanity assuredly would have been the gainer'.[50]

He travelled on over several days to Pontresina, now 'becoming as fashionable as Chamouni; and far more salubrious'.[51] Pontresina lay on the coach road over the Bernina Pass and close to the baths at St Moritz, where it had become the practice for English physicians to send their patients. The day after arriving, Tyndall was invited by a fellow Athenaeum member, Christopher Puller, and urged by his guide Jenni, to take part in an expedition to climb the Piz Bernina, the highest peak in the Engadine (and the only one above 4,000m). Leaving from Boval at 1.20am by starlight, Tyndall, Puller, and two guides worked through broken glaciers, along heavy snow slopes, over rocky spines, and along sharp arêtes with 'merciless precipices' to the right, until reaching the top just before noon. The descent was challenging but they reached Boval at 6.30pm, with a further two hours for the descent of the Morteratsch glacier; a total of nineteen hours and a remarkable achievement for someone recently arrived and unacclimatized.[52] After that his excursions were more modest, but he went up the Piz Corvatsch without a guide. As he wrote to Mary Egerton: 'To me it is surprising to see the small amount of originality developed by English climbers, Girdlestone is the only man who has shown anything of the kind. I will not say it is of the most prudent kind, but it is infinitely more creditable to him than hanging on to the skirts of a guide to whom you delegate the skill of scenting out the proper way, and the labour of leading, your own inventive powers being for the time perfectly torpid'.[53]

The British Association meeting in Edinburgh started on 2nd August. Tyndall had intended to go, not least because Thomson was presiding and had invited him to cruise with Helmholtz, Huxley, Maxwell, and Tait in his yacht *Lalla Rookh* after the meeting.[54] He would also have enjoyed the opportunity for a 'scamper' in the Highlands with Mary Egerton and her daughter. But Hirst urged him against coming, to rebuild his health, and Tyndall stressed that he had been compelled to run away from London when he did: 'The air and habits of the place become deadly to me towards the end of June, and life there is to me no life at all'.[55] So, in much improved weather, Tyndall remained for another month in the Alps, extended at the end to join the unsuccessful search over several days for the body of a student, Bodmer, from Zürich University. He stayed on despite finding Pontresina 'overflooded with people—All the hotels filled, for the quiet Alpine village of a few years ago is now a place of hotels'.[56] He wrote a charming note to Thomson wishing him every success: 'Were I as strong as Tait, and equal like him to beer and tobacco ad infiniture, I should never have come to Switzerland at all but gone to Edinburgh instead'.[57]

Had Tyndall been in Edinburgh, he would have found Thomson's Address most provoking. Reading an account in Pontresina, he wrote to Huxley: 'Thomson opened my eyes in Pontresina. Seriously I wished that theory unuttered'.[58] The theory to which he referred is Thomson's argument that as 'life proceeds from life, and from nothing but life', and because the Earth was not in its early stages suitable for life, it must have originated through 'moss-grown fragments of the ruins of another world', from beyond the Earth. Thomson ended the Address by querying whether evolution had happened in biology and nailed his colours to the mast of intelligent design, by 'one ever-acting Creator and Ruler'. Huxley declared Thomson's theory 'creation by cockshy', imagining God 'sitting like an idle boy...and slinging aerolites [at]...a planet!'.[59]

Tyndall, Thomson, and Tait were not always on opposite sides of an argument. Some months later, Tyndall found himself for once on the same side, in what Tait hoped would be a 'splendid row' over a 600-page treatise On the Nature of Comets by Zöllner, a physicist in Leipzig. The attack was not only on Tyndall's cometary theory but also on his popular writings, and Zöllner pitched into fellow Germans Helmholtz and Hofmann, as well as Thomson and Tait. Tyndall's measured response to Tait was that he would rather see Tait and Clausius as friends than himself and Zöllner. Stung by the attack on his popular writings, he revealed the motivations behind them in a letter to Helmholtz. He said that he had often been pressed by Longman to publish his lectures, but had declined until conversations at the Athenaeum, especially with the late Sir Edmund Head—once Governor General of Canada—had convinced him that what Head called 'the deplorable ignorance and apathy of the so-called culti-vated classes of England regarding science' should be addressed: 'He did not want to see these classes profoundly conversant with science, but he wished to see them competent to sympathise with it in an intelligent manner'. Tyndall was inspired to publish Heat, followed by Sound, 'to evoke sympathy for science among classes of persons whom our Universities had left wholly in the dark regarding science'.[60] Helmholtz and Clausius were both scandalized on his behalf at Zöllner's attack.

Helmholtz wrote that he should feel no need to apologize for his popular works, which were well received in Germany.[61]

Tyndall's autumn of 1871 was disturbed by a lengthy piece of work for the Board of Trade and Trinity House, but he did manage to attend the first X-Club meeting of the season on 19th October. He wrote to Francis Galton to lobby for support for Mayer as the Copley medallist of the Royal Society, pledging: 'All of us are ready to go heart & soul in for Helmholtz when his turn comes. I urge for Mayer what Sir Wm Thomson urged for Joule last year—greatness of achievement & seniority of merit'.[62] The previous year, in a diplomatic *quid pro quo*, Thomson had written to second Tyndall's proposal that Mayer be elected a foreign member of the Royal Society, expressing his pleasure at the prospect of Joule being awarded the 1870 Copley Medal, following Tyndall's nomination, which Stokes seconded. The medal was duly awarded to Joule, and Tyndall struck again in 1871 by successfully nominating Mayer, seconded by Siemens. As he explained to Clausius: 'We obtained the medal for Mayer, though Helmholtz was strenuously supported by Thomson and others' (Helmholtz received it in 1873).[63] Tyndall put his views on the record by publishing articles in *Nature* on 'The Copley Medallist of 1870' (Joule, for which Joule thanked him), 'The Copley Medallist of 1871' (Mayer), and 'Dr Carpenter and Dr Mayer'.[64] In the last of those he printed an accusatory letter from Carpenter, in which Carpenter declared that he had told Tyndall on two occasions, which Tyndall had neglected to mention, of his own early recognition of Mayer's work, and that Tyndall was not the first to recognize it publically as he had claimed.

To his disgust, Tyndall managed to rupture a muscle in November, simply by 'scampering leisurely over English fields and crossing a few English fences and walking afterwards briskly 12 or 13 miles'.[65] This misfortune laid him up on a sofa for some time, and prevented his accepting an invitation to stay with the Peels at Drayton Manor in early December. It nevertheless did not prevent him from attending his first meeting as a governor of Harrow School. Christmas was spent at the Royal Hotel in Ascot with Hirst and Debus, the two close friends of more than twenty years, even if Debus was taciturn on a long walk on Christmas Eve after a characteristic chafing by Tyndall. They returned to London from Windsor on Christmas Day, and Tyndall was pitched straight into his Christmas Lectures on 'Ice, Water, Vapour and Air'. An amusing article in the *Daily News* described a parent taking his son in the hope that he would be thoroughly bored, as he had been by 'instructive' events in his youth, but found both completely won over by Tyndall's ease, clarity, and humour. The lectures were printed as *Notes of a Course of Six Lectures on Ice, Water, Vapour and Air*, and then worked up into a book *The Forms of Water in Clouds and Rivers, Ice and Glaciers*.[66] This book would later launch a further and vitriolic stage of the long-running Forbes saga.

With the Christmas Lectures over, Tyndall's many social engagements included dinner with the Spottiswoodes, and at the Oxford and Cambridge Club in a party including a 'Hindoo' who proved 'a most intelligent man'. Sundays were spent mostly with Hirst and Debus, and on the familiar walks around Kensington Gardens and further afield. But the pressure of London got to him again in mid-January, and he went down to Eastbourne for a few days, reading George Eliot's *Middlemarch*, visiting a lighthouse where the keeper's daughter was 'one of the loveliest girls ... I have seen for some time', and watching people picking coal from a wrecked collier off Pevensey.

Then Bence Jones wrote to say that he had been let down by a speaker about the eclipse, and Tyndall would have to do the Discourse on 2nd February 1872. Back in London he dined at the Spottiswoodes before Odling's Discourse on the discovery of Indium as his thoughts turned to his own upcoming lecture, which he decided would be 'On the identity of light and radiant heat'. He spent time over many days in the laboratory perfecting his demonstrations, even managing to magnetize a ray of heat 'in the sense of Faraday's magnetization of a ray of light', but at this point, the effect was too small to be shown to the audience. Though he wrote to Lady Lyell that he was working 'night and day' in preparation for the Discourse,[67] he did have Robert Roupell (a QC and Manager at the Royal Institution for many years), Clifford, and Hirst to dinner; attended the Philosophical Club; and visited the theatre, joining the Misses Moore in Lady Coutts's box at Drury Lane for a 'capital pantomime'. He dined at the Spottiswoodes before his Discourse, on a little mutton and a couple of glasses of champagne. According to a later letter to Roscoe, Tyndall found that a glass of good dry champagne was 'important for the well-being of my brain during the lecture'.[68]

There was a great crowd in the theatre, and Tyndall did not disappoint. He started provocatively, quoting (as he had twenty years earlier) from Pope's *Essay on Man*:

> All are but parts of one stupendous whole,
> Whose body Nature is ...

deliberately omitting the final words 'and God the Soul'. Not, he declared, 'because physical science has arrived at any conclusion hostile to that clause', but 'because what the poet goes on to affirm lies outside the sphere of science'.[69] This demarcation runs throughout his writings and would take its sharpest form in the Belfast Address two years later. The Discourse centred on the relationship between light and heat, as Tyndall promised to demonstrate their connection and substantive identity to his audience in ways that had only previously been seen by experimenters in their private laboratories. He showed that radiant heat could be refracted, reflected, polarized, and focused just like visible light. Then, for the first time to a public audience, he showed its double refraction through a crystal of Iceland spar, just like light. Finally, in his *pièce de resistance*, he demonstrated the rotation of polarized radiant heat by a magnet, as Faraday had demonstrated for polarized light.

The following day, after the efforts of the lecture, Tyndall was feeling 'very shady indeed'. On the Sunday he stayed in the whole day, with Hirst and Debus to lunch. But the next week he was back on form, and his engagements illustrate a typical week when he was well: hosting Roupell, Clifford, and Hirst to dinner after a Turkish bath on Monday; dining at the Saville Club on Tuesday and at Longman's on Wednesday, with Froude and the Dean of St Paul's among the guests. The Government Grant Committee followed on Thursday, with dinner at the Athenaeum, where he also dined on Friday with Ball. At the weekend he went to see the aquarium at Crystal Palace. Given the often frenetic social whirl, in conflict with his official duties and research agenda, it is not surprising that he occasionally found it too much.

His Discourse 'On The Identity of Light and Radiant Heat' was one stimulus to Tyndall to publish a collection of his works on this topic, *Molecular Physics in the Domain of Radiant Heat*, following those on diamagnetism.[70] As he edited the papers, he worked extensively throughout February and into March on related experiments.[71] This occupied him to the exclusion of much social activity, so that apart from dining at Lady Eastlake's in late February, taking in Lady Beauclere and admiring John Ball's 'beautiful' wife, he started turning down invitations. He even forgot the X-Club meeting on 7th March, though he had been the previous evening with Spencer to the Queen's Theatre to see *The Last Days of Pompeii*.

And here Tyndall's journal ends. There is no extant journal, apart from a record of his American trip, until March 1883, just after he decided to resign his posts as scientific advisor to the Board of Trade and Trinity House.[72] His life for the next decade would be even more busy than before.

17

⊷⟐⟐⊷

GOVERNMENT SERVICE
AND EDUCATION

1871–1892

Governments since time immemorial have employed scientific and technological expertise, for war and peace. In the mid-nineteenth century, the professionalization of the military and civil service was proceeding apace, tackling inadequacies revealed by the Crimean War. Meritocratic appointments and stricter use of examinations became the norm, even if personal connections still counted. Reforms to the Royal Society from the 1840s moved science in the same direction. Tyndall was a beneficiary of these forces. In addition to his role as an examiner, for military and civilian students, he became an adviser to government on many issues. While he firmly believed that a scientific investigator should be allowed to pursue whatever researches he judged best, untrammelled by demands for public benefit, he recognized that the fruits of science had great practical potential. So did education. He thought that miners should be educated in science so that they might themselves find ways of creating safer working conditions, not just for their moral improvement.

From the mid-1860s, Tyndall juggled his desire for fundamental research with the demands of providing scientific advice to government. This public role was partly a matter of duty, partly of money. For many years, Tyndall earned more from his advisory roles to government than his salary from the Royal Institution. It was George Wynne who gave him his first break, in 1853. Wynne, then a Captain in the Royal Engineers, had become an inspecting officer for the Railway Department at the Board of Trade. He asked Tyndall's advice on locomotive boiler explosions, for which Tyndall in turn sought Bunsen's views. Two years later, now promoted to Lieutenant-Colonel, Wynne invited Tyndall to examine a boiler that had exploded at Gloucester station. Tyndall quickly came to the conclusion that the cause was electrolytic corrosion, and suggested possible remedies.[1] His letter was attached as part of Wynne's report, and Douglas Galton, Secretary to the Railway Department, proposed circulating it widely to the railway companies.[2] Tyndall had checked his conclusions with Faraday, himself a frequent adviser to government, and it was doubtless Faraday who instigated his next contribution, as a member of the Commission on Lighting Picture Galleries by Gas, which reported in 1859. The Commissioners—including Faraday, Tyndall, and Hofmann—found that gas lighting was acceptable, provided that the products of combustion were removed, as the presence of sulphur compounds in coal gas led to the generation of acids that could damage the pictures.[3]

A few years later, in a related enquiry, Tyndall was invited by Henry Cole to join the Commission on the Heating, Lighting, and Ventilation of the South Kensington Museum. The Commission was initially chaired by Thomas Graham, the last Master of the Mint.[4] The other members were Tyndall (who chaired most of the meetings in Graham's absence), Percy, Frankland, Lieutenant Colonel Scott, and Captain Donnelly. It held its first hearing on 3rd July 1865, with four more in March and April 1866, saw a total of seven witnesses including William Boxall (Director of the National Gallery), and finally reported in 1869. By this time Graham had died, so Tyndall's name appeared as the first signatory. The Commission found the situation broadly satisfactory, but pointed to the need for more research into the effects of light, air, and temperature on pigments, vehicles, and varnishes used by painters. Tyndall chaired a follow-up meeting in February 1870.[5]

While this enquiry was in progress, Tyndall's expertise in sound brought another call for advice. In 1868, he was summoned before the Select Committee on House of Commons Arrangements to advise on the acoustics of the new chamber following rebuilding after the great fire of 1834. He used his experience of speaking at the Senate House in Cambridge to suggest the importance of surfaces that dampened sound reflection, such as perforated boards or draperies, and argued for some simple experiments to identify sources of improvement.[6] In complementary roles, he was a member in 1868 of the committee advising on the building of the organ for the Albert Hall (which opened in 1871), and the following year, of a committee on the acoustics of the new lecture theatre at the South Kensington Museum.[7]

But these were all individual and time-limited enquiries. Much more demanding were his roles as Scientific Adviser to the Board of Trade and Trinity House on lighthouses, and his later parallel role as a Gas Referee, also under the Board of Trade.

* * *

Maritime capability and power has long served Britain's interests, and the network of lighthouses around the coast of the British Isles has for centuries been crucial for safe navigation. At a time of rapid development of new technologies and scientific understanding, the Board of Trade retained a scientific adviser, in the person of Faraday from

1856, to assist them in making improvements. Although the Board of Trade held overall responsibility there were separate institutions for specific geographical areas: Trinity House for England and Wales (the Commissioners often referred to as the 'Elder Brethren'), where Faraday was the adviser from 1836; the Commissioners of Northern Lighthouses for Scotland; and the Commissioners of Irish Lights for Ireland. By early 1866 the ailing Faraday was hardly able to walk without help. He envisaged Tyndall as his successor, and in January, he took Tyndall with him to see William Pigott,[8] Deputy Master of Trinity House. Both wanted Tyndall to assist him with Trinity House duties, and Tyndall was glad to oblige. He received £100 as a retainer and £100 for work to be done, lower terms than Faraday as he did not want to be seen to be taking Faraday's place. With Faraday's health increasingly failing, Thomas Farrer, Permanent Secretary to the Board of Trade, spoke to Tyndall a few months later about his offer to help Faraday with some of his duties there as well.[9] Faraday offered his resignation on 23 July, but Farrer asked for it not to be put to the Board until Tyndall had returned from Switzerland.[10] It never was, and Faraday remained formally in post until his death.

By the summer of 1867, it was clear that Faraday could no longer effectively continue. While in the Alps, Tyndall received copies of letters from Cecil Trevor, Assistant Secretary to the Harbour Department of the Board of Trade, written to the Commissioners of Northern Lighthouses and the Commissioners of Irish Lights, explaining to them the arrangements for him to succeed Faraday in his duties for the Board of Trade and Trinity House, at a salary of £400 per annum. He would be available, under certain circumstances, to the Scottish and the Irish too. There was the smell of trouble from the outset as the Commissioners of Northern Lighthouses wrote to say that they did not wish for the services of an official adviser, but the freedom to consult whomsoever they thought appropriate. They were particularly cross when, four years later, they discovered that they were paying part of Tyndall's salary at a time when Tyndall had challenged a report by their advisers, David and Thomas Stevenson. The Irish were entirely content.[11] Tyndall wrote to Trevor to accept the appointment, conscious of his debt to Faraday: 'The highest object I can set before me in the discharge of these duties is to imitate as far as in me lies the example of the high minded & celebrated man who has hitherto been your scientific advisor'.[12]

Tyndall received his first cheques in March 1868, totalling £100 for his work to the end of 1867.[13] He was soon drawn into discussions and experiments on matters such as the sizes and arrangements of lenses, the measurement of the intensities of different colours of light by photometry, and the use of different means of illumination. New types of oil and gas burners and new forms of electric lighting vied for supremacy. An electric light had first been installed at South Foreland near Dover in 1858, and new generators were in development. Money and reputations were at stake.

In early 1867, before he was formally in post, Tyndall was asked to work with James Douglass (engineer to Trinity House), Thomas Stevenson (engineer to the Commissioners of Northern Lighthouses), and James Chance (the glass manufacturer), to advise on whether lantern designs used by Trinity House were better than those elsewhere.[14] He found himself in the middle of an argument between Trinity House and the Scots, adjudicating between the reports of James Douglass and the Stevensons. His own report, sent on 20th March 1868, came down in favour of Douglass. In November, the Stevensons disputed some of Douglass's report, submitting further evidence, and the matter remained unresolved. In parallel, there was a dispute on the merits of a paraffin burner preferred by the Scots and shown to Tyndall in early 1868. The Scots supported a new burner patented by the blunt and litigious Captain Doty—who was eager to receive his recompense, and later tried unsuccessfully to extract a royalty payment of £10,000 from the Board of Trade—while Trinity House declared their own burner to be better. Based on reports by Douglass and the chemist William Valentin, compared to the evidence of the Scots, Tyndall came down for the Trinity House solution.[15]

A further issue was the question of the general superiority of gas over oil burners, which became a major point of contention. The Irish considered a gas burner, designed and patented by the Irish inventor John Wigham, to be better. But in a series of trials in 1868 and 1869, the Scots disagreed. The Irish called for an investigation by 'a thoroughly competent person'. They urged the greater power and flexibility of gas, and its economy, especially when used as a flashing light. Tyndall went to Ireland with Valentin for a week in June 1869 to observe the gas burners at Howth Baily in Dublin Bay and at Wicklow Head, and to compare gas and oil lamps. In a long report to Farrer, he came out strongly in favour of the gas lamp, on the grounds of brightness, economy, and the flexibility to increase the power rapidly in fog. He encouraged the Board to support further development in Ireland.

Tyndall's visit to the Alps that year was interrupted by the accident that delayed his return until October, but to his relief, he heard from Hirst that the Board of Trade and Trinity House had not been demanding work in his absence. Nevertheless, he did find himself writing 'a Report about a Report', concerning the benefit of gas over oil, as the dispute from the previous year between Douglass and the Stevensons continued. Tyndall pointed out significant discrepancies in the latest report by the Stevensons, as did the London Gas Referees when asked for an opinion, and reiterated his support for the use of gas in Ireland.[16] In September 1870 he went to Ireland specifically to examine the use of gas in revolving lights, again arguing strongly in its favour.[17] He followed this up with a long letter in February 1871, accompanied by a tin model, arguing that a gas flame would be cheaper than oil for revolving and fixed lights, and adding that the periodic extinction of the light would save more money.[18] At the instigation of the Board he went to Ireland again in June 1871, where he was impressed by Wigham's new annular lens with a flashing gas light, and he returned in September—though with no time to visit Emma in Gorey—for a trial with a revolving gas light at Rockabill.[19] He continued to pronounce gas superior to oil, arguing that it was more economical, with the advantage that each lighthouse, whether flashing or revolving, could emit a characteristic series of beams that uniquely identified it to mariners.

As Tyndall left for the Alps in July 1871, the question of the Scottish paraffin burner resumed.[20] When he returned he was immediately plunged into a dispute about it that affected all his other activities. His twenty-three-page report on the relative merits of the Trinity paraffin lamp and Doty's version, covering fifteen days of experiments, was sent to Trinity House in November 1871. He now came out in favour of Doty's burner—although not of Doty's apparent failure to trust his competence and independence—provided that the financial arrangements were appropriate. Though he still emphasized the longer-term potential of gas, he was not partisan. In many ways he was the epitome of the non-political, evidence-based science adviser. He ruled later on the safe storage of paraffin oils.[21] Debates over paraffin ran for years as all the lighthouse authorities explored converting to it from vegetable oils such as colza oil. Although weekends and some evenings could be spent relaxing with friends like Hirst and Debus, resolving the dispute between the rival inventors took all his working time. It seriously interfered with his research: 'These experiments have often kept me from 9 A.M. till 7 P.M. in a diabolical atmosphere. Sometimes even as late as 11 P.M.... I constantly have to think of the loss they have caused me by withdrawing me from dignified scientific work'.[22] The demands caused problems for Frankland too, who had allowed experiments at the College of Chemistry without the permission of Acton Smee Ayrton, First Commissioner of Works in Gladstone's administration, and risked being hauled over the coals.[23]

* * *

To add to his burdens, though this was a welcome one, Tyndall was appointed in October 1871 as a governor of Harrow School. He had a strong interest in scientific and technical education throughout his life, though without the messianic drive or strategic ability to influence public policy that Huxley exhibited. Tyndall's teaching experience at Queenwood College, and his statement of educational philosophy in his 1854 lecture 'On the Study of Physics', had established him early in his career as a significant force in science education. A couple of years after that lecture, in November 1856, and at the instigation of Henry Moseley, Tyndall was approached by Stephen Hawtrey, Mathematical master at Eton, asking his opinion about the introduction of a course of experimental science into the school. The major public schools at the time offered little or no teaching of science, and this was an opportunity to influence them. So, despite his commitments, Tyndall accepted an offer to give a series of lectures the following autumn. At the first, on 8th October 1857, after his return from the Alps, the boys were rowdy. But Tyndall relished the challenge, thinking of his skills honed at Queenwood: 'My old power among these young fellows will express itself before the course is ended'.[24] However, at the second lecture, the rowdiness was such that the headmaster threatened punishment. Tyndall did not support such action and suggested a less demanding lecture to engage them. Hawtrey wrote that the boys were remorseful, and hoped he would continue. He did, lecturing in November to enthusiastic applause. But, feeling that he had pitched the lectures too high, through misunderstanding the boys' capabilities, he returned £17 18s of the 60 guineas Hawtrey sent him for the six he delivered.

Tyndall's advice was soon in demand on matters of educational policy. Lyon Playfair, in 1857, sought Tyndall's recommendations of books for science teaching. Tyndall added the suggestion that equipment to demonstrate physical principles could be made at a fraction of the cost of those from manufacturers, but only if the teachers understood the principles and were instructed in the use of demonstrations. He was not impressed by the teaching force, writing: 'it is scarcely possible to conceive the amount of incompetence...which is afloat in this Country under the general designation of School Masters'.[25] The government's response to this, which Playfair expressed to Tyndall, was to pay the teachers who presented pupils for exams according to their results. He argued: 'it is the result which we wish to be attained. How this result is to be attained would be left to private enterprise'.[26] This use of State support to encourage private enterprise meshed with the predominantly laissez-faire Victorian approach to capitalism. It worked, as examination entries grew rapidly, leading to the inevitable reductions in payments per pupil over time to keep expenditure acceptable. 'Payment by results' has a long history in education. Tyndall also contributed to the training of science teachers by lecturing to them, though to a smaller extent than Huxley.

What he did not do was engage actively to advance the education of women. Even late in life, after his marriage to an intelligent woman, he could agree with Hirst, who wrote approvingly of a friend's comments: 'I told them at Girton...that I wished my daughter to pay very particular attention to three accomplishments—Horsemanship, Cookery and Whist. These secured, she might amuse herself as she pleased!'.[27] In an undated statement of his views, in what appears to be a draft letter to the editor of *Woman* magazine, he declared: 'While feeling, in common with the true womanhood of England, the reverse of sympathy for the modern development of Amazonianism,

I would willingly aid in placing the flower and fruit of human knowledge within the reach of women. I believe in their capacity to grasp and to enjoy whatever the brain of man has achieved; and to those who are striving to give their capacity for healthy exercise I would say "God Speed"!'.[28] Hardly a ringing endorsement of female potential or equality.

Although he was rarely involved directly in University policy, Tyndall did give evidence in April 1858, as the first witness, to a committee of the University of London established to consider introducing science degrees.[29] It was a plea for a physics doctorate, based on his continental experience, for which a prior science degree would not be necessary (mirroring his own case). He was adamant that degrees in science and the classics should have equal status.

One Commission and four enquiries in the 1860s reveal much about Tyndall's educational philosophy and practice. On 10th March 1866, he was appointed to a Commission following the decision to convert the Museum of Irish Industry into a College of Science for Ireland.[30] Fellow commissioners included Lord Rosse, Sabine, Huxley, Frankland, Playfair, Hofmann, and Carpenter. This Commission, with Captain Donnelly acting as Secretary, was asked to examine the purpose of the College, its course of study and its staffing. It worked speedily and reported on 9th July. Though conscious of the desire to create an institution to support a wide industrial base—wider than that of the School of Mines—including agriculture, mining, and manufacture, and the training of teachers, the Commission recommended a broad scientific curriculum over three years, leaving students to specialize later if they wished.[31] Given the membership of the Commission, full of men of science rather than industrialists, it is unlikely that any other conclusion would have been reached. The report was accepted by the Treasury.

The first of the four enquiries was the House of Lords Select Committee on the Public Schools Bill in 1865. Tyndall proposed in evidence that the science curriculum in public schools should be compulsory, amounting to 6 to 7 hours per week, covering elementary physics—including gravity, light, heat, sound, electricity, and magnetism—and the first principles of chemistry and botany. In speaking earlier to science teachers, he had emphasized the need for them to have first-hand experience of phenomena.[32] In this evidence he urged that science should be taught by demonstration, and encourage reasoning and experimental testing by the students. Questions he set for the Department of Science and Arts in 1870 illustrate his approach and helped to shape the nature of physics teaching: 'Give some proof that light itself is invisible', and 'Are clothes really warm? If not, why are they sometimes called warm? What is the real meaning and action of cool dresses and warm dresses?'.[33] Examination, he thought, should be by independent 'men of recognised eminence in science'.[34] By this he probably meant Fellows of the Royal Society. His examination load at the time—in acoustics, light and heat, electricity, and magnetism—was more than 1,000 papers, for which Hirst and Augustus Matthiessen helped out while he moderated them.

Tyndall drew on these ideas for a related enquiry for the British Association. Just before his lectures at the Royal Institution started in early 1867, Tyndall had dined at the Huxleys with two schoolmasters, Frederick Farrar (of Harrow), and James Wilson (of Rugby), and with George Griffith (Assistant Secretary of the British Association) and Hirst (now General Secretary with Francis Galton). At the 1866 British Association meeting, they and a few others had been constituted into a Special Committee to report to the meeting in Dundee in 1867 on the teaching of Natural Science in Schools.[35] Tyndall had missed the Special Committee meeting on 2nd March, but in the absence of a chemist on their sub-committee, he had drafted a paragraph on chemistry to which Wilson took exception, since in his view, it appeared to consider chemistry subordinate to physics.[36] Wilson, who had sounded out support from others on the Committee, sent a paragraph to Griffith for discussion at the Council meeting on 23rd March and suggested that Tyndall might like to submit a revised one too. He pointed out that Oxford required chemistry, as well as mechanical philosophy, for scholarships and studentships for the Natural Science School, which had been established in 1853. He thought it important that the Committee treat chemistry as they had physics, and in the eventual report, the two demonstrably had equivalent weight.

In the overall battle between a classical and a scientific education, and the appropriate balance between them, Tyndall came down firmly on the side of the scientific. Wilson thought that an education principally based on natural science would be found as defective as an exclusively classical one, arguing: 'At present science is only taught by clever & enthusiastic men, who c[oul]d teach anything. But when it is taught by everybody there will be another story'.[37] He thought that 3 hours a week would be soon achieved in most of the public schools, and would be found to be enough. This figure was one of six main recommendations in the report, which struck a diplomatic note, acknowledging the value of the classics but suggesting that some of their time, especially that devoted to Greek and Latin composition, could be given over to science. For the sciences, the report distinguished between the need for scientific information, predominantly geography, geology, and natural history, and for scientific training, which should focus on experimental physics, elementary chemistry, and botany. The report argued that both aspects should be offered, foreshadowing twentieth-century debates about education for scientific literacy and for future scientists, and that more importance should be given to arithmetic. In terms of organization, the report recommended both the compulsory system for science at Rugby and the voluntary one at Harrow. Anything more would have been too radical.

The other two parliamentary enquiries were stimulated by concerns about the superiority of education abroad, in the context of industrial competition. After the 1867 Paris Exhibition, Playfair argued that he had seen no improvement in British exhibits since the 1862 event, and put the relative decline down to better technical education on the continent. In his evidence to the subsequent Schools Enquiry Commission of 1867, Tyndall stated his view that the facilities for scientific education were far greater on the continent, and argued that England would 'find herself outstripped...both in the arts of peace and war. As sure as knowledge is power this must be the result'.[38] The Select Committee on Scientific Instruction took up this issue the following year. It made fifteen recommendations, all emphasizing the improvement needed in scientific education to counter the threat of industrial competition from abroad. Tyndall's evidence responded to questions about his lecture-based course in general physics at the Royal School of Mines.[39] He admitted that he did not focus on industrial applications, but claimed that the students would readily be able to apply the principles he taught as needed. Given his emphasis on reasoning, and the nature of his examination questions, which required more than rote learning to answer, that is a defensible position. Nevertheless, when asked if the Royal School of Mines could be developed into a technical school he gave no opinion. He did comment on the behaviour and interest of the working men, which he thought on a par with the Royal Institution audience from a 'higher stratum of society'.

Although Tyndall acknowledged the importance of mathematics, he claimed that it was surprising how much physics could be conveyed and understood without formal mathematics. Tyndall was never a mathematical physicist in the mode of Maxwell, Thomson, or Tait, which led to his denigration by some. His understanding was rooted in materiality. He conveyed that to his students, in an inspirational manner, and in the process developed a classical physics curriculum that would be recognizable to school students today. His influence extends not only to the content of the curriculum. The painstaking effort he took to design demonstrations for his lectures resulted in a multitude of classic examples that have become part of the stock-in-trade of physics teachers the world over. Likewise, his lecture notes and popular books became models of clear exposition. He had the ability to write engagingly and accessibly without compromising on scientific accuracy. But he did not write substantive school or university textbooks, nor leave a cadre of trained teachers behind him as Huxley did. He was not involved, with Huxley, Frankland, and Frederick Guthrie, in the creation of the 'Science Schools' in South Kensington, forerunners of

Imperial College. Though he apparently had plans in 1856, which he shared with Magnus, for a practical physics laboratory at Woolwich, nothing is known of them.[40] Formal education, as opposed to public education and debate, was not his priority, though he excelled in the informal environment of the public lecture. The physicist Sylvanus Thompson said that Tyndall gave him 'what we do not get at all in Guthrie, and very little in Frankland, the swing, the ease, the dash, that makes all the difference between the easy and the tedious lecture'.[41]

Nor did Tyndall found a school of research physics. Partly this reflected his desire to stay at the Royal Institution, where it would have been difficult with the facilities available for him to do so. Had he gone to Edinburgh, matters might have been different. More tellingly, it reflects Tyndall as the archetype of the individual experimenter. He consulted others, and exchanged ideas at meetings and in correspondence. Occasionally, as was the custom, he would demonstrate his experiments to others in the laboratory. Yet the laboratory to him was the place of the lone experimenter, meticulously asking his questions of Nature. Despite having been powerfully mentored himself, he was not a good mentor to others in return, and doubtless difficult to work with. He did encourage his assistant William Barrett, who became Professor of Experimental Physics at the Royal College of Science in Ireland, but that is about all. That relationship broke down, when Barrett, perhaps unfairly, felt that Tyndall was appropriating his work on 'sensitive flames'. Barrett's interest in psychic phenomena (he co-founded the Society for Psychical Research in 1882) would not have endeared him to Tyndall.

Looking back on his own education, when he gave the inaugural lecture in the new Bream's Buildings of the Birkbeck Literary and Scientific Institution in 1884, Tyndall emphasized its importance, and his debt to the Preston Mechanics' Institution (later the Harris Institute), founded on the model of the London Mechanics' Institution, which became Birkbeck.[42] He said that he did not know a higher, nobler, more blessed calling than teaching, which he saw as the combination of knowledge and the power to inform and stimulate. Stressing the duty of hard work he nevertheless contrasted the need for rigorous study with the flexibility needed for research, arguing that 'you could not call up at will the spirit of research'.[43] But the message for students, perhaps one that he did not get across to the boys at Eton, was that duty underlay all, as it had for Nelson at Trafalgar, and as Tennyson had written of Wellington.

In the end it was not Eton but its rival Harrow that occupied Tyndall's energies, though he did become involved with Spring Grove School. Officially known as the London College of the International Education Society, the school had opened in 1867 at the instigation of the Liberal politician and industrialist Richard Cobden. It had an emphasis on internationalism, to help eliminate war and promote free trade, and a science-focused curriculum. Tyndall became a director, and helped plan the science curriculum with Huxley.[44]

* * *

Tyndall's governorship of Harrow School gives an insight into the challenge of influencing educational practice at the time, particularly at such an elite establishment. His association with Harrow dates at least from the mid-1860s, when he was invited by Frederick Farrar, then an assistant master, to give a lecture to the boys. Farrar was elected a Fellow of the Royal Society in 1866, partly on the grounds of his contribution to the introduction of science teaching at the school, through voluntary botany classes. He served with Tyndall on the committee of the British Association which produced the report 'On the Best Means for Promoting Scientific Education in Schools' to the Dundee meeting in 1867.[45] Much later, when a Canon of Westminster, it was Farrar who controversially arranged for Darwin to be buried in Westminster Abbey and preached the funeral sermon. Tyndall had other connections to Harrow, through his friendships with Francis Vaughan Hawkins and Francis Galton. Hawkins had accompanied him on their attempt on the Matterhorn in 1860, and Hawkins and Galton were his companions on a walking tour of Devon and Cornwall over Easter 1861, when Tyndall gave what may be the earliest recorded description of rock-climbing on the sea cliffs.[46] Hawkins was an old Harrovian who was instrumental in drafting Statutes for the school, even before he became a governor in 1874. Galton was the brother-in-law of the Head Master, Henry Montagu Butler, and influenced his approach to science teaching in the school. They had mountaineering in common. All were at some time members of the Alpine Club, and Butler had climbed Monte Rosa in 1856.

On 29th November 1871, Tyndall found himself at the offices of the Public School Commissioners in Westminster, attending his first meeting as a governor of the prestigious Harrow School. This pupil of a rural Irish hedge school teacher could hardly believe it, writing the same day to Mary Egerton: 'I remember when a child hearing of Harrow as a place existing at almost Olympian heights. I remember my poor father telling me of the great men who were educated there; and here I am a Governor of Harrow. Were I to write a simple account of my life it would be better than many romances'.[47] The Public School Commissioners had been established following an inquiry into the nine grandest public schools, including Harrow, by the Clarendon Commission of 1861–1864. The Earl of Clarendon was a governor of Harrow from 1861–1870, but nevertheless proved determined to carry out a searching examination. To the later benefit of Tyndall he challenged the opposition of Butler to the appointment of men of science as governors. After a parliamentary battle, the Commission's report led to the Public Schools Act of 1868. This required the formation of new governing bodies, giving them increased influence over the curriculum, but it envisaged no public accountability or external scrutiny through statutory inspection. It effectively closed access to the elite public schools to all but those privileged by birth and wealth, a legacy of social divisiveness that remains today. Forster's subsequent Education Act of 1870 provided for a national system of education for the lower classes. The 'public' schools became entirely private.[48]

The new governing body for Harrow consisted of ten members: five elected by the old body and five appointed respectively by the Head and assistant masters, the Vice-Chancellors of Cambridge and Oxford, the Lord Chancellor, and the Royal Society. The old governing body had tried to elect Tyndall themselves in 1869 but were ruled out of order by the Public School Commissioners. It was not until 1871 that the Statutes and powers were agreed, at which point Tyndall was appointed by the Royal Society (Stokes got Eton, and Spottiswoode was given Westminster). He served for twenty years, until ill-health forced his resignation. Fortunately for him, the requirement that governors be members of the Church of England had been struck out in a late night Commons resolution, to the fury of the Duke of Abercorn in the Lords. Now this same Duke of Abercorn, one of the five elected members of the old body, found himself a governor with Tyndall. Neither would have imagined the connection that would develop between them later.

Tyndall was not an assiduous attendee, even though the meetings were held in London, initially at the offices of the Public Schools Commissioners, then at the offices of the school's solicitor Henry Young in Essex Street from 1877, or occasionally at Spencer House. The Earl of Verulam was the chairman during most of Tyndall's time, replaced in 1888 by Earl Spencer who had been a governor throughout. Of the first fifty-two governors' meetings up to November 1877, when Tyndall gave a lecture to the boys on fog signals, he attended just thirteen. Tyndall was not the only occasional member. There were often just four or five governors present. The bulk of the contributions came from Charles Roundell and William Stone MP, both Harrovians, and the Reverend Brooke Westcott. The Duke of Abercorn was automatically removed in 1883, having missed meetings for two years. He was promptly re-appointed by the Queen to fill up the vacancy 'caused by his Grace's non-attendance'.

As a strong Head Master, Butler was in any case capable of managing his governors astutely, and the difficulty of influencing the teaching may have affected Tyndall's attendance. It was not worth huge effort. Tyndall was one of few governors concerned with curriculum reform, and then only with science teaching. Although Butler was not interested in science, he did sense the need for change, introducing the Modern Side of the school in 1869, which included science, and initiating fundraising for science schools in 1871. He appointed the first science master,

George Griffith, in 1867, followed by Sydney Lupton in 1876. The curriculum on the Modern Side gave some boys the opportunity of an education less dominated by the classics, and by Greek in particular, but it was manifestly of lower status. Harrow was the second school after Cheltenham to offer this option, and Rugby had also started to teach science.

Reform was slow. Butler took more than three years to produce a report on science teaching requested by the governors in 1877. It was hardly challenging, requiring him to declare the subjects taught, the number of boys taught, and the time devoted to subjects. On its eventual production a subcommittee of Tyndall, Westcott, Roundell, and Stone was formed in March 1881, and planned to meet the science masters in London in May. Tyndall was unwell and the meeting did not take place until nearly a year later, on 16th February 1882 at the Royal Institution, although he did visit the school in 1881 to talk to the masters and inspect the teaching apparatus. Tyndall completed his report in March but little seems to have happened as a result. He remarked that he had tried several times 'to throw life into the department of the school, but I cannot say that my success was conspicuous'.[49]

The next opportunity for influence was the appointment of a new Head Master after the resignation of Butler in 1885 (to become Master of Trinity College, Cambridge in 1886 at the instigation of Lord Salisbury). On 1st May 1885, the governors appointed the Reverend James Welldon, then Head Master of Dulwich College. In parallel with the appointment process, the subcommittee of Tyndall, Westcott, Roundell, and Stone was reconvened, with the addition of a new governor, the historian Henry Pelham. Its task was to report on modifications to the timetable, provision for effective teaching in science, mathematics, and other non-classical subjects, and expenditure of the School Fund. Tyndall sought the advice of Helmholtz and Du Bois-Reymond, asking their views on the comparative value of a classical and a scientific education and the place of Greek.[50] Both bewailed the same issues in Germany, the poor teaching even of the classics, and the need for more education in mathematics and the sciences.[51] The new Head Master was asked for a report on the curriculum, which he produced on 30th October 1886, rather more speedily than Butler had, although Tyndall recognized the difficulty of introducing 'new forms of culture' at a time of transition and did not press hard.[52]

Welldon's report started inevitably with the place of Greek. He recognized that it could not be taught properly to all boys, but argued: 'when it is taught, it should absorb a large proportion of their time'.[53] There was a necessity, therefore, for a Modern Side in which modern languages took the educational place of the ancient. To his credit, Welldon saw the general ignorance of natural science as 'a blot upon liberal education'. He thought that some science should be compulsory for about two years, and that it should consist of botany and chemistry. A few boys might then specialize, perhaps taking physics, astronomy, or geology. At the December meeting of the governors, the physicist Tyndall was not impressed, commenting that they 'hardly touched on the main business of Welldon's scheme of education; he has practically omitted its sheet anchor, physics, making Botany take its place'.[54] Nevertheless, he did recognize the strengths of Welldon's overall approach, and commended his statement that if the boys did not distinguish themselves in science in coming years the fault would lie with the natural science masters. (Griffith, whose capability was in noticeable decline, proved difficult to sack.)

Matters dragged slowly, and on 7th February 1888, the governors approved a request by Tyndall for a full report on science teaching by the Head Master. At Tyndall's urging, there was agreement at the governors' meeting in November to spend £200 on the salary of a Demonstrator in Science. But Tyndall missed the meeting in February 1889, when the governors were informed that a Demonstrator was not necessary and a cheaper assistant had been appointed. At the March meeting, Tyndall laid out his views on improving physics teaching. The minutes record that the governors, 'in order to strengthen the position of the Head Master, and to encourage him to devise some means by which Professor Tyndall's views might be carried into effect, expressed their willingness to leave the matter in his hands'. Tyndall circulated a copy of his 1854 lecture 'On the Study of Physics' as a stimulus, but then missed the rest of the meetings in 1889. In his absence, the governors resolved to equip a laboratory in the Lower School subject to his advice. In April 1890 they finally considered the Report of the Head Master on the Natural Science teaching of the School. Welldon advised them that 'they might take it that all the Apparatus for teaching were satisfactory but not so the result of the Teaching itself', and suggested they should pass a resolution calling upon him to take steps. The governors dutifully resolved that 'steps should be taken by the Head Master for the improvement of the teaching of Science'.

That was all. The meeting was Tyndall's last, as illness took over. He wrote to Young in early 1892, explaining his non-attendance and willingness to resign. On 10th May his office was declared vacant, as he had been absent for two years.

Politically, if not educationally, Tyndall found himself at home at Harrow. The governors were largely of a Liberal mind, including Earl Spencer, a Liberal Cabinet Minister, but class interest trumped politics. When a replacement was elected for the Tory Lord Abercorn in 1885, Tyndall commented: 'It was somewhat impressive to me to see a body of men, the great majority of whom were Liberals, entirely forgetting their political predilections in the election of the new governor'.[55] They elected the Tory Sir Matthew Ridley in Abercorn's place. Whereas a majority of the governors may have been Liberal, the boys were firmly Tory. When Tyndall inadvertently projected the Tory colour on a red cloth in a lecture during the 1885 General Election, there was massive cheering: 'the boys certainly did not hide their political principles'.[56] Tyndall had in any case moved towards the Conservatives by 1885 as his opposition to Gladstone's Irish policy crystallized. Harrow moved likewise; the masters and governors increasingly identified with political conservatism after the Home Rule Bill of 1886. Tyndall's influence on Harrow was limited, but with the entrenched weight of tradition and social pressure, no other outcome was likely.

* * *

A few months after Tyndall took up his Governorship at Harrow, in early 1872, a major row blew up over the ownership and management of Kew Gardens, funded by the Board of Works but in many ways the private fiefdom of Joseph Hooker. Inevitably, Tyndall became involved. The gardens, museum, and herbarium had been built up over the years since 1840 by Hooker's father, Sir William Hooker, and by Hooker himself (who succeeded his father to the directorship in 1865), at considerable personal expense to them both on top of the public funds. Ayrton, a combative barrister, was Hooker's political master. Seeking to cut public costs, he attempted to drive through the transfer of Kew's scientific functions to the British Museum, of which Richard Owen was the superintendent of the natural history departments, leaving the gardens as a public park. Hooker had suggested in 1868 that the herbarium collection of Joseph Banks should be moved from the British Museum to Kew, threatening Owen's plans for a natural history museum in South Kensington and citing mismanagement at the British Museum as a justification. The stage was set for confrontation, and Ayrton was the man to initiate it. As Lord Derby remarked, when dragged into the debate by Tyndall: 'Ayrton's habit of harassing and ill-using his subordinates is well known'.[57]

The harassment started in 1870, soon after Ayrton's arrival in post, with what Hooker saw as the undermining of his authority with his staff. And when Ayrton appointed Douglas Galton to oversee the building of the hothouses, Hooker, who had a short fuse himself, felt his position was intolerable.[58] He resorted to a direct appeal to Gladstone, Ayrton's superior as First Lord of the Treasury. In time-honoured manner, Gladstone, through his private secretary Algernon West, prevaricated for several months before asking a Committee of the Cabinet consisting of the Marquis of Ripon, Lord Halifax, and Cardwell to examine the matter. Hooker wrote to Tyndall to say that he was summoned for interview on 13th March 1872.[59]

Quite apart from their sympathy to Hooker's personal position, Ayrton's machinations threatened the designs of the X-Club friends to create greater independence and a stronger influence for their scientific activities. Hooker's vision was for a body of the directors of the key institutions to be placed under the Lords of the Committee of the Council: 'as it is, the Lords have no scientific council that has the confidence of the public and Science is jumbled up with Art at South Kensington, with the Parks under the Board of Works, with literature at the British Museum and so forth'.[60] There was a parallel battle under way between Hooker and Richard Owen over the functions of Kew in relation to the museum. By chance there was a meeting on 26th March of The Club, of which both Owen and Gladstone were members. Tyndall used the opportunity to raise the issue with Lord Derby, who offered his help and undertook to look into the case. Lubbock stood ready to advance the cause in the Commons. In April, Lord Ripon conveyed Gladstone's verbal message following the deliberations of the Cabinet Committee, that 'Mr Ayrton has been told that Dr Hooker should in all respects be treated as the head of the local establishment at Kew: of course in subordination to the First Commissioner of Works'. This statement resolved nothing, and no answers were given to Hooker's specific points in his subsequent formal written response to the Treasury. Hooker put the questions back to the Treasury again, but no reply was forthcoming.

As the dispute gathered pace, Tyndall's series of nine weekly lectures at the Royal Institution on 'Heat and Light' demanded attention; he had tried but failed to persuade Dean Stanley or Froude to fill the first three slots. To add to the load, he had two books to complete. Writing to Tyndale on 2nd April, he said that he had spent the last six weeks like a hermit, having broken the back of his biggest book, but with the little one leaving much to be desired.[61] Publicly his star was high. *Vanity Fair* published a striking image of him with an article in April, commenting that 'social habits have taught him the scientific use of conversation, and he is one of the most welcome and expansive of table companions' (Figure 30).[62]

Lord Derby had now seen all the Kew correspondence, agreed that Hooker had a good case, and concurred that the correspondence should be asked for in Parliament. It was clear to all that such a course of action could result in the resignation of either Hooker or Ayrton, and that Ayrton might be protected by his political colleagues, notwithstanding the fact that so many people disliked him. A 'quarter-deck apology' from Ayrton to a subordinate was quite out of the question. The stakes were high. Tyndall had already petitioned, or 'memorialized', one Prime Minister, Sir Robert Peel, in 1843 over conditions of service in the Ordnance Survey. He now orchestrated the memorialization of another. In this he was aided by Huxley, who had returned in April from a long stay in Egypt to recover his health, and by others, including Spottiswoode and Lubbock, whose position as an MP in Gladstone's party was delicate. Nothing could be done formally until the House met again in late May. In early June, Lord Derby moved in the Lords for the papers to be produced, and a fortnight

Fig. 30. Tyndall as 'Man of the Day',
Vanity Fair, 6th April 1872.

later, Lubbock, who had seemed to be equivocating, did so in the Commons.[63] In the meantime, Tyndall sought signatories for the memorial to Gladstone and lined up further supporters in the Lords, as did Hooker and his colleagues. Eleven of the most eminent men of science signed the memorial, including Darwin, Lyell, Huxley, and Spottiswoode. A notable absentee was Airy, Astronomer Royal, and President of the Royal Society, who felt it was quite improper for him, as an officer under the government, to interfere. Lord Derby advised that nothing would be lost by publishing the memorial, since the signatories would guarantee interest from the press.[64] The view was that on seeing the evidence, published in full in *Nature*,[65] the press and public would back Hooker, which they largely did. There the matter stood when Tyndall left for the Alps on 25th June, telling Huxley that he would have to take over the coordination, and missing the annual excursion of X-Club members and their wives. His book *Contributions to Molecular Physics in the Domain of Radiant Heat*, an edited collection of his papers, was published just before his departure abroad, and respectfully dedicated to Bence Jones. Joule, bearing no grudges for the earlier dispute, wrote to thank him for such a handsome collection of his researches on radiation.[66]

On the way to Switzerland, Tyndall stopped off in Paris for a couple of days, dined with Jamin, and spent an hour with Pasteur, whose experiments on fermentation much impressed him. Pasteur had written the previous October in a flattering manner, sending several of his papers and a summary of his work to date. Sensing someone who could help, he wrote flatteringly that with Tyndall's 'rare genius, your brilliant imagination, your incomparable talent of explaining, you will provide a great service'.[67]

Chamonix was Tyndall's initial Alpine destination. On the first day he climbed straight up the cliffs above Montanvert, extricating himself with difficulty when he lost his way on the descent. The weather was glorious, and after an excursion to the Grands Mulets for the views, though not on this occasion to climb Mont Blanc, he hatched a plan with Thomas Kennedy, first ascensionist of the Dent Blanche in 1862, to inspect the unclimbed Aiguille du Géant. They were repulsed by the final 200 feet of smooth granite, which 'jutted above us like a perpendicular tower'. The peak remained untrodden until 1882. After a week in Chamonix, Tyndall went over the Col du Balme and on to his 'perch' in the hotel on Belalp. He settled into a routine of a regular early morning swim in a stream, followed by several hours' work, as he continued preparations for his American tour. As ever, it was a sociable place. John Ball was there when he arrived, as were various clergymen, doubtless interested in the 'Prayer Gauge Debate' stimulated by the article in July's *Contemporary Review* to which he had written an introductory note. The widowed Mary Egerton arrived later with three children. The weather in the Bernese Oberland is often less reliable than in other parts of the Alps, but it brightened for a fortnight, during which time Tyndall made excursions onto the glacier and climbed the Aletschorn for the second time, breaking his ice-axe in the process and saving his porter, who slipped and fell down an ice slope.

The Ayrton affair ground slowly to a dénouement, as correspondence flowed to and from Belalp. Without the knowledge of Hooker and his friends, Ayrton had commissioned a report from Richard Owen, highly critical of Hooker, which was published with the papers in Parliament. It was an astute tactic of Ayrton's to pit two experts against each other. Tyndall was particularly saddened by Owen's action, writing to Huxley: 'I never broke with that man, as you know, I hoped against hope that matters might in the end be healed up. But this last trick makes me feel that those who broke with him knew him better than I did. It will greatly augment his isolation'.[68]

Lord Derby raised the matter in the Lords on 29th July, in as emollient a manner as possible, meeting significant resistance from Ayrton's supporters in a sparsely attended chamber. Lubbock, whether leant-on or otherwise, had come to the view that the matter was best smoothed over informally, which was Gladstone's position, and sought to persuade Ayrton and Hooker to back down, an outcome he seemed to achieve. But on 8th August, two days before Tyndall arrived back from the Alps, Fawcett (Tyndall's ex-pupil) intervened during the third reading of the Consolidated Fund Bill, which included the funding of Kew, forcing Lubbock to speak in the Commons. Ayrton gave a robust defence against the 'scurrilous and most calumnious libel' upon him, noting that no resolution against him had been brought forward in either House. He took a swipe at the 'philosophers', eminent for their knowledge of organic and inorganic matter but not of the higher moral science of the Law nor of the relationships between Ministers and civil servants. Gladstone, attempting to stitch back together an accommodation that seemed to be unravelling, rebuked Lubbock for going into the detail of the complaints against Ayrton. He remarked, with respect to Hooker, that 'those who are not accustomed to enter into our sturdy conflicts take reproof in a much more serious manner than we who are hardened by long use are accustomed to do'. He hoped that Hooker and Ayrton would 'bury in forgetfulness the recollection of those differences'.[69]

The affair rumbled on through the autumn, until honour could be deemed to be preserved on both sides. Hooker stayed in post, though battles with Owen remained to be fought. Ayrton was moved in 1873 to the post of Judge Advocate General. He lost his seat at the 1874 General Election, and never returned to Parliament. In a postscript in 1879, two positions became vacant in The Club. Tyndall proposed Hooker, seconded by Sir James Paget. He told Hooker that Owen rarely attended and that 'even if he does there seems no reason why he and you should growl at each other'.[70] Hooker was elected. Tyndall's machinations behind the scenes in relation to Ayrton did not affect his desirability as a government adviser: at the instigation of Farrer, he was appointed in September as a Gas Referee, a position he held for the next twenty years.

Back from the Alps in 1872, Tyndall concentrated on preparing for his forthcoming lecture tour of America, taking just one day in August to go down to the British Association meeting in Brighton, where Carpenter was presiding, to hear Spottiswoode lecture to the Operative Classes of Brighton on 'Sunshine, Sea and Sky'. By early September, working daily at the School of Mines, he had established his approach for the American lectures. He wanted to do much more than 'make experiments before the American people. I want to impress upon them the means by which science flourishes, and indeed to cast a certain amount of discredit upon the very work which they have called upon me to do. I want them and England to feel in their heart of hearts the importance of the original investigator and this requires work and thought of a totally different kind from that brought into play in the arrangement of experiment'.[71] He asked Bence Jones to find him some lodgings away from the bustle of London in Folkestone, where he spent a week up until 20th

September finalizing his thoughts. Bence Jones noticed that Emily Faithfull was also due to lecture in America and teased him, as a bachelor, that 'if she should go in the same steamer you may have a chance'.[72] A conversation between them would have been fascinating to hear. She was a remarkable women's rights activist who set up the Victoria Press, hiring women as compositors against the objections of the Printers' Union. She published the feminist English Woman's Journal in the 1860s and then the Victoria Magazine, which advocated paid employment for women.

Tyndall's 'little' book, The Forms of Water in Clouds and Rivers, Ice and Glaciers, was published just before he left for America, its stimulus to future confrontation lying for now undiscovered. He spent the final weekend before his departure with Hirst and Debus, and said his goodbyes to Carlyle. Transatlantic travel was not without peril—two out of the three of Emily Faithfull's return voyages from America were booked in steamers that were wrecked on their outward journey, one with considerable loss of life—and leave-taking was poignant. His last note before departing was to Bence Jones, remembering their twenty years working together, and the more than twenty that he had known Hirst and Debus. Those two old friends accompanied him to Euston Square on 27th September to catch the train to Liverpool. He would be away for nearly five months.

18

AMERICA
1872–1873

G iven Tyndall's growing reputation, the Americans had been trying for some time to tempt him over to lecture. In November 1864, during the American Civil War, he had received an invitation from John Amory Lowell asking him to lecture on 'Heat' in the autumn of 1865 or winter of 1866 in Boston, for a fee of $1,200. But he had declined because of his research commitments; he was embroiled in the dispute with Akin.

At length, Joseph Henry, Secretary of the Smithsonian Institution in Washington, understood from Peter Lesley, Secretary of the American Philosophical Society of Philadelphia, that Tyndall 'could be induced for the gratification of your numerous American readers to visit this country provided it should not appear that you are seeking to make money by the operation'. Accordingly, he wrote in August 1871 to invite him on behalf of the Smithsonian, as 'a missionary as it were in the cause of abstract science, to vindicate its claims to popular appreciation and to government support'.[1] A further letter followed on 1 September, from a stellar group of signatories including Emerson and Agassiz, orchestrated by Lesley and Tyndale.[2]

Tyndall did not reply until 18th November, apologetic at the delay which had been caused by his returning from Switzerland only on 3rd September, and then being seized upon by the Board of Trade and Trinity House for 'a most unpleasant investigation. It quite made shipwreck of my autumn's scientific work'. But he finally accepted the invitation. He would have preferred to offer new material but regretted that he would not now have time to prepare it: 'consoling myself by the thought that I shall at all events make a first acquaintance with American audiences, and that at some future day I may make atonement for the shortcomings of the next year'.[3]

Tyndale was an important conduit in the communications, and Tyndall looked forward to spending some time with his American cousin. They had maintained their friendship across the Atlantic despite differing positions on the American Civil

War. Tyndall recorded in his diary a few months before he went to America: 'My American friends must learn somehow that I was a Southern sympathiser—not with slavery, but with bravery, during their fight'.[4] In this he was in tune with much public opinion in Britain, at least until Lincoln's Emancipation Proclamation and the growing sense that the North's victory would destroy slavery. Tyndale had fought on the Union side. Having been present at Cedar Mountain on 9th August 1862, the first battle of the Northern Virginia campaign, he was seriously wounded at the battle of Antietam on 17th September, commanding the first Brigade of the second Division of Mansfield's Corps. Antietam, of which he gave a chilling description to Tyndall, was the first battle on Union soil and the bloodiest single-day battle in American history, with some 22,000 dead, wounded, or missing. Tyndale received a ball 'exactly in the centre of the back head—on the occipital bone—thankful that my head is so hard or for the slight turning of it which glanced the ball along the skull, passing it under the large muscles back and side of neck, passing it between the jugular vein and carotid artery—grazing both and lodging the ball in front of throat: from whence it was cut out, flattened to a broadened hemisphere, a concave—convex form!'.[5] He was carried off the field under terrible fire by two of his men even though they thought he was dead. After a break to recover from his injury, and a bout of typhoid, Tyndale rejoined the fray as a Brigadier-General. He wrote to Tyndall in 1866, after a gap of nearly three years, describing his final experiences in the war, which had ended in April 1865. He had marched about 1,000 miles with the 11th and 12th Corps under Hooker, from Virginia to Chattanooga, fighting the battles of Wauhatchie, Chattanooga, Lookout Mountain, and Missionary Ridge. A further march of 250 miles to relieve Knoxville, in the ice of winter with thousands of men and even officers barefoot, led to his partial paralysis. He resigned in August 1864, just two days before his brother-in-law was killed in action, 'riddled with balls'.[6] This veteran of the war was Tyndall's generous host in America.

The planned circuit for Tyndall's American tour, negotiated under the leadership of Henry, took in the East Coast venues of Boston, Philadelphia, Baltimore, Washington, New York, Brooklyn, and Newhaven. Henry also proposed that Tyndall attend the annual meeting in San Francisco of the American Association for the Advancement of Science, founded in 1848 on the model of the British Association.[7] Unfortunately, Tyndall's commitments did not allow this, despite his desire to see 'that glorious country which has been so vividly described by Prof. Whitney, and by my excellent friend Mr. Clarence King', the geologist and mountaineer who had named Mount Tyndall in the Rockies after him.[8]

Tyndall agreed to visit America on the condition that he made no personal profit from his lectures. His mission was purely philanthropic, setting a clear moral example, and initially aimed at the founding or maintenance of a scientific or other institution in Chicago. As he told Mary Adair, hinting at her earlier rejection of his marriage proposal: 'The people of Chicago have suffered so much that they have many wants to supply before they can turn their thoughts to Science...This is one of the benefits of bachelorhood—one is free to do such things!'.[9] Tyndall planned to be back home before Christmas, but changed his mind after hearing from Edward Youmans that the political excitement in New York in the run-up to the State election on 4th November was the maddest he had ever seen.[10] To avoid a clash with the election, Tyndall asked that the New York lectures be put back and the trip extended until February.

Tyndall arrived at Liverpool's Edge Hill station from Euston in the middle of a terrific rainstorm. He was met by a carriage sent by George Warren—a wealthy shipowner Tyndall had encountered at the Eggishorn in Switzerland—and taken to Strawberry Field,[11] Warren's country house in Woolton. After breakfast, Warren drove Tyndall round the Liverpool suburbs to acquire some fine tea for the journey, and to make arrangements for the safe storage of his equipment. He accompanied Tyndall onto the tender and up to the deck of the *Russia*, built for Cunard in 1867, before bidding him a safe voyage. Tyndall's assistants, the experienced Cottrell and the young Miller, presumably made their own way on board.

Tyndall sat down to his first dinner on the *Russia*—a slice of beef—as the weather took a turn for the worse and delayed their departure. Outside, an angry Mersey echoed the latest news on the paper placards at the landing stage: the arrival of a heavy gale, the foundering of a Welsh schooner and the loss of a Liverpool ship with thirty lives.[12] Tyndall slept fitfully the first night as the *Russia*, still docked, was buffeted by the tail end of the storm. By morning the weather had subsided, and as the ship eased out of port the heavens were clear and the wind keen. On deck, Tyndall chatted to Froude, who was also setting off on a lecture tour. Travelling over with them was the Marquis of Queensbury, providing for Tyndall a melancholy echo of his failure to find the body of Queensbury's brother, Lord Francis Douglas, after the tragedy on the Matterhorn. The *Russia* had a rough crossing, accompanied by a fierce gale. As Tyndall recounted: 'the ocean ridges were high, and where they crossed each other they formed heaps tufted with foam which was caught by the wind and carried like smoke through the air'.[13] Warned away from the bow by the crew, Tyndall spent much of his time on the upper deck, watching the waves rise above the bow, curl over, and sweep the forward deck. One wave leapt the gunwale and swept along a bevy of ladies in a crash of chairs. Tyndall waded to their aid, 'to assure and rescue the frightened fair'.[14]

After eleven days, on 9th October, the travellers were woken earlier than usual to a pleasing sight: a cloudless sky and a view of the American coast. The arrival was as breath-taking as it was welcome. Tyndall wrote: 'under these favourable conditions we steamed between the forts which guard the entrance of New York harbour. I was not prepared for the view there opened out before me. It was splendid. To the left were the slopes of Staten Island, fringed with coloured foliage. The ample bay was studded with small craft flying about with their white winglike sails'.[15]

Tyndall's first impression of the city itself was not so pleasant: 'The docks are rather shabby, looking like the neglected & decayed structures of an old country, rather than the vigorous production of a new one'.[16] Once on shore, the party was quickly enfolded into the city's charismatic allure. The welcoming group for Tyndall and his assistants was led by Youmans and Appleton, who whisked Tyndall off to the Brevoort House, an 'exceedingly expensive hotel' on the corner of Fifth Avenue and 9th Street. Froude, meanwhile, was taken by Scribner, his publishers, to a lavish banquet. Such largesse did not appeal to Tyndall, who was grateful that Appleton had decided, given Scribner's action, not to organize the same for him. But there was plenty of time for sightseeing.

Over the next two days, Tyndale showed Tyndall Fifth Avenue, the 'paradise of New York capitalists', and Broadway, where 'the retail trade has here received its deification—being carried on in temples worthy of the gods'.[17] Tyndall also visited his lecture venues: the Cooper Institute in New York, and the Academy of Music in the adjacent city of Brooklyn. Youmans and Appleton took him up the Hudson River to the Military Academy at West Point. For Tyndall, it was a revelation: 'I do not think I ever saw anything more charming than the scene from this plateau. Bold hills with valley inlets clothed with woods flanked the Hudson right and left'.[18] Walking round the Academy with one of its professors, Tyndall was gratified to see his books in the library. To finish the visit, he enjoyed a glass of rye whiskey before dining with Appleton at his Bronx country house overlooking the Hudson.

Tyndall left New York for Boston with his cousin, while Cottrell and Miller travelled with Henry's assistant. The scenery impressed him again: 'the foliage at times was...indescribably brilliant; and as those wondrous trees passed me each of them started its psychological equivalent of pleasure. The monkeys from which I came must have been dandled amid such scenes'.[19] On arriving in Boston, Tyndall's impression was of a very different place to New York: 'an entirely English town as far as its aspect goes. It is massive and wealthy, and full of active life'.[20] He put up at the Revere Hotel, despite an invitation to stay with Asa Gray, the botanist and friend of Hooker and Darwin. With an eye to the local women, Tyndall remarked: 'The women here are very fine, and not of that fair and fading quality which I had thought characteristic of American ladies'.[21] Responding from across the Atlantic, Bence Jones urged matrimony: 'perhaps you will bring back an American wife. They settle those matters quickly I believe & some of them are very charming when married to Englishmen & make very good wives'.[22] Tyndall's need for a 'rare and radiant maiden' to settle down with was a standing joke between him and his close friends for many years.[23]

Tyndall gave the first of his six lectures in Boston on 15th October, in the great dingy, ugly hall of the Lowell Institute (as a competitive New York newspaper put it). Tickets were free, under the conditions of Lowell's bequest, and there was an enormous demand; 2 hours before they were distributed a queue of 500 people was already snaking from the door. When the tickets were finally handed out, they were snapped up in 20 minutes. A lively black market sprang up, with tickets trading for up to $8. At one point the organizers were urged to move the event to a bigger hall, as too many tickets were issued and some ticket-holders were turned away. But with the audience admitted—1,200 people crammed together—the event got underway. At 7.30 prompt, the figure of Tyndall, 'quite tall and thin, with his hair and side whiskers, dressed in the English style, tinged with gray', strode into the hall to take his position behind a low table about ten feet long and covered with green cloth.[24] On it stood a variety of lenses, prisms, and coloured pieces of glass, as well as a vessel half full of water, a gas burner, and an electric lamp connected to a huge battery. On viewing the spectacle, the reporter from the *Boston Journal* was underwhelmed. He had expected a more imposing sight from a lecturer with Tyndall's reputation for striking experimental displays. The hall was larger than Tyndall was used to in the Royal Institution,[25] and the curator of the Institute was concerned that the audience would become restless unless Tyndall kept his demonstrations brief. He needn't have worried. Both journalist and curator had underestimated Tyndall's ability to charm and engage an audience. One newspaper reported that Tyndall had a slight English accent. Both Hirst and Huxley found this amusing, which suggests that Tyndall retained a noticeable Irish accent at times.[26] Indeed, Charles Eliot Norton, if conceivably over-exaggerating for effect, recalled overhearing him at a dinner party a few years earlier, as he declared: 'Ah! the mountain tops, 't is there that man fales himself nearest the devine. I always sakes the mountain tops for relafe from the tile and care of the wurrld'.[27]

Normally, Tyndall would have developed a series of lectures that presented original material, as he would have done at a Royal Institution Discourse. But there had been no time before the trip to prepare; and besides, this was no ordinary course of lectures. His aim for this tour was twofold: to illustrate the growth of scientific knowledge through experiment, and to make the case for the value and support of science for its own sake, unrestrained by the motivations or needs of practical use. As a suitable topic to illustrate the development of human scientific understanding, Tyndall chose light. Although he covered reflection, refraction, diffraction, interference, and polarization in complex detail, he expressed and illustrated the phenomena beautifully using the experimental demonstrations for which he was renowned. Throughout the course of lectures he gave a historical perspective, examining how older theories are replaced by those better able to explain and predict the observed phenomena.

As an example, he gave centre stage to Young's 'undulatory' or wave theory of light, which had replaced Newton's 'emission' or particle theory. For this development of theories to take place, said Tyndall, 'the mind must possess a certain pictorial power', a scientific imagination that can go beyond observed phenomena and answer the question 'what is light?' But this is no arbitrary image, since the explanation or theory imagined must be tested by experiment. As Tyndall explained, if the phenomena deduced from the theory agree with those in the natural world it is a presumption in favour of the theory. If new phenomena arise that harmonize with the theory the presumption in its favour becomes stronger, and if the theory predicts phenomena never before seen that are then shown to exist, the truth of the theory becomes overwhelming. Tyndall highlighted the work of Thomas Young and the mathematics of Fresnel as critical in placing the wave theory on firm ground. This theory, unlike Newton's, could explain all the phenomena of diffraction and interference patterns and of the colours shown by thin films such as soap bubbles on the basis of waves of different lengths, short ones blue and longer ones red. Of course, like any theory it could be challenged in future. Outside this visible range lay the rays of 'radiant heat' in the infrared and the 'actinic rays' in the ultraviolet region. Tyndall showed the equivalence of the rays of radiant heat to those of visible light, just with longer wavelengths, and the chemical action of the actinic rays to create his famous artificial blue skies. Throughout his lectures on light, Tyndall asked the audience to picture what his experiments told us about the nature of matter, with different atoms and molecules able to absorb and emit light of different characteristic wavelengths. How were these waves of radiant heat and light transmitted? What was doing the waving, as the molecules of water move up and down while a wave is transmitted after throwing a stone into a pond? Tyndall held firmly to the ether theory, not to be displaced for many years yet, which imagined this invisible yet elastic material suffusing space and carrying the light waves with it.

Just as at the Royal Institution, the lectures were well attended by women. The Nation, taking the gender-stereotyping angle, commented that 'the instructed are satisfied with what he says, and pleased with his workmanlike way of saying it, while to the young ladies he gives gorgeous exhibitions of color, which they freely admit to be most lovely, most splendid and most nice'.[28] The more enlightened Boston Post declared the presence of so many women as 'somewhat remarkable considering the estimation put upon their capability of receiving instruction in the higher branches of learning by the dons who guard the gates of our leading institutions of learning...Their attendance here in such full numbers is no doubt a delicate bit of irony directed towards the home talent of our universities who have given them the cold shoulder in their efforts to gain instruction in their college walls'.[29]

Between lectures, Tyndall took in more sightseeing. Tyndale gave him a tour of the immediate sights, including Bunker Hill and the cemetery of Mount Auburn. He visited the birthplace of Rumford, one of the founders of the Royal Institution, taking the train to Woburn Centre and walking the 2 miles to the house. He arrived just in time for lunch: 'it was midday and the people were just sitting down to dinner. The flavour of fried onions filled the room in which the hero was born'.[30] Another of Tyndall's boyhood heroes, Ralph Waldo Emerson, tall, thin, and bald, took him out to his house, being restored after a fire. The men then visited the Old Manse in Concord, 'from the windows of which his [Emerson's] grandmother witnessed the first firing between the farmers & the troops', before travelling on to Lexington, where the first shots of the Revolutionary War were fired.[31] Engagements followed with Oliver Wendell Holmes, Henry Longfellow, and Agassiz, 'a perfect galaxy of genius'. Tyndall had already encountered Agassiz; the Swiss geologist had arrived, to much applause, at the start of the fourth Lowell Institute lecture. He had just returned from the Hassler Expedition to collect deep-sea specimens off the coast of South America to test their relationships to fossil forms. As Tyndall observed, Agassiz was much exercised by questions of evolution: 'it evidently disturbs him much'.[32]

In the final lecture in Boston, Tyndall emphasized the significance of international relations, and stressed the importance of investigation, rather than money, as the main motivation for science. The lack of government support in both England and America meant that 'the only course open to us is so to enlighten the public mind that those in a position to do so may be induced to help the work of original enquiry'.[33] He urged wealthy Americans to endow laboratories, giving scientific investigators the freedom to pursue their researches, without asking 'what is the use of your work?'. During his stay in Boston, tempting offers to lecture poured in from elsewhere: St. Louis, Cincinnati, Providence, Cleveland, and Columbus. Tyndall was forced to decline them all, as he was not due to finish in Brooklyn until late January, and had promised the students and Professors of Yale College that he would lecture in Newhaven afterwards. He was told that he could make £20,000 in two years by lecturing. Bence Jones noted that John Pepper—the popular scientific lecturer and inventor of the Pepper's Ghost illusion—was 'ghosting in America' at the same time.[34] The rewards may have been high in this Brave New World, but so was the cost of living. Tyndall's favourite wine, La Rose, was 16–18 shillings a bottle, and his total expenses during nineteen days in Boston amounted to nearly $400 (£80).

The next stop on the tour was Philadelphia, at the time the second-largest city in America after New York. It was the home of Peter Lesley, a good friend of Tyndale; they lived a few doors away on the same street. The first lecture in the city was organized for 11th November (when Tyndall witnessed the funeral procession of General Meade of Gettysburg fame),[35] and the postponement of the New York lectures gave time for sightseeing. Tyndall took an overnight train to Niagara Falls, which left an indelible impression: 'The first aspect of the Falls was rather tame, for the point of view accidentally chosen was not a good one. But their beauty and their majesty have grown upon me. I have seen them thoroughly—nobody more thoroughly, and in doing so have incurred some danger, for I went into places never visited by a traveller before'.[36] On his return the visit proved a fruitful topic for a Discourse as he made the most of his daring exploration right under the Falls. It must have made an impact, as he regaled the ill Tennyson with the story in 1888.[37]

With a population of 700,000, Philadelphia was a quarter of the size of London. Nevertheless, in his lifetime, Lesley had seen it expand from a Quaker village to a place that had experienced 'the erection of one public edifice after another...the birth of the fine arts in succession, and the planting & watering of the sciences'.[38] Despite Tyndall's preference for the independence and peace of hotels, he stayed with his cousin. His reputation was such that he found himself better known in Philadelphia than in many English cities. But even for someone used to London, he was struck by the sharp contrast between private style and public squalor: 'the houses here are wonderfully clean and luminous. As far as the Kerbstone all is beautiful—Beyond it all is—I had almost said vile. The street pavement is atrocious'.[39]

Tyndall's lectures took place in the frescoed Horticultural Hall, with its splendid new pendant chandelier, and once again they were the talk of the town. The audience, including the cream of Philadelphia society, were reserved at first—perhaps reflecting their Quaker roots—but by the third lecture their applause was enthusiastic and heated. As in Boston, the press raved about Tyndall, calling him 'the foremost disciple of pure science living', and ranking him alongside John Stuart Mill and Herbert Spencer in terms of significance.[40] To many, his lecturing style was a revelation. One newspaper gives a vivid account: 'He never reads, but holds his audience by the power of lucid and forcible extemporaneous statement. He is not what would be called a fluent or even speaker, who keeps up a continuous strain of agreeable utterance. He is not an elegant declaimer whose measured cadences are accompanied by graceful and appropriate gestures. He is irregular and sometimes hesitating in speech, and unstudied in gestures and movements...Of poetic and imaginative temperament...he is never eclipsed by his own pyrotechny, but holds the attention of his listeners closely to the question under examination'.[41] There was much appreciation for Cottrell too. Audiences and journalists alike applauded and praised his dexterous support throughout the experiments on display.

For his part, Tyndall was unrestrained in his homilies to the great and good of Philadelphia. In his last lecture, re-emphasizing the importance of fundamental research, he spoke to the audience 'in a manner which I am sure no Englishman had previously employed, and they listened to me without a murmur. I repeated to them de Toqueville's account of their poverty as regards genius in the higher sciences...that had they been alone in the world they would have found that practical science could not long be cultivated with success without abstract science but that they drew their intellectual treasury from England'.[42]

Tragedy struck in Philadelphia. Miller, Tyndall's young assistant, contracted typhoid. Fraught with worry, Tyndall ensured that the young man had the best medical attention money could buy. He even rigged up 'a respirator of cotton wool and glycerine, which would keep out all the plagues of Egypt', and visited Miller regularly, despite being advised against it.[43] As the days passed, Miller seemed to show no alarming symptoms, so Tyndall decided to proceed with the lecture tour, and left Philadelphia on 25th November for Baltimore. As Miller's replacement he took with him a 'coloured man' whom Cottrell had taught to put the vital battery together. But a week later, while he was still in Baltimore, Tyndall received the devastating news that Miller had died. He sent a telegram to Spottiswoode, who passed the news on to Jane Barnard, and she had the painful task of informing Miller's parents. The coffin carrying Miller's body eventually arrived in England, but the lack of paperwork led customs officials to suspect smuggling. They refused to let the coffin land, and Bence Jones had to complete a death certificate to obtain its release.

It was in Baltimore that Tyndall commented on the racial segregation of his audience: 'I noticed the gallery filled with coloured persons: I did not see one downstairs. Political functions are secured to the blacks, but not social freedom'. He also remarked unfavourably on the dress code of the women, who wore what he considered an unsightly bunch of clothes behind: 'I would as soon marry a negress without a hump as a white lady with one'.[44]

While he was on the other side of the Atlantic, Tyndall kept a keen interest in the goings-on at the Royal Society. According to Bence Jones and Hirst—who was about to accept an appointment as Director of Studies at the new Royal Naval College in Greenwich, where Debus would join him as Professor of Chemistry—Huxley had succeeded Sharpey as Biological Secretary of the Society, and Airy had announced his impending resignation of the Presidency. There were rumours that the Duke of

Devonshire would be his successor, although X-Club members were considering whether to put Hooker forward, having told him at a meeting in January 1873 that he would be nominated. Hooker was duly elected and served as President until 1878. Huxley, meanwhile, had been elected to the Rectorship of Aberdeen University, a role that came with significant powers, unlike similar positions elsewhere in Scotland. As Huxley put it, 'the fact of anyone, who stinketh in the nostrils of ortho-doxy, beating a Scotch peer at his own gates in the most orthodox of Scotch cities, is a curious sign of the times'. [45]

In early December, Tyndall's lecture tour moved to Washington where, he noted, 'the darks appear to muster strong'.[46] The lectures Tyndall delivered in the ornate Lincoln Hall, to an audience that included the 'President, Cabinet, Ministers, Senators, Congressmen, and all manner of minor lights', created another success, raising $2,000.[47] President Grant was represented by his son and daughter at the first lecture, but he attended the second in person on his return to Washington. As usual, Tyndall gave good value for money, speaking for up to two hours in each lecture, and—following Hirst's advice—without reading directly from a script. In this political environment, he repeated his urging of America to support original research. He avoided social engagements as much as possible while he wrote up the lectures for publication. But, fearing that he would gain a reputation for unsociability, he did meet with Sir Edward Thornton, the British Ambassador.

On Saturday 14th December, Tyndall was back in New York, and for his first few days there, he was kept busy. On the evening of his arrival, Appleton hosted a grand reception, where Tyndall met many celebrities, including the Civil War general George McClellan and the poet William Cullen Bryant. A banquet at the Century Club followed on the Monday, and on Tuesday a gathering at Hoboken, where Tyndall encountered various church dignitaries, including the Bishop of America. Tyndall's involvement in the prayer debate had caused a huge stir in America, not least since Youmans had published in his new *Popular Science Monthly* Tyndall's *Contemporary Review* article 'On Prayer', which he retitled 'Science and Religion'.[48] The Bishop of New York had been invited to the Century Club banquet but, although kind to Tyndall personally, preferred not to attend, to avoid offending 'the weaker brethren'. Throughout the tour, Tyndall was followed by the 'religious question', which he kept away from in his lectures. While some people shook their heads and prayed for his salvation, others—in Columbus, Ohio—were inspired by his words to set up a 'Tyndall Association' to discuss science.[49]

Tyndall's six lectures in New York were split into two sets of three, before and after Christmas. The organizers selected the Cooper Institute, a large, low, cellar-like room with numerous pillars supporting the ceiling. When Tyndall gave his first lecture, he found the audience made up not only of the wealthy but also the learned in literature, science, and art, 'splendid and by far the most enthusiastic I have met'.[50] Unfortunately, he also appealed to New York's thieves. When he returned to his hotel he found that his portmanteaus had been rifled and his purse—which had contained about £40 and had been left in the room against Tyndale's advice—stolen.

Just before Christmas the New York winter descended. Despite the snow and intense cold, the days were bright and bracing. Tyndall often left his hotel to visit Central Park, about 2½ miles away, where he took long walks. He watched the ice-skaters on the frozen ponds, and noticed evidence of glaciation on the rocks. On Christmas Day, after a 6-mile walk, he visited the Cooper Institute to check his equipment for the following day's lecture. Although a winter storm kept the 'fashionable element' at home, at least 800 people braved the conditions to attend. The New York run of lectures ended on 31st December, and carriages thronged the Cooper Institute despite deep snow. The six lectures yielded $8,500. After each one, the *New York Daily Tribune* printed a transcript, and sold the reprints as a series at 3 cents per copy. Demand was huge: about 300,000 were sold across America. Appleton later issued them in hardcover format. Tyndall's books were also selling well. Youmans told him that copies of *Forms of Water* were going like hot cakes on a winter's day, with one distributor buying 2,000 copies at once. When Appleton learned that King had only issued 2,000 copies of the book in England, he was surprised. His print run was 7,000. This book was the first title in the International Scientific Series, an initiative of Youmans to diffuse scientific knowledge, particularly of Spencer's evolutionism, to a wide public. It was enlightened too, in that by arrangement with the publishers Appleton, foreign authors received the benefit of copyright. Spencer, Huxley, and Tyndall formed the British Committee, which faded out at the end of the decade when the British publisher King was bought by Kegan Paul.

Tyndall moved on to Brooklyn on 4th January to repeat the six lectures, before giving two more to the students at Yale in Newhaven. By this stage, after the effort he had spent preparing the lecture course, he was already planning to reprise it in England. Brooklyn, bright and icy, was invigorating to him: 'the sleighing goes on merrily, and the sun shines from an unclouded heaven'.[51] At the lectures, his audience was intently engaged and highly serious: 'There is no display of dress; sober black is the predominating color of attire. Fashion is there but she has left her distinguishing characteristic at home'.[52]

Over the course of Tyndall's appearances in America, other lecturers realized the impact of his unique style, one that combined exhilarating exposition with traditional experiments.[53] For many, he was 'entirely altering the character of experimental teaching in these cities'.[54] Froude was receiving very different reviews for his style of lecturing, 'in a nervous, incoherent, half-distraught way that is the opposite of interesting'.[55] According to one paper, 'like almost all Englishmen, the great historian has not a very good delivery'. Tyndall's address was 'more pleasing, and not quite so English'.[56]

From Brooklyn, Tyndall and his entourage moved to Newhaven, Connecticut, to lecture on 22nd and 23rd January. The venue was smaller than those in New York and Brooklyn, but still held 1,500 people. Following his second lecture in Newhaven, he attended a banquet with the State Governor, President Porter of Yale College, and about fifty others. Tyndall was paid $1,000 for the two lectures, and immediately handed $250 to the Yale Scientific Club, which was run by students who arranged scientific lectures. On his return to New York, he finally allowed himself to socialize more, and—perhaps realizing what he'd been missing—attended three dinner parties on successive evenings. He found New York a fantastic place, 'rich with feminine beauty'. But by the end of the tour his thoughts were turning to home.

His aim was to return to England on the *Cuba*, a Cunard ship regarded as a sound vessel able to cope with the violent winter weather expected in the North Atlantic. He had heard that Bence Jones was very ill, and he was eager to see once again 'a friend who gives up all hope of seeing me in the other world, and therefore wishes to see me again in this one'.[57] There was also the challenge of money; what to do with the sizeable profits from his visit. Initial estimates put them at about $10,000 (£2,000), despite his considerable expenses (partly from the relatively expensive standard of living in America, and partly the result of Miller's treatment and the repatriation of his body).

Tyndall had discovered that his original idea to focus on Chicago was not necessary. Instead he suggested to Henry that the fund be used to support four American students for three years at Heidelberg or Berlin. One of the students would be chosen by Henry, one by Youmans, one by a third American, and one by Tyndall himself. Henry saw Tyndall's generosity as a way to make a more lasting impact than the immediate support of four students. He suggested that the money should instead be invested in perpetuity by a Board of Trustees. In Henry's estimation this would yield about $600–700 per annum to support a student in a foreign university in the pursuit of original research. Given the demand for applied science, Henry argued, as well as poor professorial salaries and a lack of government support, very few American students returning from study abroad pursued original research. Tyndall agreed with the proposal and suggested that Tyndale visit Washington to discuss the details with Henry and Youmans, and help set the plan in motion.

In this land of opportunity, one inventor wrote to Tyndall with a design for an illuminated sign based on his discoveries, and suggested that they share the proceeds of any sales of the device. According to the inventor, if the design was patented in England first it might realize £10,000 alone.[58] It was not the first time Tyndall had been approached about a patent. After the Rede Lecture, Airy had written to remind him of his question at a dinner with the Vice-Chancellor about what glass would admit solar-heat into a hothouse and prevent plant-heat from escaping: 'I am confident that…a lucrative patent might be based on this. I do not despair of seeing the "Tyndall Lights" adopted in every hothouse in the kingdom'.[59] Tyndall's response to these is not known (nor to a previous request of Pasteur's to act as his agent for a patent in England for beer-making, with a share of the profits).[60] But he was not a supporter of patents, writing later: 'I care little for patents: they are things with which I never meddled'.[61] He had no objection to the commercialization of scientific discoveries, and was friendly with many industrialists, who claimed that without patents the spirit of invention would die out. He was not motivated by money. His priority was to secure support for research aimed at understanding the natural world, and to argue for this against the prevailing opinion that scientific activity had to be for practical ends.

Tyndall returned to Philadelphia in late January, braving the snowbound roads from New York. His plan was to stay with Hector for a few days and prepare his speech for the final banquet of his tour. Organized by Alfred Mayer, this grand event was held on 4th February, the eve of Tyndall's departure, at Delmonico's restaurant in New York. It featured the cream of American scientific society. The intention had been to have Henry as host, but he was unable to attend. His substitute, Agassiz, was hauled before a legislative committee at the last minute, so William Evarts, a lawyer and President of the New York Bar Association (and later US Secretary of State under President Hayes) took the chair instead.

The banquet proved a splendid send-off, though marred by some religious controversy. In a typically forthright speech, Tyndall emphasized his call for Americans to foster original research, by private means if not by the public purse. Speaking after him was the chemist and historian William Draper, who reminded Tyndall of the American origins of the Royal Institution in the form of Rumford. Displaying the endemic racism of the time (on both sides of the Atlantic), Draper invited Tyndall back to American shores to witness 'how we have purged the African, the woolly-headed black man, of the paganism of his forefathers and are now Darwinizing him into a respectable citizen'.[62] The journalist Parke Godwin, responding to the toast to the media, expressed his support for the men of science, 'the only real magicians who are left to us', and science's ability to help improve society. But he sought to put science in its place, articulating his view that it deals only with facts, rather than with questions of personal origin or ultimate destinies. Most vocal was the Reverend

Dr Bellows, who responded forcefully, in religious terms, to the final toast to the service of science to humanity. As Mayer later put it, regretting that Henry and Agassiz had not been present to control the event properly, 'we did not anticipate that the parsons would demonize us'.[63] Mayer undertook to ensure, with the help of Barnard and Youmans, that Appleton did not publish those homilies, thus tending to 'further the objects of your visit among us instead of frustrating it as was the tendency of the speeches of the dominies'. But he did not succeed, and much of the controversy was published and debated in the newspapers. Tyndall reflected later, given the arguments that became visible after the dinner, that religious liberty was now greater in England than in America, where evangelical views were widespread.[64] Though early in his visit he had commented: 'Nothing has struck me more since I came here than the widely divergent religious views among the members of the same family and still the perfect tolerance that reigns among them'.[65]

Tyndall was astonished at the recognition he had in America. His trip had been hugely successful, enabling him to cement his relationships with many Americans and bring his work to an even wider public. Clashes with the clergy were sideshows. In a remarkable display of openness, he made public the financial accounts of his visit in a document that Tyndale referred to as the 'Tyndall Trust for Original Investigation'. The full Deed of Trust was published in the May 1873 edition of the *Popular Science Monthly*. It showed income of $23,100 from all thirty-five lectures. With expenses totalling $10,066.66, that left a surplus of $13,033.34 (more than £2,000, equivalent to at least a year of his earnings at the time) 'for the advancement of theoretic science, and the promotion of original research, especially in the department of physics, in the United States'.[66] Henry, Youmans, and Tyndale were appointed Trustees of the fund, and the money was invested in bonds yielding 7 per cent.

Tyndall's extraordinary generosity was partly intended to encourage the wealthy individuals he saw around him to make further donations. As he commented to Bence Jones, if the Royal Institution were brought to America, by his estimation it could be endowed with half a million sterling in a fortnight. His own donation in support of American science was a success. His initial aim was to support two American students to undertake fundamental research in German universities, free from American pressures to concentrate on applied research. Though he had not put it in writing, Tyndall had in mind Heidelberg and Berlin, with Bunsen and Helmholtz. Invested entirely and perhaps riskily in Lehigh Valley Rail Road bonds, the fund had grown to the equivalent of £6,000 by 1884. Tyndall decided to divide the sum equally between Columbia College, Harvard, and the University of Pennsylvania, creating three Tyndall Fellowships.[67] In December 1875 Tyndale wrote to give news that a Mr Hastings of Yale College had been chosen as the first beneficiary.[68] The scholarships ran for several decades, and until the 1960s in the case of Harvard, but no longer appear to exist.

Tyndall might have stayed on in America had he not ardently wished to see Bence Jones before he died. So when the *Cuba* sailed on 5th February, Tyndall was on board. The voyage home took fourteen days, longer than anticipated owing to the adverse weather. Tyndall arrived back in England on 19th February, and completed the circuit by meeting Hirst and Debus at Euston station.

19

FOGS AND GLACIERS

1873

Tyndall returned from America to 'six cubic feet of papers and correspondence', and settled into his new laboratory at the Royal Institution.[1] It had been installed in his absence under the direction of the ailing Bence Jones. The first visible fruit of his American visit was a Discourse on 4th April 1873, 'Some Observations on Niagara'.[2] He described his manly battle to wade out under the tumultuous cascades, before turning to the geology and the evidence that the Falls were receding northwards by erosion. A series of six lectures on 'Light', derived from his American tour, followed after Easter. But he had to fit further demand for lighthouse advice around them, which also prevented a planned trip to the Isle of Wight.

The lighthouse work was stimulated by a report sent on 10th April by the Irish Inspector of Lights to the Board of Trade. Based on experiments at Howth Baily, it concluded that gas was superior to either paraffin or colza oil. Two days later, on Easter Saturday, Tyndall wrote to the Board of Trade to say that he had proposed to Trinity House that he inspect the lights at Haisbro on the night of 21st April, with the aim of finally settling the question of the use of gas in lighthouses, by comparing gas and oil burners.[3] He wanted John Wigham there, to which the Board of Trade and Trinity House agreed. If Tyndall really thought these experiments would settle the question he was much mistaken. Wigham disputed several conclusions and processes, including the means of generating gas, to the extent that Tyndall was asked in his role as a Gas Referee to report on that too.[4] He asked Valentin to examine the system for gas generation, but it proved in need of repair and it was not until the following year that the full trial was possible. Wigham had developed a new 'triform' gas light, which could readily be increased in power if necessary. Tyndall visited Ireland in July with Richard Collinson (who became Deputy Master of Trinity House in 1875) to see it in a fixed light at Howth Baily and a revolving light at Rockabill. They were greatly impressed, and Tyndall again urged the merits of the Irish invention.[5]

In March 1874, Tyndall was finally able to compare, in their presence, Douglass's oil burner and Wigham's gas burner at Haisbro, but he did not submit his report until October after his alpine season and his Belfast Address. He came out unambiguously in favour of gas over oil.[6] But though supportive of further experiments with gas in Ireland, Trinity House were not convinced enough to replace their own oil burners with gas while the potential of electric light was also being explored. Tyndall became so disenchanted that he even offered to resign in 1875, but was persuaded to stay by Sir Frederick Arrow, the Deputy Master.[7]

The period up to the beginning of May 1873, while lighthouse affairs progressed, was the calm before a gathering storm. Tyndall invited Hirst and Huxley to lunch with the visiting Emerson, and attended the Royal Society soirée on 26th April. Hirst dined with Tyndall at the Athenaeum the next day, finding his manner chilly at first. He had noticed an increased tendency for Tyndall to feel slighted since his return from America, and hoped to see the old cheerfulness and geniality resume. It is perhaps with his closest friends that Tyndall occasionally exhibited the less admirable side of his character. His ability to charm in conversation was legendary, but he could be short and rude to his friends, giving Hirst a frigid reply to a letter saying he couldn't come to lunch. In earlier days, Jemmy Craven had written to him: 'You see you do not often laugh'.[8] Hirst suffered in relative silence, but the scratchier Debus took frequent offence. After one later incident, he refused to speak to Tyndall for more than a year.

That accessible little book, The Forms of Water in Clouds and Rivers, Ice and Glaciers, had been noticed in Scotland. It now caused a major row stretching into the autumn. The fight started with an anonymous review in Nature, which pointed out that most of the book was concerned with ice, and asked if it was necessary for Tyndall to rake up the Forbes-Rendu glacier controversy and to renew the claims of Agassiz, while ignoring objections to his ideas by Canon Moseley and James Croll.[9] Tyndall heard that Tait was intending to call him to account in a biography of Forbes, and seemed genuinely surprised that an animosity he believed had been set aside in Dundee was rearing its head again years later. The Life and Letters of James David Forbes soon appeared under the pens of John Shairp, Forbes's successor as Principal of St Andrews, Anthony Adams-Reilly, the Irish cartographer and mountaineer, and Tait. Huxley saw a copy and told Tyndall they had made an onslaught on him, as well as suggesting that they were both 'fellow conspirators against the peace & fame of Forbes'.[10] Tyndall, busy preparing his American lectures for publication, asked Hirst to buy a copy and look over it. He planned also to send it to Louis Agassiz.

The first response to the review in Nature was from Alexander Agassiz, son of Louis (who died in December, leaving his son to continue the battle for his reputation). Agassiz supported Tyndall's apportionment of the glory as 'every fair-minded investigator' would do, and reiterated the charge that Forbes had appropriated his father's idea and misrepresented their conversation.[11] Tyndall held himself aloof as he finished his book on *Light*, which he expected to take a further month, telling Huxley: 'then I shall discourse like Jeremy Taylor on justice, mercy, and judgment to come'.[12] Huxley fired the next shots in *Nature*. He denied the charge that Tyndall had any knowledge of the move to block Forbes for the Copley Medal in 1859, which Huxley claimed was done because nobody on Council knew the subject and no one had investigated rival claims. He said that Tyndall only became aware of this a year before, when he had sent over some old papers he had found.[13] This was untrue, since Tyndall had noted in his journal in November 1859 that he was aware, though after the event, of Frankland's representation about this at the Council, made at Huxley's prompting. George Forbes, son of James, defended his father against Louis Agassiz, father of Alexander, claiming, with some supporting evidence, that Forbes had made the crucial observation about the veined structure and that Agassiz had falsely claimed it for himself.[14] Shairp weighed in too, explaining that the publication of *Forms of Water*, when the *Life and Letters* was almost complete, had necessitated their adding two appendices of Forbes's statements about his glacier work, neither of which they claimed had been refuted. Tyndall commented that Shairp 'seems perfectly unconscious that he is exhibiting his friend stripped of all inner dignity'.[15] Here matters stood until Tyndall relit the touchpaper in August.

* * *

In parallel with the dispute about the relative merits of gas and oil for lighthouses, Tyndall was drawn into the problem of providing effective warnings to ships in fog. It led him to significant findings in acoustics and meteorology. While he was in America, a committee from Trinity House, led by Sir Frederick Arrow, had visited Canada and the United States to observe their system of fog signalling. Tyndall went down to Dover with Major Elliott of the American Light House Board on 19th May 1873, to observe the fog signals at the South Foreland lighthouse—the site was chosen because it had been electrified and so had the necessary steam generators. Work on sound signals for navigation now took up a significant amount of his time, as he travelled to and from Dover.

Tyndall directed experiments over a full year to test different approaches, starting with brass horns, a locomotive whistle, and a steam whistle. Most of the observations were made steaming up and down offshore on one of the Trinity House vessels—the *Irené*, *Triton*, *Galatea*, or *Palmerston*—with the enthusiastic participation of Richard Collinson. On the second day, 'through the prompt courtesy of General Sir A. Horsford', the eighteen-pounder gun at Dover Castle was fired, proving distinctly more audible than the horns and whistles. It would have been easy to draw the conclusion that such would always be the case. Tyndall did not. His persistence in repeating experiments that already appeared to be clearly answered led him to important findings about atmospheric acoustics. Hearing from Sir William Armstrong that the loudness of a gun depended on its shape, he added a howitzer and a mortar to the arsenal.

It was after ten days of experiments, giving conflicting results, that Tyndall imagined why sounds heard in one place on one day might be completely inaudible on a different day, even in calm conditions. He suggested that jumbled masses of water vapour produced by evaporation could create 'acoustic clouds'. These clouds, invisible to the eye, prevented sound from passing through. He drew the analogy with a powdered crystal. The crystal was transparent to light, but its jumbled powder transmitted no light because of reflection at the boundaries of the particles of powder. In the same way, the irregular boundaries between water vapour and air caused internal reflection of sound, which was unable to penetrate through.

In America, with the help of Henry, Tyndall had been able to see the siren, an air-driven sound generator, in action at Staten Island and Sandy Hook. The autumn series of experiments, starting on 8th October 1873, incorporated a siren made available by the Americans. Its advantage over the horns was immediately apparent, and Tyndall wrote to Henry to say how impressed he was with it.[16] He also observed that after heavy rain and a hailstorm the various sounds became more audible, counter to the received wisdom. It was as if the rain had cleared the invisible acoustic clouds, making the air more homogeneous. He recalled his winter in Chamonix in 1859, when he could hear sounds clearly through heavy snow across the glacier. Drawing a parallel with light, Tyndall realized that just as waves of higher frequency were scattered more easily than lower, so sounds of higher frequency were scattered better by an inhomogeneous atmosphere, leaving lower notes more audible. Hence the use of low frequencies for effective foghorns. After thirty-seven days of experiments, the pressure of his other duties drew him away. He was conscious that he had not yet had the opportunity to test the fog signals in fog itself, a crucial further objective. He was aware, though, that the idea that sound would be attenuated by fog in the same way as light might well be wrong.

Widespread thick fogs fortunately descended in the middle of December. Tyndall had experiments carried out on the coast, and made several days of observations himself by the Serpentine, with the help of Cottrell operating a whistle and an organ pipe. He co-opted two 'very intelligent policemen' to assist them, listening to the distant Westminster bell. This confirmed his view that fog, far from impeding the transmission of sound, might aid it. The objective overall was to produce a sound powerful enough to endure any loss by scattering, yet still retain enough for transmission. For that the siren seemed best in most situations, but he also recommended further experiments on guns. A mammoth seventy-seven-page report, submitted in May 1874, summarized more than forty days of observations which had ended in February.[17] Considering this of public interest, Tyndall published two articles 'On the Atmosphere in Relation to Fog Signalling' in the *Contemporary Review*.[18] His demonstration of 'sounding flames' arose directly from this work.

The Board of Trade approved a substantial expenditure for fog signals based on this report: £35,600 for the English coast alone in 1874–1875, out of a total lighthouse budget for England of about £300,000. By 1879 there were twenty-eight sirens in place, referred to by one newspaper as Tyndall's 'demoralised and brutalised sirens, those pretty foghorns that make one on land doubt whether hearing is a blessing, but at sea doubt whether there is a more useful faculty' (Figure 32).[19] In the meantime, Tyndall had expressed himself satisfied with the work of the War Department on a bronze fog signal gun, and recommended the acceptance of the Department's offer to make it, suggesting a trial of bronze against iron or steel.[20] Towards the end of 1874, the Royal Gun Factories were ready to trial the guns and different forms of muzzle at Woolwich. The results were not clear cut, although the shape of the muzzle seemed

Fig. 31. Cartoon of Tyndall's sound experiments. Artist unknown, but possibly Huxley.

more important than the material of the gun. Tyndall additionally recommended trialling gun-cotton against powder, and experiments in early 1875 showed powder to be more effective, price for price. Work continued on this for several years,[21] generating sufficient public interest to be reported in the *Times*.[22] Tyndall concluded that gun-cotton and cotton-powder rockets would be effective sound signals, especially for rock lighthouses and where mounting a siren or gun would be impossible.[23]

As the lighthouse work was under way in mid-1873, Tyndall received a singular honour. On 18th June, a few days after the annual outing of the X-Club and their wives to Skindles Hotel in Maidenhead, he received a DCL from Oxford University. He had received an LLD from Cambridge as long ago as 1865, but Oxford was more orthodox. When the degree had been mooted in 1870, to Tyndall, Darwin, and Froude among others after the Marquis of Salisbury had been made Chancellor, Tyndall understood that Dr Pusey, the High Church Anglican, had objected. Out of respect for his religious feelings the matter was not pressed.[24] The Universities Tests Act (allowing Roman Catholics, non-conformists, and non-Christians to take fellowships at the Universities of Oxford, Cambridge, and Durham) had been passed in 1871, which gave Tyndall a fairer wind this time. Dr Huertlet, the Professor of Divinity, protested on the grounds of Tyndall's denial of the credibility of miracles and the efficacy of prayer, contrary to the bible which the university professed to acknowledge, but the objection was not sustained. When Tyndall added a third LLD in 1886, from

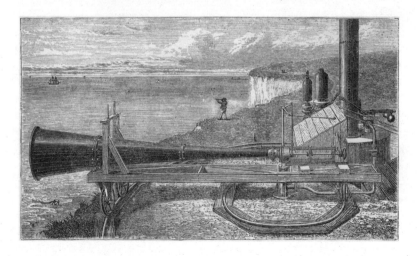

Fig. 32. A siren of the type Tyndall encouraged.

Trinity College Dublin, there was also one dissentient in the Senate, on the grounds that he was not a Christian.

Clausius wrote to congratulate Tyndall on the doctorate. He mentioned in the same letter that he had received Maxwell's seminal book A Treatise on Electricity and Magnetism, and was pleased that he had not taken offence at a previous criticism Clausius had felt it necessary to make, writing: 'he seems very different from Tait'.[25] He was probably unaware of the sarcastic poems about Tyndall that Maxwell circulated among his friends. Tyndall seems not to have taken offence at these poems, even though one, published in Nature, cast an implicit slur on his morality.[26] His relative lack of self-awareness, and apparent inability to detect irony, perhaps gave him some measure of protection. He even sent Maxwell a poem in June 1871, Brave Hills of Thuring.[27] It was a homage to Luther, written in Luther's room at the Wartburg years before.[28] Perhaps he wanted to demonstrate to Maxwell that he was sensitive to religious beliefs. He respected Maxwell, and wanted to get on with him. Indeed, he had earlier nominated him for the Philosophical Club, to Maxwell's pleasure.[29] But Maxwell held back. He was uncomfortable with Tyndall's religious outlook and his confrontational style, writing to a friend: 'the only man I know who can make everything the subject of discussion is Dr. Tyndall. Secure his attendance and that of someone to differ from him, and you are right for a meeting'.[30] Tait and Maxwell ridiculed Tyndall's popular lectures and lack of mathematical capability, and the pressure to 'Tyndallize' their language. For the North British, the essence of physics could not be communicated to those unversed in mathematics. But the ridiculing of popular lectures sounds like sour grapes: Maxwell's poor lecturing seems to have contributed to his departure from King's College London in 1865.

Tyndall had intended to go to Switzerland early in July, but Trinity House duties intervened. On his return from an Irish visit he spent an afternoon with Hirst, going over his response to The Life and Letters of James David Forbes, in an article that he titled Principal Forbes and his Biographers.[31] Hirst thought the article 'a very able one', but 'at the expense of force and pithiness I induced him to strike out a passage or two relative to Forbes's religious exercises'.[32] Tyndall finally left for the Alps with Hirst on 1st August, just after the publication of his Six Lectures on Light: Delivered in America in 1872–1873.[33] Back in England, the as yet unseen weapon of his article waited to be read in the August issue of the Contemporary Review.

Tyndall and Hirst were in Paris for three days, calling on friends including Jamin and Moigno, and taking a trip to Sèvres by steamer. On 4th August, after a visit to Anna's grave in Montmartre, they attended the Académie des Sciences, placed as guests of honour by the President Joseph Bertrand. They dined afterwards at Jamin's with a

multitude of guests including Pasteur. Leaving Paris on 5th August, they made a leisurely journey via Dijon to Ouchy near Lausanne on Lake Geneva, to Sierre, and
finally to Brig on the banquette of a diligence. On a hot day, they walked up to Belalp,
while Tyndall improved noticeably with every foot of altitude.[34] They took walks on
the next two days to Nessel and across the Aletsch glacier to Rieder Alp to call on the
two daughters of Mary Somerville, admirers of Tyndall.

Then Lady Claud Hamilton arrived with her children Louisa, Emma, and Douglas.
It appears that the Hamiltons had got to know Tyndall in London soon after his
return from America. They had several friends in common, including the Lubbocks,
Birkbecks, and Mary Egerton, and had been planning to engineer an introduction to
Louisa for some time. They attended Royal Institution lectures (Louisa was present at
Tyndall's Discourse 'On Niagara'), but it was the Birkbecks and Lubbocks who seem
to have made the direct connection. Louisa had met the Lubbocks at least by 1870
when she remarked, before going on to visit the Herschels, that the charm of Lady
Lubbock was 'the enormous interest she throws into everything & every body that
comes near her'.[35] She was particularly struck by Nelly Lubbock's awareness of the
barriers women faced in society. Two years later she visited Norway with Mary
Birkbeck, sister of John Lubbock.[36] Mary had married Robert Birkbeck, a cousin of
the mountaineer and founder member of the Alpine Club John Birkbeck, whom
Tyndall knew well. In early 1873 the Hamiltons tried to arrange a dinner for Tyndall
and the Birkbecks, but it fell through because Tyndall—who they referred to as the
'G. T'., possibly the 'Great Tyndall'—had gone to the Isle of Wight. So the Birkbecks
invited Louisa with them to Tyndall's box at the Albert Hall on 9 April to hear Bach's
St Matthew Passion.[37] Louisa accepted, despite having been to a performance the previous evening. Thus was a seed planted.

Lady Claud Hamilton capitalized by writing to Tyndall just before he went to the
Alps to ask if he would look after her party while they were there. He was hardly
encouraging in response, telling her simply where he would be found in Paris and
Switzerland. He replied: 'I never took charge of a lady or a party of ladies in my life
and I fear you would find me a poor hand at looking after the small needs & comforts
of a ladies party...I am a wild savage and fear my ability to look after the smaller
odds & ends'.[38] Did he have any designs on Lady Claud's daughter Louisa at this
stage? He was with them every day (Hirst commented: 'Tyndall of course is the life
and soul of the party'), and a much later letter hints that he was quickly smitten.[39]
They all visited the Massa Gorge, climbed the Sparrenhorn, and explored the Aletsch
glacier. Hirst left for Italy while Tyndall went on to the Eggishorn, perhaps still with
the Hamiltons, and to Zermatt on 2nd September, arriving back in England a week
later. He had been thrown together with the Hamiltons for a fortnight or so, but Hirst

was not there at the end to bear witness. To Mrs Steuart he simply wrote that he had been with Lady Claud Hamilton, two of her charming daughters, and her son.[40]

Tyndall had returned with the aim of going to the British Association Meeting in Bradford, because he wanted to support the President, Alexander Williamson. The previous year, he had heard from Spottiswoode that Joule had been invited to be President and promised that 'if he accepts I will go all the way to Bradford to support him'.[41] In the event Joule was not well enough and Tyndall was again asked to preside, despite having asked Spottiswoode to make sure that he was not invited instead. He turned it down, for the second time, pleading to Galton that 'though my presiding there would certainly be the greatest public distinction of my life, the nature of my engagements compel me to forego the distinction'.[42] Tyndall was later instrumental in securing a pension for Joule, by raising the issue with Lord Salisbury.[43] Shortly after the meeting he received one honour he did not turn down: with Huxley, Hooker, and Airy he was made a Knight of the Order of the North Star by the King of Sweden.[44]

* * *

By the time Tyndall returned to England, the *Contemporary Review* article on Forbes had done its work. To increase the firepower on Tait and his colleagues, Tyndall had even had a pamphlet of the article printed. Huxley thought that 'nothing could be better than the tone you have adopted. I did not suspect that you had such a shot in your locker as the answer to Forbes about the direction of the "crevasses" referred to by Rendu. It is a deadly thrust'.[45] Tyndall was relieved: 'I am right glad to hear that the reply to Forbes pleases you. Imagine my self denial in permitting that thing to remain unanswered for a dozen years'.[46]

He was back in England just in time to see Tait's biting response in the 11th September issue of *Nature*.[47] It was a combination of ad hominem attack and a critique of remarks of Tyndall's on the reflection of rainbows.[48] Incendiary as it was, according to a later statement by Lockyer, the editor of *Nature*, it even had some personal remarks removed by him. Tait charged that: 'Dr Tyndall has, in fact, martyred his scientific authority by deservedly winning distinction in the popular field. One learns too late that he cannot "make the best of both worlds"'. He then criticized *Light* for comments on Wheatstone, and by implication Thomson, which he asserted were a slur on men who devote themselves to practical and commercial applications of science: 'the contrast between the utter contempt for money showed by their censor and the (implied) opposite which is condemned unworthy of scientific men, is brought out with all the flow of word-painting and righteous indignation which Dr Tyndall so abundantly possesses. Besides, the monstrous doctrine is inculcated that men who devote themselves to practical applications are men incapable of original research'.

Tyndall's friends rallied round. Hooker sympathized after his experience with Ayrton, and Mary Egerton was angry that he could be accused of disrespect to men like Wheatstone. Helmholtz told him that he regarded Tait as a scientific bully and that he blamed William Thomson 'for encouraging the mad vanity of Scotchmen'.[49] Tyndall was goaded to respond, in an article the following week, in which he said that the question of priority was Agassiz's word against Forbes's (and implicitly that Tyndall believed Agassiz); that criticisms of his popular writing were 'mere ignoble spite'; and that he had been vindicated over the award of the Copley Medal to Mayer, which Tait had opposed. He accused Tait of not possessing the 'manhood' to acknowledge a committed wrong.[50] At this point Lockyer called a halt, having filled his column inches with lively material and printed Tyndall's retraction of his comments about lowering himself to Tait's level by reflecting on his manhood.[51] Lockyer also published in this issue Tyndall's clarification of what he meant with respect to rainbows, having held it over to allow the main dispute centre stage.[52]

It was not just Tait with whom Tyndall had to deal in the continuing Forbes dispute. While Tyndall was on a Sunday break from fog signal work at Lubbock's country house in October, Nelly Lubbock found out from George Allen, Ruskin's publisher, that Ruskin was about to make a bitter attack on him. Tyndall complained to Carlyle that this was most unmerited, 'for over and over again when in private conversation he has been subjected to far graver attacks than he could possibly direct against me, I have defended him'.[53] Carlyle, Pollock, and others urged him to ignore Ruskin, though Nelly Lubbock 'was in favour of instant war with him' and, as he put it to Hirst, 'really seems to consider that my pluck needs a stimulus to make me "vindicate my honour"'.[54] Huxley's response was that 'men don't make war on either women or eunuchs and I hope you will let Ruskin have his squall all to himself'.[55] Tyndall let it go, for now.

What should we make of the relative merits of the theories of Tyndall and Forbes? Who was right about the physics of glacier motion? Opinions at the time were partisan. In an early article in 1861, John Ball had attempted to steer a diplomatic line in the *Edinburgh Review*, writing: 'Forbes established the law of glacier motion; Tyndall showed how the glacier is enabled to obey that law'.[56] Even he, who had once disputed Tyndall on the point, agreed that Tyndall's explanation of the veined structure was correct, and Forbes's wrong. Hopkins had eventually produced his piece for the *Quarterly Review* in 1863, following his earlier article in *Fraser's*. He came out strongly for Tyndall over Forbes's 'vague and misty' viscous theory.[57] But the challenges did not go away. Tyndall had been alarmed by an experiment of William Mathews in 1870, in which he showed that a plank of ice could bend under its weight between supported ends without breaking.[58] Mathews argued that plasticity must be admitted alongside sliding, and fracture and regelation. He suggested that more importance should be given to Forbes's views. One impetus for Tyndall's visit to Pontresina in 1871 was to meet this challenge from Mathews. Having missed the British Association meeting in Edinburgh, he responded at the Red Lions and in *Nature*.[59] Though he could replicate the results to an extent, he also found fracturing. He expressed the hope 'before long to return to the subject', but he does not seem to have done so, even after further criticism from Forbes's supporters.

Ice is an unusual material. The extent to which it can 'flow', in Forbes's terms, depends in a complex manner on temperature, pressure, and the rate at which it is put under strain. Taking bulk ice, the upper part of a glacier, to a depth of about 100m, will tend to fracture under strain and close up under pressure: Tyndall's fracture and regelation. The lower part, in a sufficiently deep glacier, will move without fracturing: Forbes's viscous flow. The glacier can also slide along its bed, as Tyndall understood, for which the hydrostatic pressure of water at the bottom may be important, effectively lifting it over asperities. They were both partly right.

20

THE BELFAST ADDRESS

1873–1875

Tyndall was driven away from London at the end of September 1873 by a steam engine and roller crushing bricks outside his bedroom. He fled to Folkestone, taking Spencer's *First Principles* with him, and stayed at the West Cliff Hotel 'where Spencer once deprived me of a night's sleep by asking me just as I was retiring to bed whether I believed in the externality of matter'. He wished that Huxley had not persuaded him to accept the Presidency of the British Association in Belfast, writing: 'they do not want me. Well in return I may be less tender in talking to them than I otherwise should have been'.[1] Given the increasing but unforeseen demands of government work, he continued to agonize over whether to give up the Presidency. But Spottiswoode, Hooker, Huxley, and Sharpey strongly objected, fearing not just the embarrassment to the Council but the misconstruing of his motives, which could be seen as giving in to pressure from Tait or from the vocal Presbyterians in Belfast. Hooker suggested he should not worry about it until next midsummer and then decide. 'As to the Address', he advised, 'make up your mind to choose a scientific subject, that will demand no great mental strain, remembering that it is not for the address that men are chosen Presidents'.[2] That advice was to be comprehensively ignored, and when Tyndall discovered that Tait, who had been a professor in Belfast, was republishing in Belfast newspapers 'the slanderous attack of Ruskin upon me', it was clear that he could not resign.[3]

The year of Tyndall's notorious Address started with the fruits of his work on sound. Even his Christmas Lectures, ending on 8th January 1874, were on 'The Motion and Sensation of Sound'. Tyndall had been writing up his findings in acoustics for the Royal Society in parallel to his report for Trinity House, which would be laid before the House of Commons. He asked Stokes for advice on some theoretical calculations, and told him that he thought there had been no papers in the *Philosophical Transactions* on the effect of atmospheric conditions on sound since one by William Derham in

1708. Herschel and others had quoted it, but he had now shown that it needed serious correction. (Derham mistakenly thought that rain and fog absorbed sound.)

The preliminary account of his work was given to the Royal Society on 15th January.[4] He followed it with a Discourse the following day 'On the acoustic transparency and opacity of the atmosphere'.[5] An engaging version of it was written up in Nature from his lecture notes.[6] Tyndall described the experiments which had occupied him for a substantial part of the year, from mid-May to early July and from early October to the end of November, extended opportunistically by exploiting the thick London fog in December. He explained finding how the acoustic condition of the air changes all the time, massively influencing the effective distance and the relative audibility of guns, horns, and whistles. This was not just due to wind effects. The result which stunned him was his discovery, on 3rd July 1873, that acoustic transparency was not related to optical transparency, as everyone had assumed. It was a beautiful day, with perfect visibility, but the sound suddenly became much less audible.

Tyndall's explanation was that acoustic opacity is related to the condition of water vapour in the air. Though invisible, an 'acoustic cloud' of water vapour was there, reflecting echoes, and the conditions affected high- and low-pitched sounds differently. A clinching finding was that sound was more audible after heavy rain that had cleared the air of vapour, and it was still audible during the rain. It was the state of the vapour that caused the deadening of the sound.

What Tyndall had explored, which was new, was the intensity rather than the speed of sound in the atmosphere, overturning in the process ideas that had stood since Derham's paper. So the practical solution was simply to use the most powerful sound possible, and he favoured the siren. Not one to underestimate the significance of his work, he claimed: 'thus, I think, has been removed the last of a congeries of errors which for more than a century and a half have been associated with the transmission of sound by the atmosphere'.

Tyndall's main lectures at the Royal Institution in 1874 were on the subject of 'Physical Properties of Gases and Liquids', a weekly series of six before Easter. Once these were over, he planned to go as early as possible to Switzerland to prepare his Belfast Address. In that lecture period, before the beginning of June, he tied up loose ends while maintaining some social activity: a luncheon on 25th March at Sir Joseph and Lady Whitworth's new residence in Berkeley Square; hosting 'a brilliant reunion' in his rooms after Spottiswoode's Discourse on 17th April; and attending Hirst's first dinner party at Greenwich on 28th May, where the guests included the Huxleys, Spottiswoodes, and Admiral and Lady Key.[7] He did some further testing on his respirator, which he communicated to the Royal Society.[8] The outstanding work concerned his substantial paper on sound for the Philosophical Transactions, lighthouse duties, and the lingering Tait controversy. The acoustics paper was read at the Royal Society shortly before he went to Switzerland, and submitted to the tortuous publication process.[9]

True to form, Tyndall could not let go of the controversy with Tait over Forbes. He wrote to Clausius, stretching the truth a little, about Tait's attacks on his 'most innocent of little books—a boy's book of the glaciers—in which I actually ignored myself and gave all possible credit to Agassiz, Forbes, and others', and about the row in *Nature* and the refusal of other magazines to publish Tait's articles. He went on: 'there is a mad creature here who writes upon Art.—Ruskin—a beautiful imaginative creature, more a woman than a man, and he wrote a scandalous article against me, and in favour of Forbes, in an obscure magazine of his own'.[10]

The next salvo fired from Scotland in the war for Forbes's posthumous reputation was the reprinting of Rendu's book *Glaciers of Savoy*, with commentary by Tait. James Coxe suggested ignoring Tait, but Huxley advised that, though he should do nothing in a hurry, an answer in July would be effective.[11] But Tyndall was unable to wait. Just as in the previous year his 'Principal Forbes and his Biographers' had been published as he left for Switzerland, so too this year the equally combative 'Rendu and his Editors' appeared in the *Contemporary Review* as he departed.[12]

The beginning of June is a lovely time to arrive in the Alps. Many of the lower passes are still blocked by snow, and as it retreats the flowers emerge in profusion. Tyndall was isolated on Belalp: 'myriad groups of these violets over the Bel Alp bursting through the dead brown grass left behind by the retreat of the winter snow—no eye to see their beauty—for even the sheep & goats have not yet reached this height. I am the only dweller (visitor) in the place; but the very solitude saved me from loneliness'.[13] Tait intruded on this solitude, writing that in the *Contemporary Review* article Tyndall had falsely attributed a quote to him.[14] Tyndall forwarded the accusation to Huxley, with a letter that indicates a distinct loss of his sense of proportion in the affair.[15] He told Huxley that, having so far backed off any real attack on Forbes's character, he might now 'give the world a picture of the man's doings as they then appeared to me...In fact a psychological analysis of more than ordinary interest could be founded on the study of the character of Forbes'.[16] Huxley had to break the news that Tait had a point, and sensibly advised his friend not to proceed with any character attack, as 'the more severely true your portrait might be, the more loud would be the outcry against it'.[17]

Tyndall decided to make no response until Belfast was done. In that respect, Huxley was keen to know where Tyndall's thoughts were leading: 'I wonder if that Address is begun, and if you are going to be as wise and prudent as I was at Liverpool. When I think of the temptation I resisted on that occasion,... "I marvel at my own forbearance!" Let my example be a burning and a shining light to you. I declare I have horrid misgivings of your kicking over the traces'. Huxley was not mistaken in his misgivings.

The weather in the Alps that June was poor. Indeed it was only the weather that prevented Emily Peel from a proposed ascent of Mont Blanc. She teased Tyndall, suggesting that 'when worthy of the distinction you must enroll me amongst your Alpine Clubbists, if such an honour is ever accorded to one of the weaker sex!'.[18] (Women were allowed to join the Alpine Club in 1974, a hundred years later. Unless one includes Tschingel, William Coolidge's canine companion, who was made an Honorary Member in 1869.) Tyndall spent his time reading 'quite a library of books', walking and thinking over his arguments. He told Huxley: 'I have read again those early articles of yours on the Origin of Species: they are very fine, and must have been most opportune'.[19] Further reading included the work of Spencer, who was to feature in the Address though, like many others since, Tyndall found Spencer's writing difficult: 'Spencer's Psychology is also on the whole, a noble piece of work. I wish I had time to boil it down: but the wretch is tough'.[20] He caught a cold, and he suffered from severe toothache for the third time in his life—hindrances to serious thought.

By early July he was feeling better, not least perhaps because he had seen an article in the *Scotsman* on 2nd July on his side in the Forbes controversy (it was by James Coxe).[21] He started in earnest on the Address, writing to Debus and Hirst to line them up as reviewers at the end of the month. He told Hirst that he had been looking at some land to purchase on Belalp and build a house, a scheme he had mentioned to Carlyle some years before. Huxley wrote in suspense about what was being cooked up for Belfast: 'Shall I not see the address? It is tantalizing to hear of your progress and not to know what is in it'.[22]

Tyndall left the hotel on Belalp on 8th August to return to England, his last few days brightened by Lady Edith and Lady Emily Quin, the 'two charming daughters' of the late Lord Dunraven, whom he had known well.[23] At the last minute he sent sections of the Address to friends including Spencer, Huxley, and Darwin. Huxley approved, and Darwin wrote that he had been 'most deeply gratified by what you say. As far as rather hasty reading suffices, I have not one word of criticism to make'.[24] Spencer, typically, did not have time to read it.

* * *

Tyndall set off from London with Huxley and Debus on the morning of 17th August, putting up at Salt Hill near Kingstown before travelling on to Belfast on 19th August, the opening day of the Meeting. Many of Tyndall's colleagues and friends went to support him as President, among them Mary Egerton and the Hamiltons. Even the Belfast people were, he wrote, 'showing warm kindness & cordiality: proving that they have not wholly set their hearts upon home rule'.[25] Ironically, the Mayor of Belfast had earlier objected 'to the choice of an Englishman as President of the Belfast meeting'.[26] Knoblauch, who was lodged with local gentry in the Association's customary style, and Wiedemann, who had preferred to stay as Tyndall's guest, were both present. Many good friends were absent. Hirst was ill. Clausius was unable to come, and neither were Du Bois-Reymond nor Bunsen. Du Bois-Reymond was completely occupied with the building of his new laboratory (funded by the French reparations after the Franco-Prussian War). He wrote that it would be 'by far the most spacious, sumptuous, glorious place for scientific work which was ever conceived of'.[27]

Tyndall stood up in the Ulster Hall to give his Address on the evening of Wednesday 19th August.[28] It was an Address of which any orator would be proud. He started with a tour of the roots of atomic theory, and of philosophical thought from Greek and Roman times through the Middle Ages. Here, he argued, Arabic culture kept scientific knowledge and approaches alive, when elsewhere 'natural events, instead of being traced to physical, were referred to moral causes'. After describing the challenge to Scholasticism from Copernicus and then Bruno, burnt at the stake, Tyndall ran through Galileo, Newton, Bacon, and Descartes, who prepared the ground for the Enlightenment. He gave credit to Gassendi, the Catholic provost of Digne, for the reintroduction of atomic theory, commenting in a curious aside—which could be interpreted as an anti-Catholic statement designed to appeal to Protestants—that Gassendi 'assailed superstition and religion, and rightly, because he did not know the true religion'. In a type of imaginary Socratic dialogue between a follower of Lucretius and the theologian and philosopher Bishop Butler, he addressed the bishop's arguments about the 'distinction between our real selves and our bodily instruments', a distinction Tyndall did not perceive. He finally put into the bishop's mouth the words: 'You cannot satisfy the human understanding in its demand for logical continuity between molecular processes and the phenomenon of consciousness'.

This was Tyndall's position on the limits of scientific understanding, and of materialism, a dualistic view of knowledge. But the following and final part of the Address starkly staked out the territory over which, in his view, scientific understanding does lay legitimate claim.

He gave a picture of modern scientific knowledge; of the developing understanding of the vastness of geological time compared with the 6,000 years of the theologians, and of Darwin's theory of evolution by variation and natural selection, with its lack of any need for 'special Creation', and the inevitable consequence of natural selection for the growing complexity and diversity of life, the trend to 'higher forms'. Then followed other 'grand generalisations' of modern knowledge: the 'doctrine of the Conservation of Energy' and Spencer's principle of continuous interaction between the organism and its environment. He stated the essence of the materialist position, though he would substantively qualify it at the end of the Address:

> The polarity of magnetism and electricity appealed to the senses; and thus became the substratum of the conception that atoms and molecules are endowed with definite, attractive and repellent poles, in the play of which definite forms of crystalline architecture are produced. Thus molecular force becomes structural. It required no great boldness of thought to extend its play into organic nature, and to recognize in molecular force the agency by which both plants and animals are built up...Believing, as I do, in the continuity of Nature, I cannot stop abruptly where our microscopes cease to be of use...By a necessity engendered and justified by science I cross the boundary of the experimental evidence, and discern in that Matter which we, in our ignorance of its latent powers, and notwithstanding our professed reverence for its Creator, have hitherto covered with opprobrium, the promise and potency of all terrestrial Life.

In other words, life is inherent in matter. It does not need importing from outside. This may seem radical, indeed it was thoroughly radical and offensive to religious orthodoxy. He may not have known that Newton himself had written: 'We cannot say that all Nature is not alive',[29] though he criticized Newton and Boyle for what he saw as their 'clock-maker' view of God. As Lucretius himself had put it: 'since we perceive that the eggs of birds change into live chicks...it is evident that the sensible can be produced from the insensible'.[30] In modern terms one can describe Tyndall's view of life as an emergent property of matter in complex forms. In a beautiful image, he wrote elsewhere: 'Life is a *wave* which in no two consecutive moments of its existence is composed of the same particles'.[31]

In the final part of the Address, Tyndall tackled head-on, in the Protestant religious citadel of Belfast, the might of the religious establishment. It would have been almost unimaginable twenty years previously. Even now it was to create a storm of major proportions, stimulating a flurry of pamphlets, and attacks in the periodical press.[32] Yet in the midst of his attack, he gave religions what he saw as their due:

> grotesque in relation to scientific culture as many of the religions of the world have been and are—dangerous, nay, destructive, to the dearest privileges of freemen as some of them undoubtedly have been, and would, if they could, be again—it will be wise to recognize them as the forms of a force, mischievous, if permitted to intrude on the region of objective knowledge, over which it holds no command, but capable of adding in the region of poetry and emotion, inward completeness and dignity to man.

Having acknowledged the territory over which religion could legitimately act, though not to the satisfaction of those who saw religion as a matter of faith and revelation rather than feeling and emotion, Tyndall delivered his credo:

> The impregnable position of science may be described in a few words. We claim, and we shall wrest, from theology the entire domain of cosmological theory. All schemes and systems which thus infringe upon the domain of science must, in so far as they do this, submit to its control, and relinquish all thought of controlling it.

The choice of words is critical. Tyndall was attacking theology and the formalized dogma of religion, and empirical claims about nature made on religious grounds. He was not attacking the religious sense itself, the needs of which he thought might be met in time by poetry rather than organized religion. His was not some mechanistic, value-free science. His vision, Carlylean and Spencerian in its influence, was of science underpinning human development: 'The lifting of the life is the essential point'. This vision recognizes the moral source of much scientific inspiration. Tyndall argued: 'Science itself not unfrequently derives motive power from an ultrascientific source. Some of its greatest discoveries have been made under the stimulus of a non-scientific ideal'. He exhorted his listeners not to settle for intellectual compromise: 'As regards these questions science claims unrestricted right of search'. But even then, the Mystery might remain. He ended: 'Here, however, I touch a theme too great for me to handle, but which will assuredly be handled by the loftiest minds when you and I, like streaks of morning cloud, shall have melted into the infinite azure of the past'.

The Belfast Address is a statement of the relationship between science and the theology of organized religion, and of the limits of science. Though Tyndall did not see science and the religious sense in conflict, he did see potential conflict with organized religion and theology, especially with Catholicism. Indeed, stimulated perhaps by Draper's *History of the Conflict Between Religion and Science*, which was published in 1874, he wrote a short supportive preface for the English edition of Andrew White's *The Warfare of Science* in 1876, directed at the 'intelligent Catholics of England and Ireland'.[33] But Tyndall's position was not that of an atheist, as he affirmed explicitly to Mrs Steuart.[34] He wrote also to Olga Novikoff—the journalist and, according to Disraeli, 'MP for Russia in England'—at the start of what was to be a long correspondence and friendship: 'I do not know a man for whose opinion I would give a halfpenny who would avow himself an Atheist. While rejecting the Anthropomorphic notion of superstitious people I said as plainly as words could make it, that inscrutable Power lay behind it all'.[35] He stressed in a lecture in Manchester shortly after the Address that atheism would be an impossible answer to whether there is a greater power in the universe but 'only slightly preferable to that fierce and distorted Theism which I have reason to know reigns rampant in some minds'.[36]

Nevertheless, he caused a storm in religious quarters: 'Every pulpit in Belfast thundered at me. Even the Roman Catholics who are usually wise enough to let such things alone come down upon me'.[37] Cardinal Cullen, the Archbishop of Dublin and

Primate of Ireland appointed three days of prayer to keep infidelity out of Ireland, as Tyndall gleefully reported to Youmans.[38] Abbé Moigno, Tyndall's Catholic French translator, wrote sadly, contrasting the insight and instruction of Tyndall's work on *Light* with the 'clouds of Belfast'.[39]

Tyndall's view of the nature and role of poetry is revealed by the Belfast Address and his similar writings. On one level, he imagined poetry eventually replacing the emotional and, in the modern sense, spiritual needs currently met by religion, once people recognized the falsities of revealed religion and dogmatic theologies.[40] For him, the Book of Genesis was a poem, not a scientific treatise.[41] But he also saw poetry as a means of imagination, of looking through immediate experience to concepts and meanings beyond. In a creative sense, the scientific imagination was no different. Experimental findings could not lead to theories and explanations without the scientific imagination; the poetic and scientific imaginations were complementary. Poetry played its legitimizing function too. Tyndall quoted from Tennyson's *Lucretius* in the Address, and gave prominence to Wordsworth's pantheistic *Tintern Abbey* in some subsequent published versions. His meaning was both validated and extended through the cultural legitimacy and value of poetry.[42]

A week later Tyndall's duties at the British Association were done. He returned home via Kingstown for a couple of days of lighthouse experiments. The Address was published by Longmans and sold briskly. As he told his American publisher Appleton, 'the third thousand was called for in three days'.[43] In the preface to the first thousand, written from the Athenaeum on 15th September, Tyndall emphasized his position on atheism, saying that material atheism offered him 'no solution to the mystery in which we dwell, and of which we form a part'. For the seventh thousand, in December, he added a more substantial preface. He acknowledged that 'as an experience of consciousness [religion] is perfectly beyond the assaults of logic'. But it 'will have to bear more and more, as the world becomes more enlightened, the stress of scientific tests'. He was unequivocal about the rightful domain of science, asserting that in relation to the organic world, 'we must have the freedom [from religion] we have already won in the inorganic'.

Hirst observed that Tyndall was 'being attacked on every side for his Belfast Address'.[44] Many people, wilfully in pursuit of their own agendas, or otherwise, misunderstood his position on materialism and the limits of science. Others disapproved of this commandeering of the President's position to make such a statement. But there was support too, including from older hands like Lyell. The Address reached an international audience, as far away as New Zealand. Govi, Professor of Physics at Turin, wrote to say how grateful he was and to describe his difficulties with those in power after his own address on 'The Laws of Nature' in Turin in 1869, and the attacks that followed on him from Moigno.[45]

Closer to home, Mary Adair wrote in the aftermath: 'Though I do not in all you say agree with you, yet I am content to attribute much of this to my inferiority of mind—my brain being unequal to such a wide grasp of truth as your's. But I can see that much of the clamour is simply due to the ignorance and wilfulness of those who make it; and I fought many battles on yr behalf last autumn in Scotland' (Plate 18).[46] Clausius wrote that he had read the Address with great interest.[47] The perceptive Duke of Argyll, staunch Christian, understood Tyndall well, writing later: 'I have taken no part in the outcry about your Belfast Address, because I thought it greatly misunderstood, and that its tendency is rather to spiritualise matter, than to materialise Thought'.[48] Tyndall, grateful for his words, replied: 'More than once or twice…I have asked myself whether, if I were to present myself at the Duke's door, I should find it closed against me. That question can arise no more. In fact among those whom I esteem I do not think I can count a single "averted eye" '.[49] Tyndall was now a major, and controversial, cultural figure, a position acknowledged when he appeared in William Mallock's satire *The New Republic* in 1876, thinly disguised as 'the sentimental and effusively eloquent' Mr Stockton. Fellow targets included Huxley, Carlyle, Ruskin, and Jowett.

* * *

As if one controversy was not enough, Tyndall initiated a second in November, which became known as the Tyndall Typhoid Controversy. The *Times* published a long letter from him on how to combat typhoid, stimulated by an outbreak in Over Darwen and the publication of William Budd's book *Typhoid Fever*.[50] Tyndall asked if typhoid could be generated anew: 'is it produced by the decomposition and putrefaction of animal and vegetable substances, or must the matter producing it have had previous contact with an infected body?'. Pointing out that typhoid was killing 15,000 people each year and causing 150,000 to pass through its protracted miseries, answering the question was critical since it would determine 'our mode of attack upon this enemy of mankind'. He praised Budd's observations and inferences of a germ theory of the disease, and highlighted identification of 'the very organism which lies at the root of all the mischief', recently published by Klein in Germany.

Not for the first time in a historical *résumé*, Tyndall did not do full justice to predecessors of Klein. Huxley privately pointed out that in his British Association Address in 1870 he had referred to the work of Chaneveau and John Burdon-Sanderson in proving that living 'microzyms' are the efficient agents of contagion, and that Cohn had subsequently demonstrated that these minute living particles grow and multiply under suitable conditions, out of the body.[51] Tyndall was attacked by the medical community for appearing to downplay the importance of cleanliness, a particular emphasis of the 'miasmic' theory of disease, and wrote a follow-up letter to the *Times* to clarify his position: 'I neither deny nor doubt that filth, foul air, and the gaseous products of animal and vegetable decomposition are things hurtful to health, or that they are capable...of causing serious disease, and even death. What I do doubt and deny is that of themselves they ever produce a contagious fever'.[52] He added some words acknowledging Klein's predecessors.

Tyndall's difficulty was that the ground rules for germ discovery were not yet apparent nor agreed. In his interventions in the spontaneous generation debates of the early 1870s he had tried to designate particular experiments as crucial, but others differed. He would fight this battle again later in the decade. Accepted causes of disease were multifactorial and some found it difficult to countenance the idea that microscopic and apparently structureless objects were disease-specific, and alone responsible. Environmental conditions surely played a part. Acceptance of a theory of minute germs of fixed species became easier after Robert Koch's isolation of the anthrax bacillus in 1876, but even then it was not complete.[53]

It was a common pattern for Tyndall to give a Discourse in January and to spend time at the beginning of the year with friends. In 1875 the regular Huxley New Year gathering was followed by lunches and dinners with Hirst and Debus, by the X-Club meeting on 7th January, and dinner at Sir Joseph Whitworth's on 11th January. Tyndall's Discourse on 15th January, following the reading of his closely-related paper at the Royal Society, was 'On acoustic reversibility'.[54] Here Tyndall claimed to have solved the 1822 mystery of the firing of guns at Villejuif and Montlhéry, about 12 miles apart. Those at Villejuif heard everything, those at Montlhéry almost nothing, even though a very slight wind was blowing towards them. Many years earlier, Tyndall had seen Count Shaffgotsch demonstrate his ability to extinguish a gas flame in a tube by the sound of his voice at the right pitch and volume. Tyndall had achieved the reverse of igniting a flame with his, when he discovered it was beats rather than unison which caused the effect. Now, using a flame, and directing sound through a tube to the bottom of the flame, which was the most sensitive part, Tyndall showed that the sound was effective in disturbing it when the source was at a distance from a screen between it and the flame, and the flame close behind the screen. In the reverse position, effectively making the area around the sound acoustically opaque, it was ineffective. So the atmosphere at Montlhéry was diacoustic, but it was acoustically opaque at Villejuif. Tyndall put this down to the inhomogeneity of the atmosphere at Villejuif, created by the air currents arising from the multiple fires of the nearby city of Paris.

A few days before Easter, after his series of seven lectures on 'Electricity', Tyndall went to Folkestone with Lady Claud Hamilton and, one must presume, with her family, though Tyndall makes no reference to this. By April he was writing to Louisa with a hint of intimacy in response to a letter (now lost): 'It would have been a delight to be with you in Marburg...Your letter brings vividly to my mind my first visit to the

Wartburg. I felt my heart tingling with the associations of the place, as you now feel yours. When you return put me in mind of sending you a scrap of doggerel that I wrote there. It will at all events show you how similar our thoughts & reflections have been'.[55] Tyndall was working hard after Easter, not least because Longman had asked for new editions of four of his books: *Heat, Sound, Light,* and *Fragments of Science.* To relieve the pressure he spent a few days at the West Cliff Hotel in Folkestone at the end of April, where Hirst visited him, and about a week on the Isle of Wight in May. Tyndall heard that his hero, his 'General' Carlyle, had turned down a knighthood. He wrote: 'I had a conversation some time ago with Lady Derby in regard to the honour which Her Majesty proposed to bestow upon you. The mission of the Countess was to move you: but she rejoiced to find you immovable'.[56]

Tyndall is generally perceived as an unremitting champion of the importance of 'blue skies' research, untrammelled by commercial interest or objectives. This he was, at a time when the support of original research for its own sake through public money was limited, and private patrons less evident than a century before. It was important to make that argument, a view shared by his X-Club colleagues. Those, like Gassiot, who could afford to finance their own research were relatively few, and the nature and organization of research was changing as the groundwork of nineteenth century physics and chemistry was laid.

But Tyndall was never an opponent of applied or commercialized research, despite the occasional rhetorical flourish that might suggest otherwise. He was a good friend of industrial magnates like Joseph Whitworth and William Armstrong, and was welcomed into their country houses. He showed this in his next Discourse on 4th June, with the Duke of Devonshire in the chair, talking 'On Whitworth's planes, standard measures, and guns'. Doubtless in part preparation, Tyndall had dined with Whitworth and Hirst at the Athenaeum on 20th May and spent the weekend of 29th May at Whitworth's country residence Stancliffe Hall in Derbyshire (Whitworth was by this time largely retired, and looking after his quarry business locally). In the Discourse, Tyndall described the significance of the development of Whitworth's tools for reproducible mass production, including of the rifle bullet and barrel, and the breech-loading gun.[57] At the end, without naming him though the identification was obvious, he praised the work of Armstrong, Whitworth's great competitor, and offered his services should Armstrong 'desire to place an account of his labours before the members of the Royal Institution'. Some years later, Tyndall revealed that Whitworth had wanted him to become a Director of his company. But he declined, explaining to Hirst: 'I gathered that he looked to the aid I could render him with my pen, and I frankly told him that if I ever had to speak about him or his inventions, I must do so as a perfectly free and disinterested man'.[58]

Tyndall dined with Hirst, Debus, and Roupell at the Athenaeum on 18th June and left for the Alps the following day. The first week's weather at Belalp was atrocious, rain splashing through fog, driving almost all the visitors away. Confined indoors much of the time, Tyndall was able to revise *Fragments of Science* and absorb his reading matter of Lange's *History of Materialism*, and Frohschammer's *Freiheit der Wissenschaft* and *Das Recht der Eigenen Überzeugung*.[59] He had been introduced to Frohschammer, the Catholic philosopher and priest who challenged religious authority, by Olga Novikoff.[60] While the effort in writing and revising his books, and his other commitments, were substantial, he had finally begun to think that he had 'enough of money for the evening of my life. What with my books, and my pluralities and the kindness of my friends at the Royal Institution, I have gradually slipped into the position of a millionaire!'.[61] He told Huxley that his books had made him around £1,000 in the past year.[62] Despite the weather, Tyndall enjoyed a few days with Canon Liddon, whom he had not known before: 'probably he had pictured me, before he saw me, as a creature with hoofs and horns. But we parted very cordially, and I am to dine with him when I return'.[63] Liddon was Canon of St Paul's and Professor of the Exegesis of Holy Scripture at Oxford. A highly popular preacher he was a traditional Anglican and a firm supporter of Pusey. Another of the theological conservatives had been charmed by Tyndall.

Cottrell joined Tyndall at the end of June for some glacier observations. The weather brightened temporarily, and though Cottrell's ice skills left something to be desired, they staked and measured several lines, 'learning thereby something perfectly definite regarding certain appearances in the Aletsch glacier'.[64] Lady Claud Hamilton and her family were due in mid-July, and Tyndall planned to head with them for Grindelwald. He was reading around germ theories, soon to be the focus of his research. He told Hirst that if the antivivisectionists were to succeed in hampering or extinguishing such researches it would be a calamity for the human race.[65] The Hamiltons arrived but the weather remained vile, as it was across Europe, with serious flooding and even lives lost in England. On 24th July, Tyndall and the Hamiltons moved on from Belalp to Grindelwald, finding Wiedemann there. With Louisa, her sister and brother, Tyndall climbed the Wetterhorn. They were on the summit at 9am and had 'an unspeakably glorious view'.[66]

* * *

A third controversy in so many months now arose unexpectedly, this time between Tyndall and the United States Light House Board, and specifically with Joseph Henry.[67] Henry took issue with Tyndall's criticism of the Board, which Youmans had published in *Popular Science Monthly*. At the root of this was a disagreement about Tyndall's concept of an 'acoustic cloud', which Henry did not see as the principal cause of the abnormal phe-nomena of sound, and the fact that Tyndall appeared not to have referred in his report to the previous researches of the United States Light House Board on the subject of fog signals. This offended the Board and the members of the Philosophical Society of Washington. Henry wrote while Tyndall was in the Alps in 1875, to thank him for send-ing him the *Lectures on Light* and the third edition of *Sound*. But he had read the preface to the latter 'with mingled feelings of gratification and sorrow. With gratification on account of the kind manner with which you have spoken of me, for I value very highly your friendship and good opinion; and with sorrow on account of the annoyance which apparently my dissent from your views has occasioned you'.[68] Major Elliott had returned from his visit to France and Britain to report that that they were far ahead of the Americans, which led to Elliott's resignation, 'and subsequent attacks on our Light House system tending to its entire disorganization'.

In a report justifying the American system, Henry had been critical of Trinity House and of Tyndall's conclusions. He felt that Tyndall had not listened carefully enough to his paper given to the Washington Philosophical Society, when Tyndall was present as guest of honour, in which he stated his opinion that fog did not materially interfere with the propagation of sound, and gave his account of the explanations of abnormal phenomena, which he put down to the existence of upper air currents in the opposite direction. Tyndall seized on other evidence of acoustic cloudiness to defend his position, and dashed a letter off to *Nature*.[69] Commenting on a letter from his cousin Hector Tyndale to Henry, updating him on the progress of the Tyndall Fund,[70] Tyndall said he was sorry to have criticized Henry but maintained he had 'gratuitously provoked it' and was wrong.[71]

Tyndall had to be back in England to hand over the Presidency of the British Association in Bristol on 25th August 1875. The incoming President was the engineer Hawkshaw, whose history with Tyndall went back to their railway days, and who Tyndall saw as a soothing choice: 'All pious souls will have a calm year; for Sir John Hawkshaw, my successor, is a great Engineer, and instead of talking heresy he will give some account of the achievements of his own profession'.[72] The great and good were assembled in the Colston Hall, with Mary Adair too. Tyndall adroitly avoided any further controversy in a short and humorous introduction to one 'not likely to be caught up into atmospheric vortices of speculation about things organic or inor-ganic, about mind or matters beyond the reach of the mind...I have looked forward for some time to the crowning act, still in prospect, of his professional career, intended, as many of you know, to give our perturbed spirits rest in crossing the Channel to visit our fair sister France'—this a reference to Hawkshaw's vision of a

Channel Tunnel between England and France. Hawkshaw's speech, as Tyndall had foreshadowed, was worthy, but it was partly inaudible. The *Bristol News* reported that 'the audience was not very demonstrative except at the close'.[73]

Tyndall left Bristol before the end of the Meeting, having largely kept away from social interactions by accepting no invitations and staying in lodgings. In the weeks afterwards, issues connected with Belfast and glaciers remained in flux. For the new sixth edition of *Fragments of Science*, Tyndall included a specific response to the attacks on his Belfast Address by James Martineau, the religious philosopher and Unitarian, which had been published in the *Fortnightly Review*.[74] He had produced an earlier clarification and restatement of his views, 'Apology for the Belfast Address', soon after the Address itself.[75] In the world of glaciers, George Forbes returned from the Sandwich Islands, where he had been observing the transit of Venus, to discover that Tyndall had 'again reopened the glacier controversy'. He asked icily for a copy of the publication.[76]

With his commitments under control, and controversies playing out, Tyndall was ready to start his experiments on the 'floating matter' of the air. They would occupy him for several years and constitute a major research programme.

21

⬩⟷⟸ ⟹⟷⬩

FLOATING MATTER
OF THE AIR

1875–1876

Tyndall's researches into the decomposition of vapours by light, which required optically pure air, had led him to examine the nature of the 'floating matter' in the air. He had found it to be organic.[1] The question was whether this organic matter contained 'germs' that could propagate by continuous succession, or whether life could arise by spontaneous generation in putrefying matter, and disease be produced from the 'miasma' emitted from dirty surroundings. After all, a belief in spontaneous generation appeared to many people to be entirely compatible with ideas of evolution and of continuity in nature. Even Huxley had admitted that 'abiogenesis', the development of living organisms from inorganic matter, seemed to have happened at some point in the past. The argument was whether 'heterogenesis', the development of organisms from decaying organic matter, happened in the present.

One person who believed it did was Bastian. Germ theories had been extensively debated at a meeting of the Pathological Society on 6th April 1875, when Bastian, arch-proponent of spontaneous generation, had admitted the coexistence of bacteria with contagious disease. But he argued that they were pathological products of the disease, spontaneously generated, not the contagion itself. His key evidence was the observation of putrefaction in hermetically sealed flasks heated to boiling point, when germs were known to be destroyed at lower temperatures.

Tyndall traced the proof that living germs are the cause of putrefaction and contagious disease to Schwann's work in 1837, which had established that all livings things are composed of cells. It was the discoveries of Schwann and others, Tyndall said, on which Lister had built to develop his antiseptic surgery, and Pasteur his work on fermentation and a germ theory of disease. Now, some five years after Tyndall's own discovery, Pasteur's ideas continued to be contested. Much medical opinion, including that of Bastian, still considered the possibility of spontaneous generation and other causes of disease and decay. Tyndall thought that his ability to make optically pure air would enable him to bring clear experimental evidence to settle the question. On 10th September 1875, he closed the doors of the wooden case he had designed to hold test tubes containing experimental samples (Figures 33a and 33b). A beam of light revealed the floating matter, but three days later, the track had vanished. The floating matter had settled to the bottom, which was smeared with glycerine to capture it. His simple yet clever piece of apparatus was ready to serve its purpose.

Tyndall put urine into eight tubes, boiled them for five minutes, and placed them in the box with no stoppers or filtration of the air. A further eight tubes were treated likewise, but left outside the box. A week later all the tubes in the box were still bright and clear, and those outside all turbid. Looking through a microscope, Tyndall saw what he described as swarms of bacteria. He asked Huxley to come over and see his demonstration of 'the practical correctness of your statement that bacteria germs exist "in myriads" in the atmosphere'.[2] Throughout October, November, and December, he extended the experiments by using a wide range of substances in hundreds of tubes—

Fig. 33a. Tyndall's 'floating matter' box.

342

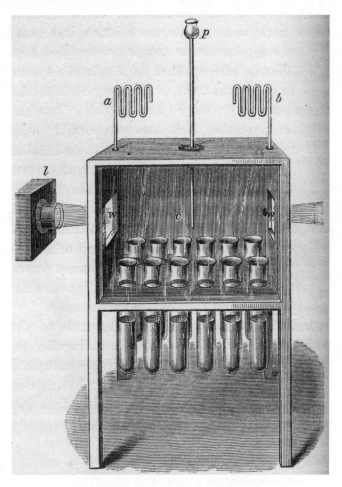

Fig. 33b. Floating matter chamber.

infusions of mutton, beef, haddock, turnip, hay, sole, liver, hare, rabbit, pheasant, and more—and by hermetically sealing the tubes to keep air out until he broke them open to the atmosphere. Though the microscope was of value, it was the use of the beam of light that Tyndall regarded as more powerful; it could reveal the presence of matter in the air or in liquids that was invisible to the microscopes of the day. It gave evidence that germs could exist where no microscope could detect them.

All manner of experiments followed. Tyndall let laboratory air into the closed box containing tubes that had been clear for days, observing with his light beam that floating matter had been introduced and that putrefaction set in after a few days. He used different strengths of infusions, different temperatures, and different times of treatment, to counter the objections of those supporting spontaneous generation. He used calcined (dried) air, he used filtered air, and he left tubes open all over the Royal Institution: in his bedroom, on the roof, in the theatre, or in the basement. In the box, the infusions remained sterile and clear. Outside it, putrefaction always set in. Puzzled when he once seemed to find life in tubes that had been clear for days in the box, he realized that the tiniest drop from the end of an unclean pipette used to take out the sample had been the source of error. On the few later occasions in which a hermetically sealed tube did appear to generate life, as identified by Huxley with a microscope, minute cracks or breaks in the glass were detected. Scrupulous experimental care was needed, and in Tyndall, the field had found its exponent.

The difficulty now for Tyndall was proving a negative. As he had done with Magnus and the effect of radiant heat on aqueous vapour, Tyndall had to identify the sources of error and meet objections one by one. Finding that Bastian had emphasized the importance of warmth, Tyndall had tubes taken to his favourite haunt, the Turkish Bath in Jermyn Street, and put in the care of 'an intelligent assistant'. The same results were obtained. Conscious of statements about the scantiness of bacteria in the air, he thought that observations outside London would be instructive. Darwin accommodated samples in his study at Down, Lubbock at High Elms, Hooker at Kew, Siemens at Sherwood near Tunbridge Wells, Rollo Russell at Pembroke Lodge in Richmond Park, and a certain Miss Hamilton at Heathfield Park in Sussex. All putrefied on exposure to the air.

In early October, in one of those 'what ifs?' of history, he nearly discovered the antibacterial action of *Penicillium*. Noting its growth on the surface of a turnip infusion, and the death of the bacteria that had developed in the tube, he put the action down to the physical barrier of the layer of mould preventing oxygen from reaching the bacteria.[3] He wrote to Huxley: 'The most extraordinary cases of fighting and failing and conquering between the bacteria and the penicillium have been revealed to me'.[4] Sometimes the bacteria seemed to win, sometimes the *Penicillium*. This was a Darwinian 'struggle for life', or 'survival of the fittest', in the phrase coined by Spencer in 1864. The idea that *Penicillium* might secrete an anti-bacterial chemical did not occur to him. Huxley, in a parallel series of experiments, showed that lack of oxygen was not the sole cause, but likewise failed to draw the conclusion.[5]

Tyndall argued that we did not entirely know whether the floating matter consisted of germs, particles of dead organic dust, or mineral matter. But if it gave rise to specific organisms, then just as we would deduce from the growth of cresses or grasses from soil that seeds of them were present, so we should deduce from these experiments that the organisms were the offspring of living seeds. As he put it later with respect to disease: 'Is it, then, philosophical to take refuge in the fortuitous concourse of atoms as a cause of specific disease, merely because in special cases the parentage may be indistinct?'.[6] For Tyndall, stimulated by the work of Pasteur and Budd, it was the conflict of the germ theory of contagious disease with the doctrine of spontaneous generation that was the critical issue, and he praised the work of Burdon-Sanderson in identifying contagious disease with the presence of self-multiplying organic forms 'of the lowest vegetable kind'.

Tyndall realized that he could 'map' the floating life of the atmosphere by making a grid of a hundred tubes filled with infusions and letting the germs fall in. Eventually, over many days, the tubes all showed evidence of growth of bacteria or mould, but at different speeds and in different ways. Tyndall deduced that the distribution of germs was not uniform in quantity or quality, which he thought might help explain the unpredictable behaviour of dressed wounds and the progress of an infectious disease (ignoring the possibility of different responses by individuals, and thereby giving further ammunition to the medical community to criticize his competence). He calculated that many millions of germs must be present in a single room, a figure he thought might be a substantial underestimate, but would buttress Huxley's statement in his 1870 British Association Presidential Address that myriads of germs are floating in our atmosphere.

* * *

After the summer break, Tyndall picked up lighthouse business in parallel. He was fortunate to have the assistance of Cottrell, who could oversee experiments and make observations on putrefaction at the Royal Institution when he was away. Laid up with a cold on 5th November, which prevented him going to dinner with Lecky,[7] he wrote to William Thomson about the use of sound in signalling. It was Thomson who had resurrected Babbage's idea of 1851 that distinctive signals could be used to identify different lighthouses, and had been involved in having a self-signalling apparatus installed at the entrance to Belfast harbour. Thomson later wrote a long letter to the *Times* in 1879 urging this. He seems to have first suggested the use of Morse code signals in an article 'Lighthouses of the Future' in *Good Words* in March 1873, although it met resistance from sailors when originally made, and from the Stevensons when they reported on the idea in 1880.[8] Tyndall was scathing about the Navy's use of horns and whistles and told Thomson he considered the siren, preferably driven by steam rather than air, to be the best option.[9] Like Tyndall, Thomson considered that the Navy could use short and long blasts for code signalling.[10] Earlier in the autumn,

Tyndall had written to Wigham to urge on him the importance of making experiments on the glare of his gas lights, when the light itself was invisible through fog, because he thought it important to show whether fog flashes could be intercalated with sound signals.[11] By 2nd December, the day of the monthly X-Club meeting, Tyndall could report to the Board of Trade that the sudden glare produced by the flash of a gun, when employed as a fog signal, had frequently been seen when the sound of the gun was unheard. But it still needed proper experiments to test its efficacy as a fog signal and he reiterated his proposal for testing it at South Foreland.[12]

Meanwhile at the Royal Institution, management business put Tyndall in a difficult position. An explosion connected with the work of Williams, John Gladstone's assistant, had risked setting the whole building on fire. He was concerned, as Superintendent of the House, about safety in the chemical laboratory, which was Gladstone's territory, but had been unable to gain access to the laboratory or see Williams. He asked Gladstone to consider closing the laboratory until he could give it his personal attention.[13] Gladstone acknowledged that Tyndall must have access at all times, but also wanted to prevent any unnecessary interference with those engaged in the Chemical Laboratory during working hours.[14] Tyndall bridled at the response and the implication of 'unnecessary interference', pointedly expressing the difficulty of resolving the issue when Gladstone was so rarely in the building.[15]

A year and a half later, matters came finally to a head; a special meeting was to be convened in March 1877 to discuss Gladstone's position, but he resigned to save embarrassment. The Managers held Tyndall in high regard and offered an increase of £200 in his salary in 1875. Tyndall wrote to Spottiswoode (the Secretary since 1873, following his spell as Treasurer) to say that with his existing salary, government work and writing, his income was more than sufficient. He asked if £85 could be made available instead, £35 to increase Cottrell's salary to a more reasonable £150 and £50 to pay his second assistant, which until now he had been paying from his own pocket.[16]

With the demands of his experimental work and of continuing lighthouse business—the work even overlapped as Tyndall used a cloudy infusion of bacteria in a glass cell to simulate fog—his social life diminished. He turned down an invitation from Olga Novikoff, saying the most he could manage was to visit Darwin on 16th October 'whose health is delicate, and whose years are advancing', and Mrs Grote '(wife of the historian) who is 83'.[17] In early December he went down to the Hamiltons at Heathfield Park for a few days. He found it knee-deep in snow: 'the trees snow-laden—their beauty, particularly that of the cedars, is simply enchanting'.[18] Even here he had to work, as it was his turn to give the Christmas Lectures, and he turned down invitations from Emily Peel and from Olga Novikoff, to whom he felt the need to apologize for inadvertently being rude to her.[19] He even turned down dinner with the Franklands, though he agreed to call by after dinner to see Mrs Dawes, widow of Richard Dawes, of whom he wrote: 'My memory of him is one of the brightest I possess; and few men of his day did so much good in the world'.[20] Perhaps picking up the educational theme of Dawes, he found time to write to the *Times* on the subject of 'The Vatican and Physics', quoting from a letter from the Bishop of Montpellier to the Deans and Professors of Montpellier. In this letter, the Bishop made clear that the Church of Rome held herself to be the depository of total truth and that 'even in the natural order of things, scientific or philosophical, moral or political, she will not admit that a system can be adopted and sustained by Christians, if it contradict definite dogmas'. Tyndall saw no possible reconcilement nor compromise between the liberty of Rome to teach mankind and the liberty of the human race: 'One liberty or the other must go down. This, in our day, is the "conflict" so impressively described by Draper, in which every thoughtful man must take a part'.[21]

Relations with Olga Novikoff were mended as she wished him Happy Christmas on Christmas Eve and he replied from the Lubbocks' on Christmas Day, relaxing with his godson and playing hockey (Nelly Lubbock praised his skills to Hirst), before returning to his 'grinding work'.[22] His six Christmas Lectures, on 'Experimental Electricity', were aimed at showing what could be done for science teaching in schools with simple and inexpensive apparatus. He wrote them up as a book, *Lessons in Electricity*, which he dedicated to Five Young Friends in the approximate order of their ages: Hugh Spottiswoode, Henry Huxley, Rolfe Lubbock, John Clausius, and Reginald Hooker.[23] To their great delight, he sent them copies complete with apparatus when the book was published later in the year (John Clausius's broke in transit and was mended by his father).

He found himself also required to give a lesson to the Duke of Argyll, who wrote asking for his view on the conceptual identity of heat and light.[24] Tyndall gave a reply that reveals his metaphysical approach to the question. Arguing that heat and light are not identical simply by virtue of the different periods of their waves, he maintained that to understand the question those waves in the ether had to be traced back to the vibrating atoms and molecules of matter that caused them. In that sense, taking the Sun as an example, all the waves transmitted by the ether were heat waves; they all produced heat when absorbed by matter. But some of them, reaching the retina, produced a psychical or subjective effect, and this we call light.[25] In modern terminology we would say that all waves carry energy, some of which are perceived by us as light.

The Duke and Tyndall met and corresponded over many years. The Duke, who declined the Presidency of the Royal Society in 1868,[26] published two substantial books in which he sought an integration of science and religion: *The Reign of Law* in 1866 and *The Unity of Nature* in 1884. Late in both their lives he wrote to ask if Tyndall had ever examined the behaviour of motes, or finely suspended matter, in air under high pressure, since he thought it might affect their distribution.[27] Tyndall thought it would make no difference, unless the floating motes had portions impermeable by the dense air, in which case they would tend to rise, but went on: 'I have been sometimes surprised to hear naturalists speak of the impossibility of life at the bottom of a very deep sea. If the elastic fluids within our own bodies were not in equilibrium with the atmosphere outside our bodies, we should be either crushed or blown up; and in the case of the deep sea organisms, the only requisite of life would be the establishment of equilibrium between the elastic and non-elastic fluids within the organism and those without it'.[28] A perceptive remark on biology from a physicist.

As the Christmas Lectures ended, as packed as ever, Tyndall revealed the initial results of his work on putrefaction. First to see the evidence, on 13th January 1876, were the Fellows of the Royal Society. They examined tubes that had been hermetically sealed for several months. No signs of life were visible. Hirst asked the question 'where is Bastian?' to which the reply from another Fellow was 'he is nowhere'. A week later Tyndall gave a Discourse, 'On The Optical Condition of the Atmosphere in its Bearings on Putrefaction and Infection', and brought the work to more public attention.[29] Bastian wrote to the *Times* after this Discourse to express his readiness to 'vivify' boxes of the type Tyndall had constructed. Tyndall responded combatively, his anger unabated since Bastian's earlier response in the *Times* in 1870: 'The dialectic dust of Dr. Bastian's letter I leave to the slow, sure process of self-subsidence'.[30] An abstract of the Discourse appeared in the *British Medical Journal* on 29th January, which Clausius for one read with great interest, writing: 'It covers a new and highly important area within the already large scope of your investigations'.[31]

Tyndall's challenge to Bastian and the believers in spontaneous generation raised the profile of the debate. He realized, as he expressed to Darwin, that after Bastian's response in the *Times* he would have to go into the whole history of Bastian's work, dealing with his logic as well as his experiments: 'I was disposed to deal with him in the tenderest manner; but his recent exhibition in the Times shows me that a far different treatment will be needed'.[32] Indeed he appeared in the *Times* a second time to rebut Bastian's claim that he had not used high enough temperatures to reveal spontaneous generation.[33] Articles followed in the *Lancet* and the *British Medical Journal* on 12th February, and in *Nature* a few days later.[34]

Pasteur was grateful for Tyndall's intervention and the publicity and weight they would carry, corroborating his earlier findings which dated back to his seminal paper of 1862, but using different methods. He could not understand how Bastian could maintain his belief in a force resident in the amorphous part of dust suspended in the air, rather than in the organized part formed by corpuscles identical in appearance to those of the germs of the organisms of the infusions.[35] This stimulated a correspondence of several letters of mutual admiration between Pasteur and Tyndall. As the significance of this research became apparent, Tyndall used the opportunity of a new edition of *Fragments of Science* to highlight Budd's earlier work.[36] His paper in the *Philosophical Transactions* was summarized in journals right across the world.[37]

* * *

On the day after his Discourse, Hirst was the first person to whom Tyndall wrote to announce his remarkable personal news. He was to be married to Louisa Hamilton, eldest daughter of Lord Claud and Lady Hamilton, and niece of the Duke of Abercorn, the current Lord Lieutenant of Ireland (and fellow governor of Harrow School). The humble Irish surveyor had stormed the citadel of the aristocracy. The news was not a complete surprise to Hirst, who had had inklings for several months. But he realized its likely impact on him, and though he went to the door of the Royal Institution to congratulate his friend, he found himself unable to go in. He walked round Hyde Park, and then to the Athenaeum, where he found Tyndall dining alone: 'the first emotions once allayed I was able to listen with deep interest to his account of how what had just happened had all been brought about'.[38] Tyndall then wrote to tell his sister and many of his close friends at home and abroad. Both Hirst and Debus were profoundly affected by the marriage, 'for it was the breaking up of a little round table which they had almost ceased to regard as possible'.[39] The writer and politician William Kinglake took a more sanguine view. He teased Olga Novikoff on 'the devotion of that gay Lothario, Tyndall', and suggested that his approaching marriage would 'clip his wings for flirtation'.[40] Tyndall had a reputation.

Tyndall had known Louisa for several years. She was just over thirty and he now in his fifties. The age difference did not perturb her sister Mary, who wrote to her other sister Emma: 'As to ages, I think so little of that. I could never fancy Louisa marrying anyone who was not a little older than herself... It seems as if it would be one of those very delightful unions which are as rare as they are happy'.[41] Lady Claud had been matchmaking for at least several months, but it was only in January that the relationship was formalized between them, after Louisa had giving substantial help in writing up Tyndall's Royal Society paper.[42] They went out for a long walk at Heathfield, where Tyndall disclosed his family background and plans at length, and unromantically intimated that he wanted to leave a formal proposal until after his Discourse.[43] He was less than gracious about Louisa in letters to friends despite the fact that he was completely besotted with her. To Spencer, as one bachelor to another, he wrote: 'She is not beautiful—she is not even handsome—save from the transcendent glow from the soul which sometimes illuminates her countenance. She is strong, well-rounded, tender, deep, true—with a power of self-renunciation such as I have rarely if ever seen on earth...Go and do thou like-wise'.[44] Spencer, his failed relationship with George Eliot well in the past, never did. To Miss Moore he wrote: 'She is not wealthy; but she has no expensive tastes—does not dream of carriages or evening dresses—is wonderfully helpful—and in short in every way fitted to be the wife of a poor man'.[45] The Hamilton family were renting Heathfield Park, and when Lord Claud died in 1884, his personal estate was small. Tyndall did not marry an heiress.

James Coxe had wisely advised Tyndall to be sure to let her know in advance that he was a 'pagan'.[46] Tyndall revealed Louisa's position on religion to Carlyle: 'Without losing an iota of a woman's reverence, or a woman's tenderness she has reasoned herself into great freedom on religious matters—You and your friend Froude being at the bottom of many of her thoughts on this subject; so that on this score no jar can occur between us'.[47] He told Carlyle that Louisa had already begun to help him, work-ing willingly from morning till night. He later said with respect to her cooking: 'Louisa is fond of experimenting. Had she been a natural philosopher, her perpetual sorting might have revealed some new truth. In her cooking some of her experiments fail, but others are remarkably successful'. She was frequently in the laboratory with him, taking notes, and helping write an account of 'our recent observations for the Philosophical Magazine'.

Despite the use of 'our', Louisa's name does not appear on any paper. Tyndall held the view that women were not capable of real scientific creativity but could be expected to understand anything revealed by the savants. She would have wanted more scope, commenting with respect to her earlier life on her want of higher work and the impossibility of getting it. It was why she found Nelly Lubbock such a kindred spirit. Despite the example of Louisa, and so many of his female friends, Tyndall remained continuously surprised by the intellectual capabilities of women. He should have known better with Mary Somerville, a superb mathematician and scientific writer. But on receiving a copy of one of her books, *Molecular and Microscopic Science*, he wrote to convey his 'astonishment' at the extent of her knowledge.[48] His 'scientific' belief in the inferiority of women blinded him to the evidence in front of his own eyes. Having said that, it was not only over women that Tyndall sought to assert his scientific authority. His books and lectures all conveyed a superior expertise that he communicated to a relatively inexpert pub-lic, an authority derived in particular from his uncompromising yet accessible lec-ture performances.[49]

Tyndall lived in the Royal Institution, and Louisa was to join him there following their marriage. As he put it to Bunsen: 'She belongs to the very cream of the English aristocracy, and still her tasks and habits are as inexpensive as those of a village

Fig. 34. John and Louisa Tyndall seated in 1876.

maiden... and she is about to share the modest rooms which have been in turn occupied by Davy, Faraday, and myself, at the Royal Institution'.[50]

These modest rooms, a suite of four on the second floor, now had to be suitably furnished. A group of more than 300 subscribers, led by the Duke of Northumberland, gave a silver plate and 300 guineas. The Moore sisters, Harriet and Julia, provided a chair and a dark red sofa, the colour chosen by Louisa. Agatha Russell and her brother Rollo gave *Milton's Works*. Pollock gave candlesticks. Darwin contributed a tea-set, writing at the same time to urge Tyndall to complete his work on spontaneous generation: 'I feel very strongly that the whole subject is not made clear until some light is thrown on the question how men like Burdon Sanderson and Wyman of Boston and Dr. Child often succeeded in getting bacteria in infusions which they had boiled for a long time'.[51]

The choice of venue for Tyndall's marriage to Louisa was noteworthy. Tyndall joked to Huxley that they discussed the registry office, Moncure Conway's free-thinking South Place Chapel and Gretna Green, before settling on Dean Stanley, 'as the man most likely to simplify reasonably the marriage service'.[52] So the ceremony was to be in Henry VII's Chapel in Westminster Abbey, perhaps some compensation for Lord Abercorn who, as Tyndall remarked to Tyndale, had 'attacked my "materialism", and will no doubt be surprised to find his niece carried off by the "materialist"'.[53] The religious venue did not please Spencer, who regretted that Tyndall had not chosen a registry office.[54] He felt unable to attend, as Tyndall had half anticipated.[55] The ceremony took place at 2pm on 29th February, giving them thereafter a leap year anniversary.

The fact that Dean Stanley agreed to officiate at all is a sign of the respect this broad church Anglican had for Tyndall, hardly known as a church-goer. That he went through with it is the more remarkable, since his beloved wife was desperately and painfully ill. She died the following day, and her illness may have influenced the decision not to have any wedding breakfast after the ceremony.

In his capacity as best man, Hirst called for Tyndall at the Royal Institution and took him in a hansom cab to Westminster Abbey. A small group of friends, including Huxley, Hooker, Frankland, Spottiswoode, Busk, Pollock, and Carlyle, gathered in the Jerusalem Chamber before going on to the chapel. Carlyle, now in his eighties, strolled in on Tyndall's arm, so slowly that they arrived at the altar at the same time as Louisa. According to Hirst, the manner in which Tyndall uttered the responses was characteristic, the clear ringing tone with which he pronounced the words 'I will' contrasting markedly with the subdued voice in which he had to invoke the 'Father, Son and Holy Ghost', even though Stanley had 'left out much that was objectionable in the ceremony itself'.[56] Carlyle, now on Hirst's arm as Louisa was on Tyndall's, shakily signed the register, with the Marquis of Lansdowne, Hirst, Huxley, Pollock, and Lord Claud Hamilton adding their signatures. Tyndall promoted his father to the 'Rank or Profession' of 'Gentleman' on the entry (Plate 19).

There followed a family dinner with Lord and Lady Claud Hamilton, the sisters Emma and Mary, and brother Douglas. The newly-married pair, with Louisa suffering from a heavy cold, then left for Heathfield. They had seventeen days there with the house to themselves. It was wild, stormy weather, but they managed to get out for a couple of hours daily. While indoors, with Louisa's help, Tyndall continued to work on his paper on germ theory and spontaneous generation for the *Philosophical Transactions*.[57] To Spottiswoode, he wrote: 'The "Fragments" are well nigh ended; the Royal Society paper is well nigh ended; my Kensington examination papers well nigh ended; a Report to the Trinity House ended'.[58] Immediately on his return from honeymoon he lectured at Sion College on bacteria and germ theory.[59]

Tyndall's lectures at the Royal Institution, a series of seven on 'Voltaic Electricity', did not start until the end of April, well after Easter. He was able to fulfil other commitments, including government work, and to host a dinner for Sylvester to mark his departure to America to take up a position at Johns Hopkins University. In May, he demonstrated the siren to Queen Victoria at the opening of the Loan Exhibition of

Scientific Apparatus at South Kensington (Figure 35).[60] The Queen did not entirely enjoy this occasion. She wrote in her journal: 'We saw all kinds of musical instruments & many very curious things, but it was very bewildering & one hardly knew how to get away from each'.[61] Some years earlier, on being introduced to some 'celebrities' at tea with the Dean of Westminster, the Queen had been unamused by Tyndall himself, recording in her diary: 'Professor Owen, charming as ever Professor Tyndall, (not very attractive) who has a great deal to say'.[62] In a delightful postscript to his work on sirens two years later, Tyndall received an unusual request from Darwin, who thought he had noticed that 'certain sensitive plants were excited into movement, by a prolonged note on the bassoon and apparently more by a high than a low note'.[63] Tyndall arranged for a siren to be sent to Down House, but to no effect. Darwin wrote: 'The plants, ill-luck to them, are not sensitive to aerial vibrations. I am ashamed at my blunder, and very much more so at having vainly given you great trouble'.[64]

At the end of May, the Tyndalls visited the remote Lochaber region of Scotland, to see the famous Parallel Roads of Glen Roy. These narrow horizontal terraces, high on the mountainsides, were originally thought to be the work of human hand but by the early nineteenth century, were considered to be natural features. How were they

Fig. 35. Professor Tyndall demonstrating a fog-horn to Queen Victoria and her entourage.

formed? Early ideas suggested that they were sea or more probably lake beaches, though the barriers that must have created them were not all obvious. Darwin had visited in 1838, after returning from the Beagle expedition, since it seemed that the Roads might help confirm his theory of land oscillating up and down, which he saw as a means of explaining speciation; a marine origin rather than the accepted lake theory. Despite no evidence of marine debris, Darwin submitted his first paper to the Royal Society, giving his marine explanation. It was immediately challenged by Agassiz with his theory of ice-ages and the suggestion that the marks were the result of glacial lakes. But that theory also suffered from problems and the dispute continued. Much later, in 1861, Thomas Jamieson provided convincing evidence that the glacial theory was correct, and Darwin gave up his own.[65]

Tyndall's reason for visiting, following an earlier 'solitary pilgrimage' in 1867, was a combination of sightseeing with his new wife, his interest in glacial action, and the commitment to deliver the final Discourse of the season on this topic. He returned to give the Discourse in short order on 9th June, and started by making clear that he disagreed with the cause of the Roads put forward by Henry Rogers in 1861 on the only previous occasion that the topic had been presented at the Royal Institution.[66] Rogers had proposed a massive inundation from the Atlantic. Tyndall's exploration with Louisa brought distinctive evidence of glacial action. He linked this with an explanation for the formation of deep glaciers at this spot, given their geography in relation to Ben Nevis and the prevailing wind, using his knowledge of the thermal properties of water vapour. This influence of the prevailing wind, and patterns of rain and snowfall (of which Jamieson had been aware), allied to the possibility of ice, as well as mountain barriers holding in the glacial lakes at times, seemed to explain all the Roads. With a model made by Cottrell, Tyndall demonstrated the formation of each lake and its Road by removing glass plates to represent the various barriers. Jamieson, to whom Tyndall had paid tribute, was delighted with the report of the lecture and the model.[67] He wrote to say that he considered Tyndall had settled the matter.[68]

In Louisa, Tyndall had found someone to share his love of the Alps. They left from Charing Cross on 9th July 1876, crossing from Folkestone to Boulogne, breakfasting at Lausanne the following day and reaching Belalp by mule to stay at Klingele's hotel. Klingele welcomed them with a triumphal arch ending in two tubs of evergreens and flowers. They would be together on Belalp for nearly two months. Their affectionate behaviour in public prompted perhaps jealous comment from the American mountaineer Meta Brevoort, who remarked that they were seen 'kissing on one of the spurs of the Sparrenhorn'.[69]

An early mission was the identification of some land on which they could build a chalet, 'with a view to the quiet of life, as contrasted with the rows of hotels'

Fig. 36. The Tyndalls in front of the Belalp Hotel.

(Plate 21).[70] The site they were allowed to explore is in a stunning position on Alp Lusgen, above the cluster of chalets and hotels. Behind is the hillside leading up to the Sparrenhorn, an easy peak which commands a breath-taking view across to the Aletschorn and the other mountains of the Bernese Oberland. In front, towards Italy, is the mountain range across the Rhône valley. To the left is the end of the huge Aletsch Glacier, the longest in the Alps, and to the right, distant views of the Weisshorn and Matterhorn. A more suitable location for Tyndall could hardly be imagined. Having identified the spot, they mapped out the ground with stones and marked the boundary with stakes after finding a 'troupe of peasants' dancing on their ground. The exact location was a challenge: 'useless rocks' counted within the land measurement and they did not want to ask for too much in case the Commune turned them down. One of the useless rocks seemed to stand in the way of a view of the Weisshorn from the western window, so they stood on each other's backs to check the sight line.

As their plans took shape they would 'sleep, eat, walk, read, write, conduct researches on Spontaneous Generation, and worship the rising and the setting Sun in an entirely unequivocal and umbrageous manner'.[71] For more than a month the weather was glorious. The eclectic reading included Pasteur's *Études sur la Bière*, Martineau's essay *Modern Materialism*, Carlyle's *Essay on Burns* and *On German Literature* and Lewes's *Problems of Life and Mind*. They took walking trips on the glacier, several with De la Rive's family, climbed the Sparrenhorn, and, in a noteworthy early ascent for a woman, the Aletschorn. At 4,195m (nearly 14,000ft) it is the second highest mountain in the Bernese Oberland, and was first climbed in 1859 by Francis Fox Tuckett with Tyndall's guide Bennen. The Tyndalls climbed it with Christian Lauener and Moritz. Tyndall suffered from considerable sunburn, while Louisa sensibly wore a fourfold black veil.

Fog signal experiments continued while Tyndall was away, but he was occupied with his own writing and researches, correcting the proofs of *Lessons on Electricity* and carrying out experiments. They had brought out eighty tubes of infusions, of which fifty-five remained unbroken in transit, to examine the influence of clear mountain air. On one occasion Louisa opened tubes on the summit of the Sparrenhorn while Tyndall did so down on the glacier. Meanwhile Bastian had presented a paper to the Royal Society and published an article in *Nature* on 10th August. But in Paris, according to Tyndall, Pasteur had given Bastian 'a stunning blow' at the Académie des Sciences.[72] He told Pasteur that he would resume his inquiry into Bastian's results on his return home, and wrote to the *British Medical Journal* to say that he would soon be following up Pasteur's and Bastian's findings.[73]

At the end of August, Louisa left for Pontresina, where her family were staying. Tyndall was strongly tempted to go with her, but stayed at Belalp to concentrate on his work and supervise the initial building of their chalet.[74] Had he gone he would doubtless have climbed the Piz Languard, as she did, and walked with her on the Rosegg glacier with Helmholtz, who told her how much Tyndall's use of glycerine to trap the floating matter was admired by the Germans (in fact the technique was due to Pouchet).

The Tyndalls met in Basle on 11th September on the way home, dropping off Harriet Hooker with her brothers in Boulogne (Harriet had come out with them to avoid having to attend the remarriage of her father), and were back in London two days later. On the day of their arrival, the British Association Meeting finished in Glasgow.

It gave Tait the opportunity to make another personal attack on Tyndall for his comments fourteen years before on Mayer, to the disgust of Hooker, Spottiswoode, and others.[75] Helmholtz was equally disgusted by Tait's *History of Science*: 'how he twists and perverts scientific sentences, to write to please theologians, which I find completely unworthy'.[76]

Tyndall shrugged it all off, and settled down to married life.

22

CONTAMINATION
1876–1878

Tyndall's work on spontaneous generation took a challenging turn in the autumn of 1876. In his work of 1875, he had mostly used animal and vegetable juices extracted by distilled water and not made artificially acidic, neutral, or alkaline. He had tried to repeat the experiments of William Roberts on 'superneutralised hay infusions', but found that the time required to sterilize them was just five minutes, against Roberts's claim of 3 hours. Unable to follow this up, he did not mention the discrepancy in his published paper. When he returned from the Alps, he took up these loose ends, finding that recent results by Ferdinand Cohn seemed to support Roberts, though there were also inconsistencies between them.

What followed, starting on 27th September 1876 with the first 'hay infusion' experiment, tested Tyndall's patience and experimental skill over the next eight months to the limit, and led to a major discovery. Louisa was with him frequently in the laboratory, taking notes and writing in her diary: 'there is nothing I enjoy more than the stillness of that little room filled with the presence of my beloved'.[1] Social engagements were not ignored. In October they had six of the young Huxleys to lunch, visited Sir Fowell Buxton in his brewery, and went with the Darwins and Hookers for a weekend with the Farrers at Abinger Hall near Dorking.

While the experiments progressed, Tyndall finalized his lecture for Glasgow on 'Fermentation, and its bearings on surgery and medicine', and travelled up with Louisa to deliver it. It was a major statement of his position against Bastian and the damage that he considered Bastian was doing to the medical profession, though he did not name Bastian in the talk. Careful to solicit advice, he sought comments on his draft from Huxley, Hirst, and Pasteur (though not from the medical community). Pasteur was unable to read English, nor to find anyone to translate the text in time, so he could not reply until after the lecture had been given. He was impressed by Tyndall's 'remarkable exposition'. But he advised him not to despise Bastian, and urged him not to publish anything on the contested subject of urine, Bastian's speciality, without coordinating with him first, avowing: 'I will do the same in regards to you. We cannot be in a disagreement'.[2]

Tyndall took a more vigorous line. He granted that Bastian was clever, but thought he was now beginning to see the 'grossness of his own blundering…I think caution will be required on your part in dealing with him, for he is sure to convert any expression of approval on your part to purposes which you would never think of applying it. Trust me when I say that the fight's for victory and not for truth'.[3] It was hardly the view of a disinterested seeker after scientific truth. But Tyndall's belief in the importance for public health of establishing a clear understanding of germ theory, and his desire, shared by Huxley and other influential men of science, to see off the challenge of spontaneous generation to their understanding of Darwinian evolution, overcame any scruples. He encouraged the work of others who also sought to combat Bastian, including William Dallinger, who worked on the life cycles of small organisms.

In the lecture, which was published in the *Fortnightly Review* for November, Tyndall ranged over the state of knowledge, the role of animal experimentation, and the importance of public education and cooperation essential to combat disease.[4] He started by comparing the deliberate fermentation of beer by introducing yeast, perhaps by his brewery-owning friend Sir Fowell Buxton, with the fermentation of grape juice to produce wine. In the second case, the yeast came automatically, as a contaminant from the skins of the grapes and twigs of the vines. There was no spontaneous generation. Uncontaminated juice would never ferment. The trick of the brewer or wine producer was to support fermentation by the right organisms and keep out all those others, floating in the dust of the air, that would spoil the ferment. The same applied to the human body. Tyndall described his accident in 1869 and the infection of his wound. In the very same room at Belalp where he had lain injured, his wound open to the air, he had this year opened sterile hermetically sealed infusions which then swarmed with germs after two days.

The conclusion was obvious to him. The germs from the air had entered his wound to cause the infection, and he praised the initiative of Lister in introducing antiseptic practices into surgery to combat this problem. He argued that each organism gives rise only to its own type, as Pasteur had shown with ferments. The same was true for disease: 'Sow typhoid virus, your crop is typhoid—cholera, your crop is cholera'. But we now needed to understand how contagious and infectious diseases took root and spread. Taking aim at the anti-vivisectionists he urged: 'while abhorring cruelty of all kinds, while shrinking sympathetically from animal suffering…no greater calamity could befall the human race than the stoppage of experimental enquiry in this direction'.[5] Tyndall rehearsed the history of the subject, praising the contribution of

Burdon-Sanderson for his exploration of splenic fever (anthrax), and brought the name of Koch for the first time before the British public. He described Koch's observation, confirmed by Cohn, of the 'spores' or 'seeds' of the 'contagium' alongside the rods or filaments of the mature organism, and his proof by inoculation into mice, guinea-pigs, and rabbits that these transmitted the fever.

Tyndall's belief that there were 'germs' of bacteria was mistaken (unless one counts a spore as a 'germ'), but that did not affect the substance of his argument or his findings. Knowledge of the mechanisms of these fermentations and diseases was vital, to prevent massive financial loss and to alleviate human and animal suffering. In three years in one district of Russia alone, 56,000 deaths by splenic fever among horses, cows, and sheep, and 528 human deaths had been recorded. Pasteur's discovery of heat treatment to preserve wine, his means of preventing contamination of vinegar, and of avoiding disease in silkworms had saved huge sums of money. Tyndall argued that public education and cooperation was needed to roll back ignorance, now that we understood the essential nature of these diseases and how to help prevent them, if not treat them. Carlyle told him that his article in the *Fortnightly Review* was 'the only thing that he had read for the last six months which gave him any pleasure'.[6]

Louisa returned to London after the lecture, while Tyndall went over to Greenock to catch a steamer to inspect a lighthouse on Arranmore, an island off Donegal on the north-west coast of Ireland. It was a laborious trip, from which he returned to resume his experiments with Cohn flasks. He also returned to continuing disputes over spiritualism, and attended an event at Carpenter's house with Huxley. As Huxley put it: 'I would not trust the Virgin Mary herself if she professed to be a medium. But we shall see what we shall see'.[7] Interest had been stimulated by the arrival in July of the American slate-writing medium Henry Slade. Ray Lankester, present at one of his séances, had snatched the slate from him, finding writing on it when there should have been none. Branding him a fraud and denouncing him to the *Times*, he prosecuted him for obtaining money under false pretences. Tyndall wrote in support. Slade lost the case in October, despite the support of Alfred Russel Wallace as a witness. He was sentenced to three month's imprisonment with hard labour, overturned subsequently on a technicality.

The middle of November saw the departure of Hirst, whose health was failing, for Egypt. He had been given leave of absence from Greenwich for four months, and was accompanied by Tyndall and Debus to Charing Cross station, where Louisa was waiting too. It was difficult for Hirst: 'the leave taking upset me, so tender and so sad it was'.[8] Later on, as Hirst remained unwell, Tyndall's deep friendship led to a generous offer to relieve him of his responsibilities at Greenwich by paying him £500 per annum, the sum he had been receiving at the University of London. Tyndall revealed that he had settled £8,000 on Louisa, leaving her secure, and that he was saving £1,000 annually from his combined income of more than £2,000 from his various salaries, books, and investments. Expecting to work at the Royal Institution for a further ten years, an accurate estimate as it turned out, he would still have accumulated £20,000 at retirement: 'all that Louisa and I need to render us as comfortable as our hearts could desire for the rest of our lives'.[9] Hirst was extremely touched, but graciously declined. Instead he sought an extension of his leave.[10]

* * *

The main focus of Tyndall's work throughout the autumn was on germs and spontaneous generation. It was interspersed with Gas Referee business and three days at the end of November at the South Foreland, comparing the existing Holmes light with the new magneto-electric machines of Gramme and Siemens, which Tyndall thought were impressively powerful, smaller, and much cheaper.[11] Tyndall's first experiments with hay infusions in late September, in which he found that organisms still grew after boiling, supported the results of Roberts and Cohn. He concluded that the germs had not been wholly killed by their boiling. In further sets of similar experiments, with open tubes plugged with cotton wool or with hermetically sealed tubes, results were contradictory. Sometimes organisms grew, supporting Roberts and Cohn, sometimes they did not. The variable results were worrying, but Tyndall could sense a reason, derived from his understanding from Cohn of the existence of bacterial spores. He noticed that the infusions that were sterilized by just 5 minutes' boiling were from fresh hay mown in 1876. The rest were from hay mown in 1875 or earlier. He thought that desiccation might be important. Germs with dried, and desiccated surfaces might not be wetted easily, and thereby resist boiling. So he started experiments with old hay and new hay, both from his father-in-law at Heathfield and with dried hay from a London merchant. The results were confusing, but he did notice that with the dried and undried new hay, though not the old hay, a certain proportion remained sterile after treatment.

Suspicion fell on contamination from the chambers. He had new ones constructed and ran a new set of experiments. Now even the new London hay gave problems. In the previous year, hay infusions of all types had been sterilized by 5 minutes' boiling. This result no longer held. Thinking that substances that could not be desiccated like hay would be unable to resist boiling, he went mushroom-picking with Louisa at Heathfield and made infusions from the fungi. In the closed chamber many tubes remained clear, but some did not. Worse was to come, when infusions of cucumber, beetroot, and other vegetables also succumbed to decay.

Tyndall began to suspect the atmosphere in which he worked. He tried making up tubes in remote rooms, but to no effect. Unlike the previous year, all samples eventually decayed. In December, he managed to devise a system that excluded all but a tiny fraction of air from getting into his tubes. Here he had more success. Though not complete, it confirmed to him that external contamination was the problem. But whether it was just in the Royal Institution, or wider, was unclear. There had been many disease epidemics in London that autumn. He described to Hirst the challenge of the previous two months of experiments: 'The results obtained, and constantly obtained, are entirely discordant with those of last year. Did Bastian know them he would assuredly raise a note of triumph'.[12] But he was confident he would succeed in the end.

Such was the position as Christmas approached, as he juggled experimental demands with social commitments. At the X-Club meeting in December, there was much talk of the extra £4,000 that the government had placed in the hands of the Royal Society following a report of the Devonshire Commission, a large increase in the usual annual £1,000 to support research. He wrote to Hirst, concerned at how this sum would be disbursed: 'A good deal of heartburning is likely to flow from this same gift. It is not one into the need of which we have firmly and naturally grown, so that it will have to be managed instead of healthily assimilated'.[13] Any increase in State funding had to be administered carefully.

Tyndall's new family welcomed him for Christmas, which he spent with Louisa at a snowy Elton Hall near Peterborough, the seat of Louisa's uncle the 5th Earl of Carysfort, her mother's younger brother. Christmas Day was 'dull and gray with occasional spittings of frozen matter, neither snow nor hail but occupying the borderland between both'.[14] The family all went to church, leaving Tyndall in peaceful solitude. The next day he returned south to his cucumber infusions, soon to be threatened by shortage of supply—they had vanished from Covent Garden market and he had to scrounge from friends (Figure 37). Back in London for the New Year, Tyndall was with the Huxleys, as he had been every year for the last twenty-three, with the exception of his American trip. Only Tyndall and Spencer were there of the usual English crowd. There were several Americans, including Alexander Agassiz, who was going on to Edinburgh to inspect the *Challenger* collection of deep-sea samples: HMS *Challenger* had returned from a four-year voyage that effectively laid the foundations of oceanography. Tyndall was amused to hear of the precautions taken to prevent him from being insulted by Tait.[15]

The infusion experiments continued from the beginning of the year, while Louisa recorded 'sweetest little words of love between exhausting the tubes and boiling of infusions'.[16] The experiments were interspersed with public and social engagements: several days in January with the Gas Referees; with Dean Stanley for an event with Grove and the Lockers; dinner for the Huxleys; and a Sunday at High Elms with the Lubbocks.

Needing uncontaminated air, Tyndall sought help from Hooker, who made available a newly built laboratory at Kew. Here Tyndall started with new apparatus in

Fig. 37. Cartoon of Tyndall's germ experiments. Artist unknown, but possibly Huxley.

early January 1877.[17] At last the results of 1875 were repeated. The clear tubes were shown as proof to members of the Royal Institution, 'including many eminent fellows of the Royal Society'. The problem seemed to lie in the contaminated air in the Royal Institution. Tyndall had a shed erected on the roof, disinfected it, and required his assistant to wear new clothes. Again the tubes remained clear, whereas in the laboratory 8 yards from the shed they did not. He commented: 'If anything were needed to illustrate the extraordinary care necessary on the part of physicians and surgeons, both as regards the clothes they wear and the instruments they use, such illustrations are copiously furnished by the facts brought to light in the inquiry'. The problems had coincided with the use of old hay, which had presumably brought the infection with it. Tyndall reasoned that such resistant germs might exist in other places too, asking: 'May not the surgeon have to fight sometimes against enemies like those here described?'.

Tyndall had examined alkalized and sterilized urine samples, finding in support of Pasteur and Roberts, but in opposition to Bastian, that there was no sign of spontaneous generation.[18] *Nature* published a report of the papers by Roberts and Tyndall in February 1877.[19] Bastian was meeting considerable resistance. After a damning referee's report, the Royal Society Council decided in January not to publish a paper he had submitted. He was effectively shut out by Huxley.[20] The argument between Bastian and Pasteur centred round Bastian's claim to the Académie that he had discovered the necessary and sufficient physicochemical conditions for spontaneous generation of certain types of microscopic organisms. The process involved neutralizing urine by using a solution of potassium hydroxide deprived of all germ organisms and exposing the mixture to a temperature of 50°C. Pasteur carried out a series of experiments which he believed showed Bastian wrong on all points.[21] In passing, Tyndall asked Pasteur's advice on the transport of beer to India, as his friend Richard Strachey was reporting enormous losses by deterioration. Pasteur replied that heating bottled beer to 50–55°C would avoid the problem. He despaired that breweries were not following this procedure, even though he had tried to work with them: 'You know very well that for us, men of science and research, it is not the fight for material gain that interests us and drives us to arms…Sometimes I hesitate in the pursuit of my scientific work and I ask myself if it would not be better to devote myself to a complete and absolute triumph of creation of a procedure that will make beer unalterable'.[22]

With the success at Kew, Tyndall felt able to go public with his findings while he took up the challenge of identifying the nature of the contamination and how it could be combatted. So during the weekend of 13th and 14th January, Tyndall and Louisa wrote up some of his paper at Heathfield while the others were at church. It was read to the Royal Society by Huxley on 18th January. The paper summarized his difficulties throughout the autumn, and his identification in the hay of dried and hardened germs that were extremely resistant to boiling.[23] He promised a fuller account in due course. The public version of his paper was given as a Discourse to a crowded house at the Royal Institution the following evening, after which the Kew tubes were inspected, and reported in detail in the *British Medical Journal*.[24]

It was another social time afterwards. For the weekend following the Discourse, the Tyndalls went to Oxford to stay with Jowett. On the Sunday they heard Jowett preach on good manners, and talked alone with him afterwards on Darwinism. It was a revealing conversation. Jowett recorded that Tyndall 'admitted that Darwinism was not distinguishable from Lamarck & said that in the case of man the dying off of barbarous races before civilised ones was a proof of the struggle for existence'.[25] It is not clear whether Tyndall meant by this some Lamarckian concept of pangenesis, which Darwin himself had introduced as an addition to natural selection in his 1868 book *The Variation of Animals and Plants under Domestication*, and which Tyndall had referred to in his Liverpool lecture in 1870. It is conceivable that he imagined a force, inherent in matter itself, that led to complexity and specialization.

Like Huxley, Tyndall had his own reservations about Darwin's mechanism of evolution by natural selection, yet he vigorously defended the theory of evolution itself. When the eminent German physician and biologist Rudolf Virchow gave an influential speech in Munich a few years later challenging the theory on the grounds that it had not been experimentally proved (and that belief in it would lead to the evils of Socialism), Tyndall leapt into action. He penned an article for the *Nineteenth Century* in which, while acknowledging the lack of formal experimental proof, he defended it and its further discussion on the grounds of its 'general harmony with scientific thought', in other words of its broad explanatory power.[26] To Jowett, Tyndall added that both Darwin and Hooker had 'remonstrated' with him on his theological tendencies. He continued that while 'Faraday found God to be a necessity of his nature, Darwin felt no such necessity and he was quite as good a man'.[27] The theologian and Tyndall explored the question of the source of moral behaviour. Tyndall cited Fichte, arguing 'that men will act as strongly from notions of morality as of religion—that "religion only completes their dignity"'. A few weeks later, Knowles called about discussions at the Metaphysical Society as to whether morality was dependent on religion. In response, Tyndall looked out his journals with accounts of the positive influence of him and Hirst on the boys at Queenwood, when the religious teachers had lost all control of them. Their own moral character was all the power they needed, and it was a character that many others respected. The American abolitionist and freethinker, Moncure Conway, recollected: 'It would be impossible to find more affectionate and tender-hearted and benevolent men than Darwin, Huxley, Tyndall, and other eminent unbelievers. Freethinkers have as much devotion as the orthodox, though it is lavished on other human beings'.[28]

On his return from Oxford, Tyndall found a letter from Hooker about the Scientific Instruments Exhibition, and the extensive pressure to which he had been subjected, as President of the Royal Society, to support the creation of a museum based on it. Hooker revealed to Tyndall, who was neutral about the scheme, that he felt he had no choice but to back it after the lobbying by, among many others, De la Rue, Spottiswoode, Siemens, Frankland, Lubbock, and Huxley. But he now thought he had been mistaken, given the lack of interest from the public and from the instrument makers themselves, who were supposed to benefit. In addition, hints of government funding had proved illusory.[29] Tyndall replied that he thought the loan collection had been a success. He thought that a permanent museum might be of value, but that it was best developed slowly like the initiatives at Kew and South Kensington.[30]

Back at the bench, Tyndall tried boiling samples for several hours, finding that 4 hours or more resulted in complete sterilization, even though in 1875 just 5 minutes had sufficed. It appeared that there was a difference between the 'stubborn germs and the soft and sensitive organisms which spring from them'. Tyndall, like Huxley, believed that germs existed in the air. Burdon-Sanderson and Cohn did not, and Bastian used this 'doctrine' to argue for spontaneous generation. Since it would not be expected that all the germs would develop at the same time—some might be drier and harder than others for example—Tyndall saw a means of dealing with them, which he tried for the first time on 1 February. The trick was to kill all the susceptible germs by a first brief heating, then catch those that had resisted but were now developing, with a further brief heating. By experimenting with different intervals between heating, Tyndall was able to sterilize infusions completely. The infusions only developed life when it was inoculated from outside. Tyndall had invented a technique of sterilization by discontinuous heating, later referred to as 'Tyndallisation' as a counterpoint to 'Pasteurisation'. By 14th February he had enough evidence to go public, after an evening listening to the Hunterian Oration at the Royal College of Surgeons given by Sir James Paget in the presence of Prince of Wales. Stamping his priority on the discovery, he wrote immediately to Huxley, who published the letter in the *Proceedings of the Royal Society*.[31]

Pasteur kept Tyndall up to date on his correspondence with Bastian, though Tyndall thought he was giving Bastian's experimental skills too much credibility.[32] Pasteur took some exception to this charge, but acknowledged that he now saw that Bastian lacked good faith. He told Tyndall that his presentation to the Académie in response would pull no punches: 'all will see that Bastian is defeated, very much defeated'.[33] In this speech, on 29th January, Pasteur challenged Bastian to an independent review. Bastian accepted; the Académie set up a Commission to referee the proposed experiments; and Bastian agreed to go to Paris in late July.[34] Pasteur hoped the Royal Society might do likewise, but Tyndall replied that their position was clear, as they had declined to publish Bastian's paper 'On the Deportment of Neutralised Urine' in the *Philosophical Transactions*.[35] He warned Pasteur of his recent finding of germs that could withstand long boiling.

The pre-Easter period was replete with dinners, including at the Chemical Society in honour of Frederick Abel, at Sir James and Lady Paget's—where Tennyson read a large part of *Harold*—and at the Spottiswoodes, where Tyndall showed experiments to the Chinese Minister, Zeng Jize. Later, the Tyndalls were at Spottiswoode's for breakfast with the Emperor of Brazil, Dom Pedro II. They spent Easter at Heathfield, where Lord Claud was in difficulty with his finances; the only horse had to go. There had been differences the previous autumn between Lord Claud and Sir Charles Blunt, who was leasing Heathfield to them, and the lease had to be given up. The Hamiltons' London house at 19 Eaton Square (bought with money from Lady Claud's Trust) was sold, its mortgage repaid, and a house purchased at 83 Portland Place with the residue of £9,500 and £500 from Lord Claud.[36] After Easter, and until they went away to the Alps at the beginning of July, Louisa's impact was felt at the Royal Institution. She was now the châtelaine, hosting dinners which Hirst, Debus, and members of her family attended, as well as Tennyson, the Froudes, and Dean Stanley.

The last Discourse of the year took place on 8th June, given by Tyndall on the subject that had occupied him so much over the previous eight months, 'Researches on the Deportment and Vital Persistence of Putrefactive and Infective Organisms from a Physical Point of View'.[37] Giving a summary of all his findings, including the method of sterilization by discontinuous heating, he ended with a ringing assertion that there was not a shadow of evidence for spontaneous generation: 'the method of nature is that life shall be the issue of antecedent life'. Though he did not mention it, this still left the origin of life itself as a mystery.

Tyndall's paper, which had been read to the Royal Society in May, was published in the *Philosophical Transactions*, with a note in the *Proceedings* critical of Burdon-Sanderson, in which Tyndall took exception to Sanderson's conclusion that bacteria could be borne out of something having no structure in the morphological or anatomical sense, but merely in the molecule or chemical sense; opening a chink of light for those who believed in spontaneous generation.[38] Tyndall emphasized the limiting power of the microscope and the error of assuming that structure that could not be seen in a microscope did not exist. Du Bois-Reymond was impressed by the work, seeing Tyndall in the theory, and Lister in the practice as the key figures.[39] Tyndall informed him that Lister was moving from Edinburgh to King's College London, adding: 'His presence will strengthen the righteous'.[40] In 1880, Tyndall was instrumental in securing a Royal Medal for Lister (and the Copley Medal for Sylvester). As Du Bois-Reymond later put it: 'We say "listern" as a verb to designate the use of the carbol-spray while bandaging a wound. I do not hesitate to proclaim Lister the greatest benefactor of mankind since Jenner's wonderful invention'.[41]

The Tyndalls went to see Carlyle on the eve of their departure for the Alps—'The dear old man seemed much inclined to come with us to Switzerland'—and they left England on 5th July in the company of Louisa's sister Emma.[42] They stayed in Paris at the Grand Hotel de St James, for Emma's sake, finding the bill tremendously high, called on Pasteur and then travelled south in a leisurely manner, staying in Lausanne overnight before going on to Brig. There they requisitioned two mules and four porters for the carry up to the hotel at Belalp. The load included sixty hermetically sealed tubes of various boiled infusions. Some fifty of them survived the journey from England. They opened twenty-three in a shed containing fresh hay and twenty-seven at the edge of a cliff overlooking the glacier, with a gentle wind blowing from the mountains. Tyndall reported to Huxley that after the elapse of some time, twenty-one of the tubes opened in the shed filled with organisms. All those opened above the purer glacial air remained clear.[43]

Their chalet on Alp Lusgen above Belalp should have been ready in mid-July. But snow prevented work before June; there were disputes over contracts and bills; and there was drunkenness among the workmen. While sorting out the workforce for the completion of the chalet, there was other work to do and the social crowd in Belalp to engage. On their arrival they found the Brabazons there, and had an excursion to Zermatt, walking most of the way from Visp, when they heard that the Emperor of Brazil was in residence. They went with the Emperor and Empress onto the glacier below Riffelberg and up to the spectacular viewpoint of Gornergrat.

Bastian followed Tyndall to Switzerland, in mind if not in body, as they exchanged letters in the *Times*.[44] Pasteur reported on Bastian's visit to Paris in mid-July. Bastian had left before the Commission reported, but not before Pasteur had clearly indicated to him the source of his error, that of failing to sterilize his vessels before using them. Bastian undertook to go back to London and do this, admitting in the process that he did not always generate bacteria in his experiments. He saw the production of bacteria as the positive evidence, while Tyndall and Pasteur asserted that the opposite was the case.[45] Effectively, Bastian was defeated; by Tyndall's and Pasteur's experiments and the collective weight of the scientific establishment, represented by Tyndall, Huxley, and their friends. Tyndall's compelling advocacy of the existence of germs in the air or water, too small to be seen by a microscope, yet whose presence could be 'revealed' by his light beams, coupled with his belittling of Bastian's experimental skill compared to his own, had prevailed. Bastian gave up work on the subject for many years, yet continued to maintain in later life his belief in spontaneous generation, alongside evolution.

Tyndall's intervention in the whole debate had been crucial in two respects: his pro-germ pronouncements over dust and typhoid in the early 1870s and his role later in the decade in the general abandonment of theories of spontaneous generation and chemical or non-bacterial theories of disease and sepsis.[46] Did his aggressive manner hinder the speedier take-up of his ideas about germ theory? At the margin, perhaps, though the issues were so complex that there is a virtue in having someone to focus the debate. Nevertheless, Sir William Jenner could say in his Presidential Address to the Clinical Society in 1875, that he refused to dismiss the spontaneous generation of typhoid, despite Tyndall lending the 'weight of his great name' against it, a 'weight which would...be greater...if he had himself studied the subject on which he has, I am sorry to say, addressed the public in a strain calculated to check unprejudiced individual enquiry'.[47] Jenner was by no means alone in his condemnation of Tyndall's style. A biographical sketch declared that he had 'a vehement and almost an arrogant aggressiveness in him which must interfere with the clearness of his views'. It saw him as 'assuredly one of the most impatient of sages, one of the most intolerant of philosophers', and argued that his temper unquestioningly tended to weaken his authority.[48] But Tyndall's efforts at least earned him a foreign doctorate of medicine, from the University of Tübingen. It was an honour of which he was proud, giving it prominence on the title page of his collected papers *Essays on the Floating-Matter of the Air in Relation to Putrefaction and Infection*.[49]

* * *

Tyndall, now well into his fifties, was not in the best of health in the Alps. His hands were shaky for writing, and he felt pressurized by commitments he had made. He had promised James Knowles an article on spontaneous generation for his new monthly magazine *Nineteenth Century*, which he produced after comments from Huxley.[50] He confessed to Huxley that aspects of germs still perplexed him. He saw the spores and the bacteria derived from them as physiologically distinct, but had toned down that claim in the article since Huxley's view was different. In parallel he had promised Morley, editor of the *Fortnightly Review*, the text of his address to be given to the Birmingham and Midland Institute, of which he had accepted the Presidency.

By mid-August though, he was writing to Hirst from 'one of the most charming little rooms you ever saw', looking through an open window along the Aletsch glacier.[51] The chalet was nearly ready. A fortnight later Louisa wrote that they had given up their hotel rooms to Helmholtz and moved into their 'own little Paradise—very rough & unfinished it is but so peaceful & lovely! The first night J & I occupied it entirely by ourselves, the maids room not being ready'.[52] They were able to enjoy their new home for three weeks before returning to England at the end of September.

Tyndall's first significant duty on returning was his lecture on 1st October as President of the Birmingham and Midland Institute, 'Science and Man', published immediately in the *Fortnightly Review* for November.[53] It tackled the whole sweep of natural law, consciousness, religious belief, poetry, and free will. Tyndall argued that the practice of science demanded personal concentration. Those who had influenced it most profoundly, like Faraday, Darwin, Mayer, and Joule, had habitually sought isolation. But it was nevertheless important that the man of science shared the fruits of his labours as far as possible with the public. Tyndall contrasted Boyle's picture of the universe as a machine with Carlyle's as a tree; the one with organization and direction from outside, the other from inside. He sided with Carlyle, arguing that the order and energy of the universe are inherent. He gave a historical picture of the development of scientific knowledge, built continuously on previous discovery like an expanding circle, occasionally punctuated, in an image by Emerson, by some thinker of exceptional power to create the extended field of a new circle. Joule's proof of the mechanical equivalent of heat, embodied in the doctrine of Conservation of Energy, was one such advance, which led inexorably to the conclusion that the animal body was equally a machine; the forces of the human body were identical with those of inorganic nature. But how were these forces put into action? Who or what was it that sent and received the messages, travelling through the nerves at the speed that Helmholtz and Du Bois-Reymond had explored, to move a muscle? Were we not forced to imagine a free human soul? Tyndall answered no, while acknowledging that we did not know the causal connection, if any, between the objective and the subjective, between molecular motions and states of consciousness. We could present to our minds no picture of the process by which consciousness emerges, nor how consciousness could produce motion.

Tyndall thought that ideas of God and soul were changeable. They offered themselves freely to the poet who understood his vocation, finding 'local habitation' for thoughts woven into our subjective life but which refuse to be mechanically defined. But, he asked, are the moral and intellectual processes known to be associated with the brain subject to the same inexorable laws found in physical nature. Is the will of man, in other words, free? Are we, or are we not, the complete masters of the circumstances which create our wishes, motives, and tendencies to action? Not entirely, Tyndall believed. Our inheritance lays down certain constraints, the development of which, after Darwin's *Origin of Species*, now seemed widely accepted even in the clerical world. (Darwin much appreciated the manner in which Tyndall referred to him.)[54]

So if we are not the masters of the circumstances in which motives and wishes originate, in what sense can our actions be said to be the result of free will? Tyndall saw the solution in our moral responsibility, which might be influenced and exercised under the possibility of reward or the threat of punishment for the public good, but was a choice nevertheless. Some claimed that religious sanctions were necessary for moral life. Indeed, the fear that atheism and materialism would lead inevitably to the moral degradation of society was a potent factor behind the response of the religious community to the Belfast Address. Tyndall gave the example in contradiction of George Holyoake, widely seen as an 'atheistic' leader, yet espousing a morality of which no Christian need be ashamed. There was now a growing repugnance, he claimed, to invoke the supernatural in accounting for the phenomena of human life, and a growing drive to seek in the interaction of social forces, including the sense of duty, the development of man's moral nature. For Tyndall, character was not determined by religion, nor could religion make good defects of character.[55]

Tyndall was careful where he talked about religion. A few months after this lecture, the Duke of Northumberland, President of the Royal Institution since 1873, wrote to complain to Spottiswoode, the Secretary, about comments Tyndall had made about the Bible's account of the appearance of humans on Earth. He had described these as 'myths and stories', and added that the theological notions of Count Rumford, of the wisdom of the Creator to provide conditions on Earth for human life, were based on erroneous arguments.[56] Tyndall's response highlights the tightrope he walked between his personal views and his official positions. As he wrote to Spottiswoode: 'The clearly understood law of the Royal Institution is that neither religion nor politics shall be introduced into our lectures...Never in a single instance have I infringed it during the twenty four years of my connexion with the Institution'.[57] The offending passages had been added into *The Forms of Water*, subsequent to his delivery of the Christmas Lectures on the subject, but the notes handed out to those attending, which formed the basis of the book, did not contain the offending remarks. Tyndall told Spottiswoode that when publishing anything controversial he deliberately stripped himself of his Oxford, Cambridge, and Edinburgh degrees, and of all authority derived from his connection with the Royal Institution. Instead, he wrote from the Athenaeum Club. He had not done this on the title-page of *The Forms of Water* as he had reason to believe that the views expressed 'were to all intents and purposes held by many of the most enlightened clergymen of the Church of England'. Within the walls of the Royal Institution, he was obedient to its laws, but claimed intellectual freedom outside. Though even as an outsider, he wrote: 'my desire has been to act the part of a conservative rather than that of a Destructive, by gradually preparing the public mind for inevitable changes which without this preparation might take a revolutionary form'.[58] For this reason, his affiliation with the Royal Institution is also absent from *Fragments of Science*.

In the autumn, Tyndall spent time with Louisa at Heathfield. The germ experiments were largely over, and he had no substantial lecture commitments until the Christmas series. There was space for other public duties. In mid-October, for most of a week, he was with Collinson at Shoeburyness and Dungeness for sound experiments, grabbing a short weekend with Louisa at Spottiswoode's afterwards. He had committed to give a special lecture at the London Institution on 10th December on 'Microscopic organisms, their Genesis and Work in the World'. He was not well—an unspecified illness of several days prevented him and Louisa taking up an invitation to Wilton House from the Earl of Pembroke—and though Huxley offered to take the lecture off his hands, Tyndall wanted to get it out of the way.

Continuing illness interfered with activities before Christmas. He managed Philosophical Club dinners in November and December, and private dinners with Lister and at the Spottiswoodes to meet Ménabréa, the Italian Minister, as well as British Association business and Royal Society elections. But he retired to Heathfield in mid-December, missing the Harrow Governors and Royal Society Club before returning to hear Burdon-Sanderson lecture on septicaemia. On the day of the lecture he sent the Royal Society a short paper 'On Schulze's Mode of Intercepting the Germinal Matter of the Air'.[59] This is a delightful paper illustrating Tyndall's ability to visualize or 'imagine' the behaviour of the germs. Schulze had claimed that air drawn through a tube of sulphuric acid in a particular way resulted in sterilization. Others, including Tyndall, had not always replicated this finding, and Tyndall emphasized the importance of wetting the germs for them to be captured and destroyed. He described separately to Huxley his reason for the differences between the speed of transmission of germs in water and in air. Once wetted, germs could develop or be destroyed with greater ease.

Christmas Day found the Tyndalls in London, surrounded by snow, as he prepared for his Christmas lectures, 'On Heat—Visible and Invisible'. It was the first time he had given them since his marriage. With Louisa as host, he made full use of his rooms for entertaining guests, including Dean Stanley, Spencer, the Pollocks, and Lady Ashburton. The Tyndalls went down to Heathfield after the lectures, where they stayed for nearly a fortnight. Tyndall was tired, and they turned down the invitation from Lady Russell of a weekend at Pembroke Lodge. At Heathfield, Tyndall spent a day on Gas Referee correspondence and a day at sound experiments with Louisa, using guns and pistols from the house-top and tower, giving them 'lots of opportunities for little talks'.[60] More sound experiments followed, requiring Tyndall to spend several days away on the coast.[61]

Marriage brought a new rhythm to Tyndall's life, but it was interrupted by ill-health. In 1878, for the first and only time in his tenure at the Royal Institution, he gave no lecture course. When he was well enough, Louisa energized his social life. Tennyson dined twice with them, with his sons Hallam and Lionel, and the Tyndalls went to Lionel's wedding at the end of the month.[62] Other guests in February included Matthew Arnold, Dean Stanley, the Speddings, the Russells, and the Grant Duffs. Hosts encompassed Siemens, Fitzjames Stephen, Lewes, and Lady Stanhope, and there was an official dinner at the College of Surgeons. It was evident to Louisa just how popular Tyndall was. After a dinner of The Club, Lord Arthur Russell told her that the discussion was entirely owing to her husband, who drew all the elements together into one united conversation. At Lady Stanhope's, with Sir William and Lady Harcourt, Lord and Lady Derby, and Lady Agatha Russell, Louisa enjoyed the party 'and seeing how everyone likes my John'.[63] Lord Lansdowne was disappointed that they could not take up an invitation to Bowood as he wanted to get to know Tyndall properly.

Since mid-January, Tyndall had been plagued concerns about his health. His doctor, Andrew Clark, suggested there might be something wrong with his heart, possibly only muscular, caused by some strain such as coughing. He urged rest. No catching of trains, going up of mountains, no emotion of any kind. Louisa confided to her diary that it was 'a heavy blow. It made us serious and intensely loving for many days'.[64] By March his state was such that Clark advised him against going out to dine with the Tennysons, leaving Louisa to go on her own, and he did not even come down from his rooms for Lord Rayleigh's Discourse on 15th March on 'The Explanation of Certain Acoustical Phenomena', which was a pressing interest. He was not completely incapacitated, attending a meeting of the Gas Referees and hosting General and Mrs Wynne and others to dinner, and in mid-March, the symptoms suddenly disappeared. He still felt low, but Clark agreed that he could lecture after Easter if he was able to have a long rest afterwards. Tyndall asked Spottiswoode if he could find a substitute for his first two lectures.[65] But before he could do so, he had a Discourse to give, on 'Recent Experiments on Fog-Signals'.

In the familiar pattern, Tyndall presented the work at the Royal Society first, on 21st March, and at the Royal Institution the next day.[66] It was a summary of observations and experiments over the past five years. One discovery he highlighted was that gun-cotton rockets could act as powerful fog signals. He suggested that they might also be an effective means of naval communication, because they could be fired from a ship. Drawing attention to the aerial echoes regularly observed, he explained that they were not caused, as some suggested, by reflection from sea waves, since they were produced when the sea was flat. They came instead from the atmosphere, in the direction from which the sound was emitted, and Tyndall believed they came from meeting air at different temperatures. He challenged Joseph Henry's conclusion that because no echoes were heard when a siren was pointed upwards, aerial echoes did not exist. Tyndall observed that air would tend either to ascend or descend directly, depending on the difference between air and sea temperatures, so that the sound going vertically would not be crossing the streams and was, therefore, not reflected. Tyndall modelled this in the laboratory with his sensitive flame, stimulated by a sounding reed. Gas jets lit between them, to disturb the air, immediately stilled the flame, as the aerial reflection caused acoustic opacity. But if the flame was placed behind the reed and the jets lit in front, it was immediately stimulated, imitating the effect of aerial echoes.

A few days after the Discourse, the Tyndalls went for three weeks to Farringford, Tennyson's house on the Isle of Wight, which he had lent them. Tyndall was in need of rest, suffering from the sleeplessness that plagued him for years. In 1875 he had written to Olga Novikoff, saying that if nature had given him the capacity of resting his brain by sleep 'I might have made my mark in the world. But as it is I am like a climber in the mountains who has to depend upon a broken leg. At rare intervals I feel what I might have been had the power that rules human destinies been propitious to me'.[67] But he could not entirely let go of his commitments, although on doctor's orders he gave up the lectures and missed the monthly meeting of the X-Club. His examination duties at South Kensington were picked up by Debus. Tyndall declined to accept any fee for his 'poor services', though Debus insisted on sending Tyndall a cheque for £41.[68] The weather at the beginning of their visit to the Isle of Wight was sharp, with snow, hail, and a cutting wind. By early April it had turned: 'Sun, sky, trees, flowers and birds conspire to make the place lovely'.[69] He corrected the proof of his paper on fog signals with Louisa, and a new edition of *Fragments of Science*. He was glad to hear from Spottiswoode of the success of James Dewar's Discourse on 29th March, writing: 'He is a genuine man; and has had to endure a good deal at the hands of men who are the reverse of genuine'.[70] Dewar had joined the Royal Institution as Professor of Chemistry the previous year. Tyndall later described him as 'thoroughly honest, but a thoroughly combative Scotchman', not long before they fell out.[71]

The Tyndalls returned just before Easter, catching a fleeting glimpse at Brockenhurst station of the Tennysons going the other way. His intermittent pulse had recurred during his stay at Farringford, and he sought further breaks from London. Hirst described him as looking 'haggard and very old, and subdued'.[72] On Good Friday, Tyndall went to High Elms for several days before going on to the Siemens at Sherwood. He wanted if possible to hear Spottiswoode's Discourse on 3rd May and to attend Jessie Huxley's wedding the next day, as well as a dinner at the Royal Academy. In the event he had to give them all up, again missing the X-Club, apologizing to Huxley that Louisa would have to attend the wedding without him (he later stood godfather to Jessie's son Noel).[73] Several of the Huxley children (there were seven surviving children at this point) went down with diphtheria shortly after the wedding. Tyndall, now in Folkestone with Louisa for a week, and very anxious for them, offered two bedrooms at the Royal Institution for the unaffected children.[74] Huxley urged him not to go to Dublin for the British Association, writing: 'If you don't I declare I will try if I have enough influence with the Council to get you turned out of your office of Lecturer and superseded'.[75] Spottiswoode was President, and had asked him to give the Working

Men's Lecture. Tyndall had agreed, though it would substantially reduce his time in the Alps, but suggested that his welcome might be less than enthusiastic: 'you know that the Archbishops, Bishops & Clergy of Ireland have banished me from their soil with bell, book & candle-light, as St. Patrick banished the snakes & toads'.[76]

The need for Tyndall to regather his strength led to an early departure for the Alps on 5th June; he missed the X-Club meeting the next day for the third time in a row. He would be away for nearly four months. On arrival at Brig, he walked slowly up to Belalp, an ascent of 5,000ft, without noticeable effect. After a week he found himself improved in health and strength, but it was a temporary recovery.[77] Though the chalet, with its roaring pine fire, was comfortable, his ill-health returned. Completing the chalet was not straightforward, hampered by an unusual amount of snow and broken promises from his builder, who he dismissed and replaced by a rival from Brig, risking legal action. But by 8th July, the carpets were being laid throughout and Tyndall was also rallying, though he had given up the idea of going to Dublin.[78]

By mid-July he was feeling well enough for an expedition with Louisa up the glacier towards the Aletschorn, and the following day they climbed the Sparrenhorn. Smoking before going to bed seemed to be helpful, enabling him to dispense with his customary brandy and water. They even gave two dinner parties, Lecky being the guest at the first: 'We had soup, fish, choice mutton, roast chicken, a delicious salad, and three or four different kinds of wine. Louisa works like a slave, or rather like an angel. She has a great deal to do in educating our native servant, who though quick and intelligent, has no notion of our habits of neatness and cleanliness essential to the comfort of an English household'. He continued: 'Louisa and I have renewed our marriage contract for a thousand years. She wished to make it a million, but I thought it just possible that I might be tired of her at the end of a thousand, so refused to extend the time'.[79]

At the beginning of August, they were still in the hands of masons, carpenters, and painters, finishing off work that should have been completed the previous year. Yet they were comfortably installed in front of their fire, as patches of inclement weather alternated with stretches of a week or more of fine conditions. The irregular pulse had subsided, and to add to the long mountain days, Tyndall decided to build a path from the chalet down to the hotel, from where their supplies came. A fierce thunderstorm swept away most the gravel of the first attempt, necessitating repairs and much stone-cutting. Like the previous year, Tyndall had taken out a hundred hermetically sealed flasks of cucumber and turnip infusions, of which eighty survived the journey. This time the stimulus was a paper by Downes and Blunt, who argued that light, especially the 'actinic' rays, could destroy or arrest bacterial growth, though their experiments were not consistent.[80] Tyndall sent his 'Note on the Influence Exercised by Light on Organic Infusions' to the Royal Society just before Christmas.[81] He had found no effect.

Snow fell copiously at the end of September as the Tyndalls planned to return for the wedding of Louisa's sister Mary. One last sound experiment necessitated Tyndall walking nearly a mile from the chalet in thick mist, while Louisa sat in the snow and rang a bell.[82] It is an image that encapsulates the companionship of their marriage.

23

⟡

ELECTRIC LIGHTS AND MINING ACCIDENTS

1879–1886

Spottiswoode succeeded Hooker as President of the Royal Society on 30th November 1878, thereby finding himself 'at the same time President of the great travelling Congress, and of the old Stationary Palladium of British Science'.[1] He resigned as Secretary of the Royal Institution in consequence. Before doing so, he suggested that Tyndall give a Discourse on 'The Electric Light', which was delivered on 17th January 1879.[2]

Spottiswoode had a stake—he was chairman of the English Edison Company—but the topic was of wide public interest. Edison's incandescent light, promised since 1878 but not to be delivered until 1881, sought to compete with the Jablochkoff candle (an arc lamp). Both were in competition with gas lighting for public spaces and potentially in homes.[3]

Louisa's opinion of the Discourse may have been biased—'my darling quite surpassed himself'—but the first evening was so popular that Hirst failed to obtain a seat, attending a hastily-organized second performance on the Monday. As the *Saturday Review* put it, quite apart from Tyndall's ability to draw an audience, the topic of the electric light was like that of the telephone, the general public suddenly brought face to face with a scientific discovery brilliant with mingled promise and performance.[4] Tyndall described witnessing Ruhmkorff's machine for generating steady current in 1853 with Biot and Magnus, and the advice of Faraday towards the end of 1859 about the introduction of the magneto-electric light in lighthouses. He referred to his own report to Trinity House of May 1868, recommending its further use, and the rush of developments to make electric lights for different places and purposes. Tyndall was of the opinion that the electric light would finally triumph over gas, though he was 'not so sure that it will do so in our private homes'. Anxious to avoid influencing the share market, he confined himself to 'a general statement of opinion'. (Shares in gas companies had plunged in September 1878 after Edison's claim that he was close to a cheap electric lighting system for homes.) He praised the instrument-makers he had known and worked with—Darker and Becker in England, and Sauerwald and others on the Continent—those who by mechanical ingenuity could 'turn to a special account facts and principles already known to the scientific man'. With a model, he illustrated Edison's 'alleged mode of electric illumination', a design for the circulation of electric current in a city, which he imagined like the circulation of blood, branching for use and being returned eventually to the beginning.

The end of this Discourse is revealing about Tyndall's views on the commercialization of scientific discoveries. He contrasted the investigator and discoverer, whose object is scientific and who cares little for practical ends, with the 'mechanician', whose object is mainly industrial: 'It would be easy, and in many cases probably true to say, that the one wants to gain knowledge, while the other wants to make money; but I am persuaded that the mechanician not infrequently merges the hope of profit in the love of his work. Members of each of these classes are sometimes scornful towards those of the other'. Yet he was aware of the impact of the applications of science on science itself, for example in improving instruments to extend scientific observations, and he was explicit that 'the amelioration of the community is also an object worthy of the best efforts of the human brain'. But he argued that the nation must bear in mind that these fruits derive from 'long antecedent labours', driven 'by the operation of a purely intellectual stimulus'. He mentioned Gauss and Weber developing the electric telegraph to communicate between the observatory and the laboratory in Göttingen, an intriguing forerunner of the development of the internet at CERN for scientific communication. He declared: 'There exists no category of science to which the name applied science could be given. We have science and the applications of science, which are united as the fruit is to the tree'. And the researcher was not to be managed or directed: 'We have amongst us a small cohort of social regenerators—men of high thoughts and aspirations—who would place the operations of the scientific mind under the control of a hierarchy which should dictate to the man of science the course that he ought to pursue. How this hierarchy is to get its wisdom they do not explain. They decry and denounce scientific theories; they scorn all reference to ether, and atoms, and molecules, as subjects lying far apart from the world's needs; and yet such ultra-sensible conceptions are often the spur to the greatest discoveries'.

Given the popularity of this Discourse, Tyndall managed to squeeze it into the extended and revised sixth edition of *Fragments of Science*, published at the end of June. The *Saturday Review*'s recognition of the public interest in the electric light and the telephone was highlighted a few months later—after his course of eight weekly lectures on 'Sound'—when Tyndall gave a demonstration of Edison's new invention of the loud-speaking telephone, or phonograph, through a wire between the Royal Institution and Piccadilly Circus.[5] Poems were read and repeated back, and songs sung from Piccadilly Circus to Albemarle Street, clearly heard by the audience in the Royal Institution. William Preece had given a Discourse on the remarkable invention of the telephone itself in early 1878, and an early demonstration of the phonograph (later called the gramophone). In the presence of Tennyson, Tyndall spoke 'Come into the garden, Maud', which was recorded and played back to the audience.

The invention of the phonograph caused enormous interest. Du Bois-Reymond only realized it was not a hoax when he saw a photograph in the *Graphic* of Tyndall turning the handle (Figure 38).[6] Du Bois-Reymond himself was an early adopter of the telephone, writing: 'I am going to establish telephone-wires in my Laboratory instead of the intended speaking-tubes, which are sometimes so very difficult to lead through an intricate building'.[7] The telephone soon became the subject of a patent dispute when the Post Office authorities proposed to sweep it into their net, arguing that it was covered by patents for the electric telegraph. Tyndall was lobbied by Lord Anson in early 1880 about what he saw as the injustice of the Postal Telegraph

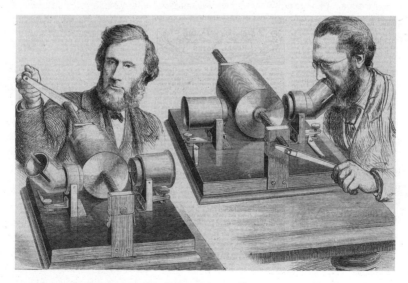

Fig. 38. Tyndall operates the phonograph as William Preece speaks, *The Graphic*, 16th March 1878.

department towards the Edison Telephone Company of which he was a Director. Tyndall promised to see Mr White, the Manager, after lectures, though he refused to take any active part, in what Louisa described as 'his own, high beautiful way'. The issue did not go away. A few weeks later, Tyndall was persuaded to sign an affadavit that the absence of his name after his lecture on the subject did not mean he was against the Edison Telephone Company. He rarely intervened in such matters, but was sufficiently exercised despite, as he put it to Helmholtz, 'My repugnance to taking an oath; my dislike of patents; and my disapproval of Mr Edison's mode of bringing his inventions forward', to write to Du Bois-Reymond and Helmholtz for support for his argument that this was unfair, a view shared by Thomson, Stokes, Rayleigh, and others.[8] Du Bois-Reymond agreed, and expressed himself 'rather astonished at a proceeding more in accordance with continental despotism and protective spirit, than with the institutions of a free country'.[9]

* * *

A month after the Discourse, Tyndall's official duties increased when he was appointed to the Royal Commission on Accidents in Mines. Mines were dangerous places in the 1870s: around a thousand miners were killed each year. Although the number of deaths had remained relatively constant while employment and tons of coal extracted doubled, the effect on public opinion could not be ignored. Not least was the impact of major disasters. In the few months leading up to the convening of the Royal Commission, 267 people had been killed at Abercarne, Monmouthshire, on 11th September 1878; 63 at Dinas, Glamorganshire, on 13th January 1879; and 62 at Leycett, Staffordshire, on 21st January 1879. The Commission's job was to inquire whether 'the resources of science furnish any practicable expedients not now in use, and are calculated to prevent the occurrence of accidents, or limit their disastrous consequences'.[10]

The government turned to its men of science to help. Of the nine members appointed, five were Fellows of the Royal Society. Three represented the employers. One of these was the remarkable Sir George Elliot, who began work as a pit boy at the age of nine, became a union leader during a strike in 1831, a pit manager and eventually owner of many mines; the archetype of Victorian social mobility. The ninth, representing the employed, was the equally remarkable Thomas Burt MP, trade unionist, and one of the first working-class Members of Parliament. The chairman was Warington Smyth, Tyndall's erstwhile colleague at the School of Mines. They were joined by Frederick Abel, the leading British expert on explosives, who later invented cordite with James Dewar.[11]

The Commission sat for the first time on 31st March. Tyndall probed their first witness, Joseph Dickinson, an inspector of mines in Lancashire, about whether there was any evidence of association between atmospheric pressure and the occurrence of explosions, about the use of naked lights and illumination, and the causes of fatalities apart from explosions. He attended the next ten sessions, up to 10th June 1879, but was an infrequent attendee thereafter, managing six of the remaining eighteen meetings, largely owing to his absence in the Alps in the summers of 1879 and 1880.

This was a group of men determined to get to the bottom of what could be done. In the first phase of work they interviewed sixty-nine witnesses, including mine inspectors, colliery managers, mining engineers, and the eminent William Siemens, asking 14,431 questions in twenty-nine sessions. Their examination was exhaustive, from collapses of walls and roofs (the most common cause of deaths) to the emission of fire-damp (methane) and other gases, explosions, the influence of coal dust, methods of illumination, ventilation, blasting, signalling, and life-saving after accidents. They visited twenty mines to observe the situation at first hand. Tyndall looked forward to mine visits, on which he was occasionally accompanied by the intrepid Louisa. He told Tyndale: 'Louisa, who purposes accompanying me, is already beginning to prepare her Collier's clothes'.[12]

Tyndall asked fewer questions than most Commissioners, though reading the transcript gives a direct feel for how he actually spoke; quite similar to his writing in letters, articles, and books. He was incisive, to the point, and patient without being condescending, recognizing the practical experience of the men before him. He pressed witnesses on the physics and chemistry of the phenomena they described, probing evidence of any relationships between changes in atmospheric pressure and explosions, the pressures of gases in coal, the effect of dust on explosions, and friction as a cause of fires. Many of his questions were about lighting. Some explored the possibility of electric illumination; in an aside during one session, Elliot thanked him for all he was trying to do to benefit mankind in this respect and offered him any access to his mines that he might need for experiments. Other questions concerned the subject of safety lamps, which became a major preoccupation of the Commission. Tyndall also saw the education of miners as critical, and asked about its provision: 'When you speak of the education of miners, you mean, of course, a practical education, which could not be got upon the surface. But do you think it would be at all advisable to give them experimental instruction as to the gases which they have the most cause to dread?'. He challenged witnesses who did not have systematic experimental evidence for their views, asking one: 'But for inferences so grave, do you not think that irregular observations are out of place?'.

The Commission published an interim report of 539 pages in 1881, summarizing the evidence they had heard and seen. Their findings pointed to the need for extensive further experiments which would take some time: on questions such as the pressures of gas in strata; the use of different safety lamps in diverse gas and colliery conditions; the development of a more sensitive gas detector than the Davy lamp; on coal dust (Faraday and Lyell had first suggested its significance for explosions in a report to the Home Secretary in 1845); on electric lighting; and on methods of bringing down coal without using sparks or flame. The two main causes of death in mines were falls of the roof and walls, and explosions of fire-damp. Both were intimately connected with the illumination of workings, hence the Commission's emphasis on safety lamps and the possibilities of electric lighting.

In the second phase of work, the Commission held just one evidence session, on 28th November 1882, interviewing five witnesses predominantly about safety lamps. The Commission's priority was experimental work, much of which related to the lamps; more than half their final 401-page report was devoted to experiments on 250 different lamps, including the well-known brands of Davy, Clanny, Stephenson (the 'Geordie'), Mueseler, and Teale. Other lines of experiment included the extent of flame projection in blown-out powder-shots in the presence and absence of coal dust (under Abel at the Royal Engineers in Chatham), the use of lime cartridge and patent wedges to bring down coal without explosions, and the use of Edward Liveing's gas indicator to detect gases to which safety lamps were not sensitive. William Galloway, whom Tyndall had heard read a paper at the Royal Society on coal dust explosions, was invited to experiment with using water in conjunction with explosives and to trial water cartridges. The aim was to reduce sparking and keep the temperature below that at which gases could combust and explode.

These experiments took time, to the extent that a question was asked in Parliament in 1884 as to why the report was not yet published. The Home Secretary, Sir William Harcourt, replied that he hoped it would be presented before the end of the session.[13] But it was not until 15th March 1886, after more than a year of drafting and editing, that the report was finished. Tyndall was not able to attend the final meetings, having suffered a breakdown after his Discourse on Thomas Young. He had worked hard to try to make the report readable: 'It is a most cumbersome document. Tried to lighten its heaviness mainly by diminishing its verbosity'.[14] He was not optimistic about its fate: 'I fear that notwithstanding all our labours the Report will be a lame affair. As I told the Trinity House some years ago, I entertain a certain contempt for committees and commissions. Problems of moment are solved by individual investigation'.[15] This is a view he held throughout his life; first expressed in a letter to Faraday in 1851.[16]

Solutions to problems might be suggested by individual investigation, but putting the solutions into practice and changing laws requires politics. Following the report, Lord Salisbury's Conservative Government passed the Coal Mines Regulation Act in 1887. It tightened up employment and safety practices across the board. No boy younger than twelve or girl, or woman of any age, was to be allowed below ground (previously several hundred boys aged between ten and twelve had been employed, especially to mine thin seams). No boy, girl, or woman was to move railway waggons, and animals had to be able to pass without rubbing the roof. Many other rules were introduced about hours and breaks, and wages were not to be paid at a public house. The scientific evidence led to extensive modification of the thirty-one General Rules in force and the addition of eight new ones, particularly concerning safety lamps. No explosive substance was to be stored in a mine. The work of the Commission proved to be a positive step in the decades-long process of changing societal norms towards employment and safety practices. For the most active members of the Commission, the commitment was substantial: 217 meetings were held in London or in colliery districts between their first meeting on 27th February 1879 and 27th March 1886. Most members attended well over a hundred, although Tyndall managed just fifty-seven, beating only Sir George Elliot (fifty-four) and the Earl of Crawford and Balcarres (eleven). The chairman, Warington Smyth, attended 213. He was rewarded with a knighthood.

* * *

Much earlier than this, not long after the Discourse on the Electric Light in 1879, a Select Committee of the House of Commons was set up to examine whether Municipal Corporations should be authorized to adopt schemes for electric lighting.[17] Lyon Playfair was appointed chairman. He asked if Tyndall would help to instruct the Committee 'on the principles by which the mechanical motion of a machine is transformed into light—and no one can put this more clearly than you'.[18] Tyndall was the first witness and gave evidence on 25th April 1879. Playfair guided him through a series of questions—which he had asked Tyndall to suggest—in which Tyndall replayed his discourse on the workings of the electric light, complete with props of magnets, coils, and diagrams. Tyndall's evidence also touched on the aesthetics of the Jablochkoff arc lighting, and comparisons with gas. The Committee could not come to a view on the relative economy of gas and electricity for lighting, but recognized that the technology would develop and that electricity might also be used as a source of power. They recommended that Municipal Corporations had powers to use the electric light for public illumination, and that they would, as need arose, have powers for future development. The subsequent Electric Lighting Act of 1882 allowed public electricity supply.

The Tyndalls planned to set off for the Alps in early June 1879, and Louisa scoured the Cooperative Stores for the many items they would not be able to obtain out in Switzerland. The journey was becoming easier: 'It is wonderfully convenient to be able to start from London at 7.40 A.M. to reach Paris at 6.5 P.M., extricate luggage, and then drive straight to the Lyons station, travel all night and reach Lausanne at 9.25 in the morning. The same afternoon, between four and five, we reach the little town of Brieg in the valley of the Rhone, which is at the foot of our own particular mountain'.[19]

By early July they were well ensconced on Alp Lusgen, enjoying lovely weather as the snows melted, snow in which Louisa buried their bacon and butter to 'stultify' the bacteria. Tyndall wrote to his mother-in-law with news of them walking across the glacier and up to Rieder Alp, seeing the wild rhododendron for the first time that year. Louisa built stone cairns to aid their return. Tyndall awaited the arrival of his books from Brig, not least to complete the revision of *Heat*, last printed in 1875 and which had been out of print, causing him a loss of several hundred pounds; the sixth edition appeared in 1880. He read a paper of Heinrich Buff's which contested his results on radiant heat. He thought Buff 'a good man, but not up to so difficult a piece of work', though he took it sufficiently seriously to pen a response for the Royal Society.[20] He told Lady Claud that they were reading *Records of a Girlhood* by Fanny Kemble, the famous actress: 'By the way I met her once here at the Bel Alp, and have regretted since that I did not seek to establish a closer acquaintance with her. She is a most remarkable woman'.[21] He described the furniture in the chalet to Hirst, and Louisa's appropriation for the 'Salon' of a walnut writing table intended for his study: 'So that while she wallows in literary luxury I find myself at the present moment in front of a blotched and spotted deal table furnished with a single drawer'.

For most of June and July, the carpenters were busy, creating two extra rooms in the loft, used the previous year for storing firewood, which Tyndall hoped, though not with great conviction, that Hirst would occupy. (He never did.)[22] After the fine conditions in June, the weather deteriorated for a while, with a diminished number of visitors to Belalp and news that hotels in Zermatt had sent away some of their servants owing to the lack of work.[23] But it then turned glorious from the middle of August until early September. Lord Claud Hamilton came to stay for nearly a fort-night, and was taken on vigorous excursions each day by his daughter and Tyndall. Eventually the extended break was up, as Tyndall planned to return in early October for a visit to Wales with the Accidents in Mines Commission.

From 1879 to 1880, Tyndall served on the Royal Society's Council for the fourth time, when he was also a Vice-President. He never served as one of the senior officers, although Hooker, Spottiswoode, and Huxley all became President. Tyndall's temperament was perhaps not suited to that office, but the possibility was mooted in 1883 on the death of Spottiswoode. Huxley approached him, even though he agreed with Hooker that Tyndall lacked judgement and had made so many enemies that he might lose an election. Tyndall replied that he would see himself 'd—d' first, so it never came to the test.[24] He was an individualist by nature, not an institutional politician or diplomat, but he did use his position on Council to advance the interests of friends and colleagues.

One significant occasion was the earlier appointment of Spottiswoode as Treasurer in 1870, a post he held until he became President in 1878. The position had become vacant when W. A. Miller died shortly after the British Association meeting in Liverpool. He had arrived there in an unstable mental state in which he considered himself, as Hirst put it, 'ordained to come to Liverpool to combat the Heresy of Tyndall and Huxley. He grew wilder and wilder and Dr Inman his physician found it necessary to have him put under restraint'.[25] Huxley immediately suggested Lubbock as his successor, and approached him by letter. But Tyndall, backed by Hirst, doubted Lubbock's ability to be able to commit fully to the role and proposed Spottiswoode. Gassiot also coveted the post, and Tyndall wrote an honest but not entirely diplomatic letter to him, to explain his support for Spottiswoode as a younger man rather than Gassiot—now seventy-two—who he had initially understood would not accept the post in any case.[26] Gassiot replied in a hurt manner, perhaps misunderstanding that Tyndall was acting so as not to go behind his back.[27] In the event, Sabine, the President, proposed Gassiot while Tyndall proposed Spottiswoode, in opposition to two of his important early patrons. He sweetened the pill for Sabine by suggesting that, as a previous Treasurer, he would be ideally placed to instruct the younger man in the role. The votes were equal, and the decision was postponed for a week, when Spottiswoode prevailed by ten votes to six after an X-Club meeting. Gassiot was bitter, both at Tyndall and Huxley, and made his views publically known, though it did not cause a long-standing rift. Tyndall stayed with Gassiot on the Isle of Wight in 1875, altercations over the Treasuryship long forgotten.

Medals were a further opportunity for influence. So it was perhaps not entirely by chance that Clausius was awarded the Copley Medal in 1879, for which he thanked Tyndall, whom he suspected of having a hand in the decision.[28] But Tyndall did not always prevail. In 1857, though he had carried Frankland for a Royal Medal, his proposal of Bunsen for the Copley Medal was not successful (though Bunsen was awarded it in 1860).[29] Nor did he succeed in 1873, when he urged Huxley to support his friend Du Bois-Reymond, writing: 'He has never had the slightest recognition from the Royal Society, though I question whether another man of his rank and merit exists who has failed to receive such recognition'.[30] Helmholtz was given the Copley Medal that year, and Du Bois-Reymond was never awarded it.

This year it was Tyndall's turn to give the six Christmas Lectures, on the subject of 'Water and Air', ending on 8th January 1880, and followed by the first X-Club meeting of the year. He managed them all, but he was again not well, the start of another period of ill-health that lasted many weeks. Even so, it did not prevent official engagements for the Harrow Governors, Trinity House, and the Royal Commission on Mines, nor a lecture on electricity to the Working Men's College in February. The event was so crowded that Thomas Hughes, who was to have presided, could not get in.[31]

Since his time in Marburg, Tyndall had admired the works of Goethe. But there was one exception, and Tyndall devoted a Discourse in March to 'Goethe's "Farbenlehre"', his 'Theory of Colours'.[32] He sent the text to Carlyle in advance, hoping it would meet with his approval, not least since Carlyle had presented him with a copy of the work sent to him by Goethe himself.[33] As he mentioned at the beginning of the Discourse, it was through the writings of Carlyle that his 'reverence for the poet had been awakened'. But in the *Farbenlehre*, Goethe seemed 'wrong in his intellectual, and perverse in his moral judgements', and Tyndall had put it aside until Carlyle gave him the copy and asked him to examine it. He found it a difficult read. Goethe's thoughts were 'strange to a scientific man', although Goethe considered it one of his most important works. Based on extensive experiments, and touching on questions of physiology, philosophy, mathematics, natural history, aesthetics, and sensuous and moral effects, it nevertheless lacked, in Tyndall's view, any clear scientific definition. Goethe had stigmatized Newton's approach as pedantry, and sought to overthrow his ideas. For Goethe, colour was concerned with the mixing of darkness and light, and however many physicists explained to him that his results could be understood according to Newton's theory, he refused to believe them.

Tyndall saw Goethe's approach as mistaken scientifically, 'trying to make the outcome of theory its foundation', though he had understood the importance of a turbid medium for the production of colour. Nevertheless, in holding fast to this, and rejecting Newton's idea that white light was a composite of different colours, he was profoundly mistaken. One valid challenge to Newton he did have, when Newton claimed that achromatic lenses could not be produced, refraction being always accompanied by colour. Though this had been shown to be wrong by Dolland, and did not overturn Newton's theory itself, Goethe pressed it at every opportunity. Indeed, as Tyndall pointed out, Newton's 'theory' of the constitution of light is in reality no theory but a demonstrable fact, depending for its explanation on either the particle or wave theories of light.

The essence of the disagreement was one of metaphysics. Goethe claimed that though phenomena may be observed and classified, deductions must be drawn by every individual for themselves: 'in knowledge, as in action, says Goethe, prejudice decides all, and prejudice, as its name indicates, is judgement prior to investigation'. Tyndall used this to assert the existence, apart from the individual, of an objective truth for science and the failure of Goethe to recognize it. He thought that Goethe should have stuck to poetry. But even then, he thought, one aberration could not detract from his many other achievements. The episode exemplified the protests made against the cold mechanical model of dealing with aesthetic phenomena employed by scientific men: 'the emotions of man are older than his understanding, and the poet who brightens, purifies and exalts these emotions may claim a position in the world at least as high and as well assured as the man of science'.

After Easter, Tyndall gave a course of six weekly lectures at the Royal Institution on 'Light as a Mode of Motion', and was soon off to the Alps again. The Tyndalls arrived on Alp Lusgen on 11th June, for a sociable season. For the last three weeks in July, the weather, which had been inclement since their arrival, turned sunny and fine, though Tyndall could not make full use of it, as he was laid up with a foot problem which an American doctor treated. The injury prevented him from accompanying the Marquis of Salisbury's family, who were staying at the hotel, onto the glacier.[34] It did give him time to prepare his lecture for the Glasgow Sunday Society in October, to which, in retrospect, he wished he had not committed himself. As with all his special lectures, it weighed heavily on his mind.[35] Pollock wrote to say that they hoped to be able to stay with the Tyndalls in early August unless a Bill on the law of partnership required him to be in London. He was also embroiled, at Tyndall's instigation, in a dispute between Lady Claud Hamilton and her brother, the fifth Earl of Carysfort, about an inheritance. It had started the previous summer, as letters and telegrams flew between Belalp and London. Tyndall tried unsuccessfully to broker a rapprochement, but relevant papers were found too late in the office of the Earl's solicitor to prevent a legal case.[36] Pollock had to report: 'Our move in Hamilton v. Carysfort has come to nothing…The case will have to be fought out in due form'.[37] In the end it subsided amicably, and family relationships were not ruptured.

The Pollocks were eventually able to come out in mid-August, staying at the same time as Maysie Ashburton and her maid.[38] Pollock brought his ice-axe, though with no great expectation of using it after seven years' idleness. Lubbock, still devastated by the loss of his wife the previous year, also enjoyed their hospitality at the end of August.[39] Nelly Lubbock had died unexpectedly, and none of her friends was able see her before she was taken away. She was one of the remarkable free spirits of Tyndall's circle. Too free for the staid Hirst, who destroyed her letters to him because they were so outspoken. He regarded her as racy, wayward, and capricious: 'There was no-one there strong enough even to temper or rather check her caprice'.[40] Few letters from her to Tyndall survive, though she once told Hirst that he was the dearest friend she had in the world, excepting her husband.

Lubbock eventually married again, but his house was never the same for his friends. The same fate befell Hooker. Frances Hooker died a month after Tyndall's Belfast Address, in the arms of twenty-one-year-old William, her eldest son, while his father was away in London. Tyndall wrote to Mary Egerton: 'it was a fearful trial for the child, for there was a good deal of convulsion. She died in two hours'.[41] Harriet Hooker asked Tyndall to spend the subsequent Christmas with them, writing: 'we will try and not make it a very dismal time, so please will you come. Papa would like it so much'.[42] Tyndall gladly did as he was asked. Eighteen months later, Tyndall was one of the first to be told by the botanist Hooker of his impending remarriage, to the aptly named Hyacinth, widow of Sir William Jardine. Hooker's joy was not unalloyed, as he wrote: 'My house never can be to myself or friends what it was for so long and so rapturously. I can only hope that you, my dear old friend and yours will not find it less a welcome one, however less attractive'.[43]

The Tyndalls finally left Alp Lusgen on 6th October, missing the British Association meeting in Swansea at which Ramsay presided—the *Saturday Review* noted that his Address covered unexceptionable geology, after Tyndall's 'Belfast harangue of 6 years' previously'—but he was back in time for the first X-Club meeting of the season.

The period up to Christmas 1880 was busy with social and official engagements. A weekend in November found the Tyndalls at High Elms with the Lubbocks, going down in the train with Sir John, the Birkbecks, and Lady Wade. Lubbock had been defeated at Maidstone in the General Election of 1880, but elected soon afterwards for the University of London. This election resulted in a landslide for the Liberals, following Gladstone's Midlothian Campaign speeches, and Gladstone became Prime Minister for the second time, replacing Disraeli. A week after the High Elms visit the Tyndalls hosted Frederick Pollock, Fitzames Stephen and his daughter, and Spedding and his daughter, dining on soup Julienne, turbot with Dutch sauce, saddle of mutton, turkey boiled with celery sauce, widgeon with watercress, apple tart, and chocolate pudding. The same menu entertained the Galtons, de la Rues, and Birkbecks later in December, after they had hosted the Miss Moores, Mary Adair, Spencer, Frederick Pollock, and Louisa's father. They spent a weekend in December at Foxwarren with the widowed Emily Buxton, the eldest daughter of Sir Henry Holland, and other guests including Lady Herschel, all of whom apart from the Tyndalls went to church on the Sunday.

On his official duties, Tyndall spent a whole day in December 1880 in London for the Gas Referees, afterwards dining with his father-in-law at the Telegraph Engineers. A meeting of the Gas Referees a few days later was followed by him taking the chair at the Society of Arts for a paper by Price Edwards on the use of sound for signals.

The *Daily Telegraph* reported the event, including mention of Tyndall's reference to wind and clouds as a great hindrance to fog signals, and the need to find signals that could enable differentiation between lighthouses. This prompted Major-General Henry Babbage, son of Charles Babbage, to write to Tyndall about his father's invention in 1857 of the occulting light as a means of achieving this, and his intention to publish a Memoir of his father, including his treatment at the hands of the Trinity Brethren.[44] Tyndall recalled Babbage showing him the light in his garden, and looked forward to his account 'of the life & action of one of the most remarkable men of the century'.[45]

Tyndall had never been a stalwart of the radical crowd, people like Chapman and Lewes, with whom Spencer was much closer. But he did know many of them. George Eliot, for many years the partner of Lewes, died just before Christmas. Spencer telegraphed Tyndall to say that she wished to be buried in Westminster Abbey, that he had telegraphed the Dean, and that John Cross (George Eliot's husband), who did not know Tyndall, hoped he would join in the pressure.[46] Tyndall wrote a 'very striking letter' to Dean Stanley, though George Eliot's denial of the Christian faith, and unconventional but monogamous life with Lewes, prevented interment at Westminster. Huxley called to ask Tyndall not to put too much pressure on, for fear of arousing criticism of her past, and Tyndall was relieved when he heard from Stanley that the funeral would be at Highgate. Cross and Eliot had only married earlier that year; Cross had survived a jump into the Grand Canal in a fit of depression while on their honeymoon in Venice. Cross wrote with gratitude for Tyndall's intervention and invited him to the funeral at Highgate. He attended on 29th December, on a pouring wet day, to see her buried next to Lewes in the area reserved for religious dissenters or agnostics.[47]

Tyndall was now around sixty years old, and still an active researcher. His next significant investigations, which proved to be his last, were stimulated by a piece of new technology.

PART V

1880–1893
POLITICAL AFFAIRS

24

-*⟹ ⟸*-

HINDHEAD
1880–1883

It was the invention of the photophone that inspired Tyndall's final significant research. Alexander Graham Bell invented this wireless 'telephone' in February 1880, with his assistant Charles Tainter, based on a discovery by Mr A. C. Brown in London in 1878. By contrast to the telephone, which used modulated electricity carried over a wire, the photophone used light, modulated by a plane mirror made of flexible material, to transmit the sound signals. This invention is the basis of modern fibre-optic communication, and Bell considered it a greater achievement than the telephone. He even proposed calling his new daughter Photophone until dissuaded by his wife.

At the suggestion of Spottiswoode, Tyndall made his laboratory available to Bell for experiments on 27th November 1880.[1] Ten days later the physicist Shelford Bidwell—who produced the entry on magnetism for the 1911 edition of the *Encyclopaedia Britannica*, writing Tyndall out of its history in the process—brought Tyndall his own development of the photophone. Tyndall noted that 'its construction was much simpler and its action vastly stronger than that of Professor Bell's instrument'.[2] Nevertheless, it was Bell who patented it, and whose invention stimulated Tyndall to make experiments that same day on the effect of intermittent beams of radiant heat on gases.

Bell had shown that sounds could be obtained by the action of an intermittent beam of light on solids. Tyndall imagined that this was caused by rapid changes of temperature causing rapid changes in shape and volume, hence emitting sound. He thought the same ought to apply to gases, so that rapid expansion and contraction of an absorbent gas would be accompanied by sound. Good absorbers of heat should produce loud sounds, the strength being a measure of absorption. Using a rotating disc with slits, to provide the intermittency of the beam, and listening to the sound through a rubber tube, his success was immediate. As he expected, the better a vapour or gas was at absorbing radiant heat according to his previous experiments, the louder the sounds generated. To his particular delight, given the strife with the late Magnus, water vapour gave him a clear sound. He discovered that sounds could be obtained from very small amounts of good absorbers, which he thought might even be used to detect fire-damp in mines.

Working quickly with the help of Louisa, Tyndall submitted a paper to the Royal Society on 3rd January 1881, before going on to Dewar's Christmas Lecture and dinner with the Huxleys. His paper 'Action of an Intermittent Beam of Radiant Heat upon Gaseous Matter' was read at the Royal Society ten days later.[3] Ever the showman, Tyndall made the sound audible to the assembled Fellows through an india-rubber tube twelve feet long. He used the occasion to give a firm statement of his belief in his scientific approach and the robustness of his findings. He said that many people, including Magnus and Buff, had disputed his conclusions about the absorption of radiant heat by gases on particular grounds, but 'occupied for the most part with details, they have failed to recognize the stringency of my work as a whole, and have not taken into account the independent support rendered by the various parts of the investigation to each other. They thus ignore verifications, both general and special, which are to me of conclusive force'. One could not have a clearer statement of the strength and validity of a well-supported scientific theory. Tyndall saw the strength of Darwin's theory of evolution in similar terms.

With a new technique to exploit, Tyndall was like a man possessed. He extended his experiments in January and February 1881 to the vapours of eighty liquids, for which Louisa spent days tabulating boiling points. The research was fitted round public commitments—at the Accidents in Mines Commission, Gas Referees and Harrow Governors—as Tyndall's annual course of lectures at the Royal Institution, this year six lectures on 'Paramagnetism and Diamagnetism', was not due to take place until after a late Easter. His second paper, 'Further Experiments on the Action of an Intermittent Beam of Radiant Heat on Gaseous Matter. Thermometric Measurements', was given to the Royal Society on 24th February.[4] The experiments were laborious, and Tyndall announced that many weeks would elapse before he could present the full results. He gave his completed findings as the Bakerian Lecture nine months later.

In the first part of 1881, while radiant heat and sound experiments continued, the social life of the Tyndalls flourished. They were at Tennyson's annual party and at dinner with him a few days later, in company including the Duke of Argyll, the Leckys, Dean Stanley, and Browning. Guests at Lady Ducie's dinner included the Duke of Argyll again and his wife and daughter, Lord Lansdowne, and Lord and Lady Elcho.

One particularly sad event intruded on these engagements. The Tyndalls had driven to the ailing Carlyle in January to give him two bottles of sixty-year-old brandy. They called again several times over the next weeks but were unable to see him. He died on 5th February, just as Tyndall returned from several days in Wigan. Tyndall took a 'mournful journey to the North' for the burial at Ecclefechan on 10th February.[5] It was a dour ceremony on a dour day, with rain overhead and thick slushy snow on the ground. 200 peasantry and children of the neighbourhood were there and 100 round the grave. The body was brought in by Mr and Mrs Alexander Carlyle, Tyndall, Froude, and Lecky, and the funeral was soon over. The coffin was lowered by tradition without word or ceremonial of any kind, only many wreaths of beautiful flowers thrown over it. Not a soul in Ecclefechan spoke to Tyndall, Froude, nor Lecky. They walked about and got some lunch at a small inn, then left for London.[6]

Not long after his death, reports emerged that Carlyle was not only incurious, but hostile to science. Tyndall wrote to the *Times* to put the record straight.[7] As a frequent visitor to Carlyle's house, and as host of Carlyle's visits to the Royal Institution, his experience was the opposite. Carlyle would engage with him deeply: 'His questions were always pertinent and his remarks often profound'. He compared Carlyle's approach to physics favourably to Goethe's, and claimed that Carlyle had even come to terms with Darwin's theory of man's descent, despite what he saw as its tendency to weaken the ethical element in man.

At the end of March, the Tyndalls had a break of several days together in Folkestone—Tyndall had not been well during the week—part of it spent correcting *Essays on the Floating-matter of the Air in relation to Putrefaction and Infection*, which was published later in the year. They returned to find that Louisa's Aunt Augusta, who had just been buried at Elton, had left Louisa's mother £15,000. Given the state of Louisa's parents' finances, this must have been welcome.

Royal Institution commitments now beckoned. William Thomson had given a Discourse in early March on 'Elasticity viewed as possibly a Mode of Motion', paying explicit tribute both through the title and in the talk to Tyndall and his 'beautiful book', *Heat a Mode of Motion*. Shelford Bidwell lectured about the photophone a few weeks later, and then Helmholtz was in London to give the Chemical Society's Faraday Lecture at the Royal Institution. His topic, appropriately, was 'On the Modern Development of Faraday's Conception of Electricity'. It was the fifth Faraday Lecture, instigated as a memorial to Faraday in 1867 and 'confined to foreigners and British colonists'. Helmholtz was well entertained by Tyndall's now extended family, with a lunch before the event with the Busks, Louisa's mother and sister Emma. Tyndall gave a Discourse the same week, on 'The Conversion of Radiant Heat into Sound'. Louisa thought it 'first rate'—indeed two people in Hirst's hearing pronounced it the best he had given—and a large party afterwards included the Duke of Argyll, the Spottiswoodes, and the Helmholtzes.[8] Anna Helmholtz wrote on her return to Germany to say that she had been particularly delighted to meet them both.[9]

For the late Easter break, the Tyndalls made a visit to the area in which they eventually made their home. They spent several days in Surrey, roaming about Hindhead, the Hog's Back, and the Devil's Punch Bowl, staying at 'a modest little hotel, in a singularly wild and romantic position near Hind Head', before returning on Easter Monday.[10] After that, they went there almost every weekend to house-hunt, staying at the Royal Huts Hotel. Part of the attraction of Hindhead, apart from the wild country and the opportunity for long walks which they both loved, was the social life, with the Tennysons and Palmers living nearby.

But Tyndall's sleeplessness returned with a vengeance, and he was frequently unwell between Easter and his departure for the Alps in June. He managed to give all six weekly lectures on 'Paramagnetism and Diamagnetism' between the end of April and early June, although illness after one of them forced him to cancel dinner at the Athenaeum with John Morley, Mori Arinori (the Japanese Ambassador),[11] and Arthur Russell. Feeling the strain, Tyndall wrote to De la Rue from Hindhead in early June, warning him that he might not be well enough to attend the Managers' meeting at the Royal Institution. He asked the Managers to vote in the interest of the Institution that he be requested to take complete rest for twelve months.[12] The Managers proved supportive, resolving that he should do just that.[13] Tyndall thought he would not need so long to restore his sleeping power, and with it his working power and general well-being. But he did ask that he be released from the Christmas Lectures. His experimental work, which he had been obliged to halt after Easter, had to be completed in the autumn in time for the Bakerian Lecture, and he felt he could not accomplish both.[14]

The Tyndalls set off for the Alps on 17th June. A pass from the French Ambassador sped their transfer between stations in Paris, and they were on Alp Lusgen the next day. A donkey arrived with beer soon afterwards. Sleep still eluded Tyndall at times, but he strengthened quickly, telling De la Rue that he expected to be fully able to resume duties on his return in October.[15] The masons, plasterers, and painters left, having made good the damage of the winter. It was sociable in their chalet and at the hotel, and with three servants they were well supported. Later in the summer, Louisa's mother came to stay for a full month. By mid-July Tyndall was sleeping and feeling well, having banished what he considered to be his first ever attacks of asthma by the expedient of smoking cigars.[16] The weather throughout June and July was wonderful; almost continuous sunshine, broken two or three times by thunderstorms which freshened the air. Tyndall wrote a graceful article for the *Pall Mall Gazette*, describing their purchase of a donkey, the 'concrete villainy of Nature', and a touching encounter with a ptarmigan.[17] They saw the Great Comet of 1881, the first to be properly photographed, from the summit of the Sparrenhorn.

Dean Stanley died while Tyndall was in the Alps, although the *Times* reported him as present at the funeral. Very little could drag Tyndall from his Alpine fastness in the summer months, and he also missed the Golden Jubilee meeting of the British Association in York, presided over by Lubbock. He excused himself to Lord Houghton on the grounds of ill-health, despite claiming that Lubbock, before he took up politics, had no more intimate friend than him.[18] He was the only one of the X-Club missing from the celebrations, although he did send in a paper to Section A, 'On the arrestation of infusiorial life', which the General Committee ordered to be printed *in extenso*.[19] Stimulated by the days of unbroken sunshine, and following up some observations of Downes and Blunt made in 1877, Tyndall had decided to experiment again on the effect of solar light on tubes of infusions that had been kept sterile for three years at Alp Lusgen.[20] He broke the ends to let in the air and kept half in the dark and half in sunlight, finding that exposure to sunlight delayed decay but did not prevent it. He made no claim that sterilization could be achieved, merely that he had not observed it.

Towards the end of their stay, now into early October, Tyndall worked on the Bakerian Lecture as summer turned to autumn and heavy snow fell. As became his practice, he had been ministering to the sick, and had dressed the wounds of one boy with terrible burns every day for five weeks, after a passing surgeon from St Bartholomew's Hospital had been called away. He then offered to pay for the boy's education in Brig.

The Tyndalls left on Saturday 8th October, halting in Vernayaz overnight, and were back in London on the Monday, not even stopping in Paris for the Electrical Exhibition. Du Bois-Reymond was disappointed that Tyndall was not there, telling him that it was a dazzling exhibition but that the scientific results were poor: 'The Frenchmen made a sorry show as scientific men this time'. Tyndall told him he had decided not to attend, fearing that its fascination would interfere with his duties. He added, after Du Bois-Reymond had complained that his work was not understood nor awarded medals: 'There cannot be a doubt of it that you occupy a far higher position in the appreciation of scientific men than the great majority of those who have been awarded medals. In cases like yours it is the difficulty of defining the work so as to render the claims intelligible to a mixed committee that constitutes the real obstacle'.[21] Du Bois-Reymond may not have known that Tyndall had tried unsuccessfully to get him that recognition.

In the run-up to the Bakerian Lecture, Tyndall still had experimental work to complete. He started turning down social invitations, dining occasionally with close friends such as Hirst, Debus, and Spencer, often at the Athenaeum or at the Reform Club. He managed one country weekend, at High Elms, from which they all went to Down House on the Sunday to pay the elderly Darwin a visit. Darwin died a few months later, on 19th April 1882 (Emerson followed eight days afterwards), and it was quickly agreed that he should be buried in Westminster Abbey. The pall bearers included Huxley, Hooker, Lubbock, Spottiswoode, Wallace, and the Duke of Argyll. George Darwin had sent personal tickets, but Tyndall was not feeling well, and he and Louisa arrived too late to enter.[22] They remained by the door at the back. Louisa reported that 'my love sank down on a seat his face buried in his hands and there remained through all the service of which we heard distant snatches'.[23]

Tyndall had returned in generally fine health after taking the Swiss air. Louisa gave a glimpse of his playfulness, as she returned late one evening: 'Presently I thought I discerned a little black mass curled up at the foot of the bed, which jumped up with the sudden exclamation "Bow wow" and gave me such a loving welcome'.[24] Tyndall described trying to wake up Louisa: 'I kissed her several times, then bit her nose and ears, then kissed her again and giving her a shake, then coiling my arms round her lifted her up, placed her head against my breast and turning my back to the pillow, drummed vigorously with my naked feet against the matched boarding'.[25]

On the day of the Bakerian Lecture, 24th November 1881, the Tyndalls walked together to Burlington House (Plate 22), looking through the blindless windows together at the apparatus and wan lights inside, and at the wise heads gathering there, animatedly conversing in dumb show. Then Louisa left him; as a woman and member of the public she could not go in.[26] This was his fourth and last Bakerian Lecture—since 1855, his first lecture, no one else has given as many—and it was another triumph. At times the audience 'loudly applauded. Unlike the habit of that grave assembly'.[27]

The presentation was a *tour de force* of his previous work, restating his convincing arguments against the claim by Magnus that water vapour was not a good absorber of radiant heat, which he had spent so many years combatting.[28] Tyndall repeatedly stressed the importance of the sustained and specific experimental knowledge and skill that he had built around his apparatus. Someone less skilled, less conscious of all the possible sources of error, could not have achieved his results. Not even Magnus could repeat the experiments, even after he had seen them with his own eyes in London. But objections to this key finding of Tyndall's, critical for understanding meteorology, were still current in 1880. So, he explained, 'with a view to my own instruction, and to the removal of uncertainty from other minds, these researches on radiant heat were resumed in November, 1880'.

In particular, Tyndall had sought to meet the challenge from Magnus and others that thin films of vapour adhering to his experimental tubes or rock salt plates were responsible for his results. He used yet another innovative experimental design in which he focused the radiation with a mirror, so that it did not touch the sides of the experimental tube on passing through. He had carried out these experiments twelve years previously, though without publishing them.[29] On repeating them with further experimental improvements, he found all his earlier results confirmed. He had a strong desire to show that the same amounts of liquid and vapour had the same absorption, an idea he had held since 1864, and which would demonstrate clearly that absorption was a property of the molecules alone. This he had succeeded in doing in a series of intricate experiments.

In the final part of the lecture, Tyndall drew on the significance for meteorology of the absorption and emission of radiant heat by water vapour, with a side-swipe at meteorologists who refused to accept the validity or implications of laboratory experiments. He had long argued that high enough in the atmosphere, above significant absorption by water, the solar spectrum would show a greater proportion of invisible to visible solar rays. A letter from Samuel Langley, from 12,000ft up Mount Whitney in October 1881, had verified this prediction, and there Tyndall's lecture ended.[30]

* * *

The weekend after the Bakerian Lecture found the Tyndalls scouting possible sites for their house. The following week was typical: at Harrow to see the science teaching on the Monday; at Gas Referee business at Lewisham on the Tuesday; at a meeting of the Harrow Governors on the Wednesday followed by Spottiswoode's Presidential Address to the Royal Society; and at the X-Club on the Thursday. Then they were off to Hatfield House, seat of Lord Salisbury, for two nights, joined by William Armstrong and many other guests. The Tyndalls did relax on Christmas Day. Louisa recorded: 'A very sweet, peaceful little Christmas morning spent with my Johnnie, starting with very special love'.[31]

In the New Year of 1882—interspersed with lighthouse business, Gas Referee commitments, a report on science teaching at Harrow, and a report of the Science Masters Conference held at the Royal Institution—Tyndall was still hard at work finishing his paper from the Bakerian Lecture and preparing a Discourse on the subject. Proving the physical facts was only half the challenge. In a long letter to Stokes, Tyndall revealed the struggles he was having to tie down a physical explanation for the greater absorption by compound rather than simple molecules, which he thought might be the consequence of the ether being more dense within the compound molecules.[32] Stokes responded: 'I think the molecules of the so-called elementary bodies, and to a smaller extent those of very stable compounds, are to be regarded as systems very tightly triced up, so that none but smart forces have much effect on them, whereas those of complex bodies are more loosely put together, so as to be more easily set swinging … by the vibrations of the ether'.[33]

Tyndall gave a shortened course of just three lectures in March, arranged in consequence of his ill-health the previous year, on his current research interest, 'Resemblances of Sound Heat and Light'. This reduced demand gave him more time for research and for official duties, both of which weighed heavily. As the lectures started, Tyndall managed to complete his paper for the *Philosophical Transactions*, which had been given as the Bakerian Lecture, and send it to Stokes. Stokes told him that he had been convinced all along, unlike some others, that Tyndall had been right about the identity of radiant heat and light. On the absorption of radiant heat by aqueous vapour, he admitted: 'I had indeed not examined, except very imperfectly, Magnus's work, and I had heard the opinion expressed that you had best shut up. But I could not imagine how what you had done could be upset'.[34] He gave his usual constructive and detailed comments on Tyndall's paper, and his support for a short paper on fog whistles that Tyndall had sent in January. In this neat paper, Tyndall explained the observation of a soundless zone often found some distance from the signal out at sea by interference of sound reflected from the sea surface.[35]

Tyndall was grateful for the comments. He respected Stokes, and he had felt keenly over the years the objections to his work which he knew were unfounded: 'I have sometimes been disheartened by the efforts of really eminent men to reconcile...Magnus's results and mine...Of course I knew that it would all come right in the end. But it is gratifying to have it set right in one's own lifetime'.[36] Part of the problem was that equipment was hand-made, unique, and operated in everyday environments. Reproducibility, often taken for granted today, was a major challenge.[37] Playfair had a similar response to Stokes, writing: 'I was very loth to doubt your previous results, but your present experiments seem to me to place them beyond all question and to establish a chief factor in the solution of many of our Meteorological difficulties'.[38] In a postscript to his work towards the end of the year, Tyndall recorded the temperature of the Earth's surface and the air above it. His results demonstrated clearly the role of water vapour in reducing heat loss from the surface by radiation.

The Tyndalls spent almost every weekend in March at Hindhead, and were twice with the Tennysons. Tennyson read *Maud* in the back drawing room, followed on request by *St Agnes Eve* and *Sir Galahad*. Tyndall even gave up the Civil Engineers' Dinner one day, despite Armstrong being President, excited about the prospect of getting their 'own little home to take refuge in', and obtaining recommendations from local labourers for a builder and well-sinker.

Personal life was ecstatic. Louisa divulged that she 'champood him with my red flannel gloves, to the edification of us both', and recorded Tyndall's declaration: '"My glorious wife!", said the dearest one with glistening eyes, "I never dreamed I could have found so glorious a love!"'.[39] He told Hirst later: 'Well, I have many blessings. Did you ever hear Louisa laugh one of her real musical laughs? That is a blessing'.[40]

At the end of March, Tyndall attended the meeting to establish the Association for the Advancement of Medicine by Research. It had been stimulated by the passing of the Cruelty to Animals Act, which threatened to hinder or prevent some research that required vivisection.[41] It was not a large meeting, but attended by 'the most eminent of the medical profession and others of high distinction'. Tyndall was asked to propose a resolution.[42] Though sensitive to cruelty and suffering, he was a strong advocate of research using animals. He reiterated this support in a letter in the Times, bringing to public attention Koch's recent discovery of the tubercule bacillus, which had been announced on 24th March to the Physiological Society of Berlin.[43] As Tyndall reminded his readers, TB was a far bigger scourge than either the plague or cholera. Koch claimed that one seventh of the world's population died of it, and one third of those in middle age. He described the animal experiments that Koch had carried out to prove the existence and mode of infection of the bacillus, which he argued could not have been made in any other way. He ended: 'however noisy the fanaticism of the moment may be, the common sense of Englishmen will not, in the long run, permit it to enact cruelty in the name of tenderness, or to debar us from the light and leading of such investigations as that which is here so imperfectly described'. In 1905, a few years after its inception, Koch was awarded the Nobel Prize for this discovery.

Tyndall extended his arguments in an article the following year on 'Methods and Hopes of Experimental Physiology' for the Pall Mall Gazette, aware that it might bring the anti-vivisectionists down on him. Pointing to the now widespread acceptance of the wave theory of light and the theory of evolution, both after initial resistance, he recollected how only a few years ago the germ theory of communicable disease had been held by just a few people. Now it was widely accepted. Nevertheless, immunity was a challenge to explain on this theory. Tyndall suggested that just as trees needed minerals to grow, perhaps the first infection removed necessary nutrients for the germ to succeed a second time. The converse of this was that minute amounts of some substances, like silver nitrate, might prevent the growth of germs, such as the current exploration of arsenic to combat malaria using human subjects in Italy. Tyndall argued that a proper trial would require some to be given the potential remedy and some not, then for both groups to be deliberately infected with malaria. But that would doubtless result in prosecution for murder, so what was to be done? Malaria could be transmitted to cows, so despite his objection to animal cruelty, Tyndall argued in support of a trial in cattle. He wrote: 'I am neither a vivisectionist nor an anti-vivisectionist, and cruelty to animals is abhorrent to my nature', but he thought those who opposed these experiments would range themselves 'on the side of the enemies of the human race'.[44]

The Tyndalls staked out the position of their 'iron house' at Hindhead in mid-May, intended as a temporary abode. As Tyndall described it to his sister: 'We have but one room; but we have a capital little kitchen range where we make our own tea, cook our own chops, and boil our own potatoes. A tidy woman comes to us now and then to wash up and arrange matters'.[45] At the opposite end to the range was a bed, with two tables between, one for writing and one for eating.[46] Doors at each end opened to the north and south.[47]

Louisa, with Tyndall ill, went off with her cheque book to pay for the land. They had found in Mr Simmons, Chairman of the local Magistrates and introduced to them by Hallam Tennyson, a welcome source of help in ensuring that legal formalities on title were carried through. Tennyson was installed on the opposite side of the valley on Blackdown, at his house Aldworth, which he had built in 1869 (with James Knowles his architect). A few days later, they bought the bricks, and two strips of heather were dug up. Tyndall was now feeling better than for several months, and superintended the building work. The chimney was finished at the beginning of June, just as he sent the first five sheets of a revised edition of *Light* to the press. He told Hooker how much they were attracted to the area: 'This region is to me very glorious. Heather knee deep gorse lately all ablaze, but now sobering down; pinewoods with their "waterfall tones" when the wind blows through them; far horizons with finely sculptured hills intervening…If the air and scenery of this place fail to cure my wounds they are incurable…It is a great comfort to me to think that during the last year and this, I have succeeded in accomplishing a solid and difficult piece of work. So that…my latest years will not be regarded as blank and fruitless'.[48]

Hooker was building a house at the same time—he had been left £1,000 by Darwin, which he gratefully used to extend the size of the house he could build—and offered his services as landscape gardener.[49] He was somewhat alarmed by Tyndall's plans for their main house: 'You are going to build a high & narrow house…in the most exposed spot in its vicinity…You propose living in the top story—have you calculated in the vibration, or the howling wind, beating rain, & bellowing noise in the chimneys…? When a site is chosen that commands a country, the occupier of it may be…held to have regard to his neighbour's amenities. Your house will be in view from all around & be a most conspicuous object…the surrounding scenery cannot but be profoundly modified by it for miles and miles in all directions'.[50] Hooker perhaps anticipated Tyndall's subsequent disputes with neighbours.

* * *

The Tyndalls arrived at Belalp in the summer of 1882, a visit delayed by the marriage of Louisa's brother on 6th July. Tyndall was still feeling low, and suffering from a foot problem and asthma. Louisa thought he had 'suddenly and unaccountably aged'.[51]

The winter and spring rains had damaged the roads, so for the first few weeks Tyndall occupied himself with road-mending: 'for the people here have no knowledge of road-making, even on a small scale, and it is uncommonly difficult to teach them'.[52] He had not been able to do any writing. A new edition of *Sound* was required—it appeared in 1883, revised and augmented—and he told Hirst that from time to time he and Louisa talked about changing his pattern of work, now they felt 'so secure that we have no reason to fear, even at the worst, either the approach of poverty or the invasion of our independence'.[53] A whole sheep was cut up and suspended in their larder. Visitors for tea included Lord Cairns, who had been Lord Chancellor under Disraeli. Tyndall hoped he had gone away 'with the impression that I am not wholly a demon, though I refuse to believe in the Mosaic cosmology'.[54] Hirst, who had been operated on by Lister to remove a tumour, told Tyndall of Frank Balfour's fatal accident on Mont Blanc.[55] Balfour, brother of the future Prime Minister Arthur Balfour, was an outstanding young biologist, regarded as Darwin's successor, and his death was widely seen as a tragedy for science. He was killed with his guide Johann Petrus, attempting the then unclimbed Aiguille Blanche.[56]

The Anglo-Egyptian War of 1882 broke in on the peace at Belalp. This war, initiated by Gladstone's government to protect British economic interests and, according to commentators, to court popularity, started with the naval bombardment of Alexandria the day after the Tyndalls arrived on Belalp. The event was of personal significance to them since Douglas Hamilton, after a brief honeymoon, left Dublin for Egypt with the British forces on 1st August.[57] Tyndall was unimpressed by Gladstone's strategy: 'Gladstone by his unstatesmanlike conduct prior to the last election has got himself and the country into difficulty. The insubordination of the Irish Constabulary is also, without doubt, based on their calculations of the character of the man with whom they have to deal. Ever since his disestablishment of the Irish church I have foreseen and prophesied that his way of conducting legislation would be followed by disaster'.[58]

At the beginning of 1884, eighteen months later, the military and political situation in Egypt provoked Tyndall to write a letter to the *Pall Mall Gazette*, criticizing what he saw as the government's (and Gladstone's) failure to apply decisive power, but writing 'with the instinct of a Liberal'.[59] A correspondent pointed out that in his lecture in Glasgow in 1880, he had declared: 'I am a Conservative'.[60] Tyndall neatly resolved the question in his reply, which reveals his Whiggish Liberalism, by holding that 'the truest Liberal is at the same time the truest Conservative', forestalling revolution by 'securing, in due time, sufficient amplitude for the national vibrations', not waiting 'for the more or less coarse expression of the popular will and then constituting himself its vehicle'.[61] By this he meant Gladstone's policy in Egypt. Lord Dunraven pressed him to speak on Egypt at a meeting of the Patriotic Association in St James's Hall, but Tyndall wrote: 'The utmost I can do is to give my mind in a letter to the Chairman. It would never do for me to get into the whirl of politics just now'.[62] The *Times* reported this meeting on 3rd March 1884, including Tyndall's letter to Dunraven, in which he hoped that the government had now shaken off the 'cowardly incubus which has hitherto paralysed them'.[63]

Further military problems lay ahead for the British, as the war in the Sudan against the Mahdist forces continued. General Gordon, sent out to command an orderly withdrawal of Egyptian forces, was cut off in Khartoum. He defended it for nearly a year, until the city was overrun and Gordon killed on 25th January 1885, two days before the belated relieving force from General Wolseley could reach him. Tyndall knew where the blame lay: 'Gordon killed. What a reckoning of blood this man [Gladstone] has to account for! And that he should be worshipped in England when his incompetence is so plain!'.[64] After the Fall of Khartoum, in April 1885, the war came close to home again for the Tyndalls and Hamiltons when Douglas announced that he was again to be sent to Egypt. They were relieved when they heard that he was not in the advance, as he was laid up with diarrhoea.[65]

While they awaited the outcome of the Egyptian campaign of 1882, Tyndall's health and strength improved; he could now go up the Sparrenhorn in quick time. Socializing on Belalp was in full swing. The mathematician Arthur Cayley came to tea, along with Mr and Mrs Biddle, 'intelligent Americans', as discussion led to Emerson and Carlyle. The Cecils, including Lady Maud and Lord Cranborne, appeared later in August.

By early September Tyndall had finished building a short cut path from the hotel, and sent off to Longman an article on 'Atoms, Molecules and Ether Waves'.[66] This was produced at Longman's request for the first issue of his new *Longman's Magazine*, which superseded *Fraser's Magazine* and ran until 1905. It was a restatement of Tyndall's belief in the role of inspiration to generate understanding of the physical world, and the importance of atomic and molecular theory. For Tyndall the idea that science was 'organised common sense'—Huxley's words—might apply to natural history, but not to the physical and mathematical sciences. The Duke of Argyll was prompted to write to Tyndall after reading the article to argue that Tyndall's statement that atomic theory was necessary to explain the law of multiple proportions was questionable.[67] Tyndall vigorously defended the theory because it explained so much diverse experimental evidence, in chemistry and in physics.[68]

Snow fell heavily in mid-September, surrounding the chalet with layers five feet high and creating a scene that they both found glorious. As Louisa put it: 'Two hearts burning and overflowing with love, in that great, silent, cold, beautiful, unconscious, majestic, solitude of snow'.[69] Such early heavy snow was not good news for the locals. They managed initially to rescue 300 sheep on the higher Alps, leaving about 600 behind, many of which perished in the drifts or in avalanches.[70] Tyndall went halfway up the Sparrenhorn to try to help find them, and described the unsuccessful rescue attempt in a letter to the *Times*.[71]

In the midst of the snow they received a telegram from Louisa's sister Emma: 'Kebir taken. Arabi's forces totally routed. Douglas quite safe'. The Battle of Tel el-Kabir, on 13th September, was the final set-piece action of the war. General Wolseley routed the forces of Ahmed 'Urabi, losing around fifty men to some 2,000 Egyptians and capturing Cairo. The war resulted in the British occupation of Egypt and the reduction of French influence in the area.

Up on Alp Lusgen, Tyndall worked away at his speech for the Carlyle memorial. After Carlyle's death the previous year, Tyndall had been asked to speak at the unveiling of his memorial statue, a duty he inevitably felt he could not decline. The Tyndalls departed the snowy landscape on 9th October, after Tyndall narrowly escaped being shot by accident by chamois hunters recklessly discharging their weapons, with bullets whizzing close to his ear.

The day for the unveiling of the statue of the seated figure of Carlyle, 26th October 1882, was foggy early but brightened about 10am and the ceremony on the Chelsea embankment took place immediately after lunch. The enclosure was small and only a few could be present. Tyndall had to tread carefully after Froude's revelations that his hero's personal life did not always live up to his own written words. As he wrote to Farrer: 'The great man has been sadly mauled since his death—not through malice prepense I am willing to believe—but through misjudgement'.[72] He gave a moving speech, loudly cheered at its close, emphasizing the deep sympathy Carlyle had for the conditions of the working man, and the moral strength of his exhortations to aristocratic government and Radical alike. Carlyle, he said, was a 'dynamic force', a believer that ultimately 'right' or 'might' were identical, who challenged sham and who believed in the ultimate power of God. Tyndall ended by proposing a companion statue of Ralph Waldo Emerson, Carlyle's lifelong friend. Lord Houghton moved the vote of thanks to Tyndall, which Lecky seconded. Then Robert Browning moved the vote of thanks to the sculptor Boehm, seconded by George Howard.

A few years before Carlyle's death, Tyndall had revealed Carlyle's influences to him directly in a letter. The debt included his introduction to Tennyson's poems (a footnote in *Past and Present* had stimulated Tyndall to start reading them) and the moral strength to fight his battles and do his work.[73] Tyndall later intended to give a Discourse on Carlyle at the Royal Institution, but illness overtook him, and he never did. Instead, he wrote his 'Personal Recollections of Thomas Carlyle' in the Alps in 1889, and published it in the *Fortnightly Review*.[74]

* * *

As their stay at Alp Lusgen in the summer of 1882 had drawn to a close, Tyndall had started preparing for lighthouse business, knowing he would be under pressure on his return: 'This is a bad business, for I shall have to fight against injustice arising out of professional jealousy: but fight I will; and my cause is so good that I hope the result will be victory for the right'.[75] The 'bad business' was Tyndall's desire to see fair play, in his terms, between John Wigham and James Douglass, respectively the promotors of gas and oil for lighthouse illumination, as their dispute of many years continued. The Irish had learnt in October 1879 that permission had not been given by the Board of Trade to convert the Copeland Island lighthouse from oil to gas. They immediately sent a deputation to London headed by the Earl of Meath, Viscount Monck, and the Lord Mayor of Dublin to remonstrate.[76] Although the President of the Board of Trade was not able to receive them, they did succeed, after a substantial fight, in overturn-ing the Board's decision, largely on the grounds of the greater responsiveness and visibility of the gas light in fog and the huge expenditure being approved in England for trialling the electric light. But it was this installation, moved from Copeland Island to nearby Mew Island on the approach to Belfast, that brought to a head the issues over Tyndall's support for gas, and for Wigham. Negotiations had dragged on for more than a year after approval, as the process for tendering lumbered through. The Board of Trade and William Douglass, now engineer to the Commissioners for Irish Lights—and brother of James Douglass, engineer for Trinity House—seemingly put obstacle after obstacle in Wigham's way. Wigham had gone to see Joseph Chamberlain, President of the Board of Trade for the new Liberal government of 1880, to try to unlock the process. He wrote to Tyndall, who raised his concerns with Chamberlain in January 1882.[77] Farrer was not impressed that Wigham had gone to Tyndall dir-ectly to complain. Wigham's tender was nevertheless accepted and the light came into operation on 1st November 1884.

In October 1881, the Commissioners of Irish Lights had written to the Board of Trade to request comparative experiments on the relative merits of gas and oil light-houses for penetrating fog, and Tyndall was asked in January 1882, at the instigation of Chamberlain, to make recommendations. His report of 8th March,[78] based on experiments at Howth Baily, detailed the difficult line he felt he had to take to provide a fair comparison between Douglass's oil lamp and Wigham's gas burner, and his desire to do justice to Wigham's patent triform system. He dined with Chamberlain, in the company of Holland and Froude a couple of days before sending in the report, but there was no talk of Trinity House affairs. Despite his antipathy to patents he supposed that if Wigham's were legal it ought to be upheld, and his most recent advice was that it was valid. Tyndall thought the Board's preference for oil mistaken and even offered, given the expense, to forgo his salary from Trinity House so the necessary experiments could be done. He felt that Douglass, with whom he nevertheless had good relations, had unfairly prevented development of Wigham's system, promoting his own oil system, which he had patented.

Tyndall saw the struggle as a microcosm of the differences between England and Ireland: 'Those who are aware of the strength of my antagonism to all schemes tending to separate England from Ireland will be able to give due weight to the declaration which I here make, that if the treatment of the gas invention and its optical adjuncts could be regarded as a fair sample of the general treatment of Ireland by England, it would be the bounden duty of every Irishman to become a Home Ruler'.[79]

The Board of Trade now decided that a full comparison, using Howth Baily, ought to include the electric light as well as gas and oil, and involve all three Lighthouse Boards. Eventually, on 30th November 1882, after tortuous communications between the various Lighthouse Boards and other interested parties, came the first meeting of the Lighthouse Illuminants Committee, set up to oversee fair play. Farrer wrote to Vernon Harcourt to let him know that he was asked to represent the Board of Trade, and that Tyndall would attend, with the Engineers of the English, Irish, and Scottish Lighthouse Boards, namely James Douglass, William Douglass, and Thomas Stevenson.

But the 'competitors', as Tyndall put it, were Wigham and James Douglass, and he thought they should have equal rights in the arrangement. Trinity House did not see it Tyndall's way, arguing that the public good was their priority, and that seeking to put James Douglass and Wigham on an equal footing in making judgements, as Tyndall had proposed, was unacceptable.[80] There was more than a hint here of the class impropriety of pitting the gentleman adviser and engineer, Douglass, against the 'manufacturer', Wigham. Trinity House also pointed out that while Wigham's patent would benefit him personally, Douglass, who was not a manufacturer, had developed his patent at his expense and granted free use of it to all the Lighthouse Boards. The Board of Trade disagreed with this interpretation, asserting that Sir James still had a pecuniary interest and therefore should not act on a Committee with a quasi-judicial character. The Board won this argument, and, as Tyndall had previously suggested, the Astronomer Royal for Ireland, Robert Ball, represented the Commissioners of Irish Lights instead of William Douglass. Captain Nisbet represented Trinity House, and was asked to chair the Committee.

At the second meeting, the Committee agreed, on the grounds of ease of access and cost, to choose South Foreland instead of Howth Bailey for the trials. That doubtless pleased Tyndall, who had not been relishing a winter trip to Ireland. The estimated cost was substantial, some £4,000, and the Board of Trade approved it. But by now it was too late for experiments in November, the prime fog season, as the Irish Board urged, and there matters rested until the New Year.

* * *

Douglas Hamilton had returned from Egypt in early November. Tyndall went to a great dinner at the Albion given by Sir William Armstrong to Wolseley and other Egyptian commanders, and viewed the grand march-past of the Egyptian troops on 18th November from the Athenaeum, while Louisa, Emma, and their mother watched from Admiralty Gardens. A few days later he was at the Harrow Board in the morning and on British Association business in the afternoon, with dinner at Grocers' Hall the following day. Tyndall was on a committee of the Grocers' Company to decide on the wording for a substantial £1,000 prize to isolate the smallpox germ and cultivate it outside the body, and a judge of candidates for a scholarship.[81] The Tyndalls socialized with dinners at the Pollocks and with Leslie Stephen, then retired to Hindhead for a week, finding their hut surrounded by a foot of snow and walking to visit the Tennysons.

In the week before Christmas, back in London, they hosted a dinner for the Huxleys and others, and attended Lady Stanley's seventy-fifth birthday party, with the Huxleys and Browning. On Saturday 23rd December, they left again for Hindhead, where Tyndall could work on his upcoming Christmas Lectures on 'Light and the Eye'. But by early New Year 1883 lighthouse affairs weighed heavily, leading to a day of complete breakdown. Stressful also was Tyndall's final lecture, during which he was barracked by a member, August Andresen, for remarks he had made about Faraday. The police were called to the door in case it got ugly and a nasty article was published in *Vanity Fair* about it, 'a string of lies from beginning to end' as Louisa put it.[82] After this Discourse the Tyndalls went straight to Hindhead, where they wrote Tyndall's paper 'Note on Terrestrial Radiation'.[83]

The Tyndalls were now frequent travellers between London and Hindhead as plans for their permanent house developed. A neighbour offered to sell them part of his 5.5 acres, and they spent time looking at building stones and at the brick and tile kilns at Rowland's Castle. Tyndall was also planning a Discourse, given in March on 'Thoughts on Radiation, Theoretical and Practical', which was published in the *Contemporary Review* for May.[84] This started with a historical tour of the subject, including the work of Leslie, Rumford, and Melloni, up to his own discovery of the absorption and radiation of heat by gases and vapours, and the more recent work by Liveing and Dewar on ultraviolet radiation, which Liveing had demonstrated at a Discourse the previous week. It ended with Tyndall describing his recent experiments, including some on the resemblance between carbon dioxide and carbon disulphide, which he had sent in to the Royal Society only the previous day.[85] He showed that carbon disulphide, normally very poor at absorbing heat, could nevertheless absorb the heat from carbon dioxide well, indicating a strong analogy between these molecules that extended to their periods of vibration. He suggested that chemists could make more use of radiant heat for exploring molecular conditions than they had done to date. In time, this idea of infrared spectroscopy became a major analytical technique.

* * *

Tyndall now lit an explosive charge under the Lighthouse Illuminants Committee. He had written to Nisbet about the Committee, and Nisbet wrote to the Board of Trade in January 1883, seeking further clarification of the positions of James Douglass and Wigham. He commented on the lack of expertise of the Committee on electric lighting, apart from Tyndall, and the arguable under-representation of the Scots.[86] The Board replied that they regarded the three engineers as members (the two Douglasses and Stevenson), with Tyndall and Wigham, plus Nisbet (Trinity House) and Ball (Irish Lights), and Vernon Harcourt representing the Board of Trade. To balance representation, they suggested a second Scot, to be named, and John Hopkinson, who had taken over the Lighthouse Department at the manufacturers Chance Brothers, and was an electric light specialist.[87] Stevenson wrote to say that he assumed that members with patent interests would not vote on matters affecting them.[88] Ball thought that those with interests should not be members of the Committee at all.[89] Tyndall went a step further, writing from Hindhead asking to be relieved of the duty of serving.[90] While the Tyndalls watched a fox hunt in the Punch Bowl and Lord Derby's woods, telegrams arrived from Wigham and Lord Meath deprecating his apparent resignation.

Tyndall returned to London to give his paper at the Royal Society, finding a letter from Farrer asking him not to 'decline to act'.[91] Tyndall held his ground, in a letter which he copied to Lord Meath, since the arrangement did 'not bear the stamp of impartiality', and the Commissioners of Irish Lights indicated they would also withdraw under these proposed arrangements.[92] Chamberlain now took a hand, and summoned Tyndall to see him on 16th February. Tyndall wrote to Chamberlain after the meeting to suggest that the membership of the Committee should include James Douglass and Wigham. But he argued that it should exclude William Douglass and Hopkinson, who he knew was an intimate friend of James Douglass, and that James Douglass and Wigham should be non-voting members.[93] He added a private note that revealed his disquiet at the way that innovations in Ireland eighteen years ago with respect to the gas system had subsequently been treated. Chamberlain did not take kindly to being told, by a member of the Committee, who else should be the members of the Committee to which he had been invited.[94] Nor did he agree that the

Board had failed to support Wigham and the Irish. Tyndall rejected the insinuation that he was prejudiced in favour of gas, and that he had been presumptuous in recommending the membership of the Committee. He mentioned in mitigation that he had tried unsuccessfully to have James Douglass elected a Fellow of the Royal Society.[95] He reminded Chamberlain that Trinity House and William Douglass had both pressed the use of oil and that the Board of Trade would not sanction the use of gas in Ireland until Lord Meath, Lord Monck, and the Lord Mayor of Dublin had caused them to reconsider. He said he regarded the current contest as 'a struggle of single-handed talent against official power'.[96]

The Board of Trade continued to insist on the membership they had proposed in January, and the Commissioners of Irish Lights caved in.[97] Chamberlain wrote to tell Tyndall so, and to state that his strong views had emerged during 1882, such that 'an enquiry conducted under your sole direction would not give general satisfaction'. It would also need to be broadened to include the electric light, thereby extending the scope of the enquiry and constitution of the Committee.[98]

This was finally too much for Tyndall. He stood stubbornly on the moral high ground and refused to relinquish it. He penned a lengthy response at the end of March, writing the same day to Farrer and Collinson to resign his positions at the Board of Trade and Trinity House, and the salaries that went with them.[99] He returned both the cheques then sent to him for the half year to 31st March.

Tyndall had been particularly offended by Chamberlain making public what he thought was a private reply to his private covering note of 17th February about his suggestions. He thought that it had forced the 'pliant gentlemen' of the Irish Board to 'bend' rather than 'run the risk of being uprooted'. Tyndall said: 'With regard to the mixing up of "personal feeling" with a scientific question, you and I are one; but it is a mistake on your part to suppose that the question is a scientific one. The unbiased evidence of science has been given, and it was the disregard of that evidence, resulting in manifest danger to a most valuable invention, that first introduced into the discussion the warmth needed in opposing a moral wrong'. Chamberlain regretted the resignation since, as he put it, Tyndall's presence would have been the best guarantee that Wigham's system would be fully and fairly examined.[100]

As one might expect from Tyndall, his resignation was not the end of the story.

25

RAINBOWS AND LIGHTHOUSES

1883–1885

The fallout from Tyndall's resignation reverberated at Hindhead over Easter, while their architect John Penfold arrived and they spent a couple of hours choosing the site and staking out their new house. For the rest of the year, while they managed the initial contractual and building work, including sinking a well to more than 200ft, the Tyndalls alternated between London and Hindhead.[1] Rollo and Agatha Russell came looking for a house nearby, which they bought shortly afterwards. The Tyndall's were good friends with the family of the late Earl Russell, and thought they would be 'very delightful neighbours'.[2]

At one point Tyndall wondered if they would get out to the Alps at all, given commitments to the new house and continuing lighthouse arguments. At the beginning of May, he gave the first of his three weekly lectures on Count Rumford, wanting to make this man, one of the founders of the Royal Institution, better known in England. It was not well attended, but Percy Bunting, now the editor of the *Contemporary Review*, asked for an article on it.[3]

June brought sadness to the X-Club. Spottiswoode had contracted typhoid in Italy, and died on 27th June (a day after Sabine). He was the first of the X-Club to go, and was buried in Westminster Abbey. Tyndall walked past the coffin, at which he could hardly bear to look: 'We stood round the grave while the service proceeded, which service whenever I gave a moment's attention to it appeared to me unreal and untrue to the last degree'.[4] Hirst found it a great loss: 'a more high-minded, unselfish and beautiful character I never knew'.[5]

Meanwhile, the lighthouse affair entered another phase. After Tyndall's resignation, all parties had been sent copies of the correspondence between him and Chamberlain. This prompted the Commissioners of Irish Lights to protest again at the constitution of the Committee, since they had assumed that Tyndall would serve on it. But they did not withdraw their agreement to participate.[6] Farrer, writing for the Board of Trade, and grateful that they were still involved, reminded them that it was the visit to Ireland by Chamberlain in 1881 that had raised the issue of the need for experiments to compare the merits of gas and oil, and that the Commissioners had then readily concurred.[7] He stressed that it was the March 1882 report from Trinity House which had raised the further issue of electric lights and led to the establishment of the larger Committee, with its double representation from each of the Lighthouse Boards, the membership of Wigham, and the addition of electrical expertise.

When the Lighthouse Illuminants Committee, in early May, rejected a resolution proposed by Wigham and Ball that would have ensured that each illuminant was to be displayed in an enclosure of a common size to its best advantage, in the view of its protagonist, the Commissioners of Irish Lights protested again.[8] When the protest was not upheld, they informed the Board of Trade that they would withdraw from participation.[9] Tyndall had stood up for the Commissioners of Irish Lights and the Irish now stood up for him. Ball resigned, but the Lighthouse Illuminants Committee held their ground on the nature of the experiments they were prepared to sanction, arguing that any other course could lead to large, currently unquantifiable, additional costs.[10] Wigham protested that he was only being allowed to use his gas system at half its possible power against the most powerful oil light.[11]

Having resigned, Tyndall was not now directly involved, but his activity behind the scenes was extensive. He saw Wigham often, and was in correspondence with many of the participants. Arthur Russell let him know that the correspondence between him and Chamberlain had been asked for in the Commons, and it was made available on 11th May. Tyndall could now see all the official correspondence. He noticed that one enclosure, from Vernon Harcourt to the Board of Trade, seemed to be substantially redacted. He asked for an 'unmutilated' copy, and began to draft a letter to Chamberlain which he worked on for many days, sending it to Hirst for comments. It became a letter published in the *Times* of 22nd June, and pulled no punches with respect to Chamberlain's actions, as Tyndall saw them.[12]

The redacted parts of Harcourt's letter included the revelation that it was only quite recently that the Board of Trade, Richard Collinson, and perhaps the Elder Brethren generally, were aware of the pecuniary interest which James Douglass had in the recommendation of the burner which he had invented. It appeared that while keeping the Elder Brethren and the Board of Trade in the dark, Douglass was engaged in floating a company with a capital of £50,000 for the purchase of 'Sir James Douglass's patent', the excellence of the burner being certified by Hopkinson, now Douglass's colleague on the committee appointed by Chamberlain.

When they found this out, the Board of Trade did not tell Tyndall. As Tyndall put it: 'while matters were being thus arranged behind my back, Mr. Chamberlain was telling me to my face that the Illuminants Committee without my presence would resemble the play of Hamlet with the part of Hamlet cut out'. Tyndall did not have it all his own way in Parliament. Lord Dunraven initiated a debate in the House of Lords on 21st June, in which the Duke of Argyll, summing up for the government, said that 'although he was perfectly satisfied as to the purity of his motives, he could not but come to the conclusion that Professor Tyndall had made a mistake'. But eventually the Board of Trade gave up and disbanded the Lighthouse Illuminants Committee in July.[13]

The dissolution of the Committee spurred Tyndall to further action. A storm of letters hit the newspapers of London and Dublin, including the *Standard*, *Daily Telegraph* and *Daily News*, both about the statements in the House of Lords and wider matters.[14] Tyndall noted with pleasure that both the nationalists and conservatives took his side in Ireland. In August the Board of Trade sent the published letters on to Trinity House and the Commissioners of Northern Lighthouses, asking for their comments on the points of substance that Tyndall had raised: the superiority of gas, and the implication that Douglass, the Trinity House engineer, was not disinterested as he held a patent.[15] Trinity House replied on 6th December 1883, disputing Tyndall's imputations about Douglass's bias and explaining why the oil system was still preferable in some circumstances to gas, since the directions of its beams were more controllable, and indeed that electrical lights might also show advantages, as the French were thinking. They pointed out again that Douglass had made his patent free for all the United Kingdom lighthouse agencies.[16] The Commissioners of Northern Lighthouses replied on the same day, enclosing a report from Thomas Stevenson, which supported the general superiority of the gas light, but stressed that further experiments were needed, including with electric lights.[17]

* * *

Back at Hindhead, the Tyndalls received the estimated cost of their new house: 'It is a formidable sum £2,300, and no doubt this sum itself will be augmented by inevitable extras'.[18] Even at £2,300, it only cost about as much as Tyndall made in his lifetime from the English editions of *Sound*. Negotiations over buying extra land from Meeson, a neighbour, were fraught. In early August he asked £15 an acre more than the £40 he had originally said, and was ready to agree if Tyndall offered £280. Tyndall remarked that he 'obviously imagines he has got a greenhorn to deal with'. But Tyndall signed the contract with Penfold, their architect, despite remarking that his discontent with Penfold was 'very serious'.[19] He selected a builder, Thomsett, cheaper than the one recommended by his architect, thereby storing up trouble on both counts. Soon they were in Guildford looking at tiles, and Witley for bricks, choosing them at the last moment as they tried to escape to Switzerland. Louisa and he were not always together. On one occasion Tyndall sat on Haslemere station out of 'love hunger' waiting for her even though he didn't know when she would arrive. As he said: 'Though she has been but a few hours away from me I long to see her face—the cheerfulest, as Tennyson says, that he has ever seen'.

Finally, on 30th August, the Tyndalls were able to leave for Switzerland. Annie, the 'domestique', travelled second class, and they in first. It was unusually late for the Tyndalls to arrive at Belalp. The maids had aired their beds, so they were able to sleep at home, but they dined at the hotel with a party including a bishop. Tyndall engaged him in a discussion on spiritualism: 'it was of course a drawn battle between him and me'.[20]

The season was winding down. At its height there had been 115 guests at the Belalp Hotel, 40 more than the previous year, following the construction of extra accommodation by Italian workmen, who were also building the new English church. About twenty remained in the hotel in mid-September, but a week later it was down to one. Emma came to visit for three weeks, carried up in a sedan chair with her luggage on a horse (she was recuperating from an operation), and the Cobden-Sandersons spent some days with them. Tyndall wrote pieces for the Pall Mall Gazette, including 'Alpine Gossip' and 'Alpine Phenomena'.[21] He was captivated simultaneously by the scenery and the scientific opportunities. Early one morning 'a belt of lovely daffodil swathed the East and from horizon to zenith no trace of cloud was to be seen anywhere . . . then the thought occurred to me that such serenity before sunrise ought not to be allowed to pass without making some experiments on radiation'.[22] He became intrigued, too, by the formation of fog bows and Brocken Spectres (the apparently magnified shadows of an observer cast on the upper surfaces of clouds or mist).

The Tyndalls had builders to organize at the chalet, amid excursions on the glacier with the Cobden-Sandersons or roped together as they explored up to the ice-fall. Indeed, there were two possible chalets to consider. In May, Louisa had described to Tyndall her dream of a cottage in the Blindthal, a halfway house between the valley and the heights of Belalp. Today Belalp can be reached in a couple of hours from Brig by bus and cable car. In 1883 it was a walk or mule trip taking most of a day. Tyndall was somewhat alarmed—'in regard to the building of houses her hunger is simply unappeasable and I know not where it will end'—but he entered into the idea, though he called it a 'hut'.[23] They inspected the ground with Toni Walden and the brothers Bamatter, but the 'hut' was never built.

Up on Alp Lusgen, Morgenthaler installed an open fireplace in the large bedroom upstairs, Joseph Imhoff painted the railings and extended the wood house, and the walls were cemented to a higher level (Imhoff later falsified a cheque from Tyndall, to cash it at a higher value).[24] Tyndall directed operations, sitting astride the roof overseeing the cementing of the chimney, telling the mason and carpenter how to finish a window properly, and road-mending with Wiessen. Klingele, his builder, turned up drunk and tried to overcharge him. Tyndall later found him in bed, a 'drink-besotted, broken man'. Towards the end of September, after the flocks had been driven down the mountain, Tyndall agreed with Bottini and Morgenthaler the work that could be completed for the year, as the snow set in and icicles 5ft long hung down from the eaves. The water supply froze and food was scarce. They realized they might have to leave earlier than planned, and although they would have been happy to stay throughout the winter the servants would not have done so. It had been a good stay. Emma left by sedan chair with Toni Walden on 12th October, with Clements and Joseph Imhoff as carriers, and her baggage on the backs of Wiessen and young Salzmann. The Tyndalls returned alone via the Blindthal. A week later they left from Brig, meeting Klingele and Bottini at the station. Klingele had not paid Bottini, so Tyndall marched him to the bank to extract his debts.

The lighthouse affair recommenced towards the end of November. Tyndall had not been well since the end of October, and this may have tipped him again into a depressive state, 'very close to the margin of misery'.[25] He did manage to attend the Anniversary meeting of the Royal Society, hearing part of Huxley's address as the new President, and proposing the toast to Hirst as Royal Medallist (Thomson received the Copley Medal and Burdon-Sanderson the other Royal Medal).

Apart from a grand country house party with the Pembrokes at Wilton, Tyndall spent almost the whole of December at Hindhead, with one excursion to London to attend a meeting of the Gas Referees and the Alpine Club dinner, at which he sat as a guest on one side of Bonney, the President—about to be succeeded by Crauford Grove—with Huxley on the other: 'The dinner was overflowing with members and very pleasant in tone'.[26] Four years later, Tyndall gladly and graciously accepted honorary membership of the Alpine Club, at the instigation of the President, Clinton Dent, and the Treasurer, Fred Pollock.[27] He had already been elected an honorary member of the Swiss Alpine Club in 1865, the first foreigner to be so, and of the Club Alpino Italiano in 1876. The earlier experience at Wilton caused him to reflect: 'Beautiful and pleasant people on the whole but I confess I sometimes muse when I see the stone breakers by the road and ask myself "when England becomes democratic is this likely to continue?"'.[28] After Christmas Tyndall flitted between Hindhead and London for Dewar's Christmas Lectures on 'Alchemy in relation to modern science', praising him after the final one on 8th January 1884: 'This is the best course he has ever given here'.[29]

* * *

In the late summer of 1883, Tyndall had written to Stokes from Belalp about the formation of fog bows and Brocken Spectres, and Stokes had encouraged him to explore the phenomena.[30] He had also observed what he took to be a circular rainbow. When he was back at the Royal Institution, he produced clouds with a steam boiler and managed to reproduce the effects, writing to tell Stokes that it proved that the fog particles were not vesicles, as many people thought.[31] Stokes was supportive: 'How the wild and baseless theory of vesicular vapour came ever to be seriously entertained, much more to meet with such general acceptance—at least on the Continent, I think John Bull had more common sense—is to my mind one of the marvels in the history of science'.[32] This theory was advanced to explain why clouds could remain suspended, but Stokes had proved that globules, rather than hollow vesicles, could perfectly well remain suspended, given the viscosity of air.

Tyndall continued his experiments, interspersed with trips to Hindhead and meetings of the Gas Referees and the Accidents in Mines Commission, and completed a paper for the Royal Society in early December. It would appear that this paper was rejected by Stokes, presumably because there was nothing of significant new scientific interest in it, and Tyndall told Stokes that he would 'call upon Mr. White presently and take away the paper'.[33] The incident did not cause an argument, but Tyndall never published again in a Royal Society journal. Stokes came to call on him a few days later, when Tyndall showed him his artificial rainbows, before they went on to the Royal Society meeting.

Tyndall carried out further observations, many of them with Louisa, both at Hindhead and in the laboratory at the Royal Institution, before he gave the first Discourse of the year on 18th January 1884, 'On Rainbows'.[34] Starting with the theory of rainbows, tracing it from Alhazan (Ibn al-Haytham) through Descartes, Newton, Young, and Airy, he went on to describe the white bows he had seen with a light behind him in fog and mist at Belalp and Hindhead, which he had determined were rainbows. They were coloured white, as Young had explained, because of the size of the drops which produced them, not the presence of vesicles. Since the physical investigator 'desires not only to observe natural phenomena but to recreate them', Tyndall described his experiments from September aimed at doing just that, standing in a cloud of steam produced by a gas boiler. According to him: 'The audiences had hardly ever so difficult a lecture placed before them. The sou-wester and waterproof clothing and my entry into the showers obviously amused them'. He published a summary of his work in the *Philosophical Magazine* for January (Airy was intrigued, having never seen a white rainbow), an additional note on the white rainbow the following month, in which he acknowledged Hammerl's recent work which he had not seen until alerted by Dewar after the Discourse, and one on rainbows and glories in March.[35] Knowles, as editor of the *Nineteenth Century*, had told him he would welcome anything for the magazine and Tyndall obliged with an article 'On Rainbows'.[36] Shortly afterwards he met Lord Lytton, who had resigned as Viceroy of India in 1880, and showed the rainbows to Lord and Lady Lytton 'and their fair daughters'.

Tyndall's six weekly lectures on 'The Older Electricity, its Phenomena and Investigators' started on 28th February. As he prepared the previous day he complained: 'The labour of lecturing is greater than any who hear the lectures have any notion of'.[37] The time, care, and rigorous preparation he gave to his lectures underpinned his success, but they took their toll. His lectures were interspersed with trips to Hindhead and dinners: at Kent House, sitting next to Lady Ashburton; at the Pollocks—who bought Kingswood Cottage nearby—with Lecky and Freshfield and his wife, 'the latter I think more beautiful as a wife than as a maiden'; and at the Hollands, with Lord and Lady Arthur Russell, the Lord Chancellor, and others.[38]

Building progressed at Hindhead (Plate 23), where the view from scaffolding around the chimney so excited him that he had a platform built with Louisa's approval as the tiles started to go on. Tyndall hoped the house would be finished in July, reporting to Hirst: 'Our great battles with builders and bricklayers have been practically won, but minor skirmishes are always on hand. They run so readily into error if not kept by external direction in the right path'.[39] At Hindhead, inspecting the works, Tyndall was roped to Louisa on one occasion with a double sash cord, declaiming: 'Hail to the Alpine steep! The heather of Hind Head, Where Gladstone cannot steal my sleep; Nor Chamberlain my bread!'.[40] Returning once from a walk he found both doors of their hut locked, 'a precaution of Louisa's to keep out murderers'.[41]

Work was in full swing up to Easter. Tyndall selected tiles and stuffed partitions and spaces between the rafters with moss. There were songs of healthy, happy young workmen: 'With the sun glinting in upon us and the tonic air sweeping through the house, I thought of the happiness of a man working with his hands and earning thus an honest, manly livelihood, the happiest lot of all'.[42] He was not happy with some of the extras, when Thompsett quoted for a rain water tank: 'it is essential, and a very high extra it is; would have been better to include in the specification'.[43] Further potential bills also alarmed him; he went with Louisa to Regent Street, commenting: 'I look with some dread to the expense of our furnishing'.[44] They decided to have pneumatic bells installed, and chose cornices for the drawing room and hall.

One feature over which Tyndall took particular care was the lightning conductor. There were no accepted standards for their design. Tyndall had seen the impact of their poor installation in lighthouses, or the lack of any installation as at Inverary Castle, the seat of the Duke of Argyll. The Duke had asked for his advice following a fire that caused serious damage in 1877, and was urged to install a lightning conductor on the new roof.[45] Ten years later Tyndall was still concerned about ineffective installations, and wrote several letters to the *Times* to explain the need for a suitably sized flat copper plate, earthed properly to the ground or water.[46] The following year he advised Tennyson on the same basis.[47] Lightning fascinated him. Keen to understand the apparently soundless lightning flickering in the distance in Northern Italy, viewed from Belalp some 50 miles away, he would frequently telegraph the manager of the Monte Generoso hotel, to find that the thunder was tremendous in that location.[48]

* * *

Pasteur had written to ask if Tyndall could arrange for the translation into English of his son-in-law's book, *Mr Pasteur, the History of a Learned Man by an Ignorant One*.[49] Tyndall sounded out Longman and started to arrange for the translation, regretting that he did not have time to do it himself as Pasteur had hoped.[50] He asked Lady Claud Hamilton to undertake it, hoping that her eye trouble would permit it. On Good Friday in 1884, he made a start on the introduction he was to write for it. But he was not impressed by the book: Pasteur 'would be to me a more considerable man if he were above such vanities as those developed in this book...How high dear Faraday was above anything of this kind'.[51] Pasteur came to London just after Easter. Tyndall dined with him and his son-in-law at James Paget's, with Andrew Clark and Lubbock (about to remarry, though he did not mention it to Tyndall). As he continued to read, his opinion hardened. It was, he wrote, 'not a noble book. There is a somewhat vulgar eagerness to extol Pasteur at the expense of everybody else...It might be imagined from Mr Radot's account that Koch was a mere nobody, and yet it was Koch's admirable memoir, sent I believe to Pasteur by myself, which constituted the starting point for all his recent investigations'.[52]

Influential people were still sceptical of the germ theory. At the Royal Academy dinner on 3rd May, the eminent physician Sir Andrew Clark 'repeated the oft exploded fallacy of bacilli being an accompaniment, not a cause'.[53] (It did not stop Tyndall signing his certificate for the Royal Society on 11th December).[54] In late May the translation was nearly finished. Tyndall wrote: 'The antivivisectionists will be furious against Pasteur but his case is so strong that they will make themselves ridiculous'.[55]

Then on 3rd June, Lord Claud Hamilton, Louisa's father, died unexpectedly just a day after developing an 'acute inflammation of the larynx'.[56] Tyndall was very shaken.[57] Louisa left immediately for Heidelberg—meeting Clausius by coincidence on the station at Cologne—where her mother had just had eye surgery.[58] Tyndall missed the X-Club on 12th June for Lord Claud's funeral at Elton, and Louisa eventually returned at the end of June, as her mother was finally improving. Lord Claud's death stimulated Tyndall to examine his own Will, which he and Louisa discussed over several weeks, having seen from Lord Claud's Will how easily they could cause family discontent.[59] In the event, Lord Claud left just £8,000 net,[60] and Douglas wrote to Tyndall, who had forgotten that he was one of the Trustees, that the house in Portland Place must be sold immediately. The mortgage exceeded its likely value, so any deficit had to be made up out of the estate. In the event, the house was sold in early 1886 for £7,500, having been bought by Lord Claud for £10,000.[61] Lady Hamilton moved into 89 Cadogan Place around the same time.[62]

Inevitably, Tyndall's thoughts turned to the Alps, but the death of Louisa's father, the return of her convalescing mother in mid-July, and the finalization of the report of the Accidents in Mines Commission delayed his departure. He left for the Alps on 30th July shortly after signing his Will, but without Louisa, who 'saw my darling off by the early train'. She stayed behind both to look after her mother and to supervise works at Hindhead. Tyndall stopped at Loretan's in Vernayaz and walked straight up to Belalp the following day.

In Louisa's absence he settled into a routine of working on his preface for *Pasteur* and regular forays onto the glaciers, usually alone. On the Oberaletsch glacier he found himself 'landed where I quarrelled with my darling in the presence of General Bradford. It was regard for her safety that caused me to do so…I did not know her then but I know her now'.[63] As usual there were repairs to do on the chalet, especially to try to repel damp by having the floors taken up and digging a ditch at the back. He socialized with the bishop who had officiated at a service in the new chapel, with a 'pleasant Archdeacon', and extensively with Lady Cust and her family. The building of the chapel, in a deal done with Klingele, led to misunderstandings and complications that Lord Lingen, with Tyndall's advice, tried to resolve with the Bishop of London.[64]

Lady Claud Hamilton's eye operation prevented her from correcting the proofs of *Pasteur*, so Louisa painstakingly undertook the job at Hindhead, sending them out to Tyndall for checking. He asked Debus to comment on his draft preface, making clear his own view: 'As a record of scientific work it is wonderful. As a record of vanity it is almost equally remarkable…We must, I suppose, make some allowance for the love of glory innate in a Frenchman'.[65] In early September he sent the first half of his preface to the printers, commenting 'I am not sure that he will like it', and hoping it would not cause any rift between them.[66] He needn't have worried. When the book appeared in early 1885 Pasteur wrote a grateful note of thanks.[67] Once the *Pasteur* proofs and preface were out of the way, Tyndall could concentrate on a speech he had agreed to give to the Birkbeck Institute in October. He had also had an invitation to Cork, which he felt he had to decline, and heard with some surprise that he had been unanimously elected President of the Polytechnic Young Men's Christian Institute. He wrote civilly, indeed cordially, to decline that honour too, and finally left Belalp on 11th October.[68]

In December 1883 there had been a rumour that Tennyson was to be made a peer. Tyndall would seem to have disapproved of such honours on principle, and wished 'a contradiction point blank had come', which it did not.[69] The rumour was true and Tennyson was created a peer on 24th January 1884. Tyndall must have made some comment, as he received a letter from Hallam Tennyson which he thought 'heated and unreasonable. He talks of my "attacking" and "denouncing" his father, where no mortal but himself could discern either attack or denouncement'.[70] He abstained from replying, saying he had steadfastly stood by his father for nearly forty years. At Hindhead a few days later, his restraint was rewarded as Hallam and his wife turned up. They laughed at the interpretations of Hallam's letter and parted cordial friends.

* * *

Despite the abolition of the Lighthouse Illuminants Committee, the comparative examination of gas, oil and electric lights continued at South Foreland, under the auspices of Trinity House. In early November 1884, the whole affair blew up again, following Chamberlain's response in the House of Commons to a question on Tyndall's resignation. Mr Gray had raised the recent report in the *Financial News* that James Douglass had benefited from the sale of his patent rights at a time when he was also a member of a Committee that would judge his own invention. He asked if Chamberlain would set up a Royal Commission into the whole matter, which Chamberlain peremptorily declined. Tyndall wrote to the *Times* to put the history straight as he saw it: 'I resigned because Mr. Chamberlain, with an arrogance which surprised me at the time, and with an amount of misrepresentation of which I then thought him incapable, insisted on the appointment of a committee, which I, in common with some of the leading members of the Board of Irish Lights, considered flagrantly unfair'.[71] Chamberlain replied with what Tyndall termed a flimsy letter and Tyndall resolved to 'give it to him hot'. His intemperate response, deplored by his friends, according to Hirst, called Chamberlain's letter 'unveracious' as well as flimsy. Standing on his honour, and declaring himself, like Chamberlain, 'a son of the people, who in the sweat of his brain has eaten his bread', he explained the financial position he was in: 'my permanent income, after a life of toil, was £1,250 a year, with no pension in view. Two years had not elapsed before Mr. Chamberlain practically placed before me the alternative of surrendering my self-respect or of forfeiting £400 of the foregoing sum. I chose the latter, my permanent income being thus reduced to

Fig. 39. 'A Lighthouse – Not Professor Tyndall', *Moonshine*, 20th December 1884 (Joseph Chamberlain is the wave).

£850 a year. Of this sum £300 still comes to me from the Board of Trade. But were Mr. Chamberlain to behave to me, in my capacity of gas referee, as he behaved to me when I was scientific adviser to the Board over which he presides, sooner than submit to his intolerant conceit the £300 should follow the £400. Happily for me the public have read my works, thereby enabling me to eat my chop in comfort, with no fear of the President of the Board of Trade before my eyes'.[72]

Fig. 40. Comparing oil, gas, and electric lights at the South Foreland.

Wigham came to talk to him. Tyndall discovered that the Irish Board supported Douglass and that Lord Meath, the chairman, was 'in woe about their poltroonery'. He wrote a third letter to the *Times*, even longer than the second, in which he defended his defence of Wigham: 'to protect from official extinction an able and meritorious man, who had the misfortune to raise a rival at the Trinity-house, and to ruffle the dignity of the gentlemen of the Board of Trade'. He continued: 'With regard to Mr. Chamberlain's charge against me, of having indulged in "unworthy and ungenerous insinuations", those who know both him and me will laugh at such a charge. I do not deal in insinuations. When I repudiate and oppose the mean and grinding despotism which he would introduce into official life, I speak to him, and of him, not by insinuation, but with honest plainness of utterance'.[73] The unruffled Chamberlain responded again, quoting the Duke of Argyll's comment that he thought Tyndall had made a mistake.[74]

When Sydney Webb, Deputy Master of Trinity House, wrote to the *Times* on 8th December to announce the end of the South Foreland experiments (Figure 40), Tyndall reacted, arguing that he thought fair weather experiments valueless. His point was that the Elder Brethren, and indeed any committee, were not in his view able to undertake experimental work. It was for the trained experimental investigator. In relation to his work on fog signals, Tyndall paid tribute to Collinson, who had understood and supported the needs of the experimental investigator. He poured scorn on Trinity House for proposing to stop the experiments at South Foreland before the lights had been tested in the proper fogs of winter.[75] He suggested also that much public money could be saved by carrying out tests in the experimental room at Blackwall rather than at South Foreland. A further letter followed, in which he implied that the experiments were being rigged in favour of the electric light.[76]

Unable to let the issue go, he wrote again, giving testimonies from ships' captains of the power of the group flashing light at Mew Island.[77] But he now lost the goodwill of the new Irish Board, from which some of his supporters had gone, although several of its previous members and advisers lined up in print to support him. His seventh *Times* letter attacked the Board of Trade itself, and it is a measure of the forbearance of Farrer and Chamberlain that Tyndall remained a Gas Referee.[78] Others might well have sacked him, but this is a sign of a more accommodating political culture than would be the case today.

Part of Tyndall's ire was caused by the age-old clash of scientific evidence and advice with political ideology and power. He pointed out that back in 1861, in a passage he thought still applied, the report of the Royal Commission on Lights, Buoys, and Beacons had criticized: 'a department of the Government whose president changes with the Government, whose members are not selected for their knowledge of the science of lighthouse illumination, and who have not necessarily any officers specially instructed in that subject'.[79] Support came from an unexpected quarter, as Tyndall reported: 'I was glad to hear that my old enemy Tait had for once deemed me right with regard to the lighthouse question'.[80]

Tyndall missed the X-Club meeting at the beginning of December as he was in Preston, having been persuaded to distribute the prizes at the Harris Institute, (previously the Institution for the Diffusion of Knowledge and Preston's Mechanics' Institute). He met the President, Ashcroft at the station. Marquis was there, old and worn looking, and they went to Lune Street to see the huge room, the New Public Hall, in which the meeting was to assemble. More than 6,000 tickets had been applied for—according to the *Leeds Mercury* there were 5,000 present—and the editor of the *Preston Herald* told him his visit to Preston was 'almost to the level of a political event in importance'.

Tyndall took three great scientific principles as the basis for his talk, those he thought would be regarded as 'the glory of the present age'. The first was the principle of conservation of energy and the inevitable consequence, in the distant future, that the eventual radiation of heat would result in all things being brought to the same temperature. All work would cease and death set in. But, he asked, if equivalence reigns between the processes of thought and the molecular motions of the brain: 'Are we as rigidly bound in the domain of human consciousness as in the physical world?'.[81] Tyndall had no answer. All he could see was the 'absolute incongruity existing between thought—or even sensation—and molecular motion'. The second was the theory of evolution, met with cries of anguish and detestation when first enunciated, but now widely accepted by those who were so perturbed in spirit. Yet Tyndall's words betrayed a teleological belief, perhaps inspired by Spencer: 'And what is this doctrine but the realisation, in a wider sense than the poet dreamt, of the poet's faith, that "through the ages one eternal purpose runs"—that, beginning with the lowest forms of life, we find them nursed up through innumerable generations until we reach the being who stands "foremost in the files of time"—the man of to-day?'.[82] He ended with germ theory, praising Koch and Pasteur, who built on the work of William Budd and others, and Pasteur's experiments on the use of an attenuated organism to create immunity against splenic fever (anthrax) in sheep and cows.

The Tyndalls slept in their new house for the first time just before Christmas. They were not yet secure in their desired isolation. Their neighbour, Meeson, had bought five acres of land in February, intending to build in front of them. Simmons helped stymie him by offering another neighbour £120 for three acres, for which Meeson offered £100.[83] Tyndall wrote to Simmons to say that 'at whatever cost to myself, no more land than he now possesses must fall into the hands of this varlet'.[84] He offered a further £400 for Miss Balfour's land, thirteen acres of Longdown heather, and declared: 'Meeson only remains to be eradicated, and we shall be monarchs of all we survey'.[85] Relationships with the architect were not perfect either. Penfold wrote to say he thought they were not happy with him and suggested they get another architect to complete the work.[86] In the midst of finishing the house, and more letters in the *Times* on lighthouse business, Tyndall had to prepare for the six Christmas Lectures which he was due to give 'On the Sources of Electricity', while finally sending off to the printer on 20th December the introduction to *Louis Pasteur: His Life and Labours*. On Christmas Day the Tyndalls dined with the Pollocks at 59 Montagu Square, 'a joyful little evening', and then he was into his Christmas Lectures, interspersed with dinners and walks.[87]

William Bowman, now Secretary of the Royal Institution, had asked Tyndall to give the first Discourse of 1885, just a week after his last Christmas Lecture. With little time to prepare he chose the topic of 'Living Contagia', basing the lecture on Pasteur's work, which was fresh in his mind. He wrote to Koch for up-to-date information, and Koch sent him 'a few specimens that are as intensely stained as possible, two of cholera bacilli and a tissue section with tuberculosis bacilli'.[88] In response to a question from Tyndall, he followed it up with a long letter explaining that he stood firmly by his claim to have isolated the true cholera bacillus.[89] Tyndall used the opportunity to highlight again the importance of animal experimentation in preventing far wider animal and human suffering.[90] He reiterated this in a letter to the *Times*, which had been carrying correspondence on the subject of vivisection, published a few days after the Discourse.[91] As he told Hirst: 'The Duke of Northumberland was in the chair, and I feared he might object to my outspokenness, as to the necessity of protecting the experimental physiologist from the mischievous legislation. But he was more than usually cordial after the lecture, saying that he entirely agreed with me, though he would not like to be the vivisector: With this sentiment I entirely sympathised'.[92]

Tyndall was in demand for public events. He was asked to preside at the Sanitary Congress and attend many others, all of which he declined. Though he was tempted to yield to an appeal 'so earnest and urgent' from the Sunday Society of Glasgow that he felt he ought to go, he changed his mind a few days later.[93] His health was again suffering, starting in mid-January with an injured foot, and he was laid up for several weeks until early April. Part of the time was spent on the final report of the Accidents in Mines Commission. Just before Easter he managed a country house weekend at Knebworth with the Lyttons, hearing Lord Lytton reading from his new poem *Glenaveril* after dinner in the drawing room.

Tyndall had never been a political activist. But he was a visible public figure and a vocal anti-Gladstonian. His visibility inspired Charles Grant, President of the Pollockshields Conservative Association, to invite him in May 1885 to stand as a Conservative for the Glasgow constituency.[94] Tyndall telegraphed that he would respond fully in due course, but did not do so until 10th October, just before he left Belalp, claiming that he was 'suddenly broken down at the end of January, and confined, by an ailing foot, to my armchair for three months', and had waited for sufficient vigour to return.[95] He wrote a response that could be published if Grant wished, expressing his personal position: 'Were intermittent members permitted in the House of Commons—members at liberty to speak, when, by virtue of any special knowledge or for some other reason, they had anything profitable to say—I should willingly accept election to that class. But the permanent atmosphere of the House would not suit me. I belong to no party. I have not even a vote. In regard to politics I am merely an interested looker-on'. He did not mention that in December 1884, he had been asked to stand as a Liberal for the Holborn Borough. Strongly criticizing the Gladstone government's forays into the Transvaal and Sudan, Tyndall finished with his views on their Irish policy. Parnell, in Tyndall's view, had proved more than a match for the Gladstonians in his 'strength and stability of character, . . . tenacity of purpose, and definiteness of aim', his aim being 'to break up an empire which he hates, and which he has taught his followers to regard as a curse to humanity'.[96] The letter was published by Blackwood and found favour with Lord Lytton.[97] Joule was inspired to write to Tyndall in support of his denouncement of Gladstone.[98] Hirst called the letter 'a violent assault on Gladstone. In substance I agree with what he says but I regret the animosity he displays. He is afflicted with Gladstonephobia'.[99]

Tyndall was at Hindhead when he received the letter from Grant inviting him to stand as a Conservative, and he read it proudly to the Duke and Duchess of Abercorn and Lady Winterton, whose impromptu visit put Louisa to rushing around preparing the drawing room and sitting room.[100] They drank 'nearly 2 bottles of Lafitte'.[101] He was in the middle of another bout of sleeplessness, writing: 'This seems to be the hardest bout of sleeplessness that I have yet had—but I may be forgetful, and I know that matters must have gone very hard with me in 1853 in Berlin'.[102] Reading Carlyle's life, he sympathized: 'It seems almost as if I were reading my own, his sufferings from sleeplessness are so similar to mine'.[103] He managed to give his five lectures on 'Natural Forces and Energies' in May and June, but missed the unveiling of Darwin's statue in the Natural History Museum on 9th June, the day Gladstone's ministry resigned, ushering in a government by Lord Salisbury before a general election.[104] He would not have imagined it, but those lectures would prove to be his last series at the Royal Institution.

The lure and importance of the Alps for Tyndall is illustrated by the length of his visit in 1885. He went out in early July without Louisa, who joined him a month later, and stayed until mid-October, when the snows finally drove them away. Towards the end of September, Hirst wrote plaintively for news, and Tyndall quickly obliged: 'We have had a most glorious season here … By day cloudless skies, & after sunset glows which put to shame the famous ones ascribed to the dust of Krakatoa … We have just finished reading the proofs of the 4th Edition of my Lectures on Light. These will bring us in a little money, which is much needed at present'.[105]

The need for funds was caused by Tyndall's machinations to prevent Meeson or others buying up land near him in Hindhead to build on. He had discovered that thirty-seven acres of land opposite his house was for sale, and instructed Simmons not to allow it to fall into the hands of speculators. Against stiff competition, Simmons had to bid £1,900, forcing Tyndall to borrow £1,200 at 4% per annum (he paid off the last of the debt in 1888). He ruefully noted that he had overpaid by some £500–600, but had driven away Meeson, who appeared to have sold his house in October.[106] Instead of retiring, he now had to 'remain in harness for a couple of years' to clear his liabilities. Victorian society invaded Belalp. Lord and Lady Lytton arrived with their two daughters and servants, carried up the 5,000ft in five *chaise à porteurs*. Tyndall commented: 'The number of attendants must have reminded Lord Lytton of his marches in India'.[107] The Bishop of Gloucester & Bristol stayed his usual three weeks. Lord and Lady Lingen stayed for five; he had just retired as Permanent Secretary to the Treasury. They were frequently on the glacier together, dining in the hotel or climbing the Sparrenhorn. The only blot was

Lingen's attitude to vivisection: 'Here his practical wisdom seemed to fail him'.[108] At the end of September, the snow rolled in but they were warm and comfortable. By 10th October the hotel was empty, and they were alone on the mountain, eventually leaving on 16th October after stuffing the chimneys and damming off the water. Before landing at Dover, every little handbag and roll of rugs was examined for explosives. The Fenian dynamite campaign had culminated in explosions in the House of Commons chamber and Westminster Hall in January, and security was still tight.

In the autumn, Tyndall helped put the finishing touches to the report of the Accidents in Mines Commission. He dined at the Athenaeum with Spencer and Hirst—Debus had been asked, but refused, still smarting over sharp words from Tyndall at the Philosophical Club nearly a year earlier—and saw Hirst off abroad for his health. Hirst was in tears on leaving, 'and John tore Louisa away from my carriage door to prevent accident'.[109] A few days later, Tyndall was shocked to hear of the death of Carpenter, from an accident with an alcohol burner. As Hirst put it: 'Burnt to death with the alcohol from which he had so long abstained and against which he had always preached'.[110] In Tyndall's view, Carpenter was a tremendous talker, but an original and penetrative thinker.[111] His death reminded Tyndall of a near-disaster of his own, when alcohol sprayed over him from a tube he was heating, and ignited. Unable to put out the flames, he pulled off his coat and lay on the floor while his students put the coat over him, extinguishing the flames.[112]

The 1885 general election took place from 24th November to 18th December. For the first time, after an extension of the franchise and redistribution of seats, a majority of adult males could vote. Tyndall had been pressed to support William Brodrick, the local Conservative candidate, and spoke at an unruly meeting in Haslemere on 18th November, with Lord Winterton and Brodrick. It was the first conservative meeting he had ever attended, for which he gave up 'the pleasure of dining with the Benchers of the Inner Temple'.[113] Winterton claimed Tyndall had abandoned liberalism and gone for conservatism. Tyndall saw it otherwise, declaring he 'was as sound a liberal as ever I was. It was the so-called "liberals" I had forsaken. I said that the new liberalism was a bad copy of the worst form of old Toryism...I said that I had no vote, for though appealed to by both parties, I so scorned party politics that I never qualified myself to vote'.[114] Rowdy Gladstone supporters let off fireworks outside, and crackers inside the room. So offended was Tyndall by this behaviour that he penned a letter to the *Times* denouncing it, declaring: 'When, some months ago, I received circulars from the Liberal and the Conservative election agents for East Surrey, my dislike of party politics was so great that I threw both circulars into the waste paper basket. I now regret doing this; for, having seen and heard Mr. Broderick...and having witnessed the scandalous conduct of those who sought, by idiotic uproar, to make the hearing of him impossible, I should be glad to give him the practical support of my vote'.[115]

Tyndall's vote was not necessary, as Brodrick won Guildford for the Conservatives. In any case, having registered for the first time, he managed to vote Conservative in two other constituencies: for W. H. Smith (founder of the eponymous business) in the Strand Borough, and the very next day—it was possible at the time to vote in more than one constituency—in the St George's Hanover constituency, for Lord Algernon Percy (both were elected). As he put it: 'The act of voting is a novel experience to me'.[116] The Liberals, led by Gladstone, won the most seats but not an overall majority. The Irish Nationalists held the balance of power. That exacerbated divisions within the Liberals over Irish Home Rule and led to a split and another general election in 1886. The Home Rule question loomed large over Tyndall for the rest of his life.

26

THE FINAL YEARS
1886–1893

In late 1885, stretching for many months into 1886, Tyndall became very ill, more ill than he had ever been. He had been taking Bromidia every night—a toxic combination of chloral, bromide, cannabis, and henbane—which he claimed disturbed his brain 'less than brandy and water'.[1] He really should not have given the opening Discourse on 22nd January 1886, on 'Thomas Young'.[2] Jane Thompson, wife of the physicist Sylvanus Thompson, described his appearance on the day: 'He is a weird-looking, thin, stooping old man, with long grey hair hanging from a high narrow head. He has a decided Irish brogue now and then, and used curious gestures'. Tyndall broke down after the lecture, which would be his last at the Royal Institution. William Preece said to Sylvanus Thompson afterwards that he hoped, if they came to that stage, that someone would prevent them making an exhibition of themselves.[3] It was a sad and embarrassing end.

The loyal Louisa described Tyndall's state to Hirst in early February: 'For more than three weeks before his lecture he was in a low and terribly sleepless state, taking things at night to help him, which sadly poisoned him. The lecture itself was fairly well got through, though he looked pale and tired. But when it was over came a complete collapse…I at last persuaded him to let me call in Gull to see him…I still feel troubled about the future, because this downfall came not after any hard work, or imprisonment in London; but after a long and free time at Hind Head'.[4] Tyndall thought he could still give his four lectures on 'Light' before Easter. But things were still bad a month later, and Gull put a stop to the idea, suggesting that he should not think of working for a year or more.[5] The Leckys were particularly kind to him during his illness, and Debus visited, putting their differences behind him.[6]

The Tyndalls left for Hindhead at the end of March. Tyndall had remained in London since the breakdown, where treatment and support was easier to obtain, and by early April he was back at the Royal Institution, attending a Managers' meeting.

He asked for Dewar to continue as Superintendent of the House because of his illness, and the Managers asked him 'to refrain from the performance of his duties for twelve months'.[7] After this he improved gradually, and was well enough to attend a postponed X-Club meeting on 3rd June. Nevertheless, he did not recover completely. Spencer, having asked him to comment on drafts of his autobiography, implored him not to. He wrote that he was not well either, and Tyndall replied: 'I seem to be much in the same condition with yourself. I am rather clinging to life than living... The accounts from Huxley are not quite satisfactory... Hirst also I hear is by no means well. So it would seem as if we were all breaking up together'.[8]

Unlike the previous year, the general election of July 1886 seems to have passed Tyndall by. The new Liberal Unionist party, led by Lord Hartington (later the Duke of Devonshire) and Joseph Chamberlain, effectively gave Lord Salisbury's Conservatives a majority, but did not join a formal coalition. Gladstone's Liberals, who supported Irish Home Rule, and Parnell's Irish Parliamentary Party, lost. The Tyndalls left for the Alps on 11th August, postponed because of his illness. Busk died on 10th August, the second of the X-Club to go, and on the eve of his departure, Hirst brought Tyndall the news, which affected him deeply.[9] It was not until 15th October, just before he left the Alps, that he summoned the ability to send condolences to Ellen Busk, writing: 'I never knew a man possessed of a more genuine love for science in its best and broadest sense... Had he wrought more for personal ends—in other words, had he been less noble than he was—he might have been more talked of by the public; but his pleasure in working himself, and in helping others to work, was of a kind which the public can neither give nor take away'.[10] Tyndall missed the funeral at Kensal Green Cemetery, but Hirst gave news of the event: 'His four daughters were there; but, of course, not Mrs Busk herself. She bears up bravely however... The X were represented at the funeral by Hooker, Lubbock, Frankland and myself. The three remaining ones, themselves broken in health, were, as you know, far away'.[11] Busk and Spottiswoode had served the Royal Institution for thirty-eight years between them, including twenty-seven as Secretary or Treasurer. That support to Tyndall had now gone.

On his arrival at Brig, Tyndall engaged a horse for himself, in addition to two luggage horses, in case his strength proved unequal to the march upwards on foot. But he was fine, reaching Alp Lusgen in 6 hours, and the maids had the advantage of the horse.[12] He was immediately out and about: 'We crossed the glacier one day to the Rieder Furca. Owing to the shiftings of the ice, I missed my way, led Louisa into danger, & had to cut her out of it'. They were blessed with fine weather well into September, and exercise on the glacier and the mountains did Tyndall enormous good. Bowman wrote to exhort him not to work. The Bishop of Gloucester came and went, and the Burdon-Sandersons. There was good news on land. Salzmann, the President of the Commune of Naters, wrote to say that there were no more obstacles in the way of finalizing the deed for the sale of three Fischel of land to Tyndall, which had been agreed in 1877.[13] Just as with Hindhead, the purchase was made to render it impossible for anyone else to build a house nearby. To this day the chalet is in a beautiful, lone position. The Tyndalls stayed on until the middle of October again until, surrounded by snow, they set off back to England.

Soon after he returned, a misunderstanding over the Christmas Lectures led slowly but inexorably to his rupture from the Royal Institution. According to Hirst, Tyndall had wanted to give them, but Dewar, who had been most supportive when Tyndall was ill, had undertaken to replace him and had failed to respond to Tyndall's request to meet him before the Managers' meeting to discuss the question. As a result, Tyndall appeared to have been irritated, and a 'certain coolness' resulted between them.[14] Hirst tried to intercede, but without success. Tyndall stayed quiet at the Managers' meeting on 6th December and the meeting of members afterwards, considering it 'more gentleman-like to ignore what I might otherwise dislike'.[15] The Managers upheld the decision for Dewar to give the lectures. To Hirst, Tyndall said he had 'yielded much in order to avoid the occurrence of a jar in the presence of a Board before whom no jar has occurred for the three & thirty years of my connection with the Institution. The matter may end by my giving up the Institution altogether; but whatever I do shall be done in a quiet and dignified manner'.[16] There the matter rested until the New Year.

Tyndall spent most of December and January at Hindhead, coming occasionally to London for the Harrow Board, the Gas Referees and the December X-Club dinner at the Athenaeum, with just Lubbock, Hooker, and Frankland. He enjoyed the Royal Society's Anniversary Dinner, where he was asked by Stokes to propose the toast of her Majesty's Ministers coupled with the name of the Lord Chancellor. He was agreeably surprised to find that the Lord Chancellor was Giffard (now Lord Halsbury) who had defended Governor Eyre.[17]

At Hindhead he read Tennyson's *Locksley Hall Sixty Years and Afterwards*, the sequel to the poem the Drummond daughters had found so unforgiving thirty years before, finding it 'extremely fine'.[18] The snow came in, necessitating cuttings three feet deep to clear the roads. Tyndall returned the proofs of Spencer's *Autobiography* while he worked hard on the hieroglyphical researches of Thomas Young for a possible article.[19] Lady Lytton telegraphed at the end of December to say that the roads were clear of snow, and invited the Tyndalls to Knebworth. They went on 1st January 1887, though Tyndall spent all of one day in bed before returning to London in time for the X-Club meeting. Lord Lytton's views on the impact of extending the franchise doubtless chimed with

Tyndall's. Lytton wrote: 'I have long been convinced that Parliamentary Government is an institution which cannot be efficiently worked by a democracy ... The most efficient restraints were placed upon the licence of Parliamentary debate by the unwritten traditions of a House of Commons elected on a restricted suffrage. These traditions once gone, no mere rules can adequately supply their place'.[20]

Tyndall's leave of absence from the Royal Institution was nearly up in March, and he decided to resign. Before doing so, he offered a handsome donation of £200, which the Managers felt unable to accept.[21] He wrote to Pollock to insist that no fund was set up for him, by testimonial subscription or pension.[22] Hirst, writing from Biarritz, and remembering being there with Anna, offered to share his pension if Tyndall was in need.[23] The Tyndalls felt some concern at the loss of income—now £550 per annum—but they considered themselves secure, and the removal of the responsibility of lecturing was a relief. In his resignation letter, Tyndall informed Sir Frederick Bramwell, recently appointed Secretary, that even though he had returned from the Alps in October refreshed, a long spell of insomnia had knocked him back and he felt he had to go.[24] The real reasons, he told Hirst, were the demands of lecturing at specific times, the appointment of a new assistant, and his changed relationship to the chemical laboratory.[25]

Tyndall's resignation was accepted at the Managers' meeting on 21st March. Lord Rayleigh was immediately appointed as Professor of Natural Philosophy in his place, and Dewar took charge of the house. The Managers resolved that one of the annual courses of lectures should be called the Tyndall lectures. Bramwell wrote to convey the resolutions of the Managers, one of whom—Frederick Pollock—had been a Manager when Tyndall joined in 1853. Tyndall was bowled over by their good wishes: 'I hardly dare trust myself to dwell upon the "Resolutions" which you have conveyed to me. Taken in connexion with the severance of my life from the Royal Institution, and with the flood of memories liberated by the occasion, this plenteous kindness, this bounty of friendship, this reward so much in excess of my merits, well-nigh unmans me'.[26]

Huxley wrote to Tyndall in May, to say that Lockyer had told him that people would like to do a large dinner for him.[27] Tyndall acquiesced, replying: 'Lockyer's friendly initiative in this matter is all the more agreeable that it was unexpected. In fact, it is certain that I shall die in peace with all the world—except Gladstone'.[28] Arranging the event was a serious affair. An Executive Committee was established, chaired by Stokes, with Lockyer and the physicist Arthur Rücker as secretaries.

The dinner, for about 200 guests, took place on 29th June in Willis's Rooms. Stokes, as President of the Royal Society, backed by Lords Derby and Lytton, and Tyndall himself, received the guests. Tyndall sat at Stokes's right hand, with Hirst immediately in front, Debus at his side and Frankland opposite. Huxley was unable to be there. In an editorial, *Nature* highlighted Tyndall's contribution, through his lectures and books, of bringing 'the democracy into touch with scientific research'.[29] It argued that while the triumphs of applied science were readily understood, the methods and conditions of scientific thought, an essential condition for national well-being, were not. As *Nature* put it, Tyndall had done more than perhaps any other living man to show the value of knowledge for its own sake and not just because of its immediate practical utility: 'Others will rank beside or above him as investigators, but in the promotion of the great scientific movement of the past 50 years he has played a part second to none'. That analysis stands up more than a hundred years later.

Stokes gave a generous speech, touching on Tyndall's work on diamagnetism, glaciers, radiant heat, and abiogenesis, and emphasizing his contribution as a communicator.[30] He referred to the scientific stimulus of Tyndall's dispute with Magnus, recognizing that: 'socially I doubt not that they were the firmest friends'. An eloquent reply from Tyndall reinforced his view, first stated in 1854, of the importance of technical education, not least to help improve working conditions, and the importance of the freedom to study science without immediate thought of application. Tyndall claimed the laws of conservation of energy and of evolution as the two great Jubilee offerings from Science for Victoria's reign, adding a third of spectrum analysis. He finished with a mountaineering analogy: 'the hardest climb, by far, that I have ever accomplished, was that from the banks of the Barrow to the banks of the Thames—from the modest Irish roof under which I was born to Willis's Rooms'.

* * *

The final part of Tyndall's life was dominated by his increasingly strident position on Irish politics, and a humourless justification of his role in the lighthouse question. As Sir James Crichton-Browne later said: 'Perhaps one touch of humour might have saved Tyndall from some extravagances into which he ran, but it might also have made him tamer and less picturesque than he was'.[31]

Free of the Royal Institution, one of his first acts was a further letter to the *Times* on lighthouse illuminants, coupling Gladstone's 'mad and wicked' Irish policy with that of the Board of Trade towards the Irish inventor, Wigham. He contrasted the behaviour of the Board of Trade to that of Gladstone, yielding to external pressure, and bemoaned the unfair contest between Wigham and James Douglass at the South Foreland, the former provided with no support and the latter with huge resources at his disposal, while Wigham's latest and most powerful invention— the double quadriform light—was excluded.[32] He poured scorn on the idea of the Board of Trade that the limit might have been reached in illumination beyond which no further practical advantage could be gained. That was not how science worked. In that he was supported by Lord Meath, who wrote to describe the opposition of Trinity House and the Board of Trade over many years to Irish innovations.[33] His view of Chamberlain had changed, however, as Chamberlain had broken from Gladstone over Home Rule. He wrote again to the *Times*: 'You mention Mr. Chamberlain as my successful opponent...But, however that may be, he has all the admiration, on my part, which his patriotic and courageous conduct, at a crisis when patriotism and courage seem to have been discarded by the majority of the Liberal party, so deservedly inspires'.[34]

Tyndall's most outspoken pronouncement on Ireland, nailing his Orange colours to the mast, came on 8th June, stimulated by the attempts of Sir George Trevelyan to unite Liberals behind Home Rule. Using the analogy of crystallization, Tyndall observed the irresistible force bringing many Conservatives and Liberals together in the cause of the Union. He continued: 'Much as I should deplore the necessity of doing so, sooner than hand over the Loyalists of Ireland to the tender mercies of the priests and Nationalists I would shoulder my rifle among the Orangemen (Figure 41). The fear of them and the dread of them have certainly acted as a powerful restraint upon rebellion in Ireland'.[35] A letter written to Charles Grant and sent to the *Times* appeared on 26th July, and another on 9th August, declaring that 'Ulster is strong enough to protect itself. The blood of the heroes of the Reformation still stirs...and it never will submit to be ruled by the Romish priesthood of Ireland...In the name of freedom, in the name of justice...I protest against these men, among whom I learnt to read and love my Bible, being handed over to their hereditary enemies, among whom their only desire is to live in peace'.[36]

Tyndall had intended to leave for the Alps on 11th July, leaving Louisa to follow, as Douglas had been ordered to take a long sea voyage for the sake of his lungs and she wanted to see him off. In the event, Tyndall delayed departing for three days to attend a garden party at Hatfield House with Lord Salisbury and to be presented to the Queen. In his absence, the ever-faithful Hirst joined Louisa at the Royal Institution, packing up thirty years' worth of possessions.

At the hotel on Belalp, Tyndall reported the clientele 'almost to a man sound unionists'.[37] Drizzle and snow had followed a prolonged period of fine weather, which had given a bountiful season to the Swiss hotel keepers. Du Bois-Reymond's son passed through, and Tyndall was bountiful too, giving him a loan to enable him to extend his holiday.[38] The hotel was nearly empty by the end of September, but the Tyndalls were still *in situ*, revelling in their isolation. As he told Hirst, they had already eaten 'four sheep, several cocks & hens, & a considerable quantity of potatoes, carrots, and salad...In our safe hang the remains of our last sheep, and we have contracted for another, small and fat, which will carry us to the end of our stay here. I

THE SCIENTIFIC VOLUNTEER.

"If ever I have to choose . . . . I shall, without hesitation, shoulder my rifle with the Orangeman."—*See Professor Tyndall's Reply to Sir W. V. Harcourt.* "*Times*," Feb. 13, 1890.

Fig. 41. The Scientific Volunteer, *Punch*, 22nd February 1890.

have been able to help the people a good deal both in the dressing of wounds and in the administration of Gregory's powder'.[39] In recognition of his contributions, Tyndall was made an honorary citizen of the Commune that year. The Tyndalls departed long after anyone else had left Belalp, to be met at Charing Cross by Hirst and Lady Claud Hamilton.

Tyndall resumed the political fray, writing to Charles Grant that he wanted to get a statement on Gladstone's Irish policy from the 'Scientific men of Britain', who he asserted were predominantly liberal. He told Grant: 'At present science is represented in the House of Commons by that silly creature Stewart, and by good natured Roscoe who knows nothing about the Irish question. Playfair hardly counts'.[40] That proved embarrassing. Grant ignored the private nature of the letter and published it, to Tyndall's discomfort. He wrote: 'Professor Stuart I do not know personally, but with Roscoe & Playfair I have hitherto stood on terms of intimate friendship which, I fear, will now be disturbed'.[41] He tried again after Christmas, writing to Huxley to seek his support for a letter that he hoped forty or fifty men of science would sign 'to record our deliberate conviction that the Irish policy of Mr Gladstone is fraught, not only with possible danger, but with certain disaster to the British Empire'.[42] It was written in true tub-thumping Tyndall style. Hooker approved of it and said he would speak to Huxley.[43] But after they met at the Athenaeum, Huxley tactfully wrote: 'Both he and I entirely agree with what you say about Gladstone and his policy but doubt whether it will not be objected to by many as having too much the air of a personal attack'.[44]

The strain of Tyndall's retirement from the Royal Institution carried over into the removal of his effects. He wrote to the Treasurer, Henry Pollock, to outline his understanding of what belonged to whom, while Louisa looked for a place to store their furniture.[45] At the end of the month they found three back rooms on the sixth floor of a mansion in Victoria Street as a temporary home. Louisa pointed out that the height of their three dwellings increased by a common multiplier of ten, the Victoria Street flat being at 8oft, Hindhead at 8ooft, and Alp Lusgen at 8,oooft. Bramwell replied with gratitude on behalf of the Managers. He returned, as Tyndall had requested, all the diagrams which had been used to illustrate his lectures.[46] But all was not resolved, and Tyndall attended a meeting of the Managers in early December, at which he had an encounter with Bramwell. Tyndall declared to Hirst: 'He came off second best. He is like a great dunghill cock whose mere volume influences the smaller chicks around him, but who is not proof against a game spur'.[47]

In the end, given the tension between the officers of the Institution and Tyndall, Tyndall arranged personally with Lord Rayleigh which instruments and samples should remain for him and which should be removed.[48] A little later the Tyndalls stayed with the Rayleighs at Terling Place. Not a word was spoken about the Royal Institution. Tyndall liked the trees—some fine cedars, which were horribly smashed by the previous last winter's snow, and a type of elm like those he had seen at Newhaven in the United States—but he did not like the location: 'I would not live in Essex if Terling Place were given to me, and £5,000 a year to keep it up'.[49] He attended the Royal Society dinner in November, with Frederick Pollock as his guest, hearing discontent that Stokes, as President of the Royal Society, had become a member of the House of Commons. Tyndall wished he had left politics alone, as he thought he could not serve science and politics together. He was at the X-Club on 8th December. Though he did not intend it at the time, it proved to be his last.

Tyndall opened 1888 with a further salvo in support of Wigham, writing twice to the *Times* quoting the glowing reports of the new gas lighthouse on Tory Island from two experienced steamer captains.[50] He suggested that the responsible Minister in the Board of Trade, Baron de Worms, should 'liberate himself from the thrall of officials whose raison d'être, as consumers of foolscap, has been so forcibly exposed in The Times, as well as from the antagonism of officials directly interested in its extinction, and to see fair play shown to this admirable Irish invention'. At the end of April, revealing that he had shaken hands with Chamberlain at the Athenaeum, Tyndall told Hirst of his plan to tell the 'round unvarnished tale of that lighthouse question'.[51] Tyndall believed that Knowles had agreed to three articles of twenty pages each for the *Nineteenth Century*.[52] In the end, one appeared in the *Nineteenth Century*, in July 1888, and two in the *Fortnightly Review*, in December 1888 and February 1889.[53]

Wigham wrote to Sir Michael Hicks-Beach, now President of the Board of Trade, to ask for the affair to be put to a committee of Tyndall, Sir Howard Grubb, and Stokes. The Board of Trade immediately ruled out Tyndall and Grubb, as not being acceptable to all parties. After much prompting by Wigham, they asked Stokes if he would name his own small committee, as the affair meandered into 1889. Tyndall's ire was further aroused when the Royal Institution invited James Douglass to give a Discourse on lighthouses in March. He wrote to Lubbock: 'Now I have taken no part whatever in Royal Institution proceedings since my retirement. The whole spirit and conduct of the individuals, who have pushed themselves into power there have been to me, so repellent, and their behaviour to myself so unaccountable, that simple pride has compelled me to hold aloof'.[54] For the enquiry, Stokes nominated Lord Rayleigh and Grove to act with him (owing to Grove's ill health, William Thomson replaced him), and Wigham agreed to be bound by their findings.[55] Their report, in October 1890, broadly supported the overall preference for electricity. They acknowledged some benefits of gas over oil, but ruled that oil was to be preferred in general, for simplicity and economy. Even so, they felt that there were circumstances in which gas would be preferable, and that all three should for the time being remain in use. A political compromise, of which Tyndall would not have been capable.

Tyndall was in good spirits in the spring of 1888, alternating life at Hindhead with dinners in London: with the President of the Institution of Civil Engineers at the Albion in Aldersgate Street; at Fishmongers Hall to hear an 'admirable speech' from Goschen, the Chancellor of the Exchequer;[56] and the Anniversary Dinner of the Philosophical Club, a few days after another dinner party at Hindhead for the Williamsons, the Pollocks, Lady Claud, and Rollo Russell. At the end of the Dinner of the Royal Academy of Arts, in early May, Tyndall responded on behalf of 'Science', and Lecky for 'Literature'. Recalling that this dinner was one of the two celebrations that gave Faraday real pleasure—the other being for Trinity House—Tyndall drew Art and Science together: 'In this respect Art and Science are identical—that to reach their highest outcome and achievement they must pass beyond knowledge and culture, which are understood by all, to inspiration and creative power, which pass the understanding even of him who possesses them in the highest degree'.[57]

Lady Welby stayed with the Tyndalls in June. She was a philosopher, a prolific writer, and a correspondent of many of Tyndall's circle. Tyndall described her as 'a being on whose spirit the clay of human nature lies very lightly. She lives in a beautiful metaphysical haze, but it is surprising what hard, practical, and brave utterances sometimes emerge from it'.[58] He told her that he was committed to two tasks before going out to the Alps in 1888, which were the first article on lighthouses for the *Nineteenth Century*, and a biography of Faraday for Leslie Stephen's *Dictionary of National Biography*.[59]

In the biography of Faraday, Tyndall remained uncompromising on diamagnetic polarity, writing: 'The explanation of the complex phenomena of magne-crystallic action was rendered impossible to him through his rejection of the doctrine of diamagnetic polarity. Applying this principle to magnetic and diamagnetic crystals the force proper to each is always found acting in 'couples' in the magnetic field, and from the action of such couples the observed phenomena flow as simple mechanical consequences'.[60] From Belalp, in 1887, Tyndall had written to Hirst that he was preparing a new edition of *Researches on Diamagnetism and Magne-Crystallic Action*, 'which I hope will bring me in a little money, and a little repute ... The reading of the proofs has revived the fascination which the subject once had for me. It is a juice of true & faithful work of the most difficult description'.[61] The edition was dedicated to Wilhelm Weber, who was touched.[62] When Hirst received his copy he wrote: 'The controversies which once bristled around the subject have now all subsided and, victory, indisputable victory, remains to you, as a reward for your conscientious and clear-sighted perseverance ... You may count this amongst the most perfect and complete of all your successes'.[63] History, perhaps unfairly, has not given the same verdict.

As soon as he had finished the draft of his article for Stephen, the Tyndalls set out for the Alps. Tyndall spent most of the first week, in poor weather, looking after the sick, having finally persuaded them to open their bedroom windows as he could not stand the stench. Hirst heard of the death of Clausius in Bonn, and wrote immediately to Tyndall.[64] The death was a shock to Tyndall, who had a greater regard for no German man apart from Bunsen. He heard also from Fred Pollock that his father was not well (he died just after Christmas). Writing to sympathize with Hirst's poor health, Tyndall revealed his opinion of doctors: 'Your case & Pollock's case have revived my old & profound scepticism regarding doctors. There is more or less of the pretender in the best of them. Andrew Clark when he has a vigorous constitution, like that of Gladstone or of Tennyson to work upon, can often help them…But in more difficult & delicate cases, he and the rest of them show the most helpless incapacity'.[65] Nevertheless he had been impressed for years by their earning power, once remarking in astonishment in 1853 that Bence Jones earned 40 guineas for an afternoon's work.[66]

September brought beautiful weather. By mid-September the number of hotel guests had already fallen from more than a hundred to less than thirty: 'Indeed the time is approaching when we shall be quite alone. Louisa likes this solitude the best'.[67] While they were away, they had let the Hindhead house for two months to friends for a hundred guineas, which would pay for their time in the Alps. Louisa was also planning a stable, to make the house more attractive to rent. Fred Pollock informed him, with a nod to Tyndall's dislike of formal dress: 'You will have to perform exorcism, and fumigation, when you come back in October. For I have seen people wearing dress coats there…Otherwise your property has not suffered any injury that I know of'.[68] On his return home, Tyndall wrote a letter to *Nature*, 'Alpine Haze', expressing the view that the haze in the valleys below them, which reminded him of his 'actinic clouds' and which persisted even in the heat of the Sun, could not be aqueous. He suggested it was caused by pollen or other organic matter in the air.[69]

* * *

After his resignation from the Royal Institution, Tyndall spent most of the rest of his life at Hindhead, apart from his summer trips to the Alps and sporadic visits to London for a few remaining official duties or occasional dinners. He was not at the X-Club on 8th November, where Hirst learnt that Huxley was to have the Copley Medal and passed on the news.[70] Tyndall took it in a matter of fact way, telling Hirst: 'The Copley medal will, I hope, cheer Huxley, & thus be a cause of rejoicing to his friends'.[71] He told Hirst he planned to come to the Anniversary Dinner on 30th November, where he sat next to Huxley. He was still active in Royal Society affairs, lobbying Stokes in January for Foreign Members and Fellows.[72]

The New Year of 1889 brought a welcome improvement in relations with Debus. Tyndall wrote to Hirst: 'I am glad he has thawed, for he was so extravagantly wrong that I should never have broken the ice'.[73] Apart from a dash to London for the Gas Referees on 18th January, he was at Hindhead battling nausea and sleeplessness while trying to complete both the second part of the 'Story of the Lighthouses', and the biography of Faraday for Leslie Stephen, due at the end of the month. His insomnia continued into February, and he turned down to his great regret an invitation from Lubbock, whom he had not seen for some time but whose political progress he followed through the newspapers.[74] He was spending two days each month in London on official business, for the Gas Referees and occasionally the Harrow Governors, but continued to miss X-Club meetings. The Faraday article was finally completed, and Jane Barnard was grateful, writing: 'You too have the magic of making your readers believe they know all about a subject of which they know little'.[75] More remunerative was an article for the American magazine *Youth's Companion*.[76] Tyndall told Hirst: 'They offer me fabulous sums for articles.[77] How serviceable such moneys would have been when you & I were in Marburg. They come when they are not needed'.[78] Tyndall was besieged by requests, as he explained to Lubbock, who was seeking a contribution from him for a book series: 'Hardly a day passes without an amount of pressure from English men & English associations, and from American men & American Associations, to do such and such a piece of work…sums that make even your liberal terms look small are offered to me from the United States'.[79]

Tyndall had long been absent from dinners of The Club, but an invitation card reached him indicating that it was his turn to chair on 12th March. He put up at the Hotel Windsor and saw Hirst and General Wynne at the Athenaeum. Many members told him that it was purely to meet him that they had come up from the country.[80] Lord Salisbury, the Prime Minister, sat to Tyndall's left and others present included Arthur Russell, Lecky, and Hooker. Several dinners followed. He was at the Gordon Dinner at the Metropole on 22nd May, where the Duke of Cambridge took the chair. Tyndall proposed the toast to the Duke, eulogizing his hero Gordon, which was well received.[81]

But as the old order changed, the ageing and crusty Tyndall felt increasing disenchantment. He decided not to go to the anniversary dinner of the Philosophical Club: 'I used to take delight in its dinners and those of the Royal Society Club. At one time I had almost resolved to dine nowhere else. But both clubs are changed; the members have become indifferent and the meetings languish for want of simple numerical life'.[82] Hooker sympathized: 'Bramwell's being elected into the Phil. Club has polluted that once fair stream, and it makes me look with less regret at the probable dissolving of that historic body'.[83] Others were withdrawing too. Frankland declined to be President of the British Association in Leeds (a fact quoted in support of the knighthood he was awarded after Tyndall's death), and Hirst turned it down too, having been sounded out by Douglas Galton at the Athenaeum.[84]

Tyndall's irritability, and probably his drinking, must have been a trial to Louisa. Lady Welby took her courage in her hands to write to tell Tyndall how much strain she thought Louisa was under from looking after him. She suggested he learn how to use a typewriter and employ a shorthand writer for his work.[85] Tyndall was stunned, but grateful to hear her opinion.[86] Lady Welby had mentioned Louisa's white hair, which prompted Tyndall to recall some early days with her: 'And with regard to the whitened hair, that came before my day. I remember her sitting beside a theodolite I had planted among the rocks high above the Ober Aletsch glacier in 1873, when turning from my instrument at times, I looked down with pathetic interest upon her head. She wore no bonnet though a light, sprinkle of snow was falling—There were the grey hairs, which I knew could not have come from years…I must plead innocence as regard the introduction of the grey hairs'.

Servants were a continual source of problems, which contributed to Louisa's stress. They tended to find the country lonesome and leave, so that 'as regards servants poor Louisa is on her beam ends once more'. Clearly it was Louisa's problem, not Tyndall's. The expectations of the Tyndalls cannot have helped. Interviewing a Miss Nightingale, Tyndall commented: 'I thought her an extremely agreeable person—bright, pleasant to look upon and a good figure. Louisa also was well impressed by her, but thought she discerned in her signs of forwardness'.[87] Another candidate, Mary Clay, 'appeared bright and innocent with extremely little character in her countenance'.[88] They favoured Nightingale, but there was 'a certain tendency to small laughter which must of course be sobered down'.[89] The duties were thoroughly regimented: daily jobs were set out throughout the week; bells must be answered immediately; tea leaves must be saved; and rooms must be swept with plenty of tea leaves (well washed out for carpets, left brown for boards).[90]

On the eve of leaving for the Alps in 1889, Tyndall dined with Debus at the Athenaeum, all ill-feeling between them banished. By mid-July the Tyndalls were on their perch on Alp Lusgen, as Tyndall sought to repair two months of sleeplessness and the effects of the drugs he took to counter it, writing: 'the enemy that assails me is an implacable one, & the world little knows how I have had to fight against it ever since I quitted Queenwood in 1853'.[91]

It was sociable again on Belalp. Emily Peel, who had been nursing a seriously ill husband, wrote to say that her three daughters, none in great health, were at the hotel and she hoped Tyndall might receive them. The Bishop of Gloucester and Bristol, with his 'almost superstitious belief in the efficacy of air that has touched ice & snow', paid his annual visit and shared 'an infidel chop' with them before his departure. They also had Professor Seeley of Cambridge, Sir Spencer Wells, '& other people of note', including Michael Foster (since 1881 a secretary of the Royal Society) and his wife.[92] He conducted the Peel daughters and Mrs Sexton, daughter of Henry Holland, onto the glacier.[93] Louisa had to return temporarily to England in mid-August to help her mother, whose sight was still poor, oversee the building of her house in Cadogan Place, 'in that newly rising region bordering upon Sloane Street'.[94] While she was away, Tyndall entertained the actress Fanny Kemble, aged eighty, and her daughter at

Fig. 42. The Screen at Hindhead.

Belalp. Fanny took off her brooch and asked Tyndall to give it to his wife because of his help. Tyndall also met a 'very sagacious man', Dr Buzzard, who would later play a significant role in his medical treatment. Quitting the Alps in deep snow, Tyndall was back at Hindhead in mid-October, leaving Louisa with friends in Paris.

Tyndall had reached the time of his life when many of his friends were dying. John Ball went on 21st October, such a friend and colleague to Hooker, and an Alpine companion of Tyndall. Hooker asked Tyndall to write the alpine part of his obituary for *Nature* while he wrote the botanical. It was Ball on whom Hooker had particularly relied to help in the revision of the huge work, *Nomenclata Britannica*, which Darwin left to his supervision.[95] Hooker's letter contains the first mention of Tyndall's proposed 'screen' at Hindhead, a huge fence built to shield him from the stables built by his neighbour (Figure 42). Tyndall proudly noted: 'It is a two-peaked Matterhorn, with minor peaks surrounding it. We propose to call it the "Heatherhorn"'.[96] The botanist Hooker suggested that ivy might not be a suitable clothing in a landscape of heather, and a wood trellis might be advisable, with trees planted as soon as possible. He came to Hindhead in early November to inspect the site and offer further advice.

November brought disruption to the depleted X-Club, the only quarrel that ever caused a serious rift. There was a row when Huxley, in the *Times*, attacked Spencer's view that nationalization of land resulted in disaster. Spencer promptly sent in a letter of resignation. To Spencer, Tyndall regretted Huxley's approach: 'He might have remembered…that the state of your health would give him an undue advantage over you…But Huxley seems incapable of resisting the temptation to hit out from the shoulder whenever a favourable opportunity offers'.[97] He wished he could have them both in front of him to resolve their differences. He wrote a long mollifying letter urging Spencer not to react further.

At the X-Club meeting a few days later only Hooker, Frankland, Hirst, and Huxley were present. Discussing the matter in the absence of Huxley, the others felt that Spencer's withdrawal from the Club was wrong, since it would become a public matter, and decided not to read out his letter. They agreed that Huxley's approach was ill-chosen, given Spencer's circumstances, and they regretted Tyndall's absence.[98] Tyndall agreed with their view and became the peacemaker along with Hooker.[99] The spat among the septuagenarians calmed down, though Spencer did not attend the X-Club meeting in February even though he was in the Athenaeum, and did not speak to Huxley when they were both in the Morning Room together.[100] Tyndall was still absent, and Hirst reported rumblings that he was no longer committed to the X-Club, but Hirst defended him on the grounds of his insomnia. It was not until October that Huxley and Spencer met—Tyndall was the only one of the remaining seven X-Club members not present. They did not speak, but Hirst noticed that the ice had begun to melt.[101]

Belfast was the fitting venue for Tyndall's last lecture. He received a letter in December from the Reverend Thomas Hamilton, staunch Unionist and President of Queen's College, Belfast, asking him to reconsider an invitation to come to Belfast to give the opening address to the Belfast Society of the Extension of University Teaching. Hamilton asked him to speak perhaps on Carlyle or science, adding: 'Pardon me for adding that your decided utterances on the question of the Union have endeared you more than ever to us here. Of course our new Society is non-political, but—we know what we know'.[102] Tyndall agreed to lecture on 21st January 1890.[103] Although he had just finished his article on Carlyle, he decided it was not a suitable topic, as it might 'have the dynamiters down upon me were I to lecture about him in Belfast'.[104] He thought a suggestion of Louisa's might be appropriate: the 'researches of Pasteur or, rather the work of Micro-organisms, in fermentation, surgery and communicable disorders, including rabies'. Further thought rendered this problematic too: 'Pasteur will not do. I am rather sick of…Pasteur's love of adulation'.[105] He telegraphed shortly afterwards with his title: 'Our Invisible Friends & Foes', which Hamilton took to be micro-organisms 'under a more attractive name'.[106] It turned out to be timely, as an influenza epidemic was in progress, to which both his host and wife succumbed shortly after the lecture. Tyndall consulted Lister and wrote to Koch for an update on the state of the field, receiving a long and informative letter about cholera, tuberculosis—about which he was planning to publish new results—rabies, and anthrax.[107]

The appearance of Tyndall's 'Personal Recollections of Thomas Carlyle' in the January 1890 edition of the *Fortnightly Review* drew praise from many, including Hirst,

Hooker, and Lady Welby.[108] One correspondent wrote: 'The portions referring to Mrs Carlyle are affecting, to tears. I earnestly hope your article will do something to mitigate the storm of querulous abuse, which has beaten so violently upon the great man's memory'.[109] He intended to write a companion article to that on Carlyle, with the title 'Personal Recollections of Michael Faraday', though he never produced it.

The expedition with Louisa to Belfast was planned like a military campaign. The Tyndalls left London on 17th January and travelled across from Stranraer to Larne rather than going via Liverpool, given his insomnia. He asked Hamilton for no social engagements prior to the event, and for a couple of days at Larne, before he went on to stay in howling weather at the Northern Counties Hotel in Portrush.[110] (The hotel bill demonstrates Tyndall's partiality to good wine. The bottle accompanying dinner, at 10s, cost almost as much as the room, 12s 6d.)[111] The lecture was a sell-out, with extra tickets made available to accommodate some 2,500 in the Ulster Hall. The *Belfast News-Letter* drew the comparison with his 1874 Address. Then, it asserted, the theory of evolution was neither universally nor immediately accepted, but was now acknowledged to be one of the deepest truths that scientific research had yielded.[112] In this 1890 lecture, Tyndall emphasized the evidence of the contagious nature of tuberculosis, a matter then of pressing interest and dispute locally. But his remarks that alcohol was 'a friend of man' did not go down well with the temperance community, who took him to task on his lack of scientific evidence, regarding his statement merely as a needful utterance in defence of his own practice.

After the lecture, and having donated fifty guineas to the Society, Tyndall followed Hamilton's recommendation to stay at the Annesley Arms Hotel in Newcastle, County Down.[113] He was delighted with it: 'The locality is charming & the hotel one of the best country houses I have ever entered. Slieve Donard, covered with snow as it now is, is quite majestic'.[114] But it was not the end of his public speaking. The Unionists took the opportunity to invite him to give a political speech at a demonstration in the Ulster Hall, and to dine at the Ulster Club with James McCalmont (the Conservative MP for Antrim East), the Duke of Abercorn (chairman of the meeting), and Lord Londonderry. The meeting had been organized by the Belfast Conservative Association in honour of the Marquis of Londonderry, to congratulate him on the success of his administration as Lord Lieutenant of Ireland.[115]

As with Tyndall's previous declarations in Belfast, the impact of his speech reverberated. His inflammatory statement that Gladstone was 'judged objectively and in reference to acts brought about not, perhaps, by his intention, but by his mismanagement, cowardice and vacillation', and that he was 'the wickedest man of our generation', raised the ire of Sir William Harcourt, a senior political ally of Gladstone. He wrote to the *Belfast News-Letter*, calling Tyndall a 'scientific Orangeman', and his remarks 'foolish and disgraceful'.[116]

To the delight of the Belfast Unionists, Tyndall riposted that scientific men were occupied with the veracities of nature while Sir William had made it his business to illustrate the unveracity of man.[117] Harcourt sent a more conciliatory response, but Tyndall did not, reiterating in the letters page of the *Times* that: 'Here, upon the Surrey moors, my views are still the same...Would that I could make Sir William Harcourt and his colleagues share my conviction...that if Mr. Gladstone should succeed in his Irish policy, the disasters awaiting us will cause to shrink into nothingness those of the Transvaal and the Soudan!'.[118]

Tyndall's Belfast speech stimulated a note from Gladstone himself the following day, taking exception to his accusation that Gladstone had called Pitt a 'blackguard'. Having consulted Lecky, Tyndall wrote a detailed response to Gladstone, which he sent straight to the *Times* as an open letter.[119] The Gladstone charge was dealt with neatly: 'I did not say that you had "called Mr. Pitt a blackguard". My exact words, as reported in every newspaper to which I have had access, were these:- "He waited until he was 76 years old to discover that Pitt was a blackguard and the Union a crime"'. Tyndall quoted back at Gladstone his words of 1856, when he had praised Pitt's vigour and wisdom in passing the Act of Union with Ireland and abolishing the slave trade. He said that Gladstone had held this position right up to 1886 when, denied an outright majority in the 1885 election, he expressed the view that he knew of 'no blacker or fouler transaction in the history of man than the making of the Union'.

Further correspondence followed, which Gladstone sent to the papers.[120] As Gladstone generously wrote: 'My only desire is to meet you on the terms on which long ago we stood when, under my roof, you gallantly offered to take me up the Matterhorn, and guaranteed my safe return'. At the end, Tyndall told Lecky that he had received a 'charming letter' from Gladstone.[121] He had written to Gladstone: 'You 80—I being 70, undertake to conduct you up the Matterhorn next Summer, and to guarantee that you shall not dash your foot against a stone. Let us do it! I once had the pleasure of climbing the Riffelhorn with your eldest son. He will assure you that I am not uncareful'.[122]

Tyndall spoke out again against Gladstone at Guildford in May, at a meeting of the Primrose League, in his standard refrain of criticizing Sir George Trevelyan and John Morley, while praising Arthur Balfour and Joseph Chamberlain for their strength of character and steadfastness in standing against their previous chief. The speech was widely reported.[123] It followed another political speech at Haslemere in April 1890, after Arthur Russell had presided at a meeting to found a Surrey Liberal Unionist association the previous year.[124] Tyndall was not well enough to attend on that occasion—neither chloral nor sulphonal were ameliorating his insomnia—though he sent a letter which was read out and reported in the *Times*.[125] He hoped to attend the Annual Meeting of the Unionists in Dublin, to which he had been invited by Sir Thomas Butler, to his delight a Carlow man.[126] But his insomnia made it impossible, and he regretted that it also prevented him taking Louisa to meet Elizabeth Steuart, whom he had known for more than sixty years.[127] He went to Lord Hartington's dinner at Crystal Palace in May despite, according to Hirst, never sleeping more than 4 out of 24 hours, and that only by the help of laudanum, but he bridled when Hirst hinted at his conscientious attendance at 'big dinners' and neglect of all smaller ones.[128]

The Tyndalls left for Switzerland in mid-July 1890, offering Hirst the use of Hindhead while they were away.[129] The weather was inclement, and Tyndall found that he did not recover his equilibrium and strength as quickly as he expected. Hirst stayed at Hindhead for a month, while men were at work on Tyndall's screen, which Hirst thought was ugly. *Punch* agreed, in a humorous cartoon in November.[130] Tyndall felt unable to get down to serious work, taking regular walks with Eduard Sarasin, his daughter Marie, and Genevese friends who were staying, but they were prevented by the weather from excursions onto the glacier.[131] The Lubbocks also visited.[132] Louisa left at the end of September to help her mother move house while Tyndall stayed on, planning to leave on 9th October.[133] He finally did so on 23rd October in snow 3 feet deep, reaching London two days later.

But the visit to the Alps had failed to restore Tyndall's health, and he had not been able to write his recollections of Faraday or start work on a planned autobiography. Opium and chloral interfered with his digestion, and Dr Buzzard found that during the two days previous to his arrival he had taken more than his usual amount of alcoholic liquors. According to Tyndall: 'This arrested his attention, & I fear caused him to lay too much stress upon it'.[134] It does seem, though, that Tyndall drank too much at times, perhaps adding to his insomnia. Feeling unable to work, Tyndall sent in his resignation as a Gas Referee, but Michael Hicks-Beach, President of the Board of Trade, 'was so kind and earnest in his opposition to the move', that he withdrew it.[135] This meant a visit to London, where he met Hirst for luncheon. Hirst told him that Dewar had asked if he would give a Discourse on the Centenary of Faraday's birth. Tyndall asked Hirst to tell him that he could only try to produce something in writing, but had already tried and failed in Switzerland.

In the event, Tyndall was too ill to accept the invitation from Bramwell to attend the Faraday commemoration at all. He wrote to Bramwell: 'As Faraday recedes from me in time, he becomes to me more and more beautiful. Anything, therefore, calculated to do honour to his memory must command my entire sympathy'.[136] The

Prince of Wales presided on 17th June 1891, when Rayleigh lectured. Thomson, Stokes, Grove, and Bramwell also spoke, the latter reading out Tyndall's letter.

* * *

Illness worse than ever now dogged Tyndall for the remainder of his life. For several weeks, since before Christmas and over New Year 1891, he suffered from an excruciating gouty phlebitis in his left leg. Newspapers regularly reported his state of health. To Juliet Pollock he wrote: 'Thirty years ago you warned me in regard to sleeplessness. Could he have found an antidote to this bane your Brother John might have made some mark in the world. But he has been compelled to climb a mountain on a broken leg. What wonder if he has stopped midway!'.[137] He had recovered enough by the end of January to let the *Times* know that the worst was past.[138] But the leg had still not healed, and Tyndall confined himself to bed, telling Hirst: 'I have postponed the recollections of Faraday so that I may apply myself in Switzerland to the task. At present I am occupied with a new volume of "Fragments"'.[139] He thought he was better in mid-March, and considered going to the seaside to recuperate. He even got out of bed on 20th March to walk downstairs and outside, leaning on Louisa's shoulder. But Louisa noted 'manifest signs of increased evil', and Buzzard was summoned a few days later when they telegraphed 'dangerous relapse'. Buzzard told Huxley that both lungs were affected, and another doctor told Louisa that 'there was now no chance of his pulling through'; he was suffering from a combination of pneumonia and pleurisy. Spencer, too, was alarmed by the newspaper reports.

The first of April turned out to be the lowest day. To Debus, Tyndall wrote: 'For the last fortnight I have been engaged in a stand up fight with death, and though the battle I believe to be now decided in my favour, it has left me very weak'.[140] A fortnight later Louisa could report: 'For three days there has been no trace of blood in the expectoration which is a great sign that things are mending'.[141] Hirst, 'the big awkward boy of 14, whose heart you won and have ever since retained', offered to defray Tyndall's expenses of the last six months, which he thought must be considerable. The situation was still not good. Buzzard suspected a block in a deep vein, as Tyndall's right leg was now greatly swollen. They decided to seek a second opinion from Andrew Clark, which confirmed the diagnosis. Tyndall was now moved to write to the *Times* to report the situation and ask that bulletins be discontinued, 'which I feel can only worry and weary the public and my friends'.[142] Spencer recommended backgammon to aid recovery, while Tyndall asked Hirst to procure him a musical box, which he bought for £25. It played tunes such as 'Home Sweet Home', 'My Pretty Jane', and 'Bonnie Dundee'.[143] Gladstone held no grudges, sending a charming letter to Louisa.

The papers continued to provide updates of variable accuracy on Tyndall's health, to his disgust. He had visits on successive days from the Tennysons—'The Bard

looked particularly well, and Hallam & his wife the picture of loving contentment'—
and from Hooker and his wife.[144] On 8th July he was well enough to go out, and was
driven 'in triumph' round the countryside. But after seven months lying down, his
legs were stiff and weak. He started to move about slowly, leaning heavily on two
sticks. Even at this stage in July, Tyndall was planning to head out to Switzerland as
usual, though Buzzard shook his head at the idea, and by the end of July, he had practically abandoned the thought. In mid-August, after a visit from the Leckys, he sent
off his article for the *Fortnightly Review* 'On the origin, propagation, and prevention of
phthisis [tuberculosis]',[145] which he hoped 'would not bring him into collision with
the big men of the International Congress'.[146] There was still a widespread belief that
tuberculosis was a hereditary disease. Georg Cornet, at Koch's Institute for Infectious
Diseases, wrote to thank him for his support on tuberculosis and for raising awareness of prophylactic measures.[147]

Apart from updates about his health, some newspapers amused themselves by
reporting on Tyndall's 'screens'. Unfortunately for Tyndall, his neighbour, Robert
Turner, was Chairman and Managing Director of Cassell & Co., and had good contacts with the press.[148] At the X-Club on 1st October, with only Hirst, Hooker, and
Huxley present, Hooker said Tyndall was bothered by comments published in the
Pall Mall Gazette about them.[149] Hirst remarked: 'I do not wonder', and noted a further
portrayal in the *Daily Graphic* on 30th October.[150] Spencer visited Tyndall at the beginning of November, and Tyndall reported to him: 'The calumniated screen which you
looked upon with favour when you were here has emerged triumphantly from all
these terrible storms. I was very anxious about it, for it shook portentously, but I
have now steadied and strengthened it with iron plates and bolts, and with additional
braces, so that it is able to defy anything likely to assail it in the future. You know the
assault that has been made upon me by the Radical newspapers. They are mean curs,
and I may tell them my opinion of them by and by'.[151] Hirst just wished he would
treat his enemies with silent contempt.

Tyndall had not quite given up hope of going to Switzerland, but on 9th September
he told Spencer: 'A few days ago, as the rain descended, and everything looked hopeless, I wrote thus to my care-taker in the Alps. "Shut up the house, make everything
fast, and see that my firewood is well sheltered"'. Hindhead held some compensa-
tions: 'To anyone not afflicted with my hunger for the mountains and glaciers, Hind
Head, in good weather, would fill the heart with satisfaction'.[152] He had the compensation of Tennyson's company too, driving over to Aldworth and hearing a reading
of *Ode to the Duke of Wellington*. On an earlier visit, Tyndall had revealed, towards the
close of both their lives, his innermost thoughts on matter and God. He told
Tennyson: 'God and Spirit I know, and matter I know; and I believe in both', and in

response to Tennyson's belief in individual immortality he declared: 'We may all be absorbed into the Godhead'.[153]

Tyndall's final book, *New Fragments*, was published on 15th January 1892.[154] The *Saturday Review* welcomed this collection of essays as showing that Tyndall was an accomplished biographer in things human as well as scientific. The paper also recognized that Tyndall had fought for the reputation of others rather than himself—mentioning Rendu, Mayer, and Wigham—and he had now done it in this book with Rumford, Young, and especially Carlyle.[155] The book included, as the final chapter, the only poem (apart from an elegy in 1850 to his mentor Dean Bernard in the *Carlow Sentinel*) that Tyndall published under his own name, though the *Saturday Review* made no mention of it. This was 'A Morning on Alp Lusgen', which first appeared in the *Pall Mall Gazette* in 1881, a poem of transcendentalism and homage to Carlylean thought, containing the evocative lines:

> Whence the craft
> Which shook these gentian atoms into form,
> And dyed the flower with azure deeper far
> Than that of heaven on days serene?[156]

At the beginning of February, Hirst was very ill with influenza, uncharacteristically missing the X-Club meeting. Hooker visited him and told Tyndall he should expect the worst.[157] Huxley tried to see him on 13th February but was turned away. Spencer managed a visit the following day and found him desperately weak. He died two days later, of prostate cancer according to Hooker.[158] 'Dr Hirst died at 4am Maclaren', was the telegram that reached the Tyndalls at midday. Hirst was buried at Highgate Cemetery near George Eliot, but Tyndall was not well enough to attend. Debus was in Germany, and the weather prevented Huxley and Spencer from going, but Grove was there. Tyndall wrote to Huxley: 'To me his life was always more or less of a tragedy. I knew his intellectual power to be great, but I saw that power perpetually broken by imperfect health. His perfect rectitude and tenderness of heart were known to nobody so well as to me'.[159]

By March, and with great regret, Tyndall had to turn down the invitation to the Royal Academy Dinner for the second year running. He explained, with only slight exaggeration, that he had been confined to his home now for seventeen months.[160] He finally went up to London on 28th June, for the first time in two years, to see the physician Sir Alfred Garrod. It was a 'fairly clean bill of health, weight 11st 4 10, and increase not wasting. Little coughing blood not a problem. Wheezing and flushing gouty symptoms not pneumonic!!'.[161] That meant he was well enough to go to the Alps, though their departure was delayed to enable him to vote for the Unionist candidate at Haslemere in the 1892 general election.[162] Characteristically he wrote to the *Times*, claiming that 'a wave of Unionism is passing over this country, which, if it only have time to act, will sweep the pestilent question of Home Rule from the minds of men... We fight, not for Protestant ascendency, but against an ascendency far more intolerant and degrading'.[163]

Spencer was worried about Tyndall's ability to survive the journey and asked: 'Why not try my mode of travelling & have a hammock slung diagonally in a saloon carriage?'.[164] But the journey passed without problem, with a rest at Vernayaz for three days. From Brig he rode on horseback to Blatten, midway between Brig and Belalp: 'there we spent the night, and next day reached our little nest'.[165] By mid-August, after a long period of 'cloudless skies and consequent sunshine almost too warm', he was feeling better, although still taking opium and chloral at night.[166] The hotel was as busy as ever, though Tyndall saw few of the guests.[167] He kept up with the political news, pleased that though Gladstone was returned, 'he has not the power which he counted on'.[168]

Gladstone formed a minority government with Irish Nationalist support after the 1892 general election. It survived until 1895, outliving Tyndall, who was unable to experience the Conservative government of the next ten years. More worrying to Tyndall's ageing cohort was the rise of the 'socialists'. Fred Pollock wrote: 'I rather hope however that the Labour Party may turn out less important than they try to look—for though Home Rule would be most mischievous to Ireland and trouble-some to Great Britain, Socialism of the John Burns kind would be ruin to the three kingdoms... Did I tell you that the real working men on the Labour Commission speak of Keir Hardie and such like with absolute contempt?'.[169] Hooker had written to say: 'It makes one weary of life to see the spread of democracy and socialism'.[170] Tyndall's early flirtation with radicalism was all but forgotten.

By early September the weather had broken, with rain and then snow. Tyndall had not recovered his strength. He did not feel able to concentrate on anything scientific and had not even been down to the glacier. He felt too weak to ride down on his return in October, and resorted to a *chaise à porteurs*. But that was so uncomfortable that in the end he did walk, stopping half-way at Blatten for the night. He faced the added burden of disinfection at the French border as the result of a cholera outbreak. From Folkestone they travelled back in the company of Alfred Wills, now Judge Wills, with whom he had made the ascent of Mont Blanc in 1858.[171]

On the day of Tyndall's arrival in England, Tennyson was buried in Westminster Abbey. Tyndall had heard of his death while in Lausanne, but was too tired to attend the funeral—although publication of his name in the *Times* led people to believe that he was there.[172] Huxley gave him news of the event: 'I am glad to say that the Royal

Society was represented by four of its chief officers, and nine of the commonality, including myself. Tennyson had a right to that as the first poet since Lucretius who has understood the drift of science'.[173]

Throughout much of the early part of 1893, Tyndall was laid up with an ankle infection that his doctors were concerned might spread if it were not completely rested. Huxley's health was good by contrast. He gave the second Romanes Lecture in May, writing to Tyndall: 'Who would have thought 33 years ago, when the great "Sammy" fight came off, that the next time I should speak at Oxford would be in succession to Gladstone, on "Evolution and Ethics" as an invited lecturer?'.[174] Tyndall had still not mustered the energy to read the text by mid-July, but he was finally mobile enough to take up Spencer's repeated urging that he should go to the Isle of Wight for a change of scene.

The Tyndalls spent nearly a month in July at Totland Bay on the west coast of the island. Then on 31st July, they left for London, staying overnight at the Grosvenor Hotel before travelling to Switzerland. The journey out was marred by delays that unsettled the fragile Tyndall, so they rested for nearly a week at Vernayaz and about the same time at Sierre.[175] His strength slowly improved in the Alps, and they were back at the end of October, meeting Lady Tennyson by chance at the station on arriving. Tyndall returned in reasonable health but succumbed immediately to another chest infection, followed by an attack of rheumatism which prevented him from shaving himself and confined him to bed.[176] By 25th November he had still not read Huxley's Romanes lecture.

On 3rd December he wrote three letters: to William Colles of the Society of Authors—Tyndall was Vice-President—thanking him for following up a recommendation to look over the poems of Ebenezer Smith; to Du Bois-Reymond, asking for news of the health of Helmholtz (who died less than a year later); and to Wiedemann, asking for information on the work of Wroblowski, who had worked on the liquefaction of gases, since 'it seems to me that Wroblowski has not received fair play in England'.[177] Just a hint, perhaps, of the defence of an underdog against Dewar.

That night Louisa recorded: 'A darling little word, spoken in the night and scribbled down by me in pencil: "If you restore your Johnnie, he will owe to you his life—entirely to you. I don't know that you will do it; but I think you will"'.[178] Her last diary entry the following morning, 4th December, read: 'slept rather later than usual. The remaining milk taken between 1/2 past 6 and 7'.[179] Tyndall took chloral nightly to help him sleep, and magnesia every other morning. This morning there was an extra bottle of chloral by the bed. Louisa used it by mistake, giving him a magnesia-sized dose of choral. She realized immediately, telling him: 'Oh John, I've given you chloral'. Calmly, Tyndall responded: 'My poor darling, you have killed your John'.[180] As the doctors arrived, they tried desperately to make him vomit it up. But it was too late. Ten hours more, and he was dead.[181]

EPILOGUE

The inquest quickly established that Tyndall's death was accidental. There was widespread sympathy for Louisa, and newspaper correspondence about the safer design of medicine bottles. A raft of obituaries appeared. The best was by Huxley,[1] who had stood with Netty by the graveside at St Bartholomew's Church in Haslemere, behind the black-shrouded Louisa and her brother (Figure 43). Frankland, Hooker, and Lubbock were present too. Wracked by guilt, though comforted by friends and public alike, Louisa returned regularly to Alp Lusgen throughout her remaining forty-seven years. In 1911 she erected a magnificent granite memorial stone high above Belalp, with stunning views to the distant glaciers and mountains. On her death in 1940, she was buried beside her husband.

But the memory of Tyndall's achievements soon faded. Lord Rayleigh gave a Discourse 'The scientific work of John Tyndall' soon after Tyndall's death, though he

Fig. 43. The Funeral of Professor Tyndall, at the Graveside, *The Graphic*, 16th December 1893.

ignored the magnetic studies.[2] Oliver Lodge's obituary of Tyndall some years later went a step further and was little short of damning of his work in this area.[3] Lodge wrote: 'His early magnetic investigations...sadly lack the definiteness which was possible at their date'. Lodge argued that Tyndall did 'not express it as a mathematician would', a spurious criticism that could equally be applied, if not more so, to Faraday.[4] Apart from a further Discourse by William Bragg in 1927, concentrating on Tyndall's magnetic research,[5] again rather critically, no significant retrospective assessments were made. He had disappeared almost without trace.

There are three reasons for this. The first is that no 'Life and Letters' was produced after his death, unlike for contemporaries such as Darwin, Huxley, Hooker, and Spencer. Louisa, desperate to pay homage to his memory, collected a mass of letters, documents, and cuttings, now held at the Royal Institution, to write a biography. But she could never bring it to completion. It was not until 1945, five years after her death, that a semi-authorized general biography was published.[6] There has not been another one until this. In a tragic sense, having outlived him so long, she killed him twice.

The second reason is that Tyndall was an experimentalist. Perhaps one of the best of his generation, though the neatness of his recording in the early years leaves much to be desired. He was no mathematical theoretician, and it is those who make the big theoretical advances who tend to be remembered in physics. Tyndall understood that the interplay of experimentalist and theoretician was critical, but he gave pride of place to the former. For Tyndall, it was the experimentalist who opened up a region for study, probing the structure of matter and exploring its interaction with forces or energies.[7] It was the Tyndallian experimentalist whose 'imagination' generated a picture of the underlying physical relationships and laws that made sense of the observed phenomena. The theoretician, through mathematical analysis, could develop an accurate quantitative picture based on these observations and interpretations, to explain existing phenomena and predict new ones. What mattered to Tyndall was the physical understanding. Though he lacked the depth of theoretical and mathematical insight of people such as Thomson, Stokes, and Maxwell, his intuition and imagination led to rich insights and powerful interpretations.

The third reason for Tyndall's lack of recognition is that he had the misfortune to die on the cusp of revolutionary discoveries and developments in physics: X-rays, radioactivity, relativity, and quantum theory. Tyndall represented the past, a classical physics that soon seemed dated. Yet this classical physics—heat, light, sound, force, energy, electricity, and magnetism—is still the bedrock of physics curricula the world over. No one else ranged across it as Tyndall did, and no one else made contributions in so many other areas: from glaciers to bacteriology. This very generality, coupled with his reputation as a communicator (still so often seen as diminishing a scientist's significance), disadvantages him. Nevertheless, he made major discoveries, rooted in a physical theory of matter and a belief in the reality of atoms and molecules. Foremost among them is his work on the absorption of radiation by gases in the atmosphere, explaining the greenhouse effect, advancing meteorology, and underpinning our current understanding of climate change (though we must now acknowledge that the actual discovery of the absorption of heat by water vapour and carbon dioxide was made by the American Eunice Foote three years before Tyndall). Then we have his answer to that perennial question 'why is the sky blue?'. Alongside these is his instigation of the field of chemical physics, using radiation to cause chemical changes. And then the practical implications and applications: to lighthouse design and safer navigation, to safety in mines, and to wider acceptance of the germ theory of disease.

Tyndall was a skilled networker, perhaps the prime catalyst between the German and British physicists in the mid-nineteenth century. He was a great communicator, to his scientific peers and the public alike. Tyndall bridged the scientific and literary worlds, not only because of his advocacy of the need for creative imagination in science as in poetry but also because of the ethos he supported at the Royal Institution and those who spoke there. They included Ruskin, Froude, Lecky, Morley, and many other figures in literature, art, philosophy, and theology. If the stridency of his politics and the lack of a sense of humour made him rather a figure of fun towards the end, he was still impossible to ignore.

Now, more than a century later, we remember him still: a complex, often contradictory figure, who embodies the intellectual excitement, the disputes, the prejudices, the tensions, and the opportunities of that nineteenth-century world.

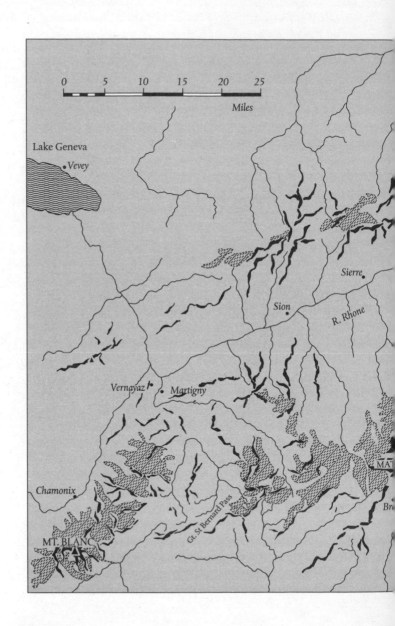

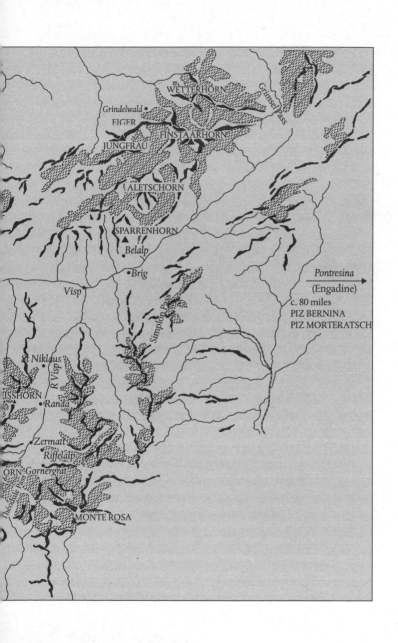

WETTERHORN

Grindelwald •
EIGER
JUNGFRAU

FINSTAARHORN

Grimsel Pass

ALETSCHORN

SPARRENHORN
▲
• Belalp

• Brig

Visp •

Simplon Pass

Pontresina
(Engadine)
→

c. 80 miles
PIZ BERNINA
PIZ MORTERATSCH

St Niklaus •

R Visp

ISSHORN
▲
• Randa

• Zermatt
Riffelalp

ORN Gornergrat
Theodule Pass

MONTE ROSA

PUBLISHER'S ACKNOWLEDGEMENTS

We are grateful for permission to quote from manuscript material held in the following archives: The Alpine Club Library; Archives Imperial College Library; The Master and Fellows, Balliol College, Oxford; Bibliothèque nationale de France, Paris; Bodleian Libraries, University of Oxford; British Library; British Science Association; Syndics of Cambridge University Library; Elton Hall Collection; Harrow School; Henry E. Huntington Library; John Rylands Library, Manchester; Library of Congress; The Trustees of the Natural History Museum, London; Royal Archive; Royal Dublin Society Archives; The Royal Institution of Great Britain; The Royal Society; Smithsonian Institution Archives; Principal and Fellows of Somerville College; St Bride Library; The Master and Fellows of Trinity College Cambridge; University College London; University of Edinburgh Library Special Collections; University of Heidelberg; University of St Andrews Library; Wellcome Library, London; Yale University.

References to individual quotes are given in the endnotes.

The publisher and author have made every effort to trace and contact all copyright holders before publication. If notified, the publisher will be pleased to rectify any errors or omissions at the earliest opportunity.

PICTURE CAPTIONS & CREDITS

Frontispiece Chalk drawing of Tyndall c.1850. Courtesy of the Royal Institution of Great Britain

Figures

22. Tyndall's journal entries noting his quick discovery of the absorption of radiant heat by gases, May 1859. Courtesy of the Royal Institution of Great Britain

23a. John Lubbock (1834–1913) in 1856. Courtesy: Lubbock Family

23b. Nelly (Ellen) Lubbock (1835–79) in 1856. Courtesy: Lubbock Family

24. Mary Somerville (1780–1872). Courtesy of the Royal Institution of Great Britain

25. The 'Matterhorn' of heat: the spectrum of radiant heat, showing its maximum beyond the red. Courtesy of the Royal Institution of Great Britain

26. The Clubroom of Zermatt in 1864. In the centre are Ball, Mathews, E. S. Kennedy,Bonney, Ulrich Lauener, Tyndall, and Wills. Lucy Walker stands by the doorway of the Monta Rosa Hotel. Roland Jackson

27. Aletschorn from the Eggishorn. Roland Jackson

28a. Tyndall's 'blue sky' tube. Paul Wilkinson/Courtesy of the Royal Institution of Great Britain

28b. Tyndall's 'blue skies' apparatus. Courtesy of the Royal Institution of Great Britain

29. Professor Tyndall lecturing at the Royal Institution, *Illustrated London News*, 56 (1870): p. 493. Courtesy of the Royal Institution of Great Britain

30. Tyndall as 'Man of the Day', *Vanity Fair*, 6th April 1872. Courtesy of the Royal Institution of Great Britain

31. Cartoon of Tyndall's sound experiments. Artist unknown, but possibly Huxley. Courtesy of the Royal Institution of Great Britain

32. A siren of the type Tyndall encouraged. Courtesy of the Royal Institution of Great Britain

33a. Tyndall's 'floating matter' box. Paul·Wilkinson/Courtesy of the Royal Institution of Great Britain

33b. Floating matter chamber. Courtesy of the Royal Institution of Great Britain

34. John and Louisa Tyndall seated in 1876. Courtesy of the Royal Institution of Great Britain

35. Professor Tyndall demonstrating a fog-horn to Queen Victoria and her entourage. Wood engraving by T. B. Wirgman (*c*.1876). Wellcome Library, London (CC BY 4.0)

36. The Tyndalls in front of the Belalp Hotel. Courtesy of the Royal Institution of Great Britain

37. Cartoon of Tyndall's germ experiments. Artist unknown, but possibly Huxley. Courtesy of the Royal Institution of Great Britain

38. Tyndall operates the phonograph as William Preece speaks, *The Graphic*, 16th March 1878. De Agostini / Biblioteca Ambrosiana

39. 'A Lighthouse – Not Professor Tyndall', *Moonshine*, 20th December 1884 (Joseph Chamberlain is the wave). Courtesy of the Royal Institution of Great Britain

40. Comparing oil, gas, and electric lights at the South Foreland. Copyright Gordon Denoon collection, St Margaret's History Society

41. The Scientific Volunteer, *Punch*, 22nd February 1890. Courtesy of the Royal Institution of Great Britain

42. The Screen at Hindhead. Courtesy of the Royal Institution of Great Britain

43. The Funeral of Professor Tyndall, at the Graveside, *The Graphic*, 16th December 1893. Michael Reidy

PICTURE CAPTIONS & CREDITS

Plates

1. Tyndall in the 1850s. Courtesy of the Royal Institution of Great Britain
2. Thomas Hirst (1830–1892). © The Royal Society
3. The Royal Institution of Great Britain in the mid-nineteenth century. Courtesy of the Royal Institution of Great Britain
4. William Thomson, later Lord Kelvin (1824–1907) in 1852. Emery Walker Ph. Sc. (Philosophical Society?), courtesy AIP Emilio Segrè Visual Archives, Zeleny Collection
5. Thomas Huxley (1825–1895). Courtesy of the Royal Institution of Great Britain
6. Faraday, Huxley, Wheatstone, Brewster, and Tyndall, c.1865 (from *English Celebrities of the Nineteenth Century*, 1875). Courtesy of the Royal Institution of Great Britain
7. Tyndall's FRS certificate. © The Royal Society
8. Henry Bence Jones (1813–1873). Courtesy of the Royal Institution of Great Britain
9. Thomas Carlyle (1795–1881). Wellcome Library, London. Copyrighted work available under Creative Commons Attribution only licence CC BY 4.0 http://creativecommons.org/licenses/by/4.0/
10. The Athenaeum c.1850. The Athenaeum
11. Tyndall in 1864, drawn by George Richmond, RA. Courtesy of the Royal Institution of Great Britain
12. The manuscript in which Tyndall set out his detailed work on the absorption of heat by gases. © The Royal Society
13. Tyndall's experimental notebook, recording experiments on the absorption of heat by gases, 16th September 1862. Courtesy of the Royal Institution of Great Britain
14. Weisshorn from the Riffel. Roland Jackson
15. Matterhorn. Roland Jackson
16. Louis Pasteur (1822–1895). Wellcome Library, London. copyrighted work available under Creative Commons Attribution only licence CC BY 4.0 http://creativecommons.org/licenses/by/4.0/
17. Tyndall in 1873. Courtesy of the Royal Institution of Great Britain
18. Letter to Mary Adair after the Belfast Address, 11th January 1875. Courtesy of the Royal Institution of Great Britain
19. Marriage certificate of John Tyndall and Louisa Hamilton. Roland Jackson
20. Louisa Tyndall reading. Courtesy of the Royal Institution of Great Britain
21. John Tyndall outside Chalet Lusgen (probably Louisa Tyndall in the doorway). The Huntington Library, San Marino, California
22. Burlington House in 1866, home of the Royal Society (1857–1968). © The Royal Society
23. Hindhead House. Courtesy of the Royal Institution of Great Britain
24. Some Fellows of the Royal Society: a key to the identities of the sitters. Wood engraving, 1885. In the front row are Stokes, Hooker, Sylvester, Huxley, Geikie, Tyndall, Cayley, Owen, Flower, and Crookes. Frankland stands third from right and Francis Galton second from left. Wellcome Library, London (CC BY 4.0) copyrighted work available under Creative Commons Attribution only licence CC BY 4.0 http://creativecommons.org/licenses/by/4.0/
25. Tyndall and Louisa in the study at Hindhead c.1887. Courtesy of the Royal Institution of Great Britain
26. Louisa at Tyndall's deathbed. Courtesy of the Royal Institution of Great Britain

NOTES

Front Matter Epigraphs

1. 6 November 1865, John William Draper to Tyndall, John William Draper Family Papers, Library of Congress.
2. Emerson, R. W., 'History', *Essays* (Boston, MA: Munroe, 1841): p. 8.

A Note on Words

1. See Ross (1962).

Introduction

1. Lee, S. (ed.) *Dictionary of National Biography* 57, (London: Smith Elder, 1899): p. 435.

Chapter 1: Irish Beginnings (*c.*1822–1844)

1. See Foster (1988) for a good account of the history.
2. McMillan, N. and M. Nevin, 'Tyndall of Leighlin', in Brock, McMillan, and Mollan (1981): p. 20.
3. He may also have had other siblings who died young, although it's not known for certain.
4. Eve and Creasey (1945): p. 2.
5. *c.*1893, Louisa Tyndall to Doctor Hutchinson, RI MS LT/14/29.
6. 16 January 1879, Louisa Journal, RI MS JT/2/23.
7. Eve and Creasey (1945): p. 3.
8. 7 October 1887, Tyndall to L. Darmstaedter, L., SB Slg Darmstaedter 1855 Tyndall Be. 10–11.
9. See Tynan, M., 'John Tyndall's Role in the Ordnance Survey of Ireland', in Cantor and Dawson (2014): p. 451–9.
10. Andrews (1975): p. 91.
11. 2 May 1841, Tyndall to father, RI MS JT/1/TYP/10/3220 (Letter 61 in TC1).
12. 25 April 1841, Tyndall to father, RI MS JT/1/TYP/10/3218 (Letter 55 in TC1).
13. 9 October 1846, Tyndall Journal, RI MS JT/2/13/151.
14. RI MS JT/8/1/2/2–10.
15. 10 January 1841, Maria Payne to Tyndall, RI MS JT/1/TYP/11/3840 (Letter 33 in TC1).
16. 26 September 1840, Tyndall to father, RI MS JT/1/TYP/10/3189 (Letter 11 in TC1).
17. 24 February 1842, George Wynne to Tyndall, RI MS JT/5/17a/41 (Letter 126 in TC1).
18. 4 January 1841, Tyndall to father, RI MS JT/1/TYP/10/3199 (Letter 31 in TC1).
19. 4 January 1841, Tyndall to father, RI MS JT/1/TYP/10/3199 (Letter 31 in TC1); 9 January 1841, Tyndall to father, RI MS JT/1/TYP/10/3200 (Letter 32 in TC1).
20. 23 January 1841, Tyndall to father, RI MS JT/1/TYP/10/3202–3 (Letter 36 in TC1).

21. 23 January 1841, Tyndall to father, RI MS JT/1/TYP/10/3202–3 (Letter 36 in TC1).

22. 5 October 1840, father to Tyndall, RI MS JT/1/TYP/10/3190 (Letter 13 in TC1).

23. 7 October 1840, Tyndall to father, RI MS JT/1/TYP/10/3191 (Letter 14 in TC1).

24. Cantor and Dawson (2016): p. 34–5, fn. 2.

25. 24 August 1841, Tyndall to father, RI MS JT/1/TYP/10/3238 (Letter 86 in TC1).

26. 21 July 1841, Phillip Evans to Tyndall, RI MS JT/1/TYP/11/3566 (Letter 79 in TC1). The trial was reported in the *Carlow Sentinel* (24 July 1841): p. 2; 16 March 1842, father to Tyndall, RI MS JT 1/TYP/10/3267 (Letter 130 in TC1).

27. 11 July 1841, Tyndall to father, RI MS JT/1/TYP/10/3231 (Letter 75 in TC1).

28. 13 October 1841, father to Tyndall, RI MS JT/1/TYP/10/3247 (Letter 102 in TC1).

29. 16 September 1841, Tyndall to father, RI MS JT/1/TYP/10/3243 (Letter 93 in TC1).

30. 18 February 1841, William Wright to Tyndall, RI MS JT/1/TYP/11/3889 (Letter 41 in TC1).

31. DeArce, McGing, and McMillan (2013–14).

32. 30 December 1841, Tyndall to father, RI MS JT/1/TYP/10/3260–1 (Letter 117 in TC1).

33. 26 January 1842, Tyndall to father, RI MS JT/1/TYP/10/3264 (Letter 119 in TC1).

34. 8 April 1842, Tyndall to father, RI MS JT/1/TYP/10/3270 (Letter 138 in TC1).

35. 29 May 1842, William Ginty to Tyndall, RI MS JT/1/TYP/10/3579 (Letter 144 in TC1).

36. 18/25 August 1842, Tyndall to father, RI MS JT/1/TYP/10/3279 (Letter 164 in TC1).

37. 15 October 1842, Tyndall to father, RI MS JT/1/TYP/10/3281 (Letter 173 in TC1).

38. 3 December 1842, Tyndall to father, RI MS JT/1/TYP/10/3283 (Letter 178 in TC1).

39. nd October 1842, Tyndall to father, RI MS JT/1/TYP/10/3280 (Letter 171 in TC1).

40. 14 February 1843, Tyndall to father, RI MS JT/1/TYP/10/3285–6 (Letter 187 in TC1).

41. 30 November 1842, John Chadwick to Tyndall, RI MS JT 1/TYP/11/3499 (Letter 177 in TC1).

42. 24 December 1842, John Lilly to Tyndall, RI MS JT/1/TYP/11/3748 (Letter 180 in TC1); 7 February 1843, William Ginty to Tyndall, RI MS JT/1/TYP/11/3593 (Letter 186 in TC1).

43. Tyndall, J., 'On Leaving Westmorland', *Preston Chronicle* (28 Jan 1843): p. 4.

44. 2 February 1843, James Sinnett to Tyndall, RI MS JT/1/TYP/11/3863 (Letter 184 in TC1); 31 January 1843, William Ginty to Tyndall, RI MS JT/1/TYP/11/3591 (Letter 183 in TC1).

45. 1 April 1843, Archibald McLachlan to Tyndall, RI MS JT/1/TYP/11/3763–72 (Letter 195 in TC1).

46. 10 April 1843, Archibald McLachlan to Tyndall, RI MS JT/1/TYP/11/3773–4 (Letter 198 in TC1).

47. 10 May 1843, Thomas Carroll to Tyndall, RI MS JT/1/TYP/11/3491 (Letter 204 in TC1).

48. 12 July 1843, Tyndall to Thomas Carroll, RI MS JT/8/2/1/20–1 (Letter 217 in TC1).

49. 25 May 1843, Joseph Payne to Tyndall, RI MS JT 1/TYP/11/3843–4 (Letter 209 in TC1).

50. See Brown (2013).

51. 3 July 1843, Tyndall Journal, RI MS JT/2/13a/iii.

52. Reid (1900): p. 66.

53. 'Personal Recollection of Thomas Carlyle', in Tyndall (1892): p. 348–9.

54. 18/25 August 1842, Tyndall to father, RI JT/1/TYP/10/3279 (Letter 164 in TC1).

55. 'Personal Recollection of Thomas Carlyle', in Tyndall (1892): p. 349.

56. 19 September 1846, Tyndall Journal, RI MS JT/2/13a/148.

57. 8 August 1843, William Ginty to Tyndall, RI MS JT/1/TYP/11/3607 (Letter 224).

58. 15 August 1843, Tyndall as 'Spectator' to Sir Robert Peel, RI MS JT/1/TYP/11/3907–10 (Letter 228 in TC1).

59. 23 September 1843, Tyndall et al. to George Murray, RI MS JT/1/TYP/10/3895–3900 (Letter 236 in TC2).

60. 14 October 1843, Tyndall et al. to Captain Tucker, RI MS JT/1/TYP/10/3893–4 (Letter 246 in TC2).

61. 7 November 1843, Henry Tucker to staff, RI MS JT/1/TYP/10/3902 (Letter 262 in TC2).

62. 17 November 1843, Tyndall to father, RI MS JT/1/TYP/10/3307–8 (Letter 271 in TC2).
63. 17 November 1843, Tyndall to father, RI MS JT/1/TYP/10/3307–8 (Letter 271 in TC2).
64. 23 November 1843, Tyndall Journal, RI MS JT/2/13a/6.
65. 24 November 1843, Tyndall to George Murray, RI MS JT/1/TYP/10/3904 (Letter 276 in TC2).
66. 30 November 1843, Tyndall to George Murray, RI MS JT/1/TYP/10/3905 (Letter 277 in TC2).
67. 6 December 1843, Tyndall to George Murray, RI MS JT/1/TYP/10/3906 (Letter 278 in TC2).
68. 'To Correspondents', *Preston Chronicle* (13 January 1844).
69. 13 January 1844, Tyndall as 'Spectator' to Sir Robert Peel, RI MS JT/1/TYP/11/3916–18 (Letter 288 in TC2).
70. 5 March 1844, Tyndall Journal RI MS JT/2/13a/22–3.
71. 12 September 1843, Tyndall to father, RI MS JT/1/TYP/10/3297 (Letter 234 in TC2).
72. 24 September 1843, William Tyndall to Tyndall snr, RI MS JT/1/TYP/10.
73. 26 September 1843, William Wright to Tyndall, RI MS JT/1/TYP/5/1811–12 (Letter 237 in TC2).
74. 23 October 1843, Tyndall to father, RI MS JT/1/TYP/10/3299–3300 (Letter 251 in TC2); 30 October 1843, Tyndall to father, RI MS JT/1/TYP/10/3302 (Letter 254 in TC2).
75. 20 March 1844, George Latimer to Tyndall, RI MS JT/1/TYP/11/3744–5 (Letter 299 in TC2).
76. 16 May 1844, Tyndall to Henry Tucker, RI MS JT/2/13a/34–5 (Letter 302 in TC2).
77. 6 July 1844, Tyndall Journal, RI MS JT/2/13a/44.
78. 25 February 1844, Tyndall Journal, RI MS JT/2/13a/19–20. It does not appear that the letter was published.
79. 30 July 1844, Tyndall to Thomas Wren, RI MS JT/2/13a/51 (Letter 311 in TC2).

Chapter 2: Railway Mania (1844–1847)

1. 1 January 1861, Tyndall Journal, RI MS JT/2/13c/1198; 27 January 1854, James Craven to Tyndall RI MS JT/1/TYP/11/3552–4.
2. There is only one clue to this, in a reference to 'un homme de stature moyen, d'une barbe rouge, et d'une activité extraordinaire', from which Hirst recognized a description of Tyndall (Hirst Journal, 28 August 1860, RI MS JT/2/32c/1552).
3. See Odlyzko (2015) for a detailed analysis.
4. 11 October 1844, Tyndall Journal, RI MS JT/2/13a/66.
5. 29 November 1844, Tyndall Journal, RI MS JT/2/13a/69.
6. 2 November 1844, Tyndall Journal, RI MS JT/2/13a/69.
7. 26 December 1844, Tyndall Journal, RI MS JT/2/13a/71.
8. 24 May 1861, Tyndall Journal, RI MS JT/2/13c/1217.
9. 12 February 1845, Tyndall Journal, RI MS JT/2/13a/75.
10. The viaduct collapsed two months later, burying fifteen workmen.
11. 3 September 1844, Tyndall Journal, RI MS JT/2/13a/61.
12. See Cantor (2015) for a discussion of Tyndall's early religious development.
13. 13 July 1845, Tyndall Journal, RI MS JT/2/13a/89.
14. 9 May 1847, Tyndall Journal, RI MS JT/2/13a/194.
15. 26 June 1847, Tyndall Journal, RI MS JT/2/13a/220.
16. 20 May 1845, Tyndall to Robert Allen, RI MS JT/1/TYP/11/3485 (Letter 321 in TC2).
17. 22 June 1845, Tyndall to Deborah McAssey, Martin Nevin, Private Collection (Letter 323 in TC2).
18. Richard Carter (1818–1895), 1896 Obituary, http://www.gracesguide.co.uk/Richard_Carter _(1818–1895) (accessed 28 December 2015).

19. 12 July 1845, Tyndall Journal, RI MS JT/2/13a/89.

20. 8 July 1845, Tyndall Journal, RI MS JT/2/13a/89.

21. 14 July 1845, Tyndall to Jack Tidmarsh, RI MS JT/1/TYP/11/3870–1 (Letter 325 in TC2).

22. 7 June 1846, Tyndall Journal, RI MS JT/2/13a/129.

23. 13 January 1846, Tyndall Journal, RI MS JT/2/13a/107.

24. 'The West Riding Union Railway', *The Times* (25 February 1846): p. 2.

25. 26 February 1846, Tyndall Journal, RI MS JT/2/13a/108.

26. 28 February 1846, Tyndall Journal, RI MS JT/2/13a/110; 9 April 1846, Tyndall Journal, RI MS JT/2/13a/114.

27. 23 March 1846, Tyndall Journal, RI MS JT/2/13a/112.

28. 5 April 1846, Tyndall Journal, RI MS JT/2/13a/113–14.

29. 30 June 1846, Tyndall Journal, RI MS JT/2/13a/131.

30. 'Select Committees on Bills', *The Times* (2 July 1846): p. 7.

31. Scrivenor (1849): p. 150.

32. 1 August 1846, Tyndall Journal, RI MS JT/2/13a/137.

33. 31 March 1847, 6 April 1847, 13 April 1847, and 22 April 1847, Tyndall to Editor, *Halifax Guardian*. The Whittington Club, a 'Bohemian experiment in Middle Class Social Reform', declined in the 1850s and closed in 1873.

34. 2 November 1846, Tyndall Journal, RI MS JT/2/13a/158.

35. 3 November 1846, Tyndall Journal, RI MS JT/2/13a/159.

36. 16 April 1847, Tyndall Journal, RI MS JT/2/13a/192.

37. See Dyde (2015) on Combe, common sense and phrenology.

38. Which celebrated Peel, Cobden, and Bright, and the repeal of the Corn Laws.

39. 4 August 1846, Tyndall Journal, RI MS JT/2/13a/139.

40. 6 August 1846, Tyndall Journal, RI MS JT/2/13a/140.

41. 15 September 1846, Tyndall Journal, RI MS JT/2/13a/147.

42. 17 September 1846, Tyndall Journal, RI MS JT/2/13a/147.

43. 'West-Riding Union Railways', *Bradford and Wakefield Observer* (22 October 1846): p. 5.

44. 9 March 1848, Tyndall Journal, RI MS JT/2/13a/302.

45. 21 February 1847, Tyndall Journal, RI MS JT/2/13a/179.

46. 8 February 1847, Tyndall Journal, RI MS JT/2/13a/177.

47. 1 June 1847, Tyndall Journal, RI MS JT/2/13a/204.

48. 10 June 1847, Tyndall Journal, RI MS JT/2/13a/208.

49. 12 March 1847, Tyndall Journal, RI MS JT/2/13a/186.

50. 11 March 1847, Tyndall to Editor of *Sligo Champion*, Tyndall Journal RI MS JT/2/13a/185. It does not appear that this letter was published.

51. 18 March 1847, Tyndall Journal, RI MS JT/2/13a/188.

52. 1 April 1847, Tyndall Journal, RI MS JT/2/13a/191.

53. 30 April 1847, Tyndall Journal, RI MS JT/2/13a/193.

54. 14 May 1847, Tyndall Journal, RI MS JT/2/13a/196.

55. 15 May 1847, Tyndall Journal, RI MS JT/2/13a/196.

56. 2 June 1847, Tyndall Journal, RI MS JT/2/13a/205.

57. Darwin (1869): p. xix.

58. 7 July 1847, Tyndall Journal, RI MS JT/2/13a/228.

59. 24 June 1847, Fanny Smith to Tyndall, RI MS JT/1/TYP/11/3866 (Letter 332 in TC2); 2 July 1847, Tyndall to Robert Allen, Tyndall Journal RI MS JT/2/13a/224–5 (Letter 333 in TC2).

60. Wat Ripton [Tyndall], 'Society', *Preston Chronicle* (10 July 1847).

61. 22 January 1847, Tyndall Journal, RI MS JT/2/13a/173.

62. 22 August 1895, John Tidmarsh to Louisa Tyndall, RI MS LT/1/43.
63. Thomson (1955): p. 247.
64. 17 June 1847, Tyndall Journal, RI MS JT/2/13a/213.
65. 10 August 1847, Tyndall Journal, RI MS JT/2/13a/243.

Chapter 3: Queenwood College (1847–1848)

1. 29 September 1850, Hirst Journal, RI MS JT/2/32b/667.
2. Thompson (1955).
3. Wilson, E., *A Practical Treatise on Healthy Skin: with Rules for the Medical and Domestic Treatment of Cutaneous Diseases* (1844); Liebig, J., *Animal Chemistry*, Gregory, W. (ed) (1842).
4. 4 September 1847, Tyndall Journal, RI MS JT/2/13a/252.
5. *Westminster Budget* (8 December 1893).
6. See Thompson (1957) for a study of Tyndall's self-education.
7. 28 April 1848, Tyndall Journal, RI MS JT/2/13a/318.
8. RI MS JT/2/35.
9. 14 February 1848, Tyndall Journal, RI MS JT/2/13a/296.
10. 29 November 1847, Tyndall Journal, RI MS JT/2/13a/275.
11. 25 May 1848, Tyndall Journal, RI MS JT/2/13a/328.
12. 20 December 1847, Tyndall Journal, RI MS JT/2/13a/285.
13. 18 December 1847, Tyndall Journal, RI MS JT/2/13a/284.
14. 27 December 1847, Tyndall Journal, RI MS JT/2/13a/287.
15. 1 February 1848, Thomas Hirst to Tyndall, RI MS JT/1/T/512.
16. Haugrud (1970).
17. 17 April 1848, Tyndall Journal, RI MS JT/2/13a/313.
18. 5 December 1847, Tyndall Journal, RI MS JT/2/13a/278.
19. 6 February 1848, Tyndall Journal, RI MS JT/2/13a/292.
20. 13 February 1848, Tyndall Journal, RI MS JT/2/13a/296.
21. 9 April 1848, Tyndall Journal, RI MS JT/2/13a/311.
22. 12 December 1847, Tyndall Journal, RI MS JT/2/13a/280.
23. 19 May 1848, Tyndall Journal, RI MS JT/2/13a/327.
24. 24 May 1848, Tyndall Journal, RI MS JT/2/13a/327.
25. 12 February 1848, Tyndall Journal, RI MS JT/2/13a/296.
26. 27 April 1848, Tyndall Journal, RI MS JT/2/13a/318.
27. 6 June 1848, Anne Edmondson to Tyndall and Edward Frankland, Tyndall Journal, RI MS JT/2/13a/331–3 (Letter 350 in TC2).
28. 8 June 1848, Tyndall to Anne Edmondson, Tyndall Journal, RI MS JT/2/13a/333–6 (Letter 351 in TC2).
29. 27 February 1854, Tyndall Journal, RI MS JT/2/13b/664.
30. 11 February 1872, Tyndall to Editor of *Woman*, Tyndall Journal, RI MS JT/2/13c/1389–90.
31. 10 December 1873, Tyndall to Martha Somerville, Oxford Bodleian Dep c.372 (MST-1).
32. 16 August 1848, Tyndall to Thomas Foy, Tyndall Journal, RI MS JT/2/13b/376 (Letter 357 in TC2)
33. 6 December 1853, Tyndall to Fanny Smith, Tyndall Journal, RI MS JT/2/13b/645.
34. 26 September 1847, Tyndall to Jack Tidmarsh, RI MS JT/1/TYP/11/3871 (Letter 336 in TC2).
35. 16 June 1848, Tyndall Journal, RI MS JT/2/13a/340.
36. 17 June 1848, Tyndall Journal, RI MS JT/2/13a/341.
37. 17 June 1848, Tyndall Journal, RI MS JT/2/13a/342.

38. 17 June 1848, Tyndall Journal, RI MS JT/2/13a/342.
39. 18 June 1848, Tyndall Journal, RI MS JT/2/13a/342.
40. In fact about 10,000 were killed or wounded.
41. 'A sketch of Paris after the late struggle', *Carlow Sentinel* (8 July 1848): p. [2]; 'Letters from Paris', *Carlow Sentinel* (22 July 1848): p. [2].
42. 1 July 1848, Tyndall Journal, RI MS JT/2/13b/349.
43. 5 July 1848, Tyndall Journal, RI MS JT/2/13b/353 (see his essay p. 352–5).
44. 16 July 1848, Elisa de Lamartine to Tyndall and Edward Frankland, RI MS JT/2/TYP/14/34 (Letter 351 in TC2).
45. 19 July 1848, Tyndall Journal, RI MS JT/2/13b/363.
46. 19 July 1848, Tyndall Journal, RI MS JT/2/13b/364.
47. 6 April 1851, Tyndall Journal, RI MS JT/2/13b/525.
48. Tyndall (1879): vol. 1, p. 396.
49. *c.* September 1848, Tyndall to Bob Allen, RI MS JT/1/TYP/12/3973 (Letter 353 in TC2).
50. See the end of Emerson's lecture 'The Uses of Natural History', given in 1833.

Chapter 4: Marburg (1848–1850)

1. 11 November 1848, Tyndall to Thomas Hirst, RI MS JT/1/T/516 (Letter 365 in TC2).
2. 17 November 1848, Tyndall Journal, RI MS JT/2/13b/399.
3. 9 February 1849, Tyndall to Thomas Hirst, RI MS JT/1/T/517 (Letter 370 in TC2).
4. 26 December 1848, Tyndall Journal, RI MS JT/2/13b/407.
5. 30 December 1848, Tyndall Journal, RI MS JT/2/13b/409.
6. 2 January 1849, Tyndall Journal, RI MS JT/2/13b/413.
7. 1 January 1849, Frankland Journal, RI MS JT/2/35.
8. Tyndall, J., 'Presidential Address before the British Association at Belfast', *Brit. Assoc. Rep.* (1874): p. lxvi–xcvii.
9. 16 April 1849, Tyndall to Thomas Hirst, RI MS JT/1/T/518 (Letter 373 in TC2).
10. 10 June 1849, Tyndall to Thomas Hirst, RI MS JT/1/T/519–1 (Letter 377 in TC2).
11. Although the Christmas tree was introduced into England by Queen Charlotte in 1800, it was the illustration in periodicals of Prince Albert's trees at Buckingham Palace in the 1840s that spread the tradition across England; 25 December 1848, Tyndall Journal, RI MS JT/2/13b/407.
12. 11 December 1848, Tyndall Journal, RI MS JT/2/13b/405.
13. 30 July 1849, Tyndall to Thomas Hirst, RI MS JT/1/T/1019 (Letter 380 in TC2).
14. 29 December 1848, Tyndall Journal, RI MS JT/2/13b/407–9.
15. 1 February 1849, Tyndall Journal, RI MS JT/2/13b/416.
16. 'The German Student' by Wat Ripton, *Preston Chronicle* (24 February 1849).
17. 26 July 1849, Tyndall Journal, RI MS JT/2/13b/442.
18. 30 March 1849, Tyndall to Thomas Hirst, RI MS JT/1/T/1018 (Letter 372 in TC2).
19. 18 May 1849, Tyndall Journal, RI MS JT/2/13b/432.
20. 1 May 1849, Tyndall Journal, RI MS JT/2/13b/428.
21. 29 May 1849, Tyndall Journal, RI MS JT/2/13b/435.
22. 10 June 1849, Tyndall to Thomas Hirst, RI JT/1/T/519–1 (Letter 377 in TC2).
23. Tyndall, J., 'Excursion into Germany' and 'A Whitsuntide Ramble', *Preston Chronicle* (19 May 1849 and 16 June 1849).
24. 14 June 1849, Tyndall Journal, RI MS JT/2/13b/439.
25. 5 February 1849, Thomas Hirst to Tyndall, RI MS JT/2/13b/417 (Letter 369 in TC2).

26. 21 June 1849 Thomas Hirst to Tyndall, RI MS JT/1/H/137 (Letter 378 in TC2).
27. 2 July 1849 Tyndall to Thomas Hirst, RI MS JT/1/T/520 (Letter 379 in TC2).
28. http://vsfp.byu.edu/index.php/Category:Truth-Seeker_and_Present_Age:_A_Catholic _Review_of_Literature,_Philosophy_and_Religion_(New_Series),_The, (accessed 4 January 2016).
29. 29 July 1849, Tyndall Journal, RI MS JT/2/13b/442.
30. 2 August 1849, Tyndall Journal, RI MS JT/2/13b/443.
31. 9 September 1849, Tyndall Journal, RI MS JT/2/13b/445.
32. 10 September 1849, Tyndall to Thomas Hirst, RI MS JT/1/T/521 (Letter 381 in TC2).
33. 18 September 1849, Thomas Hirst to Tyndall, RI MS JT/1/H/138 (Letter 382 in TC2).
34. 10 October 1849, Tyndall to Thomas Hirst, RI MS JT/1/T/862 (Letter 384 in TC2).
35. 19 September 1849, Tyndall Journal, RI MS JT/2/13b/448.
36. 21 September 1849, Tyndall Journal, RI MS JT/2/13b/450.
37. 23 September 1849, Tyndall Journal, RI MS JT/2/13b/452.
38. 26 September 1849, Tyndall Journal, RI MS JT/2/13b/454.
39. The tourist railway that is still there today was built in the 1870s.
40. 3 October 1849, Tyndall Journal, RI MS JT/2/13b/464.
41. 2 December 1849, Tyndall to Thomas Hirst, RI MS JT/1/T/524 (Letter 390 in TC2).
42. N.d. August 1851, Tyndall to Margaret Ginty, RI MS JT/1/TYP/11/3640.
43. 2 December 1849, Tyndall to Thomas Hirst, RI MS JT/1/T/524 (Letter 390 in TC2).
44. Fielding, K. J., 'Carlyle and John Tyndall: I', *Carlyle Studies Annual* 18 (1998): p. 43–53. See also Turner (1975).
45. Tyndall, J., 'Address delivered at the Birkbeck Institution', in *New Fragments* (London: Longmans, 1892): p. 237.
46. 20 October 1850, Tyndall Journal, RI MS JT/2/13b/514.
47. 21 October 1849, Tyndall to Thomas Hirst and James Craven, RI MS JT/1/T/523 (Letter 387 in TC2).
48. N.d. November 1849, Tyndall Journal, RI MS JT/2/13b/466.
49. Tyndall, J., 'Goethe and Faust', *Preston Chronicle* (29 September 1849): p. 3.
50. Tyndall, J., 'Sisters of the Rhine', *Preston Chronicle* (3, 10, 17, 24 November and 1 December 1849).
51. 30 November 1849, Tyndall Journal, RI MS JT/2/13b/469.
52. 12 January 1850, Tyndall Journal, RI MS JT/2/13b/474.
53. 18 April 1850, Tyndall Journal, RI MS JT/2/13b/486.
54. Only two other papers of Tyndall were co-authored, the next one with Knoblauch, and the first paper on glaciers with Thomas Huxley.
55. Tyndall and Knoblauch (1850a).
56. 14 January 1850, Tyndall to Thomas Hirst, RI MS JT/1/T/525.
57. Tyndall and Knoblauch (1850b); In a letter of 14 January 1850 (RI MS JT/1/T/525), Tyndall told Hirst he would send it to the *Philosophical Transactions*, possibly confusing the Royal Society's premier journal with the *Philosophical Magazine*.
58. 20 January 1850, Thomas Hirst to Tyndall, RI MS JT/1/T/142.
59. 20 January 1850, Thomas Hirst to Tyndall, RI MS JT/1/T/142. It is not clear, however, if they ever met Charlotte Brontë.
60. Tyndall, J., 'On the Death of Dean Bernard', *Carlow Sentinel* (13 April 1850): p. [4].
61. Tyndall, J., 'Man and Magnetism', *Preston Chronicle* (9 March 1850).
62. The ruler of Hesse-Cassel had come by train to Marburg the previous day. He did not even alight, as 3,000 people waited in sullen silence while the train stopped for a few minutes.

63. Tyndall, J., 'Zig-Zag', *Preston Chronicle* (20 April 1850).
64. Hirst sent Tyndall's article *Zig-Zag* to Carlyle in 1850, and was sure Carlyle had responded in *Stump Orator* (3 June 1850, Thomas Hirst/Jemmy Craven to Tyndall, RI MS JT/1/H/146).
65. 19 June 1850, Tyndall to Thomas Hirst, RI MS JT/1/T/1015.
66. 13 July 1850, Tyndall to Thomas Hirst, RI MS JT/1/T/529.
67. Huxley, L., *Life and Letters of Thomas Henry Huxley* (2 vols) (London: Macmillan, 1900): vol. I, p. 220.
68. Tyndall, J., *Fragments of Science*, 6th ed. (London: Longmans, 1879): vol. II, p. 96.
69. 12 May 1850, Tyndall to Thomas Hirst, RI MS JT/1/T/527.
70. N.d. May 1850, Tyndall to Thomas Hirst, RI MS JT/1/T/893; RI MS JT/5/16b.
71. 16 April 1850, Thomas Hirst to Tyndall, RI MS JT/1/H/144.
72. 7 Apr 1850, Tyndall to George Wynne, Tyndall Journal RI MS JT/2/13b/484.
73. 12 May 1850, Tyndall to Thomas Hirst, RI MS JT/1/T/527.
74. 12 June 1850, Tyndall Journal, RI MS JT/2/13b/494.
75. 13 June 1850, Tyndall Journal, RI MS JT/2/13b/495.
76. 14 June 1850, Tyndall Journal, RI MS JT/2/13b/496.
77. 15 June 1850, Tyndall Journal, RI MS JT/2/13b/496.
78. 15 June 1850, Tyndall Journal, RI MS JT/2/13b/498.
79. 19 June 1850, Tyndall Journal, RI MS JT/2/13b/500.
80. 19 June 1850, Tyndall Journal, RI MS JT/2/13b/501.

Chapter 5: Making a Name (1850–1853)

1. See Brock and Meadows (1998): p. 33.
2. 3 July 1850, Tyndall to Heinrich Debus, RI MS JT/1/T/357.
3. 18 July 1850, John Phillips to Tyndall, RI MS JT/1/TYP/3/976.
4. See Morrell and Thackray (1981) and McLeod and Collins (1981) for detailed histories.
5. See Higgitt and Withers (2008) and Browne (1994).
6. Tyndall (1850c).
7. 4 August 1850, Tyndall to Thomas Hirst, RI MS JT/1/T/530.
8. *The Athenaeum* (10 August 1850): p. 842; 7 August 1850, Tyndall Journal, RI MS JT/2/13b/505.
9. 7 August 1850, Tyndall Journal, RI MS JT/2/13b/504.
10. 7 August 1850, Tyndall Journal, RI MS JT/2/13b/504.
11. 7 August 1850, Tyndall to William Thomson, CUL Add 7342/T623, Kelvin Correspondence; 14 August 1850, William Thomson to Tyndall, RI MS JT/1/T/9.
12. N.d. June 1850, Tyndall Journal, RI MS JT/2/13b/501.
13. 21 August 1850, Tyndall to Thomas Hirst, RI MS JT/1/T/533.
14. 30 August 1850, Tyndall to Thomas Hirst, RI MS JT/1/T/1014.
15. 8 August 1850, George Wynne to Tyndall, RI MS JT/1/TYP/5/1841; 26 September 1850, George Wynne to Tyndall, RI MS JT/1/W/93.
16. Tyndall, J., 'The Propensities and their Equivalents', *The Leader* (29 June 1850): p. 332–3.
17. 28 September 1850, Hirst Journal, RI MS JT/2/32b/666.
18. 12 December 1850, Hirst Journal, RI MS JT/2/32b/685.
19. Tyndall, J., 'Trinkets from Deutschland', *The Leader* (16 November 1850): p. 813.
20. 25 September 1850, Hermann Knoblauch to Tyndall, RI MS JT/1/K/15.
21. 24 October 1850, Tyndall to Michael Faraday, Letter 2333 in James (1996).
22. 19 November 1850, Michael Faraday to Tyndall, Letter 2344 in James (1996).
23. 6 December 1850, Tyndall Journal, RI MS JT/2/13b/518.

24. Tyndall (1851a); 19 April 1851, Michael Faraday to Tyndall, Letter 2411 in James (1996).
25. Tyndall (1851b).
26. 23 February 1851, John de Haas to Tyndall, Tyndall Journal, RI MS JT/2/13b/522.
27. 19 March 1851, Tyndall to William Frances, St Bride Library archive, Authors' letters.
28. 5 March 1851, William Francis to Tyndall, RI MS JT/1/TYP/11/3573–4.
29. 3 March 1851, George Wynne to Tyndall, RI MS JT/1/W/94.
30. 19 August 1850, Tyndall to Thomas Hirst, RI MS JT/1/T/532.
31. From *An Essay on Man* by Alexander Pope.
32. 9 April 1851, Tyndall Journal, RI MS JT/2/13b/533.
33. 28 April 1851, Tyndall Journal, RI MS JT/2/13b/539.
34. 26 May 1851, Tyndall Journal, RI MS JT/2/13b/544.
35. 28 April 1851, Tyndall Journal, RI MS JT/2/13b/539.
36. 19 June 1851, Tyndall Journal, RI MS JT/2/13b/548.
37. 25 May 1851, Alexander von Humboldt to Tyndall, Tyndall Journal, RI MS JT/2/13b/544.
38. 30 April 1851, Tyndall Journal, RI MS JT/2/13b/541.
39. 4 May 1851, Tyndall Journal, RI MS JT/2/13b/542.
40. 11 May 1851, Tyndall, Journal, RI MS JT/2/13b/542.
41. 26 May 1851, Tyndall to Michael Faraday, Letter 2427 in James (1996).
42. Becquerel, E., 'De l'action du magnetisme sur tous les corps', *Annales de Chimie et de Physique* 32 (1851): 68–112. Becquerel, referring also to his previous results in *Annales de Chimie et de Physique* 28 (1849): p. 283, specifically contradicted Plücker's position in this paper.
43. 7 June 1851, Tyndall to Thomas Hirst, RI MS JT/1/T/542.
44. Tyndall, J., 'The Forester's Grave', *The Leader* (3 May 1851): p. 420–1; 11 May 1851, Thomas Hirst to Tyndall, RI MS JT/1/H/157.
45. 17 May 1851, Tyndall to Thomas Hirst, RI MS JT/1/T/540.
46. 1 June 1851, John Phillips to William Francis, RI MS JT/1/TYP/11/3575.
47. 22 June 1851, Tyndall Journal, RI MS JT/2/13b/547.
48. 22 June 1851, Tyndall Journal, RI MS JT/2/13b/546; from a poem by William Maccall.
49. 22 June 1851, Tyndall Journal, RI MS JT/2/13b/548.
50. 28 November 1851, Tyndall Journal, RI MS JT/2/13b/553.
51. 15 July 1851, Tyndall to Thomas Hirst, RI MS JT/1/T/543.
52. Tyndall (1851c); Tyndall (1851d).
53. *The Athenaeum*, 12 July 1851.
54. Tyndall, J., 'On Air-bubbles formed in Water', *Brit. Assoc. Rep, Notes and Abstracts of Miscellaneous Communications to the Sections* (1851): p. 26–7; Tyndall, J., 'Experiment in thermo-electricity with the monothermic pile invented by Prof. Magnus of Berlin', *Brit. Assoc. Rep, Notes and Abstracts of Miscellaneous Communications to the Sections* (1851): p. 18–19.
55. 1 October 1851, Tyndall Journal, RI MS JT/2/13b/550.
56. Tyndall (1851e).
57. 24 September 1851, Tyndall to Thomas Hirst, RI MS JT/1/T/545.
58. 30 July 1851, Tyndall to Michael Faraday, Letter 2451 in James (1996).
59. Faraday never gave open testimonials.
60. 31 August 1851, Thomas Hirst to Tyndall, RI MS JT/1/H/161.
61. 1 September 1851, Robert Bunsen to Tyndall, RI MS JT/1/B/141.
62. 8 November 1851, Tyndall to John Herschel, RI MS JT/1/TYP/2/493.
63. 4 December 1851, Thomas Huxley to Tyndall, RI MS JT/1/TYP/9/2855 (ICHC 8.5).
64. 9 January 1852, John Herschel to Tyndall, RI MS JT/1/TYP/2/496.
65. 6 November 1851, Edward Sabine to Tyndall, Tyndall Journal, RI MS JT/2/13b/550–1.

66. 20 February 1852, Tyndall Journal, RI MS JT/2/13b/558.
67. 11 January 1852, Tyndall Journal, RI MS JT/2/13b/557. He later told Hirst that he had invented 'a ducky little instrument for experiments on conduction, it is as pretty and as effectual a little affair as I have seen for some time. I look upon it with somewhat of the feelings of a father': 5 July 1852, Tyndall to Thomas Hirst, RI MS JT/1/T/552.
68. Tyndall (1853).
69. 'Notices respecting New Books', Phil. Mag. 3 (1852): p. 57–60.
70. 20 July 1852, Tyndall to William Francis, RDS 27/11.
71. Hampshire Advertiser (14 February 1852): p. 4.
72. Hampshire Advertiser (20 December 1851): p. 5
73. c.25 April 1852, Tyndall to Thomas Hirst, RI MS JT/1/T/894.
74. 24 November 1862, Tyndall to John Lubbock, RI MS JT/1/T/1038.
75. 6 October 1852, Tyndall Journal, RI MS JT/2/13b/585.
76. 2 April 1852, Hirst Journal, RI MS JT/2/32b/805.
77. 28 March 1852, Tyndall Journal, RI MS JT/2/13b/562.
78. 3 June 1852, Tyndall Journal, RI MS JT/2/13b/585.
79. 17 June 1852, Tyndall Journal, RI MS JT/2/13b/570.
80. 15 June 1852, William Thomson to Thomas Bell, RS RR/2/247.
81. Faraday, M., 'On Lines of Magnetic Force', Phil. Trans. R. Soc. Lond. 142 (1852): p. 50.
82. 11 July 1852, Tyndall to Michael Faraday, Letter 2550 in James (1996).
83. 28 July 1855, Lord Wrottesley to James Wilson, Royal Society, PP, 1854–5 (466) p. 1.
84. RS Minutes of Council 3 (1858–69): p. 24. See also Donnan (1950).
85. 2 July 1852, Tyndall Journal, RI MS JT/2/13b/573.
86. 5 July 1852, Tyndall to Edward Sabine, RI MS JT/1/TYP/3/1026–7.
87. 6 July 1852, Edward Sabine to Tyndall, RI MS JT/1/S/12.
88. 24 April 1855, Edward Sabine to Tyndall, RI MS JT/1/S/35.
89. See Hilts (1875): p. 15.
90. DeYoung (2011): p. 248.
91. 29 August 1852, Thomas Hirst to Tyndall, RI MS JT/1/H/171.
92. 24 October 1852, Hirst Journal, RI MS JT/2/32b/1081.
93. 24 October 1852, Hirst Journal, RI MS JT/2/32b/1081.
94. 29 October 1852, George Wynne to Tyndall, RI MS JT/1/T/555.
95. 29 August 1852, Tyndall Journal, RI MS JT/2/13b/581.
96. 1 September 1852, Tyndall Journal, RI MS JT/2/13b/581.
97. 2 September 1852, Tyndall Journal, RI MS JT/2/13b/581.
98. Tyndall (1852a).
99. Tyndall (1852b).
100. 4 September 1852, Tyndall Journal, RI MS JT/2/13b/583.
101. The Athenaeum (18 September 1852): p. 1010–11.
102. 4 September 1852, Tyndall Journal, RI MS JT/2/13b/583.
103. N.d. October 1853, Tyndall Journal, RI MS JT/2/13b/583.
104. 8 September 1852, Tyndall Journal, RI MS JT/2/13b/584.
105. 9 January 1853, Tyndall to Thomas Hirst, RI MS JT/1/T/558.
106. 7 October 1852, Thomas Hirst to Tyndall, RI MS JT/1/HTYP/210; 24 October 1852, Tyndall Journal, RI MS JT/2/13b/588.
107. 9 January 1853, Tyndall to Thomas Hirst, RI MS JT/1/T/558.
108. 17 October 1852, Tyndall Journal, RI MS JT/2/13b/588.

109. 21 November 1852, John Barlow to Tyndall, RI MS JT/1/TYP/1/142. Barlow may have been confused, as these were different electromagnets (Frank James, personal communication).
110. 6 January 1853, Tyndall Journal, RI MS JT/2/13b/594.
111. 25 October 1847, Tyndall to Thomas Hirst, RI MS JT/1/T/511.
112. 10 February 1853, Tyndall Journal, RI MS JT/2/13b/599.
113. Tyndall, J., 'On the Influence of Material Aggregation Upon the Manifestations of Force', *Proc. Roy. Inst.* 1 (1853): p. 254–9.
114. Frankland, E., 'John Tyndall', *Proc. Roy. Soc. Lond.* 55 (1894): p. xviii–xxxiv.
115. 11 February 1853, Tyndall Journal, RI MS JT/2/13b/600. Samuel Whitbread became a Fellow of the Royal Society in 1854.
116. See Kurzer, F., 'Chemistry and Chemists at the London Institution 1807–1912', *Annals of Science* 58 (2001): p. 163–201.
117. 22 February 1853, Tyndall to Thomas Huxley, RI MS JT/1/TYP/9/2858.
118. 25 February 1853, Thomas Huxley to Tyndall, RI MS JT/1/TYP/9/2859–61 (ICHC 8.9).
119. 18 March 1853, Tyndall Journal, RI MS JT/2/13b/602.
120. 23 February 1853, Tyndall to John Barlow, RI MS AD/03/A/02.
121. 6 March 1853, Robert Bunsen to Tyndall, RI MS JT/1/B/144.
122. 15 April 1853, Henry Bence Jones to Tyndall, RI MS JT/1/HTYP/239.
123. 5 June 1853, Tyndall to Henry Bence Jones, RI Managers' Minutes XI/17–18. Draft: Tyndall Journal, RI MS JT/2/13b/606–7.
124. Bence Jones was mistaken; the election took place on 4 July. Minutes of General Meetings of The Royal Institution RI MS AD/02/B/01/A06, p. 93; 6 June 1853, Henry Bence Jones to Tyndall, RI MS JT/1/TYP/1/683.
125. Tyndall, J., 'On some of the Eruptive Phenomena of Iceland', *Proc. Roy. Inst.* 1 (1853): p. 329–35.
126. 19 April 1853, Robert Bunsen to Tyndall, RI MS JT/1/B/145.
127. 16 June 1853, Tyndall to Thomas Hirst, RI MS JT/1/T/563.
128. 26 June 1853, Tyndall Journal, RI MS JT/2/13b/607.
129. 6 July 1853, Tyndall Journal, RI MS JT/2/13b/609.
130. 27 June 1853, Tyndall to Alexander Cumming, RI MS JT/1/TYP/1/255–6.
131. 19 November 1852, Tyndall Journal, RI MS JT/2/13b/589.
132. The firm only moved out in the 1980s and the building still stands.
133. 11 July 1853, Tyndall Journal, RI MS JT/2/13b/610.
134. 4 July 1853, Emil Du Bois-Reymond to Tyndall, RI MS JT/1/D/146.
135. 24 July 1853, Tyndall to Michael Faraday, Letter 2704 in James (1996).
136. Plücker, J., 'On Magnetism', *Brit. Assoc. Rep, Notes and Abstracts of Miscellaneous Communications to the Sections* (1853): p. 7–8.
137. 24 August 1853, Tyndall Journal, RI MS JT/2/13b/617.
138. 7 September 1853, Tyndall Journal, RI MS JT/2/13b/623.
139. 30 September 1853, Tyndall Journal, RI MS JT/2/13b/624.
140. 26 October 1853, Tyndall Journal, RI MS JT/2/13b/630.
141. De la Beche had served as a geologist on the Ordnance Survey under Colby.
142. 1 January 1854, Tyndall to Edward Forbes, Tyndall Journal, RI MS JT/2/13b/652.
143. 6 July 1853, Tyndall Journal, RI MS JT/2/13b/608.
144. 6 July 1853, Tyndall Journal, RI MS JT/2/13b/608. Tyndall's researches were not specifically directed at this question, although he did refer to the connection in his first Discourse.

145. 5 November 1853, Tyndall to Thomas Hirst, RI MS JT/1/T/933.
146. 25 November 1853, Tyndall to Thomas Hirst, RI MS JT/1/T/581.
147. 21 November 1853, Peter Gassiot to Tyndall, Tyndall Journal, RI MS JT/2/13b/639; For a full account of this episode see Jackson (2014).
148. 30 November 1853, Tyndall Journal, RI MS JT/2/13b/641.
149. 30 November 1853, Tyndall Journal, RI MS JT/2/13b/641.
150. RS Minutes of Council 2 (1846–58): p. 282.
151. 30 November 1853, Tyndall Journal, RI MS JT/2/13b/641.
152. 12 November 1853, Tyndall Journal, RI MS JT/2/13b/635.

Chapter 6: Clash of Theories (1854–1856)

1. 1 January 1854, Tyndall Journal, RI MS JT/2/13b/652.
2. 1 January 1854, Tyndall Journal, RI MS JT/2/13b/653.
3. 19 May 1857, Tyndall Journal, RI MS JT/2/13c/936.
4. 22 March 1854, Gustav Magnus to Tyndall, RI MS JT/1/M/9.
5. 20 January 1853, Tyndall Journal, RI MS JT/2/13b/596.
6. 21 January 1854, Tyndall Journal, RI MS JT/2/13b/654.
7. 12 February 1854, Tyndall to Thomas Hirst, RI MS JT/1/T/587.
8. Tyndall (1854).
9. 26 January 1854, Tyndall Journal, RI MS JT/2/13b/655.
10. 15 February 1854, William Grove, RS RR/2/251.
11. 19 February 1860, Hirst Journal, RI MS JT/2/32c/1528.
12. Forgan (1977), p. 108. The Wellesley Index lists forty-seven contributions by her between 1858 and 1886. This may underestimate her oeuvre. In addition to her novel *Hanworth*, serialized in *Fraser's Magazine* in 1858, I have identified a further novel *Ida Conway*, serialized in twelve monthly instalments in *Fraser's Magazine* in 1860–1. The Wellesley Index incorrectly attributes it to J. M. Capes.
13. 6 February 1854, Tyndall Journal, RI MS JT/2/13b/657.
14. 15 March 1854, Tyndall to Thomas Hirst, RI MS JT/1/T/588.
15. 21 March 1854, Tyndall to Thomas Hirst, RI MS JT/1/HTYP/328.
16. See Ashton (2006).
17. 4 April 1854, Tyndall Journal, RI MS JT/2/13b/671.
18. 14 May 1854, Tyndall Journal, RI MS JT/2/13b/674.
19. See Pollock, F. (ed.), *William Charles Macready, Reminiscences, and Selections from his Diaries and Letters*, 2 vols (London and New York: Longmans, Green & Co., 1875).
20. Tyndall, J., 'On some phenomena connected with the motion of liquids', *Proc. Roy. Inst.* 1 (1853): p. 446–8.
21. 20 May 1854, Tyndall to Thomas Hirst, RI MS JT/1/HTYP/340.
22. See James (2010): p. 99–101.
23. *Lectures on Education, delivered at the Royal Institution of Great Britain* (London: J W Parker & Son, 1854).
24. 2 May 1854, Tyndall to Thomas Hirst, RI MS JT/1/HTYP/338.
25. See Burchfield, J. D., 'John Tyndall at the Royal Institution', in James (2002).
26. Tyndall had the lecture reprinted in *Fragments of Science*. See Tyndall (1879): vol. 1, p. 333–55.
27. 27 May 1854, Tyndall Journal, RI MS JT/2/13b/675.

28. 27 May 1854, Tyndall to Thomas Hirst, RI MS JT/1/T/589.

29. 6 June 1854, Tyndall Journal, RI MS JT/2/13b/677.

30. 6 June 1854, Tyndall Journal, RI MS JT/2/13b/677.

31. 1 July 1854, Tyndall to Thomas Hirst, RI MS JT/1/HTYP/349.

32. 13 July 1854, Tyndall to Anna Martin, RI MS JT/1/HTYP/351–2.

33. Faraday, M., 'On Magnetic Hypotheses', *Proc. Roy. Inst.* 1 (1854): 457–9.

34. Faraday, M., 'A speculation touching Electric Conduction and the Nature of Matter', *Phil. Mag.* 24 (1844): p. 136.

35. Hesse (1961): p. 210. See also Gooding (1990).

36. 30 June 1854, Tyndall to Michael Faraday, Letter 2861 in James (1996).

37. 29 June 1854, Royal Society Minutes of Council, which implies that Sabine told him before the decision had been formally made, as the dinner was on 28 June.

38. 23 July 1854, Wilhelm Weber to Tyndall, RI MS JT/1/W/13. The letter to Weber of 3 July has not been found.

39. 5 August 1854, Tyndall Journal, RI MS JT/2/13b/680.

40. 5 August 1854, Tyndall to Thomas Hirst, RI MS JT/1/HTYP/353–4.

41. 27 July 1852, Tyndall Journal, RI MS JT/2/13b/577.

42. 22 August 1854, Tyndall Journal, RI MS JT/2/13b/684.

43. 26 August 1854, Tyndall Journal, RI MS JT/2/13b/685.

44. 2 September 1854, Tyndall Journal, RI MS JT/2/13b/688.

45. N.d. September 1854, Tyndall to Thomas Hirst, RI MS JT/1/HTYP/359.

46. It closed two years later to become the Alhambra Theatre.

47. Tyndall, J., 'On some Peculiarities of the Magnetic Field', *Brit. Assoc. Rep., Notes and Abstracts of Miscellaneous Communications to the Sections* (London: Murray, 1854): p. 16–17; Tyndall, J., 'On the diamagnetic force', *Brit. Assoc. Rep., Notes and Abstracts of Miscellaneous Communications to the Sections* (1854): p. 14–16.

48. 21 September 1854, Tyndall Journal, RI MS JT/2/13b/691.

49. 22 September 1854, Tyndall Journal, RI MS JT/2/13b/691.

50. 1 October 1854, Tyndall to Thomas Hirst, RI MS JT/1/HTYP/361–3.

51. 30 September 1854, Tyndall Journal, RI MS JT/2/13b/692.

52. 1 October 1854, Tyndall to Thomas Hirst, RI MS JT/1/HTYP/361–2.

53. 17 October 1854, Thomas Huxley to Tyndall, RI MS JT/1/TYP/9/2866.

54. 18 October 1854, Tyndall to Thomas Huxley, RI MS JT/1/TYP/9/2867 (ICHC 8.17).

55. 22 October 1854, Thomas Huxley to Tyndall, RI MS JT/1/TYP/9/2868 (ICHC 8.18); 27 October 1854, Tyndall to Thomas Huxley, RI MS JT/1/TYP/9/2869 (ICHC 8.19).

56. 20 November 1854, James Coxe to Tyndall, RI MS JT/1/TYP/1/263a.

57. 23 November 1854, Tyndall to James Coxe, RI MS JT/1/T/243.

58. 27 November 1854, Tyndall to James Coxe, RI MS JT/1/TYP/1/265.

59. 28 November 1854, Thomas Hirst to Tyndall, RI MS JT/1/H/208.

60. 3 December 1854, Tyndall Journal, RI MS JT/2/13b/702.

61. 31 December 1854, Tyndall Journal, RI MS JT/2/13b/709.

62. 25 December 1854, Tyndall Journal, RI MS JT/2/13b/706.

63. 12 December 1854, Tyndall to Thomas Hirst, RI MS JT/1/HTYP/379.

64. 29 January 1855, Tyndall to Thomas Hirst, RI MS JT/1/T/592.

65. 25 January 1855, Tyndall Journal, RI MS JT/2/13b/715.

66. 25 January 1855, Tyndall Journal, RI MS JT/2/13b/715.

67. 11 January 1855, Tyndall Journal, RI MS JT/2/13b/713.

68. Tyndall (1855a).
69. See Tyndall (1855b), Faraday, M., 'Prof. Faraday on some points in Magnetic Philosophy', *Phil. Mag.* 9 (1855): p. 81–113 and Faraday, M., 'Prof. Faraday's Magnetic Remarks', *Phil. Mag.* 9, (1855): p. 253–5. See also Williamson, A. W., 'A Note on the Magnetic Medium', *Proc. Roy. Soc. Lond.* 7 (1855): p. 306–8 and Hirst's reply, stimulated by Tyndall: Hirst, T. A., 'On the Existence of a Magnetic Medium', *Proc. Roy. Soc. Lond.* 7 (1855): p. 448–54.
70. 16 April 1855, W. H. Miller, RS RR/2/252.
71. N.d. William Thomson to George Stokes, RS RR/2/253.
72. 4 February 1855, Tyndall to Thomas Hirst, RI MS JT/1/T/593.
73. 13 February 1855, Tyndall Journal, RI MS JT/2/13c/727.
74. 6 May 1855, Tyndall Journal, RI MS JT/2/13c/744.
75. Asked by Mr Tite on 1 January. See lecture notes RI MS JT/4/7a.
76. Tyndall (1855c).
77. 9 January 1855, Tyndall Journal, RI MS JT/2/13c/713.
78. He wrote at least six reviews of science for the *Westminster Review* between April 1855 and July 1857: (*Westminster Review* 63 (April 1855): p. 551–63; 64, (July 1855): p. 255–63; 64 (October 1855): p. 557–65; 65 (January 1856): p. 254–61; 65 (April 1856): p. 596–601; 68 (July 1857): p. 268–73.
79. 'Science', *Westminster Review* 63 (1855): p. 552–8.
80. 22 February 1855, Tyndall Journal, RI MS JT/2/13b/730.
81. Tyndall, J., 'Review VIII: "Rise, Progress, and Present Condition of Animal Electricity"', *The British and Foreign Medico-chirurgical Review*, 25 (1854): p. 126–42.
82. 2 June 1855, Tyndall Journal, RI MS JT/2/13c/750.
83. See Allibone (1976).
84. 12 May 1855, Tyndall Journal, RI MS JT/2/13c/746.
85. 12 May 1855, Tyndall Journal, RI MS JT/2/13c/746.
86. 27 May 1855, Tyndall to Thomas Hirst, RI MS JT/1/T/602.
87. Tyndall, J., 'On the Currents of the Leyden Battery', *Proc. Roy. Inst.* 2 (1855): p. 132–5. See also Tyndall (1855d).
88. 21 June 1855, Tyndall Journal, RI MS JT/2/13c/757.
89. 25 June 1855, Tyndall Journal, RI MS JT/2/13c/771.
90. 25 June 1855, Tyndall Journal, RI MS JT/2/13c/771.
91. 1 July 1855, Tyndall to Michael Faraday, Letter 3000 in James (1996).
92. 21 June 1855, Tyndall Journal, RI MS JT/2/13c/760.
93. 21 June 1855, Tyndall Journal, RI MS JT/2/13c/762.
94. 21 June 1855, Tyndall Journal, RI MS JT/2/13c/762.
95. 1 July 1855, Tyndall to Thomas Huxley, RI MS JT/1/TYP/9/2875 (ICHC 8.24).
96. 18 June 1855, Henry Moseley to Tyndall, RI MS JT/1/TYP/3/884.
97. 9 July 1855, Charles Hawker to Tyndall, RI MS JT/1/TYP/3/885; 12 July 1855, Henry Moseley to Tyndall, RI MS JT/1/TYP/3/889.
98. 3 August 1855, Tyndall Journal, RI MS JT/2/13c/780.
99. 5 August 1855, Elie Wartmann to Tyndall, RI MS JT/1/W/6.
100. 8 August 1855, Elie Wartmann to Tyndall, RI MS JT/1/W/7.
101. Chandler, A., 'The French Universal Exposition of 1855', *World's Fair Magazine* 6 (1986): no. 2.
102. 1 August 1855, Tyndall Journal, RI MS JT/2/13c/779; 14 August 1855, Tyndall Journal, RI MS JT/2/13c/781.
103. 9 September 1855, Tyndall Journal, RI MS JT/2/13c/789.

104. 12 September 1855, Tyndall Journal, RI MS JT/2/13c/791.
105. 1 June 1855, William Thomson to Tyndall, RI MS JT/1/T/13.
106. Tyndall, J., 'Experimental Demonstration of the Polarity of Diamagnetic Bodies', *Brit. Assoc. Rep., Notes and Abstracts of Miscellaneous Communications to the Sections* (1855): p. 22–3.
107. 17 September 1855, Tyndall to Thomas Hirst, RI MS JT/1/T/611.
108. *The Athenaeum* (6 October 1855): p. 1157
109. 22 October 1855, Edward Frankland to Tyndall, RI MS JT/1/F/44.
110. 6 October 1855, Michael Faraday to Tyndall, Letter 3027 in James (1996).
111. 23 December 1855, Tyndall to Auguste de la Rive, RI MS JT/1/TYP/1/335–7.
112. 27 October 1855, Tyndall Journal, RI MS JT/2/13c/797.
113. 5 December 1855, Tyndall to William Grove, RI MS Gr/3a/152.
114. Tyndall (1856a); Tyndall (1856b).
115. Tyndall (1856c).
116. Tyndall (1856c): p. 137.
117. 8 March 1856, George Airy to Tyndall, Tyndall Journal, RI MS JT/2/13c/829.
118. 10 March 1856, Tyndall to George Airy, CUL RGO.6/694.
119. Tyndall, J., 'The Disposition of Force in Paramagnetic and Diamagnetic Bodies', *Proc. Roy. Inst.* 2 (1856): 159–64. Lecture notes RI MS JT/4/5/4.
120. 17 September 1856, Tyndall Journal, RI MS JT/2/13c/913.
121. Plücker (1858).
122. 10 April 1858, Tyndall Journal, RI MS JT/2/13c/1069.
123. 5 November 1855, Tyndall to Thomas Hirst, RI MS JT/1/T/935.
124. 22 February 1856, James Clerk Maxwell to William Thomson, CUL Add. MSS 7342, M95.
125. Tyndall (1870).
126. 29 December 1855, Tyndall Journal, RI MS JT/2/13c/811.
127. *Dictionary of National Biography*, 1885–1900, Vol. 3, p. 193–4.
128. 12 January 1856, Tyndall Journal, RI MS JT/2/13c/817.
129. 12 January 1856, Tyndall to Thomas Hirst, RI MS JT/1/T/617.
130. 23 April 1856, Joseph Galbraith to Tyndall, Tyndall Journal, RI MS JT/2/13c/836.
131. Tyndall to *The Times* (28 April 1856): p. 12.
132. 10 May 1856, Tyndall to Thomas Hirst, RI MS JT/1/T/622.
133. 19 April 1856, Tyndall to Thomas Hirst, RI MS JT/1/HTYP/443.
134. 6 May 1856, Thomas Hirst to Tyndall, RI MS JT/1/H/223.
135. There was a happy ending. Despite expulsion, Taylor later qualified as a doctor and inherited his father's practice in Romsey.
136. 7 May 1856, Tyndall to Thomas Hirst, RI MS JT/1/T/621.
137. 18 June 1856, Thomas Hirst to Tyndall, RI MS JT/1/H/226.
138. Tyndall (1856d). This was stimulated by the examination of several cases brought to his attention by White-Cooper, and his reading of George Wilson's treatise: Wilson, G., *Researches on Colour-Blindness* (Edinburgh: Sutherland & Knox, 1855).
139. 8 April 1856, Tyndall Journal, RI JT/2/13c/833. Tyndall published two articles 'On Binocular Vision and the Stereoscope', *Journal of the Photographic Society* 3 (1856): p. 96–102 and p. 116–21.
140. 17 November 1856, Tyndall to Auguste De la Rive, RI MS JT/1/TYP/1/343–4.
141. 17 May 1856, Tyndall Journal, RI MS JT/2/13c/843.
142. 14 May 1856, Tyndall Journal, RI MS JT/2/13c/843.
143. 6 May 1856, Tyndall Journal, RI MS JT/2/13c/841.

Chapter 7: Glacial Explorations (1856–1857)

1. 27 October 1855, Tyndall Journal, RI MS JT/2/13c/797.
2. 27 July 1856, John Herschel to Tyndall, RI MS JT/1/H/92. Tyndall had this letter published in the *Philosophical Magazine* for September.
3. Tyndall, J., 'Comparative view of the cleavage of crystals and slate rocks', *Proc. Roy. Inst.* 2, (1856): p. 295–308, at p. 305.
4. 12 June 1856, Tyndall Journal, RI MS JT/2/13c/849.
5. 15 June 1856, Tyndall Journal, RI MS JT/2/13c/849.
6. 9 August 1856, Tyndall to Thomas Hirst, RI MS JT/1/T/631.
7. 17 August 1856, Tyndall Journal, RI MS JT/2/13c/870.
8. 21 August 1856, Tyndall Journal, RI MS JT/2/13c/881.
9. 24 August 1856, Tyndall Journal, RI MS JT/2/13c/886.
10. 5 September 1856, Tyndall Journal, RI MS JT/2/13c/900.
11. 17 September 1856, Tyndall Journal, RI MS JT/2/13c/913.
12. Tyndall, J., 'Über Spalten im Gletschereise', *Amtliche Bericht über die zwei und dreissigste Versammlung der Deutscher Naturforscher und Ärzte zu Wien im September 1856* (Wien: Staatsdruckerei, 1858).
13. 26 October 1856, Tyndall to Auguste De la Rive, RI MS JT/1/TYP/1/339.
14. 7 November 1856, Auguste De la Rive to Tyndall, RI MS JT/1/D/89.
15. Tyndall and Huxley (1857).
16. 6 January 1857, Tyndall Journal, RI MS JT/2/13c/917.
17. 4 December 1856, Tyndall to Thomas Hirst, RI MS JT/1/T/634.
18. Tyndall, J., 'Observations on Glaciers', *Proc. Roy. Inst.* 2 (1857): p. 320–7.
19. 6 January 1857, Tyndall to Thomas Huxley, RI MS JT/1/TYP/9/2877.
20. Tyndall was unaware at this point that Forbes had retracted that claim in his Thirteenth Letter, though when he found out he thought the new explanation also unsatisfactory. See Tyndall (1860): p. 379.
21. 24 January 1857, Frederick Pollock to Tyndall, RI MS JT/1/P/225.
22. 4 February 1857, Charles Darwin to Tyndall, RI MS JT/1/TYP/9/2801.
23. RS RR/3/267.
24. Probably written by the mountaineer Alfred Wills (Nanna Kaalund, personal communication).
25. 'Dr Tyndall's Theory of Glaciers', *Saturday Review* (31 January 1857): p. 102–3.
26. 27 November 1856, Tyndall to Rudolf Clausius, RI MS JT/1/T/171.
27. 2 December 1856, Rudolf Clausius to Tyndall, RI MS JT/1/TYP/7/2224.
28. 6 March 1857, James Forbes to Tyndall, RI MS JT/1/F/31.
29. 7 March 1857, Tyndall to James Forbes, University of St Andrews Library, msdep7—Incoming letters 1857, no. 33.
30. 13 March 1857, Tyndall Journal, RI MS JT/2/13c/919.
31. 13 March 1857, Tyndall to Auguste De la Rive, RI MS JT/1/TYP/1/1/347. In the event, he obtained a theodolite from Murchison: Louisa draft Tyndall biography, RI MS JT/5/10/863.
32. 23 March 1857, William Hopkins to Tyndall, RI MS JT/1/H/505.
33. 13 March 1857, Tyndall Journal, RI MS JT/2/13c/919.
34. 12 April 1857, Tyndall Journal, RI MS JT/2/13c/922.
35. 18 April 1857, Tyndall to Michael Faraday, Letter 3272 in James (2008).
36. 27 April 1857, Joseph Hooker to Tyndall, RI MS JT/1/TYP/8/2545.
37. Parry (1993): p. 182.

38. 8 January 1858, Tyndall to Thomas Hirst, RI MS JT/1/T/635.
39. Tyndall (1858): p. 223–4.
40. Tyndall (1857a).
41. 2 May 1857, Tyndall Journal, RI MS JT/2/13c/931.
42. 6 May 1857, Tyndall Journal, RI MS JT/2/13c/932.
43. Tyndall, J., 'On M. Lissajou's Acoustic Experiments', *Proc. Roy. Inst.* 2 (1857): p. 441–3.
44. 23 May 1857, Tyndall Journal, RI MS JT/2/13c/937.
45. 28 May 1857, Tyndall Journal, RI MS JT/2/13c/938.
46. 22 May 1857, Tyndall Journal, RI MS JT/2/13c/937.
47. 10 June 1857, Tyndall Journal, RI MS JT/2/13c/939.
48. 12 June 1857, Tyndall Journal, RI MS JT/2/13c/941.
49. 13 June 1857, Tyndall Journal, RI MS JT/2/13c/942.
50. Tyndall (1857b).
51. 28 June 1857, Tyndall Journal, RI MS JT/2/13c/943.
52. 14 June 1857, Thomas Hirst to Tyndall, RI MS JT/1/H/233.
53. 2 July 1857, Hirst Journal, RI MS JT/2/32e/1271.
54. 1 July 1857, Tyndall Journal, RI MS JT/2/13c/945.
55. 4 July 1857, Tyndall Journal, RI MS JT/2/13c/948.
56. 6 July 1857, Tyndall Journal, RI MS JT/2/13c/949.
57. 1 July 1857, Tyndall to Juliet Pollock, RI MS JT/1/T/1197; 4 July 1857, Tyndall to William Grove, RI MS Gr/3a/154.
58. 12 July 1857, Tyndall Journal, RI MS JT/2/13c/954.
59. 15 July 1857, Tyndall Journal, RI MS JT/2/13c/957.
60. 18 July 1857, Tyndall to Michael Faraday, Letter 3320 in James (2008).
61. 16 July 1857, Tyndall Journal, RI MS JT/2/13c/958.
62. 25 July 1857, Tyndall Journal, RI MS JT/2/13c/976.
63. 27 July 1857, Tyndall Journal, RI MS JT/2/13c/977.
64. 9 August 1857, Tyndall Journal, RI MS JT/2/13c/1001.
65. 31 July 1857, Tyndall Journal, RI MS JT/2/13c/985.
66. 4 August 1857, Tyndall Journal, RI MS JT/2/13c/994.
67. 11 August 1857, Tyndall Journal, RI MS JT/2/13c/1003.
68. A tube stuffed with gunpowder, since Simond had forgotten to bring a pistol, with which Tyndall wished to repeat Saussure's experiment of firing a shot on the summit.
69. 13 August 1857, Tyndall Journal, RI MS JT/2/13c/1019.
70. 14 August 1857, Tyndall Journal, RI MS JT/2/13c/1020.
71. 17 August 1857, Tyndall Journal, RI MS JT/2/13c/1021.
72. 20 August 1857, Tyndall Journal, RI MS JT/2/13c/1028.
73. 15 August 1857, George Airy to Tyndall, RI MS JT/1/TYP/1/22.
74. 16 July 1857, Hermann Knoblauch to Tyndall, RI MS JT/1/K/20.
75. 22 September 1857, Tyndall Journal, RI MS JT/2/13c/1034.
76. 17 December 1857, Tyndall Journal, RI MS JT/2/13c/1041.
77. 31 October 1857, Tyndall Journal, RI MS JT/2/13c/1038.
78. Huxley, H., 'Observations on the Structure of Glacier Ice', *Phil. Mag.* 14 (1857): p. 241–60.
79. 14 September 1857, Tyndall Journal, RI MS JT/2/13c/1033.
80. 14 October 1857, Tyndall to Thomas Hirst, RI MS JT/1/T/964.
81. 21 October 1857, Tyndall to James Forbes, St. Andrews Letters (85).
82. 23 October 1857, James Forbes to Tyndall, RI MS JT/1/F/32. For a full discussion of this episode see Fyfe and Moxham (2016): p. 367–70.

83. Ball, J., 'Observations upon the Structure of Glaciers', *Phil. Mag.* 14 (1857): p. 481–504.
84. 15 November 1857, Tyndall to Rudolf Clausius, RI MS JT/1/T/173.
85. 8 December 1857, Rudolf Clausius to Tyndall, RI MS JT/1/TYP/7/2226–30.
86. 9 December 1857, Michael Faraday to Tyndall, Letter 3363 in James (2008).
87. 16 December 1857, William Hopkins to Tyndall, RI MS JT/1/TYP/2/612.
88. Tyndall (1858); Tyndall, J., 'On Some Physical Properties of Ice', *Proc. Roy. Inst.* 2 (1858): p. 454–7.
89. 14 October 1857, Tyndall to Thomas Hirst, RI MS JT/1/T/964.
90. 14 October 1857, Tyndall to Thomas Hirst, RI MS JT/1/T/964.

Chapter 8: Storms over Glaciers (1858–1860)

1. 16 February 1858, Tyndall to Thomas Huxley, RI MS JT/1/TYP/9/2887.
2. 2 January 1857, Tyndall Journal, RI MS JT/2/13c/916.
3. 27 July 1854, Hirst Journal, RI MS JT/2/32b/1139.
4. 19 February 1858, Tyndall Journal, RI MS JT/2/13c/1053.
5. Tyndall, J. 'On some Physical Properties of Ice', *Proc. Roy. Inst.* 2 (1858): p. 454–7.
6. 13 January 1858, George Stokes to Tyndall, RI MS JT/1/S/219.
7. 12 July 1858, Tyndall to John Murray, National Library of Scotland.
8. N.d., Joseph Hooker to Tyndall, RI MS JT/1/TYP/8/2552.
9. 4 September 1868, Tyndall to John Murray, National Library of Scotland.
10. In the event, Bence Jones was not elected Secretary of the Royal Institution until 3 December 1860, and though they had their disagreements, this would be an important partnership through to Bence Jones's death in 1873.
11. 19 February 1858, Tyndall to Michael Faraday, Letter 3397 in James (2008).
12. It was increased by £100 in 1859.
13. 6 February 1858, Tyndall Journal, RI MS JT/2/13c/1051.
14. 8 February 1858, Tyndall to Thomas Hirst, RI MS JT/1/T/642.
15. See Louisa draft biography, RI MS JT/5/11/925.
16. 2 March 1858, Tyndall Journal, RI MS JT/2/13c/1056.
17. 11 March 1858, Tyndall Journal, RI MS JT/2/13c/1057.
18. 5 April 1858, Tyndall Journal, RI MS JT/2/13c/1066.
19. 5 April 1858, Tyndall Journal, RI MS JT/2/13c/1066.
20. 6 April 1858, Tyndall Journal, RI MS JT/2/13c/1066.
21. 6 April 1858, Tyndall Journal, RI MS JT/2/13c/1067.
22. 7 April 1858, Tyndall to Juliet Pollock, RI MS JT/1/T/1202.
23. 6 January 1858, Tyndall Journal, RI MS JT/2/13c/1044.
24. 9 January 1858, Tyndall Journal, RI MS JT/2/13c/1045.
25. 15 March 1858, Tyndall Journal, RI MS JT/2/13c/1059.
26. 19 March 1858, Tyndall Journal, RI MS JT/2/13c/1061.
27. 19 March 1858, Tyndall Journal, RI MS JT/2/13c/1061.
28. 8 May 1858, Tyndall Journal, RI MS JT/2/13c/1074.
29. 5 September 1858, Tyndall Journal, RI MS JT/2/13c/1075.
30. 5 September 1858, Tyndall Journal, RI MS JT/2/13c/1074.
31. 7 May 1858, George Airy to Tyndall, RI MS JT/1/A/34.
32. 23 May 1858, Rudolf Clausius to Tyndall, RI MS JT/1/TYP/7/2231.
33. Tyndall (1859a); Tyndall, J., 'The Mer-de-Glace', *Proc. Roy. Inst.* 2 (1858): p. 544–53.
34. 24 July 1858, Tyndall Journal, RI MS JT/2/13c/1081.
35. 10 August 1858, George Airy to Tyndall, RI MS JT/1/A/35.

36. 1 August 1858, Tyndall Journal, RI MS JT/2/13c/1090.
37. 1 August 1858, Tyndall Journal, RI MS JT/2/13c/1093.
38. 2 August 1858, Tyndall Journal, RI MS JT/2/13c/1094.
39. Tyndall (1860): p. 120.
40. Tyndall (1860): p. 160.
41. 2 August 1857, Tyndall Journal, RI MS JT/2/13c/988.
42. 31 August 1858, Juliet Pollock to Tyndall, RI MS JT/1/P/193.
43. 'The Glaciers of Switzerland', *The Times* (3 September 1858): p. 10.
44. 18 August 1858, Tyndall Journal, RI MS JT/2/13c/1128.
45. Donati's Comet was the brightest in the nineteenth century, apart from the Great Comet of 1811.
46. 12 September 1858, Tyndall Journal, RI MS JT/2/13c/1152.
47. Tyndall (1858b); see 18 February 1859, Richard Owen to Tyndall, in Tyndall (1860): p. 193–4.
48. 19 October 1858, Tyndall to Thomas Hirst, RI MS JT/1/T/647.
49. See Barton (2018): ch. 2.2 and ch. 2.1.
50. 31 December 1858, Tyndall to Thomas Hirst, RI MS JT/1/T/648.
51. 17 January 1859, Tyndall to George Airy, CUL RGO.6/408.
52. 12 February 1859, James Forbes to Tyndall, St. Andrews Letters (532).
53. Tyndall (1859b).
54. Forbes, J. D., 'Remarks of a paper "On Glaciers and Ice" in the last number of the Philosophical Magazine. In a letter to Prof. Tyndall', *Phil. Mag.* 17 (1859): p. 197–201.
55. Faraday, M., 'On Regelation, and on the Conservation of Force', *Phil. Mag.* 17 (1859): p. 162–9.
56. 7 June 1859, David Brewster to Tyndall, RI MS JT/1/B/129.
57. 1 June 1859, Tyndall to Rudolf Clausius, RI MS JT/1/T/176.
58. 9 June 1859, Rudolf Clausius to Tyndall, RI MS JT/1/TYP/7/2234–5.
59. Tyndall (1859c); Tyndall, J., 'The Veined Structure of Glaciers', *Proc. Roy. Inst.* 3 (1859): p. 72–8.
60. Ball, J., 'Remarks on the Veined Structure of Glaciers', *Phil. Mag.* 17 (1859): p. 263–8.
61. 4 March 1859, Tyndall Journal, RI MS JT/2/13c/1155.
62. 3 April 1859, Tyndall to Thomas Hirst, RI MS JT/1/T/649.
63. 29 December 1896, Julia Moore to Louisa Tyndall, RI MS LT/65/19.
64. 24 July 1859, Thomas Hirst to Tyndall, RI MS JT/1/H/224.
65. 19 April 1859, Tyndall to Francis Fox Tuckett, Science Museum, Wroughton 2108/79a.
66. 17 April 1859, Tyndall to Juliet Pollock, RI MS JT/1/TYP/6/1954–8.
67. 18 April 1859, Frederick Pollock to Tyndall, RI MS JT/1/P/231.
68. Sketch of the Ascent of Scafell, Papers of Sir Edward Frankland, ex Joan Bucknell, Box 2, John Rylands Library, University of Manchester. They may have climbed Scafell Pike, the highest mountain in England, and slightly higher than nearby Scafell.
69. 16 August 1859, Auguste De la Rive to Tyndall, RI MS JT/1/TYP/1/360; 18 August 1859, Tyndall to Auguste De la Rive, RI MS JT/1/TYP/1/361.
70. The colour magenta was named after this battle, from a dye discovered in 1859.
71. 21 September 1859, Tyndall to Thomas Hirst, RI MS JT/1/T/653.
72. 31 October 1859, Hirst Journal, RI MS JT/2/32c/1512.
73. 5 September 1859, Tyndall to Thomas Hirst, RI MS JT/1/T/653 (or possibly 29 August).
74. 5 September 1859, Thomas Huxley to Tyndall, RI MS JT/1/TYP/9/2893–4; 10 September 1859, Michael Faraday to Tyndall, Letter 3639 in James (2008); 4 November 1859, Tyndall to John Barlow, RI MS AD/03/04, RI MS C93 LO/6.

75. London Institution. Minutes of the Committee of Managers, CLC/009/MS03076/007. 1851–9, London Metropolitan Archive.

76. 9 November 1859, Lyon Playfair to Tyndall, RI MS JT/1/P/117.

77. c.10 November 1859, Tyndall to Lyon Playfair, RI MS JT/1/T/1127.

78. 12 November 1859, Lyon Playfair to Tyndall, RI MS JT/1/P/118.

79. 15 November 1869, Willie Cumming to Tyndall, RI MS JT/1/TYP/1278.

80. 17 November 1859, Michael Faraday to Tyndall, Letter 3675 in James (2008).

81. 19 November 1859, Tyndall to Lyon Playfair, Tyndall Journal, RI MS JT/2/13c/1163.

82. 21 November 1859, Roderick Murchison to Tyndall, Tyndall Journal, RI MS JT/2/13c/1164.

83. 4 November 1859, Tyndall Journal, RI MS JT/2/13c/1159.

84. 12 November 1859, Tyndall Journal, RI MS JT/2/13c/1161.

85. 29 April 1859, William Thomson to George Stokes, in Wilson (1990): p. 239.

86. 7 December 1859, Tyndall to Rudolf Clausius, RI MS JT/1/T/179.

87. 20 December 1859, Escher von der Linth to Tyndall, RI MS JT/1/E/108.

88. 16 November 1859, Auguste Balmat to Tyndall, RI MS JT/1/C/35.

89. 22 January 1860, Hirst Journal, RI MS JT/2/32c/1522–3.

90. 'Discussion on Mr Preece's Paper', Journal of the Society of Telegraph Engineers 9 (1880): p. 381.

91. 18 December 1859, Balmat to Tyndall, RI MS JT/1/B/6.

92. 24 December 1859, Tyndall to Sarah Faraday, Letter 3699 in James (2008).

93. 25 December 1859, Tyndall Journal, RI MS JT/2/13c/1168.

94. 25 December 1859, Tyndall Journal, RI MS JT/2/13c/1169.

95. 26 December 1859, Tyndall to Michael Faraday, Letter 3700 in James (2008).

96. 27 December 1859, Tyndall to Michael Faraday, Letter 3701 in James (2008).

97. Tyndall (1860): p. 213. The words have echoes of his late poem 'A Morning on Alp Lusgen'.

98. 29 December 1859, Tyndall Journal, RI MS JT/2/13c/1179.

99. Tyndall, J., 'The Influence of the Magnetic Force on the Electric Discharge', Proc. Roy. Inst. 3, (1860): 169–74.

100. 28 January 1860, Tyndall Journal, RI MS JT/2/13c/1185.

101. 19 February 1860, Hirst Journal, RI MS JT/2/32c/1528.

102. 10 February 1860, Hirst Journal, RI MS JT/2/32c/1527.

103. 'Professor Tyndall on Scientific Research', Leeds Mercury (5 December 1884).

104. 30 July 1873, Thomas Huxley to Tyndall, RI MS JT/1/TYP/9/3017–18.

105. 19 February 1860, Hirst Journal, RI MS JT/2/32c/1527.

106. 19 May 1860, William Hopkins to Tyndall, RI MS JT/1/TYP/2/615; 29 May 1860, William Hopkins to Tyndall, RI MS JT/1/TYP/2/616.

107. 4 June 1860, William Hopkins to Tyndall, RI MS JT/1/TYP/2/617–9.

108. Tyndall, J., 'Remarks on Alpine Phenomena', Proc. Roy. Inst. 3 (1860): p. 269.

109. 24 February 1861, Tyndall to John Tidmarsh, RI MS JT/1/TYP/11/3872–3.

110. 'The Glaciers of the Alps', Saturday Review (21 July 1860): p. 81–3.

111. Forbes (1860): pp. xiii–xxix, and Forbes, J. D., Reply to Professor Tyndall's Remarks, in His Work 'On the Glaciers of the Alps,' Relating to Rendu's 'Théorie Des Glaciers'. (Edinburgh: A. and C. Black, 1860).

112. 29 June 1862, James Forbes to William Whewell, CUL Trinity/Add.Ms.a/204/144.

113. Hopkins, W. (anonymously), 'Glaciers of the Alps', Fraser's Magazine 62 (December 1860): p. 793–809.

Chapter 9: Radiant Heat (1859–1862)

1. Tyndall's earliest surviving experimental notebook (RI MS JT/3/46) on heat starts on 21 June 1860, with notes on the calibration of his apparatus, and with the next entry on 28 June, the day on which he left for the famous Oxford meeting of the British Association.

2. 'Scientific Worthies: IV John Tyndall', *Nature* 10 (20 August 1874): p. 299.

3. See Eve and Creasey (1945): p. 308–9.

4. Tyndall (1859d); Tyndall, J., 'The Transmission of Heat of Different Qualities through Gases of different Kinds', *Proc. Roy. Inst.* 3 (1859): p. 155–8. The Prince later helped obtain plates of rock-salt for him, important because they are transparent to radiant heat.

5. He even speculated that starlight might cause the same effect.

6. 'Sur la transmission de la chaleur de qualité différente à travers les diverses espèces de gaz, par M. John Tyndall', *Cosmos* 15 (1859): p. 321–5; 'Della trasmissione del calore di diverse qualita' attraverso al diversi gas, di T. Tyndall', *Il Nuovo Cimento* 10 (1859): p. 196; 'Sur la diathemansie des gaz, Lettre de M. John Tyndall a M. le prof. Auguste de la Rive', *Bibliothèque Universelle de Genève* 5 (1859): p. 232.

7. Foote, E., 'Circumstances affecting the Heat of the Sun's Rays', *The American Journal of Science and Arts* 22, (1856): p. 382–3. See also Sorenson, R. P., 'Eunice Foote's pioneering work on CO$_2$ and climate warming', *Search and Discovery* article #70092 (2011), and http://www.climatechangenews.com/2016/09/02/the-woman-who-identified-the-greenhouse-effect-years-before-tyndall/ (accessed 10 September 2016). Joseph Henry presented Eunice Foote's paper, but there is no evidence that he told Tyndall of it, nor that Tyndall or others in Europe heard about it. Henry did not visit England between his trip in 1837 and Tyndall's discovery. Extant letters between Tyndall and Henry date from the 1870s.

8. See 9 November 1889, Tyndall to Mrs (Leslie) Stephen, RI MS JT/1/T/1395.

9. Tyndall (1872): p. 63.

10. Tyndall, J., 'Remarks on recent researches in radiant heat', *Phil. Mag.* 23 (April 1862): p. 252–66.

11. Tyndall, J., 'Physics and Metaphysics', *Saturday Review* (4 August 1860): p. 140–1.

12. Eve and Creasey (1954): p. 83.

13. 27 October 1860, Tyndall Journal, RI MS JT/2/13c/1186.

14. N.d., Tyndall to Heinrich Debus, RI MS JT/5/11.

15. 8 November 1860, Tyndall Journal, RI MS JT/2/13c/1188.

16. See 19 August 1858, Tyndall Journal, RI MS JT/2/13c/1129

17. Tyndall, J., 'Snowdon in Winter', *Saturday Review* (5 January 1861): p. 11–12.

18. Tyndall, J., 'The Action of Gases and Vapours on Radiant Heat', *Proc. Roy. Inst.* 3 (1861): p. 295–8; Tyndall (1861a); Tyndall (1872): p. 7–50.

19. 7 February 1861, Tyndall Journal, RI MS JT/2/13c/1199.

20. Libera (1974).

21. Tyndall (1861a): p. 28.

22. Tyndall (1861a): p. 29.

23. *Report from the Select Committee on Industries (Ireland)*, PP, 1884–5 (288), p. 147.

24. 9 April 1863, Mary Somerville to Tyndall, RI MS JT/1/S/80.

25. Tyndall, J., 'Ueber die Absorption und Strahlung der Wärme durch Gase und Dämpfe; und über den physischen Zusammenhang von Strahlung, Absorption und Leitung', *Ann. Phy. Chem.* 113 (1861): p. 1–53.

26. Tyndall, J., 'De l'absorption et du rayonnement de la chaleur par des gaz et des vapeurs, et sur la connexion physique du rayonnement, de l'absorption et de la conduction', *Comptes Rendus* 52 (1861): p. 364–7.

27. For Thomson's report see RS RR/4/272–3, and for Stokes see 7 May 1861, George Stokes to Tyndall, RI MS JT/1/S/221 and subsequent correspondence.

28. 7 February 1861, Tyndall Journal, RI MS JT/2/13c/1199.

29. Barton (2018): ch. 3.2.

30. 22 April 1861, Tyndall Journal, RI MS JT/2/13c/1213.

31. Tyndall, J., 'On the Physical basis of Solar Chemistry', *Proc. Roy. Inst.* 3 (1861): p. 396.

32. Tyndall, J., 'On the Physical basis of Solar Chemistry', *Proc. Roy. Inst.* 3 (1861): p. 387–96.

33. 21 July 1861, John Herschel to Tyndall, RI MS JT/1/H/94.

34. 24 July 1861, Tyndall to John Herschel, RI MS JT/1/TYP/2/507.

35. Tyndall (1861b).

36. 29 June 1861, Tyndall to Rudolf Clausius, RI MS JT/1/T/180.

37. 14 October 1861, Tyndall to Heinrich Debus, RI MS JT/5/11.

38. RI MS JT 3/46.

39. 25 October 1861, Tyndall to Heinrich Debus, RI MS JT/5/11.

40. Tyndall (1861b).

41. 10 November 1861, John Herschel to Tyndall, RI MS JT/1/H/95; 17 November 1861, Hirst Journal, RI MS JT/2/32c/1594.

42. Tyndall (1861c).

43. 15 November 1861, Tyndall to Nevil Story-Maskelyne, NHM DF 15/2, © The Trustees of the Natural History Museum, London.

44. 26 December 1861, Tyndall to Nevil Story-Maskelyne, NHM DF 15/2; 15 November 1861, Tyndall to George Stokes, RI MS JT/T/1396.

45. 24 October 1861, Tyndall to Heinrich Debus, RI MS JT/1/T/262.

46. Brodie had resigned because of his failing eyesight. See RS Minutes of Council (1858–69), vol. 3, p. 89–90.

47. 2 November 1861, Tyndall to Michael Faraday, RI MS JT/1/TYP/12/4112–13. Letter 4049 in James (2012).

48. Tyndall, J., 'On the Absorption and Radiation of Heat by Gaseous Matter', *Proc. Roy. Inst.* 3, (1862): 404–7; Tyndall (1863).

49. 17 March 1861, Gustav Magnus to Tyndall, RI MS JT/1/M/24.

50. Tyndall's account is given in Tyndall (1872): p. 61–3. The German note was not published until 1861, hence explaining why Magnus may not have been aware of the discovery: Tyndall, J., 'Note on the transmission of radiant heat through gaseous bodies', *Die Fortschritte der Physik im Jahre 1859* 15, (1861): p. 368–9.

51. 29 June 1861, Tyndall to Rudolf Clausius, RI MS JT/1/T/180.

52. 26 July 1861, Tyndall to Rudolf Clausius, RI MS JT/1/T/181.

53. 10 December 1861, Gustav Magnus to Tyndall, RI MS JT/1/M/26.

54. 29 December 1861, Tyndall to George Stokes, RI MS JT/1/T/398; 1 January 1862, George Stokes to Tyndall, RI MS JT/1/S/228.

55. Tyndall (1862a).

56. This became the basis of a later means of measuring the amount of carbon dioxide in exhaled breath.

57. We now know that ozone molecules contain three oxygen atoms (O_3) and oxygen molecules two (O_2), but Tyndall was correct to identify ozone as a form of oxygen, which was not certain at the time.

58. 12 April 1862, John Herschel to Tyndall, RI MS JT/1/H/98.
59. 15 April 1862, Tyndall to John Herschel, RI MS JT/1/TYP/2/518.
60. 24 May 1862, Johann Poggendorf to Tyndall, RI MS JT/1/P/131.
61. 18 February 1862, Tyndall to Juliet Pollock, RI MS JT/1/T/1135.
62. Tyndall (1862c).
63. Tyndall (1862b).
64. Tyndall (1865a).
65. 17 April 1862, Tyndall to Juliet Pollock, RI MS JT/1/T/1136.
66. 22 June 1862, Geraldine Jewsbury to Tyndall, RI MS JT/1/J/28.
67. 5 May 1862, Tyndall to Hermann Helmholtz, RI MS JT/1/T/480.
68. 10 May 1862, Hermann Helmholtz to Tyndall, RI MS JT/H/41.
69. 7 May 1862, Rudolf Clausius to Tyndall, RI MS JT/1/TYP/7/2240-1.
70. 17 May 1862, Rudolf Clausius to Tyndall, RI MS JT/1/TYP/7/2242-3.
71. 23 May 1862, Tyndall to Edward Frankland, RI MS JT/1/TYP/12/3970.
72. Hopkins, W., 'On the Theory of the Motion of Glaciers', *Proc. Roy. Soc. Lond.* 12 (1862): p. 110–20.

Chapter 10: Heated Exchanges (1862–1865)

1. Tyndall, J., 'On Force', *Proc. Roy. Inst.* 3 (1862): p. 527–36.
2. 21 July 1862, James Joule to Tyndall, RI MS JT/1/TYP/2/662.
3. Joule, J., 'Note on the History of the Dynamical Theory of Heat', *Phil. Mag.* 24 (August 1862): p. 121–3.
4. Tyndall, J., 'Mayer, and the mechanical theory of heat', *Phil. Mag.* 24 (September 1862): p. 173–5.
5. Tyndall laboratory notebook (1860–2), RI MS JT/3/46.
6. 29 September 1862, Johann Poggendorff to Tyndall, RI MS JT/1/P/133.
7. N.d. September 1862, Tyndall to Thomas Hirst, RI MS JT/1/T/655.
8. 10 November 1862, Gustav Kirchhoff to Tyndall, RI MS JT/1/K/7.
9. N.d. November 1862, Tyndall to William Sharpey, Wellcome Library MS. 7777/75.
10. Fiske (1894): p. 139.
11. Ibid.
12. 11 October 1862, Tyndall to Juliet Pollock, RI MS JT/1/T/1140.
13. 28 November 1862, Tyndall Journal, RI MS JT/2/13c/1259.
14. Tyndall (1863a).
15. N.d. n.m. 1862, Tyndall to Joseph Hooker, RI MS JT/1/TYP/8/2561.
16. 18 March 1863, William Thomson to George Stokes, RS RR/5/273.
17. Tyndall (1863b): p. 205.
18. Earlier, in 1853, he had put this down to the ability of sand to conduct heat: 25 February 1853, Tyndall to Thomas Hirst, RI MS JT/1/T/560
19. 2 March 1865, Tyndall to Sir Charles Lyell, RI MS JT/1/T/1054.
20. 28 December 1862, Hirst Journal, RI MS JT/2/32c/1628.
21. 25 March 1863, Geraldine Jewsbury to Tyndall, RI MS JT/1/J/30.
22. Tyndall, J., 'On radiation through the earth's atmosphere', *Proc. Roy. Inst.* 4 (1863): p. 5–8. See also *Phil. Mag.* 25 (March 1863): p. 200–6 (Memoir XIII in Molecular Physics).
23. 4 January 1863, Tyndall to Sir John Herschel, RS Letters, HS/392a.
24. 5 January 1863, Sir John Herschel to Tyndall, RS Letters, HS/392b.
25. 15 March 1863, Hirst Journal, RI MS JT/2/32c/1633.

26. Lightman (2007): p. 490.
27. 17, 18, 30 January and 10 February 1863, Rudolf Clausius to Tyndall, RI MS JT/1/TYP/7/2248–50, RI MS JT/1/TYP/7/2251–2, RI MS JT/1/TYP/7/2253–4 and RI MS JT/1/TYP/7/2255–6.
28. Thomson, W. and P. G. Tait, 'Energy', *Good Words* 3 (October 1862): p. 601–7.
29. 10 March 1863, Tyndall to Rudolf Clausius, RI MS JT/1/T/185; Tyndall, J, 'Remarks on an article entitled "Energy" in "Good Words"', *Phil. Mag.* 25 (March 1863): p. 220–4.
30. Tait, P. G., 'Reply to Professor Tyndall's Remarks on a paper on "Energy" in "Good Words"', *Phil. Mag.* 25 (April 1863): p. 263–6.
31. Tyndall, J., 'Remarks on the dynamical theory of heat', *Phil. Mag.* 25 (May 1863): p. 368–87.
32. 24 March 1862, William Snow Harris to Tyndall, RI MS JT/1/TYP/2/480; 27 August 1863, George Airy to Tyndall, RI JT/1/A/37.
33. Thomson, W., 'Note on Professor Tyndall's "Remarks on the Dynamical Theory of Heat"', *Phil. Mag.* 25 (June 1863): p. 429.
34. Tait, P. G., 'On the Conservation of Energy', *Phil. Mag.* 26 (June 1863): p. 429–31.
35. Tyndall, J., 'Remarks on Professor Tait's last letter to Sir David Brewster', *Phil. Mag.* 26 (July 1863): p. 65–7.
36. Mayer, J. R., 'Remarks on the Forces of Inorganic Nature' (1842), *Phil. Mag.* 24 (November 1862): p. 371–7; 'On Celestial Dynamics' (1848), *Phil. Mag.* 25 (April 1863): p. 241–8; (May 1863): p. 387–409; (June 1863): p. 417–18; 'Remarks on the Mechanical Equivalent of Heat' (1851) (June suppl. 1863): p. 493–522. Mayer's 112-page book *Die Organische Bewegung in ihren Zusammenhange mit dem Stoffwechsel* (Heilbronn: Drechsler, 1845) does not appear to have been translated at the time.
37. 11 May 1864, Rudolf Clausius to Tyndall, RI MS JT/1/TYP/7/2264–6.
38. 19 September 1863, Rudolf Clausius to Tyndall, RI MS JT/1/TYP/7/2259–60.
39. 9 September 1863, Robert Bunsen to Tyndall, RI MS JT/1/B/149.
40. Tyndall (1864a).
41. 31 May 1863, Julius Robert Mayer to Tyndall, RI MS JT/1/M/80.
42. 31 October 1863, Julius Robert Mayer to Tyndall, RI MS JT/1/M/82.
43. See Neswald, E., 'Saving the World in the Age of Entropy', in Lightman and Reidy (2014): p. 15–31.
44. N.d. February 1858, Herbert Spencer to Tyndall, RI MS JT/1/S/191.
45. Tyndall, J., 'An Account of Researches on Radiant Heat'. *Proc. Roy. Inst.* 4 (1863): p. 146–50; Tyndall (1864b).
46. Tyndall (1863c).
47. 28 April 1863, Gustav Magnus to Tyndall, RI MS JT/1/M/29.
48. 25 October 1863, Hirst Journal, RI MS JT/2/32c/1658.
49. 9 October 1865, Tyndall to Sarah Tyndall, RI MS JT/1/T/1483.
50. 9 May 1889, Tyndall to Mrs Steuart, RI MS JT/1/TYP/10/3465–3465a; 11 September 1867, Tyndall to Mrs Steuart, RI MS JT/1/TYP/10/3362; 19 December 1877, Tyndall to Mrs Steuart, RI MS JT/1/TYP/10/3439; 7 October 1879, Mrs Steuart to Tyndall, RI MS JT/1/TYP/10/3451.
51. 26 April 1873, Tyndall to Henrietta Huxley, RI MS JT/1/TYP/9/2983a.
52. 29 March 1878, Louisa Journal, RI MS JT/2/23.
53. 17 February 1879, Charles Darwin to Tyndall, RI MS JT/1/TYP/9/2847.
54. 10 November 1878, Joseph Hooker to Tyndall, RI MS JT/1/TYP/8/2749; 10 November 1878, Tyndall to Joseph Hooker, Kew-Hooker-229; 11 November 1878, Joseph Hooker to Tyndall, RI MS JT/1/TYP/8/2750, see Anniversary Meeting, *Proc. Roy. Soc. Lond.* 28 (1878): p. 42–79.
55. 20 November 1863, Tyndall to Michael Faraday, Letter 4409 in James (2011).
56. 29 November 1863, Henry Bence Jones to Tyndall, RI MS JT/1/J/54.

57. 9 June 1863, Henry Bence Jones to Tyndall, RI MS JT/1/J/53.
58. Jones, H. Bence, *Report on the Past, Present, and Future of the Royal Institution, Chiefly in Regard to its Encouragement of Scientific Research* (London: Clowes, 1862).
59. 'Donation Fund (1863-72)', *Visitor's Annual Report 1873*, p. xxxiii–v, RI MS AD/12/E/03/A/1873.
60. Tyndall (1864d); Tyndall, J., 'Contributions to Molecular Physics. Being the Fifth Memoir of Researches on Radiant Heat', *Proc. Roy. Inst.* 4 (1864): p. 233–40.
61. 20 March 1864, Hirst Journal, RI MS JT/2/32c, p. 1667.
62. 30 May 1864, Tyndall to George Stokes, RI MS JT/1/T1430.
63. 24 June 1864, George Stokes to Tyndall, RI MS JT/1/S/235; 30 June 1864, George Stokes to Tyndall, RI MS JT/1/S/236; 15 September 1864, George Stokes to Tyndall, RI MS JT/1/S/237.
64. 29 March 1864, Tyndall to Henry Bence Jones, RI MS JT/1/TYP/3/794.
65. 5 April 1864, Tyndall to Thomas Hirst, RI MS JT/1/T/659.
66. 19 April 1864, Tyndall to Thomas Hirst, RI MS JT/1/T/660.
67. 27 April 1864, Tyndall to Thomas Hirst, RI MS JT/1/T/661.
68. 14 May 1864, Tyndall to Miss Roche, RI MS JT/1/TYP/11/3848–9.
69. 28 June 1864, Mrs Steuart to Tyndall, RI MS JT/1/TYP/10/3347.
70. 24 June 1864, Tyndall to Miss Roche, RI MS JT/1/TYP/11/3850.
71. 14 May 1864, Tyndall to Hector Tyndale, RI MS JT/1/TYP/5/1661–3.
72. Tyndall, J., 'A Magnetic Experiment – (Force)', *Proc. Roy. Inst.* 4 (1864): p. 317–22.
73. 14 June 1864, Tyndall to Thomas Hirst, RI MS JT/1/HTYP/532 (BL Add MS 63092, ff. 31–2).
74. 11 June 1864, Juliet Pollock to Tyndall, RI MS JT/1/P/171.
75. 16 September 1864, Tyndall to Juliet Pollock, RI MS JT/1/T/1142.
76. 3 August 1864, William Huber to Tyndall, RI MS JT/1/H/509; 1 September 1864, William Huber to Tyndall, RI MS JT/1/H/510; 21 September 1864, Felix Foucon to Tyndall, RI MS JT/1/F/35.
77. 22 September 1864, Heinrich Debus to Tyndall, RI MS JT/1/D/32.
78. 11 September 1864, Juliet Pollock to Tyndall, RI MS JT/1/TYP/6/2020.
79. 7 November 1864, Tyndall to Miss Roche, RI MS JT/1/TYP/11/3851.
80. Tyndall notebook, RI MS JT/3/47.
81. 29 October 1864, Tyndall to George Stokes, RI MS JT/1/T/1427.
82. 3 November 1864, Tyndall to George Stokes, RI MS JT/1/T/1428.
83. 8 September 1862, C. K. Akin to Tyndall, CUL Add 7656/A615.
84. N.d. October 1864, Tyndall to C. K. Akin, RI MS JT/5/12.
85. Tyndall (1864e).
86. Tyndall (1864e): p. 338.
87. 1 December 1864, Tyndall Journal, RI MS JT/2/13c/1270.
88. Akin, C. K., 'Note on Ray-Transmutation', *Phil. Mag.* 28 (1864): p. 554–60.
89. 9 December 1864, Frederick Pollock to Tyndall, RI MS JT/1/P/234.
90. 18 November 1864, Heinrich Debus to Tyndall, RI MS JT/1/TYP/7/2399.
91. Tyndall (1865b).
92. Akin, C. K., 'On Calcescence', *Phil. Mag.* 29 (1865): p. 28–43.
93. Tyndall (1865c).
94. Tyndall, J., 'On Calorescence', *Phil. Mag.* 29 (1865): p. 164.
95. Tyndall, J., 'On Combustion by Invisible Rays', *Proc. Roy. Inst.* 4 (1865): p. 329–35.
96. Akin, C. K., 'Further Statements Concerning the History of Calcescence', *Phil. Mag.* 29 (1865): p. 136–51.
97. Tyndall (1865a).

98. 3 February 1865, Tyndall to Sir William Armstrong, RI MS JT/1/TYP/1/61–2.
99. 26 February 1865, Tyndall to Rudolf Clausius, RI MS JT/1/T/191.

Chapter 11: The X-Club (1864–1866)

1. See Barton (2018) for an extensive analysis of the X-Club.
2. Tyndall had been unsuccessfully proposed for the Royal Medal for the third time the previous year, by Stokes and Wheatstone, for this work; RS Minutes of Council 3 (1858–69): p. 160. On that occasion Gassiot was awarded the Royal Medal instead. Ironically, Tyndall had unsuccessfully proposed Gassiot for it in 1861; RS Minutes of Council 3 (1858–69): p. 91.
3. 7 November 1864, Tyndall to Miss Roche, RI MS JT/1/TYP/11/3851.
4. 4 December 1864, Hirst Journal, RI MS JT/2/32c/1708.
5. 20 April 1858, Thomas Huxley to Tyndall, RI MS JT/1/TYP/9/2888.
6. 20 April 1858, Tyndall to Thomas Huxley, RI MS JT/1/TYP/9/2889.
7. 18 November 1864, Tyndall to Sir John Herschel, RS HS/17/398a; 18 November 1864, Tyndall to Henry Bence Jones, RI MS JT/1/TYP/3/793.
8. 21 November 1864, Sir John Herschel to Tyndall, RS HS/17/398b.
9. Tyndall, J., 'Vitality', *The Reader* 95 (29 October 1864): p. 545–6.
10. Tyndall, J., 'Science and the Spirits', *The Reader* 102 (10 December 1864): p. 725–6.
11. 21 March 1865, Tyndall Journal, RI MS JT/2/13c/1287.
12. Baldwin (2015): p. 25.
13. 19 January 1865, Tyndall Journal, RI MS JT/2/13c/1278.
14. N.d. June 1865, Tyndall to Thomas Hirst, RI MS JT/1/T/904.
15. 2 April 1865, Hirst Journal, RI MS JT/2/32c/1726.
16. 20 April 1865, Hirst Journal, RI MS JT/2/32c/1780.
17. 20 August 1844, Tyndall Journal, RI MS JT/2/13a/58.
18. 18 May 1865, Tyndall Journal, RI MS JT/2/13c/1289.
19. 11 February 1865, Cookson to Tyndall, Tyndall Journal, RI MS JT/2/13c/1282.
20. 18 May 1865, Tyndall Journal, RI MS JT/2/13c/1289.
21. 16 May 1865, F. Cobb to Tyndall, Tyndall Journal, RI MS JT/2/13c/1291.
22. Later known as Skindles Hotel.
23. 4 February 1866, Tyndall Journal, RI MS JT/2/13c/1298.
24. Tyndall (1865d). In this paper he states that Hooker suggested the term 'regelation', which delighted Hooker (n.d. December 1865, Joseph Hooker to Tyndall, RI MS JT/1/TYP/8/2563).
25. 16 November 1865, Rudolf Clausius to Tyndall, RI MS JT/1/TYP/7/2281–3.
26. 27 October 1865, James Sylvester to Tyndall, RI MS JT/1/S/293.
27. 4 February 1866, Tyndall Journal, RI MS JT/2/13c/1298.
28. 3 November 1865, Tyndall to Juliet Pollock, RI MS JT/1/T/1146.
29. Tyndall (1866a).
30. Tyndall (1866b).
31. Tyndall (1866c).
32. 10 February 1866, Henry Coxwell to Tyndall, RI MS JT/1/C/56.
33. Tyndall, J., 'On Radiation and Absorption with Reference to the Colour of Bodies and their State of Aggregation', *Proc. Roy. Inst.* 4 (1866): p. 487–92.
34. 10 March 1866, Tyndall Journal, RI MS JT/2/13c/1312.
35. 26 March 1866, Tyndall to Thomas Carlyle, RI MS JT/1/T/148.
36. 19 September 1857, Tyndall Journal, RI MS JT/2/13c/1033.

37. Conway (1904): vol. 2, p. 95.
38. 2 April 1866, Tyndall to Jane Carlyle, RI MS JT/1/TYP/1/190/1.
39. 3 April 1866, Jane Carlyle to Tyndall, Tyndall Journal, RI MS JT/2/13c/1329.
40. 17 April 1866, Tyndall Journal, RI MS JT/2/13c/1328.
41. 5 January 1869, Tyndall Journal, RI MS JT/2/13c/1348.
42. 8 April 1866, Hirst Journal, RI MS JT/2/32c/1781.
43. 22 April 1866, Geraldine Jewsbury to Tyndall, Tyndall Journal, RI MS JT/2/13c/1331.
44. Tyndall (1867).
45. 12 May 1866, Gustav Magnus to Tyndall, RI MS JT/1/M/31.
46. Magnus, G., 'On the influence of the absorption of heat on the formation of dew', Phil. Mag. 32 (1866): p. 111–17.
47. Tyndall (1866d).
48. 21 May 1866, Sir Charles Lyell to Tyndall, RI MS JT/1/L/52.
49. 1 June 1866, Tyndall to Sir Charles Lyell, RI MS JT/1/T/1053.
50. Tyndall experimental notebook, RI MS JT/3/48, records experiments on vibrating strings on 1 May 1866, p. 55, and 7 May 1866, p. 56.
51. Tyndall, J., 'Experiments on the Vibration of Strings', Proc. Roy. Inst. (1866): p. 685–94.
52. 15 June 1866, Hirst Journal, RI MS JT/2/32c/1787.
53. Sir John Lubbock's brother, who played for Kent.
54. 21 June 1866, Emily Tennyson to Tyndall, RI MS JT/1/T/5.
55. c.8 July 1866, Tyndall to Thomas Hirst, RI MS JT/1/T/909.
56. 31 July 1866, Tyndall to Juliet Pollock, RI MS JT/1/TYP/6/2073–4.
57. 7 August 1866, Henry Bence Jones to Tyndall, RI MS JT/1/J/56.
58. 18 August 1866, Kathleen Ginty to Tyndall, RI MS JT/1/TYP/11/3647.
59. 26 August 1866, Tyndall to Margaret Ginty, RI MS JT/1/TYP/11/3650.
60. 30 September 1866, Henry Bence Jones to Tyndall, RI MS JT/1/J/86.
61. 12 October 1866, Tyndall to Rudolf Clausius, RI MS JT/1/T/195.

Chapter 12: Eyre Affair and Death of Faraday (1866–1868)

1. See Winter, S., 'On the Morant Bay Rebellion in Jamaica and the Governor Eyre-George William Gordon Controversy, 1865–70', BRANCH: Britain, Representation and Nineteenth-Century History. Ed. Dino Franco Felluga. Extension of Romanticism and Victorianism on the Net. Web accessed 4 April 2017; and Semmell (1962).
2. 10 July 1853, Tyndall Journal, RI MS JT/2/13b/609.
3. See Barton (2018): ch. 4.1.
4. 16 November 1866, Ellen Lubbock to Tyndall, RI MS JT/1/TYP/8/2575.
5. N.d., Tyndall to Joseph Hooker, RI MS JT/1/TYP/8/2592.
6. 13 November 1866, Joseph Hooker to Tyndall, RI MS JT/1/TYP/8/2571.
7. 20 September 1866, Tyndall to Hamilton Hume, Wellcome Library MS. 7777/59.
8. 12 October 1866, Jamaica Committee to Tyndall, RI MS JT/1/TYP/8/2565.
9. 7 November 1866, Tyndall to Jamaica Committee, RI MS JT/1/TYP/8/2566–7.
10. 8 November 1866, Tyndall to Thomas Huxley, RI MS JT/1/TYP/9/2906.
11. 9 November 1866, Thomas Huxley to Tyndall, RI MS JT/1/TYP/8/2568.
12. 'The Eyre Defence and Aid Fund', Morning Post (19 February 1867).
13. 'Philosophical Institution—Professor Tyndall on Sonorous Vibrations', Caledonian Mercury (15 December 1866).
14. 15 February 1867, Thomas Carlyle to John Ruskin, in Cate (1982): p. 125.

15. N.d. December 1866, Tyndall to Sir Roderick Murchison, Science Museum, Wroughton Ms 2108/79b.
16. 2 January 1867, Tyndall to Ellen Busk, RI MS JT/1/TYP/1/186.
17. Tyndall, J., 'On sounding and sensitive flames', Proc. Roy. Inst. 5 (1867): p. 6–12.
18. Tyndall (1867a).
19. Tyndall (1867); 13 February 1867, Rudolf Clausius to Tyndall, RI MS JT/1/TYP/7/2298–300.
20. 12 July 1867, Gustav Magnus to Tyndall, RI MS JT/1/M/35; 26 January 1867, Wilhelm Weber to Tyndall, RI MS JT/1/W/15.
21. 12 July 1867, Gauthier-Villars to Tyndall, RI MS JT/1/V/16. The book appeared in both France and Germany, through Vieweg, in 1869.
22. 7 February 1867, Hirst Journal, RI MS JT/2/32c/1798.
23. 15 March 1867, Gustav Magnus to Tyndall, RI MS JT/1/M/33; Magnus, G., 'On the influence of the adhesion of vapour in experiments on the absorption of heat', Phil. Mag. 33 (1867): p. 413–25.
24. 23 March 1867, Gustav Magnus to Tyndall, RI MS JT/1/M/34.
25. Tyndall (1867b).
26. Amy Harriet Lubbock, eldest daughter of John and Nelly Lubbock, was born in 1857 and was aged nine at this point. Five years later Tyndall wrote in his journal: 'From the first moment of my meeting Harriet Lubbock: when she was a wee child—I have been wonderfully drawn towards her. So much so indeed that were it not for the immense disparate of years I might have endeavoured to be nearer and dearer to her than I now am' (26 September 1872, Tyndall Journal, RI MS JT/2/12/4).
27. 9 August 1867, Tyndall to Hector Tyndale, RI MS JT/1/T/1448.
28. 'The Paris Exhibition, 1867', The Times (1 April 1867): p. 9.
29. 7 May 1867, Hirst Journal, RI MS JT/2/32c/1802.
30. Tyndall, J., 'Experiments of Faraday, Biot and Savart', Proc. Roy. Inst. 5 (1867): p. 188.
31. See Barton (2018): ch. 6.4 and ch. 5.
32. 19 August 1867, Tyndall to Thomas Hirst, RI MS JT/1/HTYP/539.
33. 27 August 1867, Tyndall to Auguste De la Rive, RI MS JT/1/TYP/1/366.
34. Tyndall, J., 'Matter and Force' in Tyndall (1879): vol 1, p. 72.
35. N.d., Tyndall Journal, RI MS JT/2/13c/1334.
36. 29 December 1867, Tyndall to Peter Tait, University of Edinburgh GN 237 Coll-234 Gen.2169 p. 229.
37. 18 September 1867, Thomas Huxley to Tyndall, ICHC 8.56.
38. 24 September 1867, Tyndall to Thomas Hirst, RI MS JT/1/T/962.
39. 21 October 1867, Tyndall to Thornton, BL Add MS 60631, ff. 123–4.
40. 1 November 1867, Tyndall to Richard Strachey, BL Add MS 60631, ff. 116–18.
41. 22 October 1867, Henry Holland to Tyndall, BL Add MS 60631, ff. 121–2.
42. 5 November 1867, Tyndall to Lady Lyell, RI MS JT/1/T/1055.
43. 'Professor Huxley on Science and the Clergy', Saturday Review (30 November 1867): p. 691–2.
44. 13 January 1868, Tyndall to Hermann Helmholtz, RI MS JT/1/T/485; 19 January 1868, Tyndall to Rudolf Clausius, RI MS JT/1/T/196.
45. 4 December 1867, Tyndall to Lady Emily Peel, RI MS JT/1/TYP/3/956.
46. 18 November 1867, Anna Helmholtz to Tyndall, RI MS JT/H/31.
47. I am grateful to Sandy Walkington for pointing this out to me. The reference appears in Chapter XXVII. Levin, reading Tyndall's book, 'remembered his criticisms on Tyndall's self-satisfaction in the cleverness of his management of his experiments and on his lack of

philosophical views…'. He comments further on the relationship between electricity and heat. See Orwin (2009).

48. N.d., Tyndall to Juliet Pollock, RI MS JT/1/T/1200.
49. 30 November 1867, Tyndall to Hermann Helmholtz, RI MS JT/1/T/484.
50. 22 November 1867, Friedrich Vieweg und Sohn to Tyndall, RI MS JT/1/V/8.
51. 1 December 1867, Gustav Wiedemann to Tyndall, RI MS JT/1/W/46; 21 December 1867, Gustav Wiedemann to Tyndall, RI MS JT/1/W/47.
52. 30 November 1867, Gauthier-Villars to Tyndall, RI MS JT/1/V/17.
53. 7 January 1868, Jules Jamin to Tyndall, RI MS JT/1/J/8.
54. 25 March 1868, Abbé Moigno to Tyndall, RI MS JT/1/M/113.
55. 10 May 1868, Abbé Moigno to Tyndall, RI MS JT/1/M/114.
56. 2 December 1867, Tyndall to Lady Colville, BL Add MS 60634, f. 129.
57. Tyndall, J., 'On Faraday as a Discoverer', *Proc. Roy. Inst.* 5 (1868): p. 199–272.
58. 15 January 1868, Thomas Carlyle to Tyndall, RI MS JT/1/TYP/1/204.
59. 10 February 1868, Tyndall to Jane Barnard, RI MS JT/1/T/109.
60. 11 February 1868, Jane Barnard to Tyndall, RI MS JT/1/B/37.
61. 27 February 1868, Jane Barnard to Tyndall, RI MS JT/1/B/38.
62. 11 March 1868, Emily Peel to Tyndall, RI MS JT/1/P/46; 8 March 1868, Emily Tennyson to Tyndall, Tyndall Journal, RI MS JT/2/13c/1336; 16 March 1868, Prince of Wales to Tyndall, Tyndall Journal RI MS JT/2/13/1335.
63. 10 March 1868, Tyndall to Mrs Steuart, RI MS JT/1/TYP/10/3374.
64. 7 March 1868, Joseph Hooker to Tyndall, RI MS JT/1/TYP/12/4208.
65. 11 March 1868, Tyndall, Hooker and Huxley to Managers of the Royal Institution, RI MS JT/1/TYP/12/4209.
66. 12 March 1868, Henry Bence Jones to Tyndall, RI MS JT/1/TYP/2/693.
67. 13 March 1868, Tyndall to Joseph Hooker, RI MS JT/1/TYP/8/2596; n.d., Henry Bence Jones to Tyndall, RI MS JT/1/TYP/12/4208.
68. 23 March 1875, Tyndall to Frederick Pollock, RI MS JT/1/TYP/6/2153.
69. 23 March 1875, Tyndall to Frederick Pollock, RI MS JT/TYP/1/6/2153.
70. A copy stands outside the Institution of Engineering and Technology in Savoy Place, London.
71. 29 March 1868, Tyndall Journal, RI MS JT/2/13c/1337.
72. 3 April 1868, Tyndall Journal, RI MS JT/2/13c/1338.
73. 3 April 1868, RI MS JT/3/48/62; 6 April 1868, air/vapour experiments RI MS JT/3/48/63.
74. 8 April 1868, Tyndall to Thomas Huxley, RI MS JT/1/TYP/9/2919.
75. 9 April 1868, Tyndall Journal, RI MS JT/2/13c/1339.
76. 13 April 1868, Tyndall to Juliet Pollock, RI MS JT/1/T/1164.
77. 16 April 1868, Tyndall Journal, RI MS JT/2/13c/1345.
78. N.d., Tyndall to Mary Somerville, Oxford Bodleian Dep c.372 (MST-1); n.d. April 1868, Tyndall to Mrs Steuart, RI MS JT/1/TYP/10/3375.
79. 27 April 1868, Lady Emily Peel to Tyndall, RI MS JT/1/P/47.
80. 29 April 1868, Tyndall to Lady Emily Peel, RI MS JT/1/TYP/3/957.
81. 10 June 1868, Tyndall to Thomas Huxley, RI MS JT/1/TYP/9/2920.
82. 29 June 1868, Gustav Magnus to Tyndall, RI MS JT/1/M/38.
83. 14 June 1868, Sir John Herschel to Tyndall, RI MS JT/1/TYP/2/546; 11 July 1868, Tyndall to Sir John Herschel, RI MS JT/1/TYP/2/545.
84. 20 June 1868, Tyndall to Rudolf Clausius, RI MS JT/1/T/200.

Chapter 13: Prayer, Miracles, Metaphysics, and Spirits (1865–1880)

1. 11 March 1862 Tyndall to unknown, RI MS JT/1/T/1539.
2. 24 March 1862, Tyndall to Frederick Pollock, RI MS JT/1/TYP/6/1984. His actual response does not survive, but Tyndall's letter implies that he was concerned that, as drafted, it would cause unnecessary offence.
3. 30 April 1862, Alfred Wills to Tyndall, BL Add MS 63092, ff. 183–4.
4. 9 October 1865, 'Prayers against the cholera', *Pall Mall Gazette*.
5. 4 February 1866, Tyndall Journal, RI MS JT/2/13c/1298.
6. Tyndall to Editor of the *Pall Mall Gazette*, 'Prayer and the cholera', *Pall Mall Gazette* (12 October 1865).
7. 7 November 1865, Tyndall to Frances Power Cobbe, Huntington Library 8.5482.
8. 17 October 1865, 'On prayer and the cholera', *Pall Mall Gazette*.
9. Tyndall to Editor of the *Pall Mall Gazette*, 'Prayer and the cholera', *Pall Mall Gazette* (19 October 1865).
10. Tyndall, J., 'The constitution of the Universe', *Fortnightly Review* 3 (December 1865): p. 129–44.
11. Tyndall, J., 'The constitution of the Universe', *Fortnightly Review* 3 (December 1865): p. 144.
12. 10 March 1866, Tyndall Journal, RI MS JT/2/13c/1312–13.
13. 7 February 1866, Thomas Huxley to Editor of the *Spectator*, Tyndall Journal, RI MS JT/2/13c/1316.
14. 10 March 1866, Tyndall to Editor of the *Spectator*, Tyndall Journal, RI MS JT/2/13c/1314.
15. 8 March 1866, Arthur Stanley to Tyndall, Tyndall Journal, RI MS JT/2/13c/1317–18.
16. 13 March 1866, Tyndall to Arthur Stanley, Tyndall Journal, RI MS JT/2/13c/1318.
17. 22 March 1866, Tyndall Journal, RI MS JT/2/13c/1323.
18. Tyndall, J., 'Miracles and Special Providences', *Fortnightly Review* 7 (June 1867): p. 645–60.
19. 11 June 1867, Tyndall Journal, RI MS JT/2/13c/1333–4.
20. Tyndall (1879): vol. 2, p. 37. Tyndall added this comment from the 5th edition (1876) onwards.
21. Mozley, J. B., '"Of Christ alone without sin": a reply to Professor Tyndall', *Contemporary Review* 7 (April 1868): p. 481–96.
22. 19 July 1867, Hector Tyndale to Tyndall, RI MS JT/1/T/59.
23. 8 November 1860, Tyndall Journal, RI MS JT/2/13c/1189.
24. 8 May 1868, Alfred Wallace to Tyndall, NHM WP1/8/287, © The Trustees of the Natural History Museum, London.
25. Tyndall to *Pall Mall Gazette* (18 May 1868): p. 2–3.
26. Tyndall, J., 'Science and the Spirits', *The Reader* (10 December 1864): p. 725–6. Tennyson was not described as the 'Poet of Science' until after his death. See Holmes (2012): p. 659.
27. 18 December 1869, G Wheatley Kenneth to Tyndall, Tyndall Journal, RI MS JT/2/13c/1362.
28. 22 December 1869, Tyndall to G Wheatley Kenneth, Tyndall Journal, RI MS JT/2/13c/1363.
29. 22 December 1869, William Crookes to Tyndall, Tyndall Journal, RI MS JT/2/13c/1364.
30. See Gay, H., 'The Declaration of Students of the Natural and Physical Sciences', revisited: Youth, Science and Religion in mid-Victorian Britain', in Sweet and Feist (2007).
31. See Brock and Macleod (1976).
32. 11 June 1864, Tyndall to Capel H. Berger, CUL Add.5989: f.24, 1(3)–3(3).
33. See Marshall, Lightman and England (2015).
34. Spencer (1904): vol. II, p. 209.

35. See White, P., 'The Conduct of Belief: Agnosticism, the Metaphysical Society, and the Formation of Intellectual Communities', in Dawson and Lightman (2014): p. 220–41; Lightman, B., 'Science at the Metaphysical Society: Defining Knowledge in the 1870s', in Lightman and Reidy (2014): p. 187–206.

36. 12 March 1872, Tyndall Journal, RI MS JT/2/11/62, also (wrongly transcribed) RI MS JT/2/13c/1399.

37. 26 December 1869, Henry Sidgwick to Mary Sidgwick, TCC Trinity/Add.Ms.c/99/111.

38. See Lightman, B., 'Science at the Metaphysical Society: Defining Knowledge in the 1870s', in Lightman and Reidy (2014): p. 206, and Lightman (2007): p. 5–9.

39. 'The "Prayer for the Sick": Hints Towards a Serious Attempt to Estimate Its Value', *The Contemporary Review* 20 (July 1872): p. 205–10.

40. 'On Prayer', *The Contemporary Review* 20 (October 1872): p. 763–6.

41. 'On Prayer', *The Contemporary Review* 20 (October 1872): p. 777–82.

42. See Turner, F. M., 'Rainfall, plagues, and the Prince of Wales' in Turner (1993): p. 151–70.

43. 'Captain Galton criticised', *Spectator* (3 August 1872): p. 974–5 (He was not a Captain. The *Spectator* confused him with his cousin Douglas.)

44. Duke of Argyll, 'Prayer. The two spheres: are they two?', *Contemporary Review* (February 1873): p. 464–73.

45. Barton (2018): ch. 5.4. See also Barton, R., 'Sunday Lecture Societies: Naturalistic Scientists, Unitarians and Secularists Unite against Sabbatarian Legislation', in Dawson and Lightman (2014): p. 189–219.

46. 3 April 1867, Hirst Journal, RI MS JT/2/32d/1799–1800.

47. 'Glasgow Sunday Society', *Glasgow Herald* (26 October 1880).

48. Tyndall, J. 'The Sabbath', *Nineteenth Century* 8 (November 1880): p. 690–714. It was also issued as a pamphlet by Longmans, but was not a best-seller. Of the 1,000 copies printed, making Tyndall £8, some 600 were marked as 'wasted' by 1884.

49. For example, speaking at the opening of the merged Parkes Museum of Hygiene and the Sanitary Institute in 1883, and providing ammunition to help demolish that argument that running water could be safely drunk a few hours after contamination with raw sewage (16 May 1880, Tyndall to Edward Frankland, Frankland Rylands Letters, 1616).

50. See *Saturday Review* (12 May 1877): p. 595.

51. Also first curator of the Parkes Museum of Hygiene at UCL in 1877.

52. 5 May 1884, RI MS JT/2/19.

53. 9 May 1884, Joseph Hooker to Tyndall, RI MS JT/1/TYP/8/2759.

54. 15 April 1889, Tyndall to Thomas Hirst, RI MS JT/1/T/810.

55. 22 April 1889, Thomas Hirst to Tyndall, RI MS JT/1/H/403.

56. 23 August 1889, Tyndall to El Medini, BL Add MS 41295, f. 30.

57. 30 July 1888, Tyndall to Captain Toynbee, Bodleian Library, Oxford, MS. Autogr. d. 42, fol. 179–80.

58. See for example Tyndall's reference to William Hamilton and the 'Unknown and Unknowable God': Tyndall, J., 'Professor Huxley's Doctrine', *Spectator* 39 (1866): p. 188.

Chapter 14: Mountaineering in the 1860s (1860–1868)

1. Eve & Creasey (1945): p. 386.

2. He suggested that black-balling should be more difficult than was proposed, recommending one in three rather than one in five, to avoid placing power in the hands of a very small number of members, and proposed the requirement to ascend a mountain to the height

of 13,000 feet rather than reach the summit. The black ball rule was in fact introduced at one in ten until it was abolished in 1938, but the height requirement was removed. See Dangar and Blakeney (1957).

3. 16 December 1857, Tyndall to Edward Kennedy, Alpine Club Library.

4. Eve & Creasey (1945): p. 387.

5. Undated (probably December 1861), Tyndall to A. P. Whateley, in Eve & Creasey (1945): p. 389.

6. Stephen (1924): p. 189. See also Stephen (1977): p. 100.

7. 12 February 1897, Leslie Stephen to Louisa Tyndall, RI MS LT/79/8.

8. Tyndall, J., 'A Day Among the Séracs of the Glacier du Géant', Peaks, Passes and Glaciers (London: Longman, 1859): p. 39–57.

9. Eve and Creasey (1945): p. 362.

10. Tyndall, J., 'From Lauterbrunnen to the Aeggischorn by the Lauwinen-Thor', in Galton (1861): p. 305–17.

11. Vaughan Hawkins, F., 'Partial Ascent of the Matterhorn', in Galton (1861): p. 282–304.

12. Tyndall (1871): p. 23. For a slightly different wording, omitting Tairraz's scream, see Tyndall to The Times, 'The Accident on the Col-du-Geant', The Times (8 September 1860): p.8.

13. Francis Fox Tuckett found one the following year: 'Alpine Expeditions', The Times (5 August 1861): p. 10. The others placed were never found.

14. 28 November 1860, John Ball to Tyndall, Eve & Creasey (1945): p. 387.

15. 3 December 1860, John Ball to Tyndall, Eve & Creasey (1945): p. 388.

16. 2 August 1861, Tyndall Journal, RI MS JT/2/13c/1222.

17. 6 August 1861, Tyndall Journal, RI MS JT/2/13c/1228.

18. 21 August 1861, Tyndall to Michael Faraday, Letter 4047 in James (2011).

19. 18 August 1861, Tyndall Journal, RI MS JT/2/13c/1241. This passage indicates that Bennen was the first to stand on the summit. In Tyndall's published account he does not specify who reached the summit first: Tyndall (1871): p. 104. But see also 7 December 1861, Tyndall to Editor of the Athenaeum, Athenaeum (14 December 1861): p. 808. In this letter, Tyndall states that he 'ordered' Bennen to go first.

20. 18 August 1861, Tyndall Journal, RI MS JT/2/13c/1241.

21. Tyndall (1862): p. 58.

22. Bennen Führerbuch 1857–63 (1922/K15), Alpine Club Library.

23. Tyndall to The Times, 'An Alpine Observation', The Times (22 October 1861): p.10.

24. 22 August 1861, Tyndall Journal, RI MS JT/2/13c/1245.

25. 23 August 1861, Tyndall Journal, RI MS JT/2/13c/1246.

26. 23 August 1861, Tyndall Journal, RI MS JT/2/13c/1247–8.

27. 7 July 1862, RI Managers' Minutes, XI, p. 444.

28. 2 July 1862, Tyndall to Hector Tyndale, RI MS JT/1/T/1445.

29. 22 July 1862, Tyndall to Michael Faraday, Letter 4202 in James (2011).

30. 24 July 1862, Tyndall to Thomas Hirst, RI MS JT/1/HTYP/520.

31. 24 July 1862, Tyndall to Juliet Pollock, RI MS JT/6/2093b.

32. 20 August 1862, Tyndall to Thomas Hirst, RI MS JT/1/T/920.

33. 29 July 1862, Tyndall to Thomas Huxley, ICHC 8.43.

34. 1 February 1865, Clarence King to Tyndall, Tyndall Journal, RI MS JT/2/13c/1286.

35. Mount Tyndall in Tasmania is also named after John Tyndall (27 November 1863, Tyndall to Julius Haast, Alexander Turnbull Library, National Library of New Zealand MS-Papers Turnbull 0037–14), as are the Tyndall Glaciers in Chile, Colorado and Alaska. There are Tyndall craters on the Moon and Mars.

36. 1 August 1862, Tyndall to Juliet Pollock, RI MS JT/1/TYP/6/1998.

37. See Thackray (1999) for a detailed account.
38. For example, in 1855 he had strongly challenged David Forbes's views on the cause of foliation of rocks in Scotland and Norway, apparently, according to him, convincing the assembled Fellows (4 February 1855, Tyndall to Thomas Hirst, RI MS JT/1/T/593).
39. Tyndall (1862d). See also Tyndall (1864c).
40. 29 July 1863, Tyndall to Thomas Hirst, RI MS JT/1/HTYP/523.
41. 26 August 1863, Thomas Hirst to Tyndall, RI MS JT/1/H/249.
42. c.22 August 1863, Tyndall to Juliet Pollock, RI MS JT/1/T/1171.
43. 25 August 1863, Tyndall to Juliet Pollock, RI MS JT/1/T/1183.
44. Cram, A. G., *Pillar Rock* (Fell and Rock Climbing Club, 1977): p. 3.
45. 25 August 1863, Tyndall to Juliet Pollock, RI MS JT/1/T/1183.
46. 8 September 1863, Tyndall to Mrs Steuart, RI MS JT/1/TYP/10/3346.
47. 3 December 1863, Tyndall to Rudolf Clausius, RI MS JT/1/T/188.
48. 14 June 1864, Tyndall to Thomas Hirst, RI MS JT/1/HTYP/532.
49. N.d. July 1864, Tyndall Journal, RI MS JT/2/13c/1259A.
50. Tyndall (1864c).
51. 5 October 1864, Andrew Ramsay to Tyndall, RI MS JT/1/R/4; 'Mr A C Ramsay on the Erosion of Valleys and Lakes; a reply to Sir Roderick Murchison's Anniversary Address to the Geographical Society', *Phil. Mag.* 28 (1864): p. 295–311.
52. 19 July 1864, Tyndall to Juliet Pollock, RI MS JT/1/TYP/6/2011–12.
53. The mountaineer Martin Conway described in graphic detail the danger of crossing this couloir twelve years later: Conway (1920): p. 44–6.
54. 30 September 1864, Tyndall to the Editor of The Times, *The Times* (1 October 1864): p. 10. Wellcome Library MS. 7777/108.
55. 18 August 1864, Tyndall to Ellen Lubbock, BL Add MS 63092, ff. 212–3.
56. Tyndall (1871): p. 252.
57. 23 August 1865, Tyndall to Michael Faraday, Letter 4561 in James (2011).
58. 25 August 1878, Emil Du Bois-Reymond to Tyndall, RI MS JT/1/D/158.
59. Girdlestone also had an interest in science, publishing an intriguing theoretical piece on molecular motion: Girdlestone, A. G., 'On the Condition of the Molecules of Solids', *Phil. Mag.* 29 (1864): p. 108–11.
60. 31 July 1866, Tyndall to Thomas Hirst, RI MS JT/1/HTYP/536.
61. 11 August 1866, Tyndall to Rudolf Clausius, RI MS JT/1/T/237.
62. 23 July 1867, Tyndall to Juliet Pollock, RI MS JT/1/TYP/6/2082–5.
63. N.d. July 1867, Tyndall to Thomas Hirst, RI MS JT/1/HTYP/538.
64. 27 September 1867, Tyndall to Emily Peel, RI MS JT/1/TYP/3/955.
65. A single-horse carriage.
66. N.d. August 1868, Tyndall to Thomas Hirst, RI MS JT/1/HTYP/546.
67. 19 November 1868, Tyndall to Rudolf Clausius, RI MS JT/1/T/201.
68. 16 October 1868, Felice Giordano to Tyndall, RI MS JT/1/G/5.

Chapter 15: Clouds of Imagination (1868–1870)

1. 3 February 1867, Tyndall to Joseph Hooker, RI MS JT/1/TYP/8/2595.
2. Published in 1870 and 1871 under the titles 'Scientific Limit of the Imagination', Tyndall (1870): p. 52–65; and 'Scope and Limit of Scientific Materialism', Tyndall (1871): p. 107–22.
3. 5 September 1868, Tyndall to Thomas Hirst, RI MS JT/1/T/668.
4. Tyndall (1879): vol. 2, p. 89–90.

5. 23 August 1868, Sir William Armstrong to Tyndall, Tyndall Journal, RI MS JT/2/13c/1347.
6. Tyndall, J., 'Vitality', *The Reader* 95 (29 October 1864): p. 545–6.
7. See Chalmers (1996): p. 297–9.
8. Tyndall laboratory notebook, RI MS JT/3/48.
9. Tyndall, J. 'Dust and Disease', *Proc. Roy. Inst.* 6 (1870): p. 1–14.
10. 20 October 1868, Tyndall to Mrs Steuart, RI MS JT/1/TYP/10/3383.
11. 14 July 1872, John Muir to Mrs Ezra Carr, in Bade (1924): p. 335.
12. 9 October 1868, Tyndall to Charles Darwin, RI MS JT/1/TYP/9/2806.
13. 7 October 1868, Charles Darwin to Tyndall, RI MS JT/1/TYP/9/2805.
14. 20 October 1868, Tyndall to Charles Darwin, RI MS JT/1/TYP/9/2807.
15. Tyndall (1868a).
16. 14 November 1868, Tyndall to Edward Frankland, Frankland Rylands Letters, 2183.
17. 19 November 1868, Tyndall to Rudolf Clausius, RI MS JT/1/T/201.
18. 11 December 1868, Rudolf Clausius to Tyndall, RI MS JT/1/TYP/7/2309–10.
19. 3 November 1868, Tyndall to Thomas Hirst, RI MS JT/1/T/669.
20. 6 November 1868, Tyndall to Thomas Hirst, RI MS JT/1/T/670.
21. 6 November 1868, Tyndall to Thomas Hirst, RI MS JT/1/T/671.
22. 19 November 1868, Tyndall to Thomas Huxley, RI MS JT/1/TYP/9/2924.
23. 26 November 1868, Sir John Herschel to Tyndall, RI MS JT/1/H/110.
24. 30 November 1868, Tyndall to Sir John Herschel, RI MS JT/1/TYP/2/550–550a.
25. 1 December 1868, Tyndall to Sir John Herschel, RI MS JT/1/TYP/2/551–2.
26. 3 December 1868, Tyndall to Sir John Herschel, RI MS JT/1/TYP/2/555.
27. 2 December 1868, Tyndall to George Stokes, RI MS JT/1/S/244.
28. 'Water—Decay of Public Buildings', *Carlow Sentinel* (20 January 1849): p. [2]. The quoted words are from his journal draft. They are slightly different from, though with the same sense as, the published article: 30 December 1848, Tyndall Journal, RI MS JT/2/13b/409–10.
29. Tyndall, J., 'The Sky', *The Queenwood Observer* (28 February 1853): p. 6–7.
30. 2 December 1868, Tyndall to George Stokes, RI MS JT/1/T/1429.
31. 3 December 1868, George Stokes to Tyndall, RI MS JT/1/S/245.
32. 7 December 1868, George Stokes to Tyndall, RI MS JT/1/S/246.
33. 11 December 1868, George Stokes to Tyndall, RI MS JT/1/S/248.
34. Roscoe, H. E., 'On the opalescence of the atmosphere', *Proc. Roy. Inst.* 4 (1866): p. 651–9.
35. 12 December 1868, Tyndall to Sir John Herschel, RI MS JT/1/TYP/2/556.
36. 13&14 December 1868, Sir John Herschel to Tyndall, RS HS/24/243.
37. Tyndall (1869a).
38. Tyndall, J., 'Chemical Rays and the Light of the Sky', *Proc. Roy. Inst.* 5 (1869): p. 429–50.
39. 11 February 1869, Tyndall to George Airy, CUL MS.RGO.6/479.
40. 12 March 1868, George Airy to Tyndall, RI MS JT/1/A/48.
41. Tyndall (1869b).
42. Tyndall, J., 'Odds and ends of Alpine life (Part I)', *Macmillan's Magazine* 19 (1869): p. 369–85, and 'Odds and ends of Alpine life (Part II)', *Macmillan's Magazine* 19 (1869): p. 465–79.
43. Tyndall laboratory notebook, RI MS JT/3/48. Experiments were recorded on various days from 19 January to 15 March. Monday 15 February was a big day.
44. Tyndall (1869c).
45. 11 April 1869, Tyndall to Rudolf Clausius, RI MS JT/1/T/203.
46. 8 March 1869, Hirst Journal, RI MS JT/2/32c/1842; Tyndall (1869d).
47. 15 May 1869, George Airy to Tyndall, RI MS JT/1/A/49.
48. 28 May 1869, Sir John Herschel to Tyndall, RI MS JT/1/H/113.

49. 29 May 1869, Tyndall to Lady Mary Egerton, RI MS JT/1/TYP/1/389.
50. For a contemporary reflection see Proctor, R. A., 'Professor Tyndall's Theory of Comets', *Fraser's Magazine* 80 (October 1869): p. 504–12.
51. 8 June 1869, Heinrich Debus to Tyndall, RI MS JT/1/D/38.
52. 16 June 1869, Tyndall to Sir Roderick Murchison, BL Add MS 46128, f. 223.
53. 29 June 1869, Tyndall to Heinrich Debus, RI MS JT/1/T/264.
54. 5 July 1869, Tyndall to unknown, BL Add MS 63092, ff. 39–42.
55. 23 July 1869, Tyndall to Juliet Pollock, RI MS JT/1/TYP/6/2108–10; BL Add MS 63092, ff. 222–5.
56. 17 July 1869, Henry Bence Jones to Tyndall, RI MS JT/1/J/57.
57. Fiske (1894): p. 135.
58. See Spottiswoode, W., 'The Old and New Laboratories at the Royal Institution', *Proc. Roy. Inst.* 7 (1873): p. 1–11.
59. 23 July 1869, Tyndall to Juliet Pollock, RI MS JT/1/TYP/6/2108–10; BL Add MS 63092, ff. 222–5.
60. 22 July 1869, Tyndall to *The Times* (published 26 July), Tyndall Journal RI MS JT/6/5/1.
61. 16 August 1869, Tyndall to (possibly) Henry Bence Jones, BL Add MS 63092, ff. 43–4.
62. *Memoirs of Emily Elliott*, http://bartonhistory.wikispaces.com/Rev.+Julius+Marshall+Elliott+(1841–1869), p. 55–6 (accessed 15 January 2015).
63. 29 August 1869, Tyndall to Thomas Hirst, RI MS JT/1/HTYP/553–5.
64. 2 August 1869, Tyndall to Thomas Hirst, RI MS JT/1/HTYP/551–2.
65. Egerton was a Conservative MP, who served as Under-Secretary of State for Foreign Affairs from 6 July 1866 to 1 December 1868, under Lord Derby and then Disraeli. He died on 27 August 1869.
66. 11 September 1867, Lady Mary Egerton to Tyndall, RI MS JT/1/E/48.
67. 2 October 1869, Tyndall to Thomas Huxley, RI MS JT/1/TYP/9/2927–9.
68. 28 December 1867, Lady Mary Egerton to Tyndall, RI MS JT/1/E/65.
69. 29 August 1869, Tyndall to Thomas Hirst, RI MS JT/1/HTYP/553–5.
70. Tyndall (1879): vol. 1, p. 125–7.
71. Tyndall (1879): vol. 1, p. 128.
72. 10 September 1869, Auguste de la Rive to Tyndall, RI MS JT/1/D/107.
73. 13 September 1869, Tyndall to Thomas Hirst, RI MS JT/1/HTYP/558.
74. 5 October 1869, Tyndall to Thomas Hirst, RI MS JT/1/HTYP/560.
75. 22 September 1869, Tyndall to Thomas Hirst, RI MS JT/1/HTYP/556, 559.
76. Tait, P. G., 'On Comets', *Brit. Assoc. Rep. Notices and Abstracts of Miscellaneous Communications to the Sections*, 39 (1869): p. 21.
77. *Brit. Assoc. Rep.* 39 (1869): p. 175–6; *Brit. Assoc. Rep.* 39 (1869): p. 213–4.
78. 30 September 1869, Thomas Huxley to Tyndall, RI MS JT/1/TYP/9/2926.
79. 2 October 1869, Tyndall to Thomas Huxley, RI MS JT/1/TYP/9/2927–9.
80. 30 September 1869, Herbert Spencer to Tyndall, RI MS JT/1/S/98.
81. 14 October 1869, Henry Bence Jones to Tyndall, RI MS JT/1/J/62.
82. 26 October 1869, Tyndall to Thomas Carlyle, RI MS JT/1/T/149.
83. 1 November 1869, John Ruskin to Tyndall, Tyndall Journal, RI MS JT/2/13c/1350.
84. Cook and Wedderburn (1903–1912): vol. 19, p. 292.
85. 2 June 1866, Tyndall to William Stanley Jevons, Jevons Archive, John Rylands University Library of Manchester.
86. 11 March 1870, Tyndall to Rudolf Clausius, RI MS JT/1/T/204.
87. 7 November 1869, Tyndall Journal, RI MS JT/2/13c/1349.

88. November 1869, Tyndall Journal, RI MS JT/2/13c/1350–4.
89. 5 December 1869, Tyndall to Auguste De la Rive, RI MS JT/1/TYP/1/372.
90. 16 November 1869, Tyndall to Sir John Herschel, RI MS JT/1/TYP/2/569.
91. 19 November 1869, Sir John Herschel to Tyndall, RI MS JT/1/TYP/2/570.
92. 21 May 1871, Tyndall to *Daily News*, RI MS JT/1/TYP/2/576.
93. 20 November 1869, Tyndall to unknown (possibly John Tyndall at Gorey), RI MS JT/1/TYP/10/3317.
94. Tyndall described Mary Adair as the niece of Sir Charles Hamilton, but it seems that they were only distant relations.
95. 28 November 1869, Tyndall Journal, RI MS JT/2/13c/1357.
96. 28 November 1869, Tyndall Journal, RI MS JT/2/13c/1357.
97. 26 October 1869 (copy), Tyndall to Mary Adair, RI MS JT/2/10.
98. 28 November 1869, Tyndall Journal, RI MS JT/2/13c/1357.
99. 28 November 1869, Tyndall Journal, RI MS JT/2/13c/1358.
100. 28 November 1869, Tyndall Journal, RI MS JT/2/13c/1358.
101. 28 November 1869, Tyndall Journal, RI MS JT/2/13c/1358.
102. N.d. November 1869, Tyndall to George Airy, CUL MS.RGO.6/393.
103. 30 November 1869, George Airy to Tyndall, RI MS JT/1/A/51; 30 November 1869, Tyndall to George Airy, CUL MS.RGO.6/393; 1 December 1869, George Airy to Tyndall, RI MS JT/1/A/52.
104. 7 December 1871, Hirst Journal, RI MS JT/2/32d/1922.
105. Tyndall, J., 'Climbing in search of the sky', *Fortnightly Review* 13, (January 1870): p. 1–15.
106. 5 December 1869, Tyndall to Miss Moore, RI MS JT/1/TYP/3/864b–865a.
107. Possibly not completed. His notes of the course earlier in the year were published: Tyndall, J., *Notes of a Course of Nine Lectures on Light; delivered at the Royal Institution of Great Britain April 8–June 3 1869* (London: Longman, 1870).
108. Tyndall (1870b).
109. 23 October 1870, Sir John Herschel to Tyndall, Tyndall Journal, RI MS JT/2/13c/1367.
110. 5 December 1869, Tyndall to Auguste De la Rive, RI MS JT/1/TYP/1/372.

Chapter 16: Dust and Disease (1870–1872)

1. Tyndall, J., 'On Dust and Disease', *Proc. Roy. Inst.* 6 (1870): p. 1–14.
2. Tyndall to *The Times*, 'Professor Tyndall on Filtered Air', *The Times* (7 April 1870): p. 5.
3. Tyndall, J., 'On haze and dust', *Nature* 1 (27 January 1870): p. 339–42.
4. Tyndall, J., 'On Dust and Disease', *Fraser's Magazine* 1 (March 1870): p. 302–10.
5. Tyndall, J., 'The Germ Theory of Disease', *The Times* (21 April 1870): p. 8.
6. Worboys (2000): p. 279.
7. Tyndall, J., 'Pasteur's Researches on the Diseases of Silkworms', *Nature* 2 (7 July 1870): p. 181–3.
8. 17 February 1870, Tyndall to Auguste De la Rive, RI MS JT/1/TYP/1/1/374.
9. 18 March 1870, Tyndall to Josiah Whitney, Yale MS 555, Box 15, 1–4.
10. Tyndall, J., 'Death of Professor Magnus', *Nature* 1 (14 April 1870): p. 607.
11. Tyndall (1870c).
12. British Association Council Minutes 1841–68, Bodleian Library, Oxford, BAAS 20/297.
13. 20 June 1870, Tyndall to Juliet Pollock, RI MS JT/1/T/1152.
14. 16 June 1870, Tyndall to Thomas Huxley, RI MS JT/1/TYP/9/2933 (ICHC 8.81).
15. Huxley (1870): p. 161. The paragraph ends: '...the errors of systematic materialism may paralyse the energies and destroy the beauty of a life'.

16. 14 July 1870, Henry Bence Jones to Tyndall, RI MS JT/1/J/64.
17. 19 July 1870, Tyndall to Thomas Hirst, RI MS JT/1/TYP/565–6, BL Add MS 63092, ff. 67–9.
18. 1 September 1870, Rudolf Clausius to Tyndall, RI MS JT/1/TYP/7/2316–17.
19. N.d. August? 1870, Tyndall to Auguste de la Rive, RI MS JT/1/TYP/1/378.
20. Tyndall (1870).
21. Tyndall, J., 'Researches on the Deportment and Vital Persistence of Putrefactive and Infective Organisms from a Physical Point of View', Proc. Roy. Inst. 8 (1877): p. 476–7.
22. 8 September 1870, Charles Darwin to Tyndall, Tyndall Journal, RI MS JT/2/13c/1366.
23. Quoted in Lightman, B., 'Science at the Metaphysical Society: Defining Knowledge in the 1870s', in Lightman and Reidy (2014): p. 200.
24. Tyndall, J., 'On the Colour of The Lake of Geneva and The Mediterranean Sea', Nature 2 (20 October 1870): p. 489–90.
25. 7 January 1871, Henrietta Huxley to Tyndall, RI MS JT/1/TYP/9/2944.
26. Tyndall, J., 'On the colour of water and on the scattering of light in water and air', Proc. Roy. Inst. 6 (1871): p. 189–99.
27. See Braun, C. L. and S. N. Smirnov, 'Why is water blue?', J. Chem. Edu. 70 (1993): p. 612–14.
28. 'Professor Tyndall on the London Water Supply', The Times (31 January 1871): p. 4.
29. 15 February 1871, Abbé Moigno to Tyndall, RI MS JT/1/M/118; 26 February 1871, Jules Jamin to Tyndall, RI MS JT/1/J/15.
30. 15 March 1871, Henri Regnault to Tyndall, RI MS JT/1/R/10.
31. 6 July 1871, Tyndall to Emil Du Bois-Reymond, RI MS JT/1/T/403.
32. In the election in Marylebone, Huxley, though elected was beaten into second place by Elizabeth Garrett. She would doubtless have been unimpressed by his remarks to the electors that 'girls should be trained to comprehend decency, order and frugality', 'The Education Act', The Times (10 November 1870): p. 4.
33. 29 November 1879, Hirst Journal, RI MS JT/2/32c/1888.
34. 15 March 1871, Lord Hatherley to Tyndall, Tyndall Journal, RI MS JT/2/13c/1376.
35. 2 March 1875, Olga Novikoff to Tyndall, RI MS JT/1/N/22.
36. Autobiography of William Joseph Dibdin F.I.C., F.C.S., Analytical Chemist, 1850 –1925, http://www.guise.me.uk/dibdin/wjdibdin/autobiography/sect9.htm (accessed 5 August 2016).
37. Tyndall (1871).
38. 24 May 1871, Vieweg und Sohn to Tyndall, RI MS JT/1/V/9.
39. Tyndall earned about £1500 in his lifetime from the English editions alone.
40. Tyndall gave evidence on 17 April 1877. PP 1878 (C2036). Copyright Commission, p. 314–16.
41. PP 1878 (C2036). Copyright Commission, p. viii.
42. Tyndall (1871a).
43. 7 October 1871, Tyndall to Lady Emily Peel, RI MS JT/1/TYP/3/963.
44. 23 June 1871, Tyndall to George Stokes, RS MC.9.220.
45. 2 June 1871, Rudolf Clausius to Tyndall, RI MS JT/1/TYP/7/2322.
46. Tyndall, J., 'On Dust and Smoke', Proc. Roy. Inst. 6 (1871): p. 365–76.
47. Worboys (2000): p. 97.
48. Nature published a description of the trial: Tyndall, J., 'Foul air in mines and how to live', Nature 5 (7 March 1872): p. 365–6.
49. 22 February 1871, Tyndall to Charles Darwin, RI MS JT/1/TYP/9/2809.
50. 23 July 1871, Tyndall to Thomas Hirst, RI MS JT/1/HTYP/582.
51. 12 October 1871, Tyndall to Mrs Steuart, RI MS JT/1/T/1434.
52. 23 July 1871, Tyndall to Thomas Hirst, RI MS JT/1/HTYP/582; BL Add MS 63092, ff. 74–5.

53. 31 July 1871, Tyndall to Lady Mary Egerton RI MS JT/1/TYP/1/391–4. Similar words of Tyndall were quoted in a review of Girdlestone's *The High Alps Without Guides*: 'The Literary Examiner', *The Examiner and London Review* (9 July 1870): p. 436.

54. 9 April 1871, William Thomson to Tyndall, RI MS JT/1/T/18.

55. 31 July 1871, Tyndall to Lady Mary Egerton, RI MS JT/1/TYP/1/391–4.

56. 4 September 1871, Tyndall to Juliet Pollock, RI MS JT/1/T/1154.

57. 2 August 1871, Tyndall to William Thomson, CUL Add 7342/T627, Kelvin Correspondence.

58. 4 October 1871, Tyndall to Thomas Huxley, ICHC 8.90.

59. 11 August 1871, Thomas Huxley to Joseph Hooker, ICHC 2.177.

60. 13 April 1872, Tyndall to Hermann Helmholtz, RI MS JT/1/T/489 (not sent until 18 June).

61. 23 June 1872, Hermann Helmholtz to Tyndall, RI MS JT/1/H/48.

62. 1 November 1871, Tyndall to Francis Galton, UCL–13, Galton Papers, UCL Library Services, Special Collections.

63. 13 November 1879, Tyndall to Rudolf Clausius, RI MS JT/1/T/218.

64. Tyndall, J., 'The Copley Medallist of 1870', *Nature* 5 (21 December 1871): p. 137–8; Tyndall, J., 'The Copley Medallist of 1871', *Nature* 5 (14 December 1871): p. 117–20; Tyndall, J., 'Dr Carpenter and Dr Mayer', *Nature* 5 (21 December 1871): p. 143–4.

65. 29 November 1871, Tyndall to Lady Mary Egerton, RI MS JT/1/TYP/1/395–6.

66. Tyndall (1872a).

67. 21 January 1872, Tyndall to Lady Lyell, RI MS JT/1/T/1057.

68. Roscoe (1906): p. 130–1.

69. Tyndall, J., 'The Identity of Light and Radiant Heat', *Proc. Roy. Inst.* 6 (1872): p. 417–21.

70. Tyndall (1872).

71. See RI MS JT/3/48/238–61.

72. Tyndall's journals for 1883–1891 are RI MS JT/2/15–22. Louisa's journals cover the wider period 1876–1894, RI MS JT/2/23–31.

Chapter 17: Government Service and Education (1871–1892)

1. Tyndall suggested the use of different metals or design, or sacrificial anodes, to reduce the possibility of corrosion. This was more science-based than practical for the railway companies, who continued to employ the same materials. The main message that these enquiries established was the need for more frequent detailed examinations of boiler interiors to detect serious thinning before it led to failure. Michael Bailey, personal communication.

2. *Two reports of an inspecting-officer of the Railway Department to the Lords of the Committee of Privy Council for Trade, on the explosion of boilers of locomotive engines, one of which occurred at the Gloucester station of the Great Western Railway, On the 7th February 1855, and the other at the Greenock station of the Caledonian Railway, on the 5th April 1855, together with a copy of a circular issued by their Lordships to the railway companies on the subject.* PP 1854–5 (1951); 6 June 1855, Tyndall Journal, RI MS JT/2/13C/751.

3. *Lighting picture galleries by gas. Report of the commission appointed to consider the subject of lighting picture galleries by gas.* PP 1859 Session 2 (106).

4. The office, once held by Isaac Newton, was abolished after Graham's death in September 1869. The duties were incorporated with those of the Chancellor of the Exchequer.

5. *Report of the Commission on the Heating, Lighting, and Ventilation of the South Kensington Museum.* PP 1868–9 (4206).

6. *Report from the Select Committee on House of Commons (Arrangements).* PP 1867–8 (265).

7. 'Metropolitan Improvements: Albert Hall of Arts and Sciences', *The Standard* (12 December 1868); 'South Kensington Museum', *Daily News* (21 April 1869): p. 5.

8. 4 February 1866, Tyndall Journal, RI MS JT/2/13c/1303.

9. 9 May 1866, Michael Faraday to Tyndall, Letter 4580 in James (2011).

10. 23 July 1866, Michael Faraday to Thomas Farrer, Letter 4584 in James (2011); *c*. 30 July 1866, Thomas Farrer to Michael Faraday, Letter 4586 in James (2011).

11. *Board of Trade (scientific adviser). Copy of correspondence showing the mode in which the remuneration to the scientific adviser to the Board of Trade and the Trinity House is charged.* PP 1872 (258).

12. 4 August 1867, Tyndall to Cecil Trevor, Board of Trade (scientific adviser). *Copy of correspondence showing the mode in which the remuneration to the scientific adviser to the Board of Trade and the Trinity House is charged.* PP 1872 (258).

13. 16 March 1868, Board of Trade to Tyndall, NA MT 10–128, file H2727/3 (possibly a draft).

14. *Lighthouse lanterns. Copy of correspondence relative to the comparative advantages of the lighthouse lanterns adopted by the Corporation of the Trinity House and the Commissioners of Northern Lighthouses.* PP 1870 (5).

15. 13 May 1871, Tyndall to Robin Allen (Trinity House), Lighthouses (mineral oils). *Copy of correspondence between the general lighthouse authorities, and the Board of Trade, relative to proposals to substitute mineral oils for colza oil in lighthouses.* PP 1871 (318).

16. 6 November 1869, Tyndall to Cecil Trevor, *Further papers relative to a proposal to substitute gas for oil as an illuminating power in lighthouses.* PP 1871 (c.282).

17. 8 October 1870, Tyndall to Board of Trade, *Further papers relative to a proposal to substitute gas for oil as an illuminating power in lighthouses.* PP 1871 (c.282).

18. 7 February 1871, Tyndall to Board of Trade, NA MT 10/131 H64B 2309–2313.

19. 16 February 1871, Board of Trade (Cecil Trevor) to Tyndall, NA MT 10/131 H64B 2318; 29 June and 5 July 1871, Tyndall to Cecil Trevor, *Further papers relative to a proposal to substitute gas for oil as an illuminating power in lighthouses, Part III.* PP 1875 (c.1151).

20. See *Lighthouses (mineral oils). Copy of further correspondence relative to proposals to substitute mineral oils for colza oil in lighthouses (in continuation of parliamentary paper, no. 2, of session 1872).* PP 1872 (2) (264).

21. 22 June 1872, Tyndall to Robin Allen, *Lighthouses (mineral oils). Copy of further correspondence relative to proposals to substitute mineral oils for colza oils in lighthouses.* PP 1873 (3) (378).

22. 16 October 1871, Tyndall to Henry Bence Jones, RI MS JT/1/TYP/3/803.

23. 15 November 1871, Edward Frankland to Tyndall, RI MS JT/1/F/53.

24. 8 October 1857, Tyndall Journal, RI MS JT/2/13c/1035.

25. 4 April 1857, Tyndall to Lyon Playfair, RI MS JT/1/T/1125.

26. 7 January 1857, Lyon Playfair to Tyndall, RI JT/1/P/115.

27. 17 June 1890, Tyndall to Thomas Hirst, RI MS JT/1/T/824; 16 June 1890, Thomas Hirst to Tyndall, RI MS JT/1/H/422.

28. RI MS JT/3/23/26–9.

29. *Report of the Committee Appointed to Consider the Propriety of Establishing a Degree or Degrees in Science, and the conditions on which such a Degree should be conferred, with the evidence taken before the committee* (University of London, 1858): p. 1–9.

30. *Thirteenth report of the Science and Art Department of the Committee of Council on Education.* PP (1866), p. 22–3.

31. *Fourteenth report of the Science and Art Department of the Committee of Council on Education.* PP (1867), p. 22–3.

32. Tyndall, J., 'Elementary Magnetism: A Lecture to Schoolmasters', Tyndall (1879): vol. 1, p. 395–420.

33. Examination Papers for Science Schools and Classes, May 1870, Science and Art Department of the Committee of Council on Education (London: Spottiswoode, 1870), p. 46–7.

34. *Report from the Select Committee of the House of Lords, on the Public Schools Bill.* PP 1865 (482), p. 310–11.

35. Report of 1867 Meeting of British Association for the Advancement of Science, p. xxxix–liv.

36. 14 March 1867, James Wilson to Tyndall, RI MS JT/1/W/77.

37. 15 March 1867, James Wilson to Tyndall, RI MS JT/1/W/78.

38. *Schools Inquiry Commission. Report relative to technical education.* PP 1867 (3898), p. 10.

39. *Index to the report from the Select Committee on Scientific Instruction.* PP 1867–8 (432) (432-I), p. 118–21.

40. 20 June 1856, Gustav Magnus to Tyndall, RI MS JT/1/M/19.

41. Thompson and Thompson (1920): p. 21. Thompson was even more impressed by Huxley, adding 'As a speaker he beats Tyndall hollow'.

42. 'Professor Tyndall at the Birkbeck Institution', *The Times* (23 October 1884): p. 10.

43. Tyndall (1892): p. 241–2.

44. Bibby (1956). The science teaching was not a great success according to Hooker, whose son Charles attended the school.

45. The report was sent by the President, the Duke of Buccleuch, to the Lord President of the Privy Council, and published by the House of Commons on 11 March 1868. PP 1868–9 (137).

46. 30 March 1861, Tyndall Journal, RI MS JT/2/13c/1208.

47. 29 November 1871, Tyndall to Mary Lady Egerton, RI MS JT/1/TYP/1/395–6.

48. See Tyerman (2000) for an excellent history.

49. Notes n.d., RI MS JT/3/44.

50. 30 May 1885, Tyndall to Hermann Helmholtz, Archiv der Berlin-Brandenburgischen Akademie der Wissenschaften 477; 30 May 1885, Tyndall to Emil du Bois-Reymond, SB Slg Darmstaedter 1855 Tyndall Be. 1–58, 21–22.

51. 7 June 1885, Hermann Helmholtz to Tyndall, RI MS JT/1/H/63; 2 June 1885, Emil du Bois-Reymond to Tyndall, RI MS JT/1/D/171.

52. Notes n.d., RI MS JT/3/44.

53. Harrow School, Report of the Head Master on Studies, 30 October 1886, Harrow Archive R2/8.

54. 22 December 1886, Tyndall Journal, RI MS JT/2/19.

55. 15 December 1885, Tyndall Journal, RI MS JT/2/18.

56. 4 December 1885, Tyndall Journal, RI MS JT/2/18.

57. 27 March 1872, Lord Derby to Tyndall, RI MS JT/1/TYP/8/2608.

58. See Endersby (2008): p. 282–310; Drayton (2000): p. 211–20; MacLeod, R. M., 'The Ayrton Incident: A Commentary on the Relations of Science and Government in England, 1870–1873', in Thackray and Mendelsohn (1974): p. 45–78.

59. 12 March 1872, Joseph Hooker to Tyndall, RI MS JT/1/TYP/8/2606.

60. 8 April 1872, Joseph Hooker to Tyndall, RI MS JT/1/TYP/8/2617.

61. These were Tyndall (1872) and Tyndall (1872a) respectively; 2 April 1872, Tyndall to Hector Tyndale, RI MS JT/1/T/1452.

62. RI MS JT/6/4a.

63. 25 June 1872, Sir John Lubbock to Joseph Hooker, RI MS JT/1/TYP/8/2670.

64. 19 June 1872, Tyndall to Thomas Huxley, RI MS JT/1/TYP/9/2962.

65. 'Mr. Ayrton and Dr. Hooker', *Nature* 6 (11 July 1872): p. 211–16.

66. Some 1,000 copies were printed, netting Tyndall £140. Perhaps unsurprisingly for such a technical work, largely of reprints of papers in the *Philosophical Magazine* and the *Philosophical Transactions*, there was no second edition.

67. 3 October 1871, Louis Pasteur to Tyndall, RI MS JT/1/P/4.
68. 5 September 1872, Tyndall to Thomas Huxley, RI MS JT/1/TYP/8/2712 (ICHC 8.127).
69. 'The Royal Gardens, Kew—Dr. Hooker, and the First Commissioner of Works', HC Deb 08 August 1872 vol. 213 cc756–757, *HANSARD 1803–2005* (online, accessed 2 October 2016).
70. 17 February 1879, Tyndall to Joseph Hooker, RI MS JT/1/TYP/8/2751.
71. 9 September 1872, Tyndall to Henry Bence Jones, RI MS JT/1/TYP/3/804.
72. 22 July 1872, Henry Bence Jones to Tyndall, RI MS JT/1/J/81.

Chapter 18: America (1872–1873)

1. 14 August 1871, Joseph Henry to Tyndall, RI MS JT/1/H/75.
2. 1 September 1871, Joseph Henry to Tyndall, RI MS JT/1/H/76.
3. 18 November 1871, Tyndall to Hector Tyndale, RI MS JT/1/T/1451.
4. 6 February 1872, Tyndall Journal, RI MS JT/2/13c/1389.
5. 23 November 1862, Hector Tyndale to Tyndall, RI MS JT/1/T/52.
6. 15 May 1866, Hector Tyndale to Tyndall, RI MS JT/1/T/58.
7. 11 May 1872, Joseph Henry to Tyndall, Smithsonian RU 33, Vol 28, Page 769.
8. 28 May 1872, Tyndall to Joseph Henry, RH 3965, William Jones Rhees Papers, The Huntington Library, San Marino, California; Clarence King was a remarkable man who lived a double life for his last thirteen years with an African-American woman (and former slave), pretending to be an African-American Pullman car attendant, despite his blue eyes and fair skin.
9. 8 June 1872, Tyndall to Mary Adair, RI MS JT/1/T/87.
10. 19 September 1872, Tyndall to Peter Henry, Henry E. Huntington Library 62.256.
11. The place later made famous by the Beatles.
12. 29 September 1872, Tyndall to Thomas Hirst, RI MS JT/1/T/693.
13. 20 October 1872 (my dating), Tyndall to Thomas Hirst, RI MS JT/1/T/999.
14. Ibid.
15. Ibid.
16. Ibid.
17. Ibid.
18. Ibid.
19. Ibid.
20. Ibid.
21. 23 October 1872, Tyndall to Henry Bence Jones, RI MS JT/1/TYP/3/811.
22. 8 November 1872, Henry Bence Jones to Tyndall, RI MS JT/1/J/71.
23. See e.g. 10 March 1855, Tyndall to Hector Tyndale RI MS JT/1/T/1442. The 'rare and radiant maiden' is a quote from Edgar Allen Poe's poem 'The Raven', ii. 5.
24. 'Professor Tyndall on Light and Heat', *Boston Daily Globe* (16 October 1872), RI MS JT/6/7/D.
25. I am grateful to Bernard Lightman for pointing this out.
26. 1 January 1873, Thomas Huxley to Tyndall, ICHC 8.131.
27. Norton and Howe (1913): vol. 1, p. 313.
28. 'Our English Lecturers', undated, cutting from the *Nation*, RI MS JT/6/7/58.
29. 'Professor Tyndall's Second Lecture', *Boston* Post (18 October 1872), RI JT/6/7/1.
30. 5 November 1872, Tyndall to Henry Bence Jones, RI MS JT/1/TYP/3/812.
31. 20 October 1872 (my dating), Tyndall to Thomas Hirst, RI MS JT/1/T/999.
32. 5 November 1872, Tyndall to Henry Bence Jones, RI MS JT/1/TYP/3/812.
33. 26 October 1872, Tyndall to Joseph Henry, RH 3970. William Jones Rhees Papers, The Huntington Library, San Marino, California.

34. 26 October 1872, Henry Bence Jones to Tyndall, RI MS JT/1/J/70.
35. 11 November 1872, Tyndall Journal, RI MS JT/2/12/41.
36. 5 November 1872, Tyndall to Thomas Hirst, RI MS JT/1/T/694.
37. Tennyson (1895): iv, p. 195.
38. 27 October 1872, Peter Lesley to Tyndall, RI MS JT/1/L/17.
39. 23 November 1872, Tyndall to Thomas Hirst, RI MS JT/1/T/696.
40. 'Prof. Tyndall on "Light"', undated, cutting from *Philadelphia Press*, RI JT/6/7/12.
41. 'Professor Tyndall', *Philadelphia Age* (11 November 1872), RI JT/6/7/12.
42. 23 November 1872, Tyndall to Thomas Hirst, RI MS JT/1/T/696.
43. 23 November 1872, Tyndall to Thomas Hirst, RI MS JT/1/T/696.
44. 25 November 1872, Tyndall Journal, RI MS JT/2/12/44.
45. 1 January 1873, Thomas Huxley to Tyndall, ICHC 8.131. Other nominees included the Duke of Argyll and the Marquis of Huntly.
46. 3 December 1872, Tyndall Journal, RI MS JT/2/12/49.
47. 17 December 1872, Tyndall to Henry Bence Jones, RI MS JT/1/TYP/3/817.
48. Tyndall, J., 'Science and Religion', *Popular Science Monthly* 2 (1872): p. 79–82.
49. 8 December 1872, T. Meldenhall to Tyndall, RI MS JT/1/M/98.
50. 17 December 1872, Tyndall to Hector Tyndale, RI MS JT/1/T/1472.
51. 11 January 1873, Tyndall to Thomas Hirst, RI MS JT/1/T/699.
52. 'Tyndall and His Audiences', undated, cutting from a Brooklyn newspaper, RI MS JT/6/7/56.
53. 11 January 1873, Tyndall to Thomas Hirst, RI MS JT/1/T/699.
54. 11 January 1873, Tyndall to Thomas Hirst, RI MS JT/1/T/699.
55. Undated, newspaper cutting, RI MS JT/6/7/11.
56. 'Professor Tyndall', *Morning Journal* (23 January 1873), RI MS JT/6/7/57.
57. Bence Jones was to die on 20 April; 3 January 1873, Tyndall to Henry Bence Jones, RI MS JT/1/TYP/3/821.
58. 4 January 1873, W. Tisdale to Tyndall, RI MS JT/1/T/35.
59. 21 April 1866, Sir George Airy to Tyndall, RI MS JT/1/A/39.
60. 10 August 1871, Louis Pasteur to Tyndall, RI MS JT/1/P/3.
61. 28 March 1883, Tyndall to Joseph Chamberlain, RI MS JT/8/3/9 (draft); PP 1883 (168) (263) Lighthouse Illuminants, p. 51.
62. 'THE TYNDALL DINNER.: Honors to our Distinguished Visitor A Notable...', *New York Times* (5 February 1873): p. 5.
63. 1 March 1873, Alfred Mayer to Tyndall, RI MS JT/1/M/75.
64. Spencer (1904): vol. II, p. 246.
65. *c.*18 October 1872, Tyndall Journal, RI MS JT/2/12/13.
66. 'Professor Tyndall's Deed of Trust', *Popular Science Monthly* 3 (1873): p. 100–1.
67. 27 November 1884, Tyndall Journal, RI MS JT/2/16.
68. 26 December 1875, Hector Tyndale to Tyndall, RI MS JT/1/T/79.

Chapter 19: Fogs and Glaciers (1873)

1. 19 March 1873, Tyndall to Emily Peel, RI MS JT/1/TYP/3/968.
2. Tyndall, J., 'Observations on Niagara', *Proc. Roy. Inst.* 7 (1873): p. 73–91. See also Tyndall, J., 'Niagara', *Macmillan's Magazine* 28 (1973): p. 49–62.
3. 12 April 1873, Tyndall to Cecil Trevor, NA MT 10/220. Illumination of lighthouses by gas. H2075.
4. 19 May 1873, Cecil Trevor to Tyndall, *Further papers relative to a proposal to substitute gas for oil as an illuminating power in lighthouses. Part III.* PP 1875 (c.1151).

5. 25 July 1873, Tyndall to Cecil Trevor, *Further papers relative to a proposal to substitute gas for oil as an illuminating power in lighthouses. Part III.* PP 1875 (c.1151).

6. 16 October 1874, Tyndall to Cecil Trevor, NA MT 10/220. Lighthouses, Illumination by means of gas. H7863.

7. McLeod (1969): p. 21.

8. 17 May 1849, James Craven to Tyndall, RI MS JT/1/TYP/11/3530–3.

9. 'Tyndall's Forms of Water', *Nature* 7 (27 March 1873): p. 400–1.

10. 29 April 1873, Tyndall to Thomas Hirst, RI MS JT/1/T/703.

11. Agassiz, A., 'Originators of Glacial Theories', *Nature* 8 (8 May 1873): p. 24–5.

12. 10 May 1873, Tyndall to Thomas Huxley, ICHC 8.141.

13. Huxley, T. H., 'Forbes and Tyndall', *Nature* 8 (22 May 1873): p. 64.

14. Forbes, G., 'Forbes and Agassiz', *Nature* 8 (22 May 1873): p. 64–5.

15. 30 May 1873, Tyndall to Thomas Huxley, ICHC 8.147. In a curious twist, the students of St Andrew's approached Tyndall four years later to serve as Rector. Tyndall graciously turned them down, as Robert Browning and Matthew Arnold did the same year, pleading inadequate time to discharge the duties properly. Shairp is unlikely to have been devastated by this news; 'St Andrew's Nov. 22', *The Times* (23 November 1877).

16. 25 October 1873, Tyndall to Joseph Henry, Smithsonian RU 26, Vol. 139, p. 178.

17. 22 May 1874 in Fog signals. *Copy of report by Professor Tyndall to the Trinity House, upon recent experiments with regard to fog signals.* PP 1874 (188).

18. Tyndall, J., 'On the atmosphere in relation to fog-signalling (Part I)', *Contemporary Review* 24 (November 1874): p. 819–41; 'On the atmosphere in relation to fog-signalling (Part II)', *Contemporary Review* 25 (December 1874): p. 148–68.

19. 'Our London Correspondence', *Liverpool Mercury* (6 August 1879).

20. 22 April 1874, Tyndall to Robin Allen, NA MT 10/290. Papers re Fog Signals. H8476 (HW: H4380). Also in 1875 (224) *Fog signals. Copy of correspondence with the Board of Trade respecting any fog signals authorized by the three general lighthouse authorities, or sanctioned by the Board of Trade, since 18th February 1873 (in continuation of Parliamentary Papers, no. 119, of session 1873, and no. 188, of session 1874).*

21. See Tyndall (1879): vol. 1, p. 312 and p. 314.

22. 'Military and Naval Intelligence', *The Times* (23 February 1875): p. 5.

23. 19 November 1877, Tyndall to Robin Allen, NA MT 10/290. Papers re Fog Signals. H3458.

24. N.d. July 1870, Tyndall to Rudolf Clausius, RI MS JT/1/T/239.

25. 25 June 1873, Rudolf Clausius to Tyndall, RI MS JT/1/TYP/7/2329–30.

26. Maxwell, J. C., 'A Tyndallic Ode', *Nature* 4 (1871): p. 291.

27. N.d. June 1871, Tyndall to James Clerk Maxwell, CUL Add. 7655/Vl/3 (vii).

28. A version of the poem had been published in 1849 under the pseudonym Wat Ripton: 'A Whitsuntide Ramble', *Preston Chronicle* (16 June 1849): p. 3.

29. 20 April 1864, James Clerk Maxwell to Tyndall. See Harman (1995): p. 247.

30. 19 July 1865, James Clerk Maxwell to H. R. Droop, in Campbell and Garnet (1882): p. 342–3.

31. Tyndall, J., 'Principal Forbes and his Biographers', *Contemporary Review* 22 (August 1873): p. 484–508.

32. 13 July 1873, Hirst Journal, RI MS JT/2/32e/1980.

33. Tyndall (1873).

34. 8 August 1873, Hirst Journal, RI MS JT/2/32e/1981.

35. 11 June 1870, Louisa Hamilton to Lady Claud Hamilton, Elton Hall FAM182/13.

36. 1 July 1900, Mary Birkbeck to Louisa Tyndall, RI MS LT/22/1; Mary Birkbeck letters (1872), Elton Hall FAM122.

37. 7 and 9 April 1873, Louisa and Lady Claud Hamilton to Mary Hamilton, Elton Hall FAM123. Tyndall did not own a box at the Albert Hall, nor did the Birkbecks, so he may have hired one for the evening. According to Eve and Creasey (1945): p. xxi, the box was de la Rue's and the performance an early one of Wagner. However, de la Rue did not own a box until June 1876 (John Williams, Royal Albert Hall archive, personal communication). The Albert Hall put on a series of performances of the St Matthew Passion in the week beginning 7 April. See 'Royal Albert Hall Choral Society', *The Morning Post* (8 April 1873): p. 5.

38. 28 July 1873, Tyndall to Lady Claud Hamilton, BL Add MS 63092, ff. 185–6.

39. 14 August 1873, Hirst Journal, RI MS JT/2/32e/1982; 11 Mar 1889, Tyndall to Lady Welby, RI MS JT/1/TYP/5/1780.

40. 15 September 1873, Tyndall to Mrs Steuart, RI MS JT/1/TYP/10/3410.

41. 28 February 1872, Tyndall to William Spottiswoode, RI MS JT/1/T/1301.

42. 7 March 1872, Tyndall to Douglas Galton, Tyndall Journal, RI MS JT/2/13c/1397.

43. 9 April 1878, Tyndall to Joseph Hooker, RI MS JT/1/TYP/8/2741.

44. See Huxley (1918): vol. 2, p. 186–7.

45. 24 August 1873, Thomas Huxley to Tyndall, RI MS JT/1/TYP/9/3019–20.

46. 2 September 1873, Tyndall to Thomas Huxley, RI MS JT/1/TYP/9/3021.

47. Tait, P. G., 'Tyndall and Forbes', *Nature* 8 (11 September 1873): p. 381–2.

48. Tyndall, J., 'Reflection on the rainbow', *Nature* 8 (25 September 1873): p. 432–3.

49. 21 October 1873, Tyndall to Thomas Hirst, RI MS JT/1/T/708.

50. Tyndall, J., 'Tyndall and Tait', *Nature* 8 (18 September 1873): p. 399.

51. 'Tyndall and Tait', *Nature* 8 (25 September 1873): p. 431.

52. Tyndall, J., 'Reflection on the rainbow', *Nature* 8 (25 September 1873): p. 432–3.

53. 9 October 1873, Tyndall to Thomas Carlyle, RI MS JT/1/T/151.

54. 21 October 1873, Tyndall to Thomas Hirst, RI MS JT/1/T/708.

55. 13 November 1873, Thomas Huxley to Tyndall, ICHC 8.159.

56. Ball, J. (anonymously), 'Forbes and Tyndall on the Alps and their Glaciers', *Edinburgh Review* 113 (1861): p. 221–52.

57. Hopkins, W. (anonymously), 'Voyages dans les Alpes…', *Quarterly Review* 114 (1863): p. 76–125.

58. Mathews, W., 'Mechanical Properties of Ice, and their relation to Glacier Motion', *Nature* 1 (24 March 1870): p. 534–5.

59. Tyndall, J., 'On the bending of glacier ice', *Nature* 4 (5 October 1871): p. 447.

Chapter 20: The Belfast Address (1873–1875)

1. 24 September 1873, Tyndall to Thomas Huxley, RI MS JT/1/TYP/9/3022.

2. 16 October 1873, Joseph Hooker to Tyndall, RI MS JT/1/TYP/8/2721.

3. 25 October 1873, Tyndall to Joseph Hooker, RI MS JT/1/TYP/8/2722.

4. Tyndall (1874a).

5. Tyndall, J., 'On the acoustic transparency and opacity of the atmosphere', *Proc. Roy. Inst.* 7 (1874): p. 169–78.

6. 'The acoustic transparency and opacity of the atmosphere', *Nature* 9 (29 January 1874): p. 251–3 and (5 February 1874): p. 267–9.

7. Hirst had recently become the first Director of Studies at the Royal Naval College, which had been formally established on 16 January. Admiral Key was the first President and Debus the first Professor of Chemistry. A house came with Hirst's position, enabling him to give dinner parties with his niece.

8. Tyndall (1874b).

9. Tyndall (1874c); Tyndall (1874d).

10. 26 March 1874, Tyndall to Rudolf Clausius, RI MS JT/1/T/206.

11. 3 May 1874, James Coxe to Tyndall, RI MS JT/1/TYP/1/1/299; 14 May 1874, Thomas Huxley to Tyndall, RI MS JT/1/TYP/9/3031.

12. Tyndall, J., 'Rendu and his Editors', *Contemporary Review* 24 (June 1874): p. 135–48.

13. 5 June 1874, Tyndall to Thomas Hirst, BL Add MS 63092, ff. 86–7.

14. 10 June 1874, Peter Tait to Tyndall, RI MS JT/1/TYP/9/3033.

15. 15 June 1874, Tyndall to Thomas Huxley, RI MS JT/1/TYP/9/3033.

16. 11 June 1874, Tyndall to Thomas Huxley, RI MS JT/1/TYP/9/3032.

17. 24 June 1874, Thomas Huxley to Tyndall, RI MS JT/1/TYP/9/3034–5.

18. 2 December 1874, Lady Emily Peel to Tyndall, RI MS JT/1/P/65.

19. 15 July 1874, Tyndall to Thomas Huxley, RI MS JT/1/TYP/9/3036.

20. 15 July 1874, Tyndall to Thomas Huxley, RI MS JT/1/TYP/9/3036.

21. 22 August 1874, Lady Coxe to Tyndall, RI MS JT/1/TYP/1/304.

22. 22 July 1874, Thomas Huxley to Tyndall, RI MS JT/1/TYP/9/3037–8.

23. 15 August 1874, Tyndall to Mrs Steuart, RI MS JT/1/TYP/10/3412.

24. 12 August 1874, Charles Darwin to Tyndall, RI MS JT/1/TYP/9/2834.

25. 15 August 1874, Tyndall to Thomas Hirst, RI MS JT/1/T/917.

26. 'The British Association', *The Times* (23 September 1873): p. 5

27. 31 May 1874, Emil Du Bois-Reymond to Tyndall, RI MS JT/1/D/151.

28. Tyndall, J., 'Presidential Address before the British Association at Belfast', *Brit. Assoc. Rep.* (1874): p. lxvi–xcvii. See also *Nature* 10 (20 August 1874): p. 309–19, Tyndall (1874) and Tyndall (1879): vol. 2, p. 137–203. The quotations here are all from the *British Association Report*. Other versions differ in places.

29. From a draft to the *Queries to the Optiks*, quoted in Levitin. D., 'Newton and scholastic philosophy', *British Journal of the History of Science* 49 (2016): p. 70.

30. Smith (2001): 2, 928–31.

31. 'Vitality', Tyndall (1879): vol. 2, p. 51. Tyndall may have been inspired by words of Emerson: 'Society is a wave. The wave moves onward, but the water of which it is composed does not', Emerson, R. W., 'Self-reliance', *Essays* (Boston: Munroe, 1841).

32. Lightman (2004).

33. White, A. D., The Warfare of Science (London: King, 1876): p. iii–iv. In this book, White (the President and cofounder of Cornell University) argued that interference by religion with science had invariably resulted in direct evils to both, and that untrammeled scientific investigation has invariably resulted in the highest good of both. See also Lightman, B., 'Tyndall, Draper and the Conflict Thesis', in Numbers, Hardin & Binzley (forthcoming).

34. 7 September 1874, Tyndall to Mrs Steuart, RI MS JT/1/TYP/10/3413.

35. 5 September 1874, Tyndall to Olga Novikoff, RI MS JT/1/N/17.

36. 'Crystals and Molecular Force' in Tyndall (1874, 7th thousand): p. 82.

37. 26 August 1874, Tyndall to Thomas Hirst, RI MS JT/1/T/715.

38. 15 September 1874, Tyndall to Youmans, John Fiske (ed), E.L. Youmans: Interpreter of Science for the People (1894): p. 319

39. 23 September 1874, Abbé Moigno to Tyndall, RI MS JT/1/M/126.

40. For a complementary and contemporary view, see Feinstein (2013): p. 80.

41. Tyndall, J., 'Apology for the Belfast Address', in Tyndall (1879): vol. 2, p. 212.

42. I am grateful to Gregory Tate for sight of an unpublished paper touching on these points.

43. 28 September 1874, Tyndall to William Appleton, in John Fiske (ed), *E.L. Youmans: Interpreter of Science for the People* (1894): p. 319. The fourth thousand followed on 2 October; the fifth

on 6 October; and the sixth on 20 October. Tyndall made more than £300 from the publication, and he reissued it in several other forms, as a collection with other key essays and in *Fragments of Science*.

44. 30 September 1874, Hirst Journal, RI MS JT/2/32d/2006.
45. 10 November 1874, Gilberto Govi to Tyndall, RI MS JT/1/G/11.
46. 12 January 1875, Mary Adair to Tyndall, RI MS JT/1/A/13.
47. 13 January 1875, Rudolf Clausius to Tyndall, RI MS JT/1/TYP/7/2331–2.
48. 23 April 1875, Duke of Argyll to Tyndall, RI MS JT/1/A/101.
49. 28 April 1875, Tyndall to Duke of Argyll, RI MS JT/1/TYP/1/75.
50. Tyndall, J., 'Typhoid Fever', *The Times* (9 November 1874): p. 9.
51. 9 November 1874, Thomas Huxley to Tyndall, RI MS JT/1/TYP/9/3042.
52. Tyndall, J., 'Typhoid Fever', *The Times* (11 November 1874): p. 5.
53. See Worboys (2000).
54. Tyndall (1875); Tyndall, J., 'On acoustic reversibility', *Proc. Roy. Inst.* 7 (1875): p. 344–50.
55. 25 April 1875, Tyndall to Louisa Hamilton, RI MS JT/1/T/479. The 'scrap of doggrel' would have been 'Brave Hills of Thuring', the poem he sent to Maxwell.
56. 6 May 1875, Tyndall to Thomas Carlyle, RI MS JT/1/T/153.
57. Tyndall, J., 'Whitworth's Planes, Standard Measure and Guns', *Proc. Roy. Inst.* 7 (1875): p. 524–39.
58. 28 March 1884, Tyndall to Thomas Hirst, RI MS JT/1/T/883.
59. *On the Freedom of Science* and *The Right of One's Own Conviction*.
60. 9 July 1875, Tyndall to Olga Novikoff, RI MS JT/1/TYP/3/937–8.
61. 13 July 1875, Tyndall to Eliza Spottiswoode, RI MS JT/1/TYP/3/1249.
62. 29 August 1875, Tyndall to Thomas Huxley, RI MS JT/1/TYP/9/3047.
63. 27 June 1875, Tyndall to Heinrich Debus, RI MS JT/1/T/272.
64. See 16 July 1875 observations, RI MS JT/3/35.
65. 8 July 1874, Tyndall to Thomas Hirst, RI MS JT/1/HTYP/631.
66. 26 August 1875, Tyndall to Thomas Hirst, RI MS JT/1/T/718.
67. 6 December 1874?, Joseph Henry to Tyndall, Smithsonian RU 33, Vol 52, p. 287–90.
68. 26 July 1875, Joseph Henry to Tyndall, DRAFT Smithsonian Institution Archives, Record Unit 7001 Microfilm Reel 6.
69. Tyndall, J., 'Acoustic cloudiness', *Nature* 13 (25 November 1875): p. 72–3.
70. 2 September 1875, Hector Tyndale to Joseph Henry, RI MS JT/1/T/78.
71. 22 September 1875, Tyndall to Hector Tyndale, RI MS JT/1/T/1458.
72. 9 July 1875, Tyndall to Olga Novikoff, RI MS JT/1/TYP/3/937–8.
73. 28 August 1875, enclosure in Tyndall to Mary Adair, RI MS JT/1/T/88.
74. Tyndall, J., '"Materialism" and its opponents', *Fortnightly Review* 24 (November 1875): p. 579–99. Also published in Tyndall (1879) as 'James Martineau and the Belfast Address'.
75. It was published in the 5th edition of *Fragments of Science* in 1876.
76. 13 September 1875, George Forbes to Tyndall, RI MS JT/1/F/26.

Chapter 21: Floating Matter of the Air (1875–1876)

1. Tyndall (1870b).
2. 6 October 1875, Tyndall to Thomas Huxley, RI MS JT/1/TYP/9/3047.
3. Tyndall (1881): p. 60.
4. 17 November 1875, Tyndall to Thomas Huxley, RI MS JT/1/TYP/9/3025.
5. Friday (1974).

6. Tyndall (1881): p. 22.
7. 5 November 1875, Tyndall to William Lecky, Trin Coll Dublin–13.
8. 26 January 1880, *Report by D and T Stevenson I, Lighthouse characteristics. Copy of correspondence on recent suggestions respecting the characteristics of lights in lighthouses*, PP 1880 (168). The English, Scottish and Irish Commissions were equally sceptical of the use of Morse code, but the Irish took the opportunity to promote their use of group flashing gas lights.
9. 5 November 1875 Tyndall to William Thomson, RI MS JT/1/TYP/5/1562–3.
10. 17 November 1875 William Thomson to Tyndall, RI MS JT/1/T/20.
11. 22 September 1875, Tyndall to Thomas Farrer, NA MT 10/211. Lighthouses, Illumination by means of gas. H6704.
12. 2 December 1875, Tyndall to Cecil Trevor, NA MT 10/211. Lighthouses, Illumination by means of gas. H8495.
13. 11 October 1875, Tyndall to John Hall Gladstone, RI MS JT/1/TYP/1/410.
14. 12 October 1875, John Hall Gladstone to Tyndall, RI MS JT/1/TYP/1/411.
15. 13 October 1875, Tyndall to John Hall Gladstone, RI MS JT/1/TYP/1/412.
16. 30 October 1875, Tyndall to William Spottiswoode, RI MS JT/1/T/1303.
17. 13 October 1875, Tyndall to Olga Novikoff, RI MS JT/1/T/392.
18. 8 December 1875, Tyndall to Lady Emily Peel, RI MS JT/1/TYP/3/971.
19. 12 December 1875, Tyndall to Olga Novikoff, RI MS JT/1/T/389.
20. 12 December 1875, Tyndall to Sophie Frankland, Frankland Letters, 2030, John Rylands Library, University of Manchester.
21. Tyndall, J., 'The Vatican and Physics', *The Times* (18 December 1875): p. 7.
22. 30 December 1875, Thomas Hirst to Tyndall, RI MS JT/1/H/260; 25 December 1875, Tyndall to Olga Novikoff, RI MS JT/1/T/390.
23. Tyndall (1876).
24. 31 December 1875, Duke of Argyll to Tyndall, RI MS JT/1/A/102.
25. 10 January 1876, Tyndall to Duke of Argyll, RI MS JT/1/T/98.
26. White, W., (ed) *The Journals of Walter White* (London, 1898): 31 October 1868, p. 210.
27. 2 November 1891, Duke of Argyll to Tyndall, RI MS JT/1/A/114.
28. 5 November 1891, Tyndall to Duke of Argyll, RI MS JT/1/T/102; 4 November 1891, Tyndall to Duke of Argyll, RI MS JT/1/TYP/1/103.
29. Tyndall, J., 'On the Optical Condition of the Atmosphere in its Bearings on Putrefaction and Infection', *Proc. Roy. Inst.* 8, (1876): p. 6–27.
30. Tyndall, J., 'Spontaneous Generation', *The Times* (28 January 1876): p. 12.
31. 31 January 1876, Rudolf Clausius to Tyndall, RI MS JT/1/C/30.
32. 2 February 1876, Tyndall to Charles Darwin, RI MS JT/1/TYP/9/2839.
33. Tyndall, J., 'The Germ Theory', *The Times* (4 February 1876): p. 11.
34. Tyndall, J., 'The Germ Theory', *The Lancet* 107 (12 February 1876): p. 262–3; Tyndall, J., 'Reply to Dr. Charlton Bastian's Remarks on the Development of Germs in Infusions', *BMJ* 12 (February 1876): p. 188–90; Tyndall, J., 'Professor Tyndall on germs', *Nature* 13 (17 February 1876): p. 305.
35. 8 February 1876, Louis Pasteur to Tyndall, RI MS JT/1/P/7.
36. 25 February 1876, Tyndall to Thomas Huxley, RI MS JT/1/TYP/9/3058.
37. Tyndall (1876a).
38. 24 January 1876, Hirst Journal, RI MS JT/2/32d/2024.
39. 25 December 1876, Tyndall to Hector Tyndale, RI MS JT/1/T/1460.
40. William Kinglake to Olga Novikoff, quoted in Stead (1909): p. 177.
41. N.d. *c*.1876, Mary Hamilton to Emma Hamilton, RI MS LT/9/8.

42. Late December 1875/early January 1876, Louisa Hamilton to Lady Claud Hamilton, Elton Hall, FAM123/LT14.

43. 10 January 1876, Louisa Hamilton to Lady Claud Hamilton, Elton Hall, FAM123/LT17.

44. 23 January 1876, Tyndall to Herbert Spencer, RI MS JT/1/T/1290.

45. 23 January 1876, Tyndall to Miss Moore, RI MS JT/1/T/1078.

46. 12 February 1876, Tyndall to (possibly) Constance Gordon-Cumming, RI MS JT/1/TYP/1/306.

47. 27 January 1876, Tyndall to Thomas Carlyle, RI MS JT/1/T/154.

48. 8 January 1869, Tyndall to Mary Somerville, Bodleian Libraries, University of Oxford, Dep c.372 (MST-1).

49. See e.g. Gooday (2015).

50. 29 January 1875, Tyndall to Robert Bunsen, Library of the University of Heidelberg, UB-D229/11 #1.

51. 4 February 1876, Charles Darwin to Tyndall, RI MS JT/1/TYP/9/2840.

52. 4 February 1876, Tyndall to Thomas Huxley, RI MS JT/1/TYP/9/3055.

53. 23 January 1876, Tyndall to Hector Tyndale, RI MS JT/1/T/1459.

54. 24 February 1876, Herbert Spencer to Tyndall, University of London—145.791/18.

55. 22 February 1876, Tyndall to Herbert Spencer, University of London—145.791/113.

56. 29 February 1876, Hirst Journal, RI MS JT/2/32d/2027.

57. 31 March 1876, Tyndall to Mrs Steuart, RI MS JT/1/TYP/10/3430; Tyndall (1876a).

58. 17 March 1876, Tyndall to William Spottiswoode RI MS JT/1/TYP/3/1254.

59. RI MS JT/6/4a/19/153.

60. 'Loan Collection of Scientific Apparatus', The Bristol Mercury (20 May 1876).

61. RA VIC/MAIN/QVJ (W), 13 May 1876: vol. 65, p. 103–4.

62. RA VIC/MAIN/QVJ (W), 23 March 1870: vol, 59, p. 57.

63. 4 December 1878, Charles Darwin to Tyndall, RI MS JT/1/TYP/9/2843.

64. 5 December 1878, Tyndall to Charles Darwin, RI MS JT/1/TYP/9/2844; 22 December 1878, Charles Darwin to Tyndall, RI MS JT/1/TYP/9/2845.

65. See Rudwick, M., 'The Parallel Roads of Glen Roy' (2009): https://www.darwinproject. ac.uk/sites/default/files/Rudwick_Glen-Roy-field-guide_DCP.pdf (accessed 12 August 2017).

66. Tyndall, J., 'The Parallel Roads of Glen Roy', Proc. Roy. Inst. 8 (1876): p. 233–46.

67. 20 June 1876, Thomas Jamieson to Tyndall, RI MS JT/1/J/4.

68. 28 March 1877, Thomas Jamieson to Tyndall, RI MS JT/1/J/5.

69. Clark (1959): p. 102. Earlier, Miss Brevoort had sneered: 'She is very plain and thirty if a day old. He looks old and ghastly'.

70. 24 July 1876, Tyndall to Frederick Pollock, RI MS JT/1/TYP/6/2158.

71. 24 July 1876, Tyndall to Frederick Pollock, RI MS JT1//TYP/6/2158.

72. 5 August 1876, Tyndall to Thomas Hirst, RI MS JT/1/T/720.

73. 21 August 1876, Tyndall to Louis Pasteur, CP 106–7; 25 August 1876, Tyndall to British Medical Journal 2, (September 1876): p. 321.

74. 30 August 1876, Tyndall to William Spottiswoode, RI MS JT/1/TYP/3/1257–8.

75. 18 September 1876, Tyndall to Hermann Helmholtz, RI MS JT/1/T/495.

76. 29 September 1876, Hermann Helmholtz to Tyndall, RI MS JT/1/H/54.

Chapter 22: Contamination (1876–1878)

1. 7 October 1876, Louisa Journal, RI MS JT/2/23.

2. 22 October 1876, Louis Pasteur to Tyndall, RI MS JT/1/P/14.

3. 27 October 1876, Tyndall to Louis Pasteur, CP 116–17.

4. Tyndall, J., 'Fermentation, and its bearings on the phenomena of disease', *Fortnightly Review* 20 (1 November 1876): p. 547–72.

5. The Cruelty to Animals Act was passed in 1876; see also Turner (1993): p. 2067.

6. 11 November 1876, Tyndall to Thomas Hirst, RI MS JT/1/T/723.

7. 27 October 1876, Thomas Huxley to Tyndall, RI MS JT/1/TYP/9/3064.

8. 18 November 1876, Hirst Journal, RI MS JT/2/32d/2042.

9. 24 February 1877, Tyndall to Thomas Hirst, RI MS JT/1/T/730. Indeed, Tyndall left this sum in his Will.

10. 2 March 1877, Thomas Hirst to Tyndall, RI MS JT/H/274.

11. 27 November 1876, Tyndall to Robin Allen, NA MT 10/249. Papers re Fog Signals. H7393.

12. 17 December 1876, Tyndall to Thomas Hirst, RI MS JT/1/T/727.

13. 17 December 1876, Tyndall to Thomas Hirst, RI MS JT/1/T/727.

14. 25 December 1876, Tyndall to Hector Tyndale, RI MS JT/1/T/1460.

15. 5 January 1877, Tyndall to Thomas Hirst, RI MS JT/1/T/728.

16. 2 January 1877, Louisa Journal, RI MS JT/2/23.

17. 1 December 1876, Joseph Hooker to Tyndall, RI MS JT/1/TYP/8/2735.

18. Tyndall (1876b).

19. 'The Spontaneous Generation Question', *Nature* 15 (1 February 1877): p. 302–3.

20. 14 January 1877, Thomas Huxley to Tyndall, RI MS JT/1/TYP/9/3068.

21. 11 November 1876, Louis Pasteur to Tyndall, RI MS JT/1/P/15.

22. 1 December 1876, Louis Pasteur to Tyndall, RI MS JT/1/P/17.

23. Tyndall (1877a).

24. Summarized in Tyndall, J., 'Researches on the Deportment and Vital Persistence of Putrefactive and Infective Organisms from a Physical Point of View', *Proc. Roy. Inst.* 8 (1877): 467–77; Tyndall, J., 'A lecture on a combat with an infective atmosphere', *British Medical Journal* 1 (27 January 1877): p. 95–9.

25. Jowett Papers, Journal No. 9, August 1876 to July 1877: I H30 96, Balliol College, Oxford.

26. Tyndall, J. 'Virchow and Evolution', *Nineteenth Century* 4 (November 1878): p. 809–33.

27. Jowett Papers, Journal No. 9, August 1876 to July 1877: I H30 f57, Balliol College, Oxford.

28. Conway (1904): vol. 2, p. 37.

29. 20 January 1877, Joseph Hooker to Tyndall, RI MS JT/1/TYP/8/2738–9.

30. 23 January 1877, Tyndall to Joseph Hooker, RI MS JT/1/TYP/8/2740.

31. 14 February 1877, Tyndall to Thomas Huxley, Tyndall (1877b).

32. 18 January 1877, Louis Pasteur to Tyndall, RI MS JT/1/P/24; 19 January 1877, Tyndall to Louis Pasteur, CP 142–3; 26 January 1877, Tyndall to Louis Pasteur, CP 146–7.

33. 27 January 1877, Louis Pasteur to Tyndall, RI MS JT/1/P/26.

34. 14 June 1877, Louis Pasteur to Tyndall, RI MS JT/1/P/33.

35. 15 February 1877, Louis Pasteur to Tyndall, RI MS JT/1/P/27; 16 February 1877, Tyndall to Louis Pasteur, RI MS JT/1/T/1102.

36. 14 November 1884, Claude Penley to Tyndall, Elton Hall FAM124.

37. Tyndall, J., 'Researches on the Deportment and Vital Persistence of Putrefactive and Infective Organisms from a Physical Point of View', *Proc. Roy. Inst.* 8 (1877): p. 467–77.

38. Tyndall (1877c); Tyndall (1877d).

39. 6 June 1877, Emil du Bois-Reymond to Tyndall, RI MS JT/1/D/153.

40. 9 June 1877, Tyndall to Emil du Bois-Reymond, RI MS JT/1/T/424.

41. 1 October 1880, Emil Du Bois-Reymond to Tyndall, RI MS JT/1/D/163.

42. 4 July 1877, Louisa Journal, RI MS JT/2/23.

43. 18 September 1877, Tyndall to Thomas Huxley, Tyndall (1877e).

44. Henry Bastian to Editor of *The Times*, 'Professor Tyndall on Germs', *The Times* (12 June 1877): p. 4; Tyndall to Editor of *The Times*, 'Professor Tyndall on Germs', *The Times* (18 June 1877): p. 6; Henry Bastian to Editor of *The Times*, 'Professor Tyndall on Germs', *The Times* (26 June 1877): p. 4; Tyndall to Editor of *The Times*, 'The Germ Theory', *The Times* (24 July 1877): p. 4.

45. 6 August 1877, Louis Pasteur to Tyndall, RI MS JT/1/P/34.

46. Worboys (2000): p. 278.

47. 'Clinical Society: President's Address', *Medical and Times Gazette* 50 (20 February 1875): p. 195.

48. McCarthy (1872): p. 239.

49. Tyndall (1881).

50. Knowles, one of the founders with Tennyson of the Metaphysical Society in 1869, and editor shortly afterwards of the *Contemporary Review*, had resigned that editorship in 1877, after objections by the proprietors to articles attacking theism, to found his new journal; Tyndall, J., 'Spontaneous Generation', *Nineteenth Century* 3 (January 1878): p. 22–47.

51. 17 August 1877, Tyndall to Thomas Hirst, BL Add MS 63092, ff. 101–2.

52. 29 August 1877, Louisa Journal, RI MS JT/2/23.

53. Tyndall, J., 'Science and Man', *Fortnightly Review* 22 (1 November 1877): p. 593–617.

54. 20 October 1877, Charles Darwin to Tyndall, RI MS JT/1/TYP/9/2842.

55. Tyndall (1892): p. 37.

56. Tyndall (1872a), §376; Tyndall (1872a), §318–24.

57. 11 November 1877, Tyndall to William Spottiswoode, RI MS JT/1/T/1313.

58. 11 November 1877, Tyndall to William Spottiswoode, RI MS JT/1/T/1313.

59. Tyndall (1878a).

60. 18 January 1878, Louisa Journal, RI MS JT/2/23.

61. Tyndall experimental notebook, RI MS JT/3/28.

62. Lionel died aged thirty-two on 20 April 1886, at Aden, on board the *Chusan*.

63. 27 February 1878, Louisa Journal, RI MS JT/2/23.

64. 27 January 1878, Louisa Journal, RI MS JT/2/23.

65. 18 March 1878, Tyndall to William Spottiswoode, RI MS JT/1/T/1317.

66. Tyndall, J., 'Recent Experiments on Fog-Signals', *Proc. Roy. Inst.* 8 (1878): p. 543–58; Tyndall (1878b).

67. 14 January 1875, Tyndall to Olga Novikoff, RI MS JT/1/TYP/3/951.

68. 17 August 1878, Tyndall to Heinrich Debus, RI MS JT/1/T/276; 6 November 1878, Heinrich Debus to Tyndall, RI MS JT/1/D/46.

69. 4 April 1878, Tyndall to Thomas Huxley, RI MS JT/1/TYP/9/3075.

70. 30 March 1878, Tyndall to William Spottiswoode, RI MS JT/1/T/1318.

71. 18 March 1885, Tyndall Journal, RI MS JT/2/18.

72. N.d. April 1878, Hirst Journal, RI MS JT/2/32d/2074.

73. 3 May 1878, Tyndall to Thomas Huxley, RI MS JT/1/TYP/9/3078.

74. 19 May 1878, Tyndall to Thomas Huxley, RI MS JT/1/TYP/9/3081.

75. 6 May 1878, Thomas Huxley to Tyndall, RI MS JT/1/TYP/9/3079.

76. 30 December 1877, Tyndall to William Spottiswoode, RI MS JT/1/T/1314.

77. 15 June 1878, Tyndall to Rudolf Clausius, RI MS JT/1/T/212.

78. 8 July 1878, Tyndall to Heinrich Debus, RI MS JT/1/T/275.

79. 18 July 1878, Tyndall to Thomas Hirst, RI MS JT/1/HTYP/651–52a; BL Add MS 63092, ff. 103–4.

80. Downes, A. and T. P. Blunt, 'Researches on the Effect of Light on Bacteria and other Organisms', *Proc. Roy. Soc. Lond.* 26 (1877): p. 488–500.

81. Tyndall (1878c).

82. 24 September 1878, Tyndall to Emma Tyndall, RI MS JT/1/TYP/10/3318.

Chapter 23: Electric Lights and Mining Accidents (1879–1886)

1. 21 June 1878, Tyndall to William Spottiswoode, RI MS JT/1/TYP/3/1273.
2. Tyndall, J., 'The Electric Light', *Proc. Roy. Inst.* 9 (1879): p. 1–24.
3. See Gooday (2008) for a detailed analysis.
4. 'Professor Tyndall on the Electric Light', *Saturday Review* 47 (25 January 1879): p. 105–6.
5. 'The "Loud-Speaking" Telephone', *Saturday Review* (5 April 1879): p. 423–4.
6. 29 March 1878, Emil Du Bois-Reymond to Tyndall, RI MS JT/1/D/155; 'The Phonograph at the Royal Institution', *The Graphic*, 16 March 1878, pp. 259, 262, 268 (image).
7. 6 November 1877, Emil du Bois-Reymond to Tyndall, RI MS JT/1/D/154.
8. 7 December 1880, Tyndall to Hermann Helmholtz, RI MS JT/1/T/501; 11 November 1880, Tyndall to Du Bois-Reymond, Tyndall Journal, RI MS JT/8/4; 11 November 1880, Tyndall to Hermann Helmholtz, RI MS JT/1/T/500. For a more extensive discussion of this issue see Arapostathis and Gooday (2013): p. 92–3.
9. 13 November 1880, Emil Du Bois-Reymond to Tyndall, RI MS JT/1/D/164.
10. *Accidents in mines. Preliminary report of Her Majesty's commissioners appointed to inquire into accidents in mines, and the possible means of preventing their occurrence or limiting their disastrous consequences, together with the evidence and an index.* PP 1881 (C 3036).
11. The two were unsuccessfully sued by Alfred Nobel for patent infringement.
12. 29 March 1879, Tyndall to Hector Tyndale, RI MS JT/1/T/1462.
13. 22 March 1884, *Accidents in Mines—Report of the Royal Commission.* HC Deb 21 March 1884 vol. 286 cc450–1.
14. 23 February 1885, Tyndall Journal, RI MS JT/2/18.
15. 17 July 1884, Louisa Journal, RI MS JT/2/19.
16. 4 February 1851, Tyndall to Michael Faraday, Letter 2379 in James (1999).
17. *Report from the Select Committee on Lighting by Electricity*, PP 1878–9 (224).
18. 8 April 1879, Lyon Playfair to Tyndall, RI MS JT/1/P/123.
19. 2 June 1879, Tyndall to Mrs Steuart, RI MS JT/1/TYP/10/3448.
20. Tyndall (1879).
21. 2 July 1879, Tyndall to Lady Claud Hamilton (Birdie), RI MS JT/1/T/141.
22. 27 July 1879, Tyndall to Thomas Hirst, BL Add MS 63092, ff. 105–7.
23. 18 July 1879?, Tyndall to Thomas Hirst, RI MS JT/1/T/910.
24. See Barton (1990) and 30 June 1883, Thomas Huxley to Joseph Hooker, ICHC 2.250; 2 July 1883, Joseph Hooker to Thomas Huxley, ICHC 2.254; 11 July 1883, Thomas Huxley to Joseph Hooker, ICHC 2.258.
25. 14 September 1870, Hirst Journal, RI MS JT/2/32d/1884.
26. 22 October 1870, Tyndall to Peter Gassiot, Tyndall Journal, RI MS JT/2/13c/1371–2.
27. 24 October 1870, Peter Gassiot to Tyndall, Tyndall Journal, RI MS JT/2/13c/1372–3.
28. 10 November 1879, Rudolf Clausius to Tyndall, RI MS JT/1/TYP/7/2349–50.
29. Tyndall was not then on Council. Bunsen was proposed by Williamson and Yorke. He was awarded the medal after an initial tie in votes with Louis Agassiz.
30. 15 May 1873, Thomas Huxley to Tyndall, RI MS JT/1/TYP/9/2981 (ICHC 1.101 & 8.142)
31. 'Professor Tyndall on Electricity', *The Times* (16 February 1880): p. 6.
32. Tyndall, J. 'Goethe's "Farbenlehre"', *Proc. Roy. Inst.* 9 (1880): p. 340–58. The text was published in *The Fortnightly Review* in April (Tyndall, J., 'Goethe's "Farbenlehre"', *The Fortnightly Review* 33 (April 1880): p. 471–90).
33. This copy is now in Carlyle's House in Chelsea.
34. 10 August 1880, Emil Du Bois-Reymond to Tyndall, RI MS JT/1/T/415.
35. 19 September 1880, Tyndall to William Spottiswoode, BL Add MS 53715, f. 39.

36. 23 March 1880, William S. Forster to Tyndall, Elton Hall, FAM124.
37. 8 July 1880, Sir Frederick Pollock to Tyndall, RI MS JT/1/P/138.
38. 7 August 1880, Sir Frederick Pollock to Tyndall, RI MS JT/1/P/139.
39. 1 September 1880, Sir John Lubbock to Tyndall, RI MS JT/1/L/43.
40. 20 October 1879, Hirst Journal, RI MS JT/2/32d/2091.
41. 18 November 1874 Tyndall to Mary Egerton, RI MS JT/1/TYP/1/399.
42. 29 November 1874, Harriet Hooker to Tyndall, RI MS JT/1/TYP/8/2725.
43. 26 June 1876, Joseph Hooker to Tyndall, RI MS JT/1/TYP/8/2733.
44. 17 December 1880, Henry Babbage to Tyndall, RI MS JT/1/B/2.
45. 19 December 1880, Tyndall to Henry Babbage, BL Add MS 37199, f. 593. In retirement, Henry built six demonstration pieces for Babbage's Difference Engine No. 1 and an experimental four-function calculator for the mill for the Analytical Engine.
46. 24 December 1880, Herbert Spencer to Tyndall, RI MS JT/1/TYP/3/1172.
47. Sharp (1926): p. 81.

Chapter 24: Hindhead (1880–1883)

1. 25 November 1880, William Spottiswoode to Tyndall, RI MS JT/1/S/206.
2. 7 December 1880, Tyndall Journal, RI MS JT/3/48/286.
3. Tyndall (1881a).
4. Tyndall (1881b).
5. 9 February 1881, Tyndall to Canon Westcott, WT MS. 7777/25.
6. 11 February 1881, Louisa Journal, RI MS JT/2/25.
7. Tyndall to *The Times*, 'Mr. Carlyle', *The Times* (4 May 1881): p. 13.
8. 12 April 1881, Thomas Hirst to Tyndall, RI MS JT/H/287.
9. 14 May 1881, Anna Helmholtz to Tyndall, RI MS JT/1/H/38.
10. 14 April 1881, Tyndall to Thomas Hirst, RI MS JT/1/T/735.
11. Mori Arinori, an educational innovator, read Spencer's work extensively. See Swale, A., *The Political Thought of Mori Arinori: A Study of Meiji Conservatism* (Richmond: Japan Library, 2000).
12. 5 June 1881, Tyndall to Walter De la Rue, RI MS AD/03/D/04, RI MS CG3L/14.
13. 6 June 1881, RI Managers' Minutes (1874–1903), XIII, p. 267.
14. 16 June 1881, Tyndall to Walter De la Rue, RI MS AD/03/D/04, RI MS CG3L/15.
15. 1 July 1881, Tyndall to Walter De la Rue, RI MS AD/03/D/04, RI MS CG3L/16.
16. 20 July 1881, Tyndall to Thomas Hirst, RI MS JT/1/HTYP/658.
17. 'From the Alps', *Pall Mall Gazette* (22 July 1881): p. 11.
18. 20 August 1881, Tyndall to Lord Houghton, TCC, Houghton.26/79.
19. Tyndall (1881c).
20. Downes, A. and T. P. Blunt, 'Researches on the Effect of Light upon Bacteria and other Organisms', *Proc. Roy. Soc. Lond.* 26 (1877): 488–500.
21. 28 November 1881, Emil Du Bois-Reymond to Tyndall, RI MS JT/1/D/166; 1 December 1881, Tyndall to Emil Du Bois-Reymond, RI MS JT/1/T/418.
22. 25 April 1882, George Darwin to Tyndall, RI MS JT/1/TYP/9/2851.
23. 26 April 1882, Louisa Journal, RI MS JT/2/26.
24. 17 October 1881, Louisa Journal, RI MS JT/2/25.
25. 6 January 1884, Tyndall Journal, RI MS JT/2/19.
26. A proposal in 1867 by the chemist Stenhouse, to allow members of the public into meetings, was rejected by the Council. 21 March 1867, RS, 'Minutes', p. 361.
27. 24 November 1881, Louisa Journal, RI MS JT/2/25.

28. Tyndall (1882a).
29. They are referred to in Tyndall (1872): p. 394.
30. Tyndall (1882a), p. 353–4.
31. 25 December 1881, Louisa Journal, RI MS JT/2/25.
32. 17 January 1882, Tyndall to George Stokes, RI MS JT/1/T/1411.
33. 17 January 1882, George Stokes to Tyndall, RI MS JT/1/S/268.
34. 21 March 1882, George Stokes to Tyndall, RI MS JT/1/S/270.
35. Tyndall (1882b).
36. 24 March 1882, Tyndall to George Stokes, RI MS JT/1/T/1413.
37. See Gooday, G., 'Instrumentation and Interpretation: Managing and Representing the Working Environments of Victorian Experimental Science', in Lightman (1997): p. 409–37.
38. 15 August 1882, Lyon Playfair to Tyndall, RI MS JT/1/P/126.
39. 5 March 1882, Louisa Journal, RI MS JT/2/26, 14 March 1882, Louisa Journal, RI MS JT/2/26.
40. 24 December 1888, Tyndall to Thomas Hirst, RI MS JT/1/T/801.
41. The Association for the Advancement of Medicine by Research developed from the Science Defence Association, established in 1881 with Darwin's support. See Richards. E. 'Redrawing the Boundaries: Darwinian Science and Victorian Women Intellectuals', in Lightman (1997): p. 135; and French (1975).
42. 28 March 1882, Louisa Journal, RI MS JT/2/26. Tyndall did not become a member of the Council (Huxley and Hooker did), which met for the first time on 20 April 1882. The Council was chaired by Sir William Jenner, who had criticized Tyndall's style some years earlier. Minute book of the Association for the Advancement of Medicine by Research, WT MSS 5210, p. 2.
43. Tyndall to *The Times*, 'Tubercular Disease', *The Times* (22 April 1882): p. 10.
44. Tyndall, J., 'Methods and Hopes of Experimental Physiology', *Pall Mall Gazette* (30 October 1883).
45. N.d. n.m. 1882, Tyndall to Emma Tyndall, RI MS JT/1/TYP10/3320.
46. 7 January 1884, Tyndall to Rudolf Clausius, RI MS JT/1/T/225.
47. According to 'On Rainbows'.
48. 22 May 1882, Tyndall to Joseph Hooker, RI MS JT/1/TYP/8/2756–67.
49. 20 May 1882, Joseph Hooker to Louisa Tyndall, RI MS JT/1/TYP/8/2755.
50. 12 June 1882, Joseph Hooker to Tyndall, RI MS JT/1/H/501.
51. 10 July 1882, Louisa Journal, RI MS JT/2/26.
52. 25 July 1882, Tyndall to Thomas Hirst, RI MS JT/1/HTYP/659–60; BL Add MS 63092, ff. 113–6.
53. 25 July 1882, Tyndall to Thomas Hirst, RI MS JT/1/HTYP/659–60; BL Add MS 63092, ff. 113–6.
54. 25 July 1882, Tyndall to Thomas Hirst, RI MS JT/1/HTYP/659–60; BL Add MS 63092, ff. 113–6.
55. 28 July 1882, Thomas Hirst to Tyndall, RI MS JT/H/290.
56. Gos (1948): p. 125–34.
57. Douglas married Lady Frances Hely Hutchinson on 6 July 1882, after an attempt by Lady Claud Hamilton to match-make with Maysie, daughter of the second Lady Ashburton, had come to naught. His great-grandson, Sir William Proby Bt, is the current occupant of Elton Hall.
58. 11 August 1882, Tyndall to Heinrich Debus, RI MS JT/1/T/281.
59. 'Professor Tyndall on our Responsibilities in Egypt', *Pall Mall Gazette* (8 February 1884): p. 8.
60. 'Professor Tyndall's Politics', *Pall Mall Gazette* (14 February 1884): p. 2.

61. 'Professor Tyndall's Politics', *Pall Mall Gazette* (15 February 1884): p. 9.
62. 25 February 1884, Tyndall Journal, RI MS JT/2/19.
63. 'Egyptian Policy of the Government', *The Times* (3 March 1884): p. 7.
64. 6 February 1885, Tyndall Journal, RI MS JT/2/18.
65. 3 April 1885, Tyndall Journal, RI MS JT/2/18.
66. Tyndall, J., 'Atoms, Molecules and Ether Waves', *Longman's Magazine* 1 (November 1882): p. 29–40.
67. 17 November 1882, Duke of Argyll to Tyndall, RI MS JT/1/TYP/1/91–2.
68. 22 January 1883, Tyndall to Duke of Argyll, RI MS JT/1/TYP/1/93–4.
69. 7 October 1882, Louisa Journal, RI MS JT/2/26.
70. 17 September 1882, Tyndall to Heinrich Debus, RI MS JT/1/T/283.
71. Tyndall to *The Times*, 'The Weather in the Alps', *The Times* (22 September 1882): p. 8.
72. 14 November 1882, Tyndall to Thomas Farrer, WT MS. 7777/88–88a.
73. 14 April 1878, Tyndall to Thomas Carlyle, RI MS JT/1/T/157.
74. Tyndall, J., 'Personal recollections of Thomas Carlyle', *Fortnightly Review* 50 (January 1890): p. 293–309. Also in Tyndall (1892): p. 347–91.
75. 20 September 1882, Tyndall to Thomas Hirst, RI MS JT/1/HTYP/661.
76. 20 October 1879, W Lees to Board of Trade, *Copeland Island light*, PP 1880 (343).
77. 30 January 1882, Tyndall to Joseph Chamberlain, *Irish lights (Copeland or Mew Island Station)*, PP 1882 (32) (330).
78. 8 March 1882, Tyndall to J. Inglis, *Lighthouse Illuminants*, PP 1883 (168) (263), p. 8–10.
79. Ibid p. 10.
80. 28 October 1882, Trinity House to Board of Trade, *Lighthouse Illuminants*, PP 1883 (168) (263), p. 24–5.
81. The committee, of which the membership was recommended by John Simon, the reforming Chief Medical Officer from 1855 to 1876, consisted of Simon, Tyndall, Burdon-Sanderson and George Buchanan. The £1,000 prize was intended to be quadriennial, and there were three scholarships of £250 a year.
82. 12 January 1883, Louisa Journal, RI MS JT/2/27.
83. Tyndall (1883a).
84. Tyndall, J., 'Thoughts on Radiation, Theoretical and Practical', *Proc. Roy. Inst.* 10 (1883): p. 253–65; Tyndall, J., 'On Radiation', *Contemporary Review* 43, (May 1883): p. 660–73.
85. Tyndall (1883b).
86. 22 January 1883, Committee on Illuminants (Nisbet) to Board of Trade (Farrer), *Lighthouse Illuminants*, PP 1883 (168) (263), p. 32.
87. 30 January 1883, Board of Trade (Trevor) to Trinity House, *Lighthouse Illuminants*, PP 1883 (168) (263), p. 33.
88. 31 January 1883, Thomas Stevenson to Board of Trade, *Lighthouse Illuminants*, PP 1883 (168) (263), p. 37.
89. 2 February 1883, Robert Ball to Board of Trade, *Lighthouse Illuminants*, PP 1883 (168) (263), p. 37.
90. 3 February 1883, Tyndall to Board of Trade (Trevor) *Lighthouse Illuminants*, PP 1883 (168) (263), p. 38.
91. 8 February 1883, Board of Trade (Farrer) to Tyndall, *Lighthouse Illuminants*, PP 1883 (168) (263), p. 38.
92. 9 February 1883, Tyndall to Lord Meath, RI MS JT/1/T/1099; 9 February 1883, Tyndall to Board of Trade (Farrer), *Lighthouse Illuminants*, PP 1883 (168) (263), p. 39; 9 February 1883 Commissioners of Irish Lights (Armstrong) to Board of Trade, *Lighthouse Illuminants*, PP 1883 (168) (263).

93. 17 February 1883, Tyndall to Board of Trade (Chamberlain), *Lighthouse Illuminants*, PP 1883 (168) (263).

94. 20 February 1883 Board of Trade (Chamberlain) to Tyndall, *Lighthouse Illuminants*, PP 1883 (168) (263), p. 43.

95. He was eventually elected in 1887, although Tyndall did not sign the certificate.

96. 26 February 1883, Tyndall to Board of Trade (Chamberlain), *Lighthouse Illuminants*, PP 1883 (168) (263), p. 45–7.

97. 9 March 1883, Commissioners of Irish Lights (Armstrong) to Board of Trade (Farrer), *Lighthouse Illuminants*, PP 1883 (168) (263), p. 48.

98. 17 March 1883, Board of Trade (Chamberlain) to Tyndall, *Lighthouse Illuminants*, PP 1883 (168) (263), p. 49.

99. 28 March 1883, Tyndall to Board of Trade (Chamberlain), *Lighthouse Illuminants*, PP 1883 (168) (263), p. 49–53.

100. 3 April 1883, Board of Trade (Chamberlain) to Tyndall, *Lighthouse Illuminants*, PP 1883 (168) (263), p. 55.

Chapter 25: Rainbows and Lighthouses (1883–1885)

1. 12 June 1883, Tyndall Journal, RI MS JT/2/15.

2. 14 July 1883, Tyndall Journal, RI MS JT/2/15.

3. Tyndall, J., 'Count Rumford', *Contemporary Review* 44 (July 1883): p. 38–58.

4. 5 July 1883, Tyndall Journal, RI MS JT/2/15.

5. 5 July 1883, Hirst Journal, RI MS JT/2/32d/2130.

6. 5 May 1883, Commissioners of Irish Lights (Armstrong) to Board of Trade, *Lighthouse Illuminants*, PP 1883 (168) (263), p. 60.

7. 11 May 1883, Board of Trade (Farrer) to Commissioners of Irish Lights, *Lighthouse Illuminants*, PP 1883 (168) (263), p. 60.

8. 18 May 1883, Commissioners of Irish Lights (Armstrong) to Board of Trade, *Lighthouse Illuminants*, PP 1883 (168) (263), p. 3.

9. 25 May 1883, Commissioners of Irish Lights (Armstrong) to Board of Trade, *Lighthouse Illuminants*, PP 1883 (168) (263), p. 4.

10. 30 May 1883, Dr Ball to Board of Trade, *Lighthouse Illuminants*, PP 1883 (168) (263), p. 6; 8 June 1883, Lighthouse Illuminants Committee (E Price Edwards) to Board of Trade, *Lighthouse Illuminants*, PP 1883 (168) (263), p. 7.

11. 7 June 1883, Copy of protest handed in by Wigham, *Lighthouse Illuminants*, PP 1883 (168) (263), p. 10.

12. Tyndall to *The Times*, 'Professor Tyndall and the Board of Trade', *The Times* (22 June 1883): p. 8.

13. 10 July 1883, Board of Trade (Farrer) to Lighthouse Illuminants Committee, *Lighthouse Illuminants*, PP 1883 (168) (263), p. 13.

14. See RI JT/6/4d/183; 21 July 1883, Tyndall to Newspapers, 'The Lighthouse Question', *Lighthouse Illuminants*, PP 1884–5 (34), p. 3–6; 26 July 1883, Tyndall to Newspapers. 'The Lighthouse Question', *Lighthouse Illuminants*, PP 1884–5 (34), p. 6–9.

15. 25 August 1883, Board of Trade (Calcraft) to Trinity House and Commissioners of Northern Lighthouses, *Lighthouse Illuminants*, PP 1884–1885 (34), p. 9.

16. 6 December 1883, Trinity House (Inglis) to Board of Trade, *Lighthouse Illuminants*, PP 1884–1885 (34), p. 17.

17. 6 December 1883, Commissioners of Northern Lighthouses (Duncan) to Board of Trade, *Lighthouse Illuminants*, PP 1884–5 (34), p. 17.

NOTES

18. 20 June 1883, Tyndall Journal, RI MS JT/2/15.
19. 18 August 1883, Tyndall Journal, RI MS JT/2/15.
20. 2 September 1883, Tyndall Journal RI MS JT/2/15.
21. Tyndall, J., 'Alpine Gossip', *Pall Mall Gazette* (21 September 1883); Tyndall, J., 'Alpine Phenomena', *Pall Mall Gazette* (16 October 1883).
22. 9 September 1883, Tyndall Journal, RI MS JT/2/15.
23. 6 May 1883, Tyndall Journal, RI MS JT/2/15.
24. 11 November 1883, Escher to Tyndall, RI MS JT/1/E/109.
25. 29 November 1883, Tyndall Journal, RI MS JT/2/15.
26. 19 December 1883, Tyndall Journal, RI MS JT/2/15.
27. 1 May 1887, Tyndall to William Frederick Donkin, AC 2S 6 A. C. Committee Minutes 23/4/1885–15/5/1888.
28. 12 December 1883, Tyndall Journal, RI MS JT/2/15.
29. 8 January 1884, Tyndall Journal, RI MS JT/2/19.
30. 5 October 1883, George Stokes to Tyndall, RI MS JT/1/S/275.
31. 23 October 1883, Tyndall to George Stokes, RI MS JT/1/T/1414.
32. 24 October 1883, George Stokes to Tyndall, RI MS JT/1/S/276.
33. 9 December 1883, Tyndall to George Stokes, CUL Add 7656/RS1791.
34. Tyndall, J., 'On Rainbows', *Proc. Roy. Inst.* 10, (1884): p. 455–69.
35. Tyndall (1884a); 14 January 1884, Sir George Airy to Tyndall, RI MS JT/1/A/61; Tyndall (1884b); Tyndall (1884c).
36. Tyndall, J., 'On Rainbows', *Nineteenth Century* 15, (February 1884): p. 345–60; also 'The Rainbow and its Congeners' in Tyndall (1892), p. 199–224.
37. 27 February 1884, Tyndall Journal, RI MS JT/2/19.
38. 6 March 1884, Tyndall Journal, RI MS JT/2/19.
39. 28 March 1884, Tyndall to Thomas Hirst, RI MS JT/1/T/883.
40. 11 February 1884, Tyndall Journal, RI MS JT/2/19.
41. 15 March 1884, Tyndall Journal, RI MS JT/2/19.
42. 9 April 1884, Tyndall Journal, RI MS JT/2/19.
43. 3 July 1884, Tyndall Journal, RI MS JT/2/19.
44. 7 April 1884, Tyndall Journal, RI MS JT/2/19.
45. 22 October 1877, Duke of Argyll to Tyndall, RI MS JT/1/A/105; 29 October 1877, Duke of Argyll to Tyndall, RI MS JT/1/A/106; 27 October 1877, Tyndall to Duke of Argyll, RI MS JT/1/A/105; 31 October 1877, Tyndall to Duke of Argyll, RI MS JT/1/T/100.
46. Tyndall to *The Times*, 'Lightning Conductors', *The Times* (31 August 1887): p. 8; Tyndall to *The Times*, 'Lightning Conductors', *The Times* (13 September 1887): p. 13; Tyndall to *The Times*, 'Lightning Conductors', *The Times* (27 September 1887): p. 3.
47. 8 May 1888, Tyndall to Thomas Hirst, RI MS JT/1/T/780.
48. Tyndall to *Nature*, *Nature* 28 (17 May 1883): p. 54.
49. *c.* December 1883, Louis Pasteur To Tyndall, in Pasteur Vallery-Radot (ed.), *Correspondance de Pasteur, 1840–1895*, vol. 3 (Paris: Flammarion, 1940): p. 407–8.
50. 11 January 1884, Tyndall to Louis Pasteur, CP 163.
51. 11 April 1884, Tyndall Journal, RI MS JT/2/19.
52. 12 May 1884, Tyndall Journal, RI MS JT/2/19.
53. 3 May 1884, Tyndall Journal, RI MS JT/2/19.
54. 11 December 1884, Hirst Journal, RI MS JT/2/32d/2166.
55. 22 May 1884, Tyndall Journal, RI MS JT/2/19.
56. 8 June 1884, Tyndall to Mrs Steuart, RI MS JT/1/TYP/10/3457.

57. 9 June 1884, Tyndall to Joseph Hooker, RI MS JT/1/TYP/8/2760.

58. 7 April 1885, Rudolf Clausius to Tyndall, RI MS JT/1/TYP/7/2362.

59. He left his entire estate to Louisa, and appointed her sole executor, in his final Will dated 1 April 1891.

60. 25 November 1884, Claude Penley to Tyndall, Elton Hall FAM124.

61. Statement of Account Relating to Sale of 83 Portland Place, 1886, Elton Hall FAM116/29.

62. Summary of Hamilton Family Letters 1856–88, Elton Hall FAM162.

63. 31 August 1884, Tyndall Journal, RI MS JT/2/17.

64. 18 January 1886, Lord Lingen to Tyndall, BL Tyndall Papers, Vol. 1, ff. 45–8. Responsibility for the chapel was handed over to the local Catholic bishop in the Second World War. It retains not a trace of the English use.

65. 10 September 1884, Tyndall to Heinrich Debus, RI MS JT/1/T/285.

66. 2 September 1884, Tyndall Journal, RI MS JT/2/17.

67. 7 March 1885, Louis Pasteur to Tyndall, RI MS JT/1/P/42.

68. c. 4 October 1884, Tyndall to Thomas Hirst, RI MS JT/1/HTYP/669.

69. 6 December 1883, Tyndall Journal, RI MS JT/2/15.

70. 24 October 1884, Tyndall Journal, RI MS JT/2/16.

71. Tyndall to The Times, 'Mr. Chamberlain and Professor Tyndall', The Times (8 November 1884): p. 9.

72. Tyndall to The Times, 'Mr. Chamberlain and Professor Tyndall', The Times (12 November 1884): p. 8.

73. Tyndall to The Times, 'Mr. Chamberlain and Professor Tyndall', The Times (26 November 1884): p. 6.

74. Joseph Chamberlain to The Times, 'Professor Tyndall and Mr. Chamberlain', The Times (28 November 1884): p.8.

75. Tyndall to The Times, 'The South Foreland Experiments', The Times (13 December 1884): p. 10.

76. Tyndall to The Times, 'Lighthouse Illuminants', The Times (19 December 1884): p. 13.

77. Tyndall to The Times, 'Lighthouse Illuminants', The Times (29 December 1884): p. 3.

78. Tyndall to The Times, 'Lighthouse Illuminants', The Times (12 January 1885): p. 13.

79. Tyndall to The Times, 'Lighthouse Illuminants', The Times (12 January 1885): p. 13.

80. 9 January 1885, Tyndall Journal, RI MS JT/2/18.

81. 'Professor Tyndall on Scientific Research', Leeds Mercury (5 December 1884).

82. Actually: 'I the heir of all the ages, in the foremost files of time', from Tennyson's Locksley Hall.

83. 26 October 1884, Tyndall Journal, RI MS JT/2/16.

84. 2 November 1884, Tyndall Journal, RI MS JT/2/16.

85. 28 November 1884, Tyndall Journal, RI MS JT/2/16.

86. 22 November 1884, Tyndall Journal, RI MS JT/2/16.

87. 25 December 1884, Tyndall Journal, RI MS JT/2/16.

88. 26 December 1884, Robert Koch to Tyndall, RI MS JT/1/K/28.

89. 11 January 1885, Robert Koch to Tyndall, RI MS JT/1/K/29.

90. Tyndall, J., 'Living Contagia', Proc. Roy. Inst. 11, (1885): p. 161–7.

91. Tyndall to The Times, 'Surgery and Vivisection', The Times (20 January 1885): p. 4.

92. 19 January 1885, Tyndall to Thomas Hirst, RI MS JT/1/HTYP/671.

93. 19 January 1885, Tyndall to Thomas Hirst, RI MS JT/1/HTYP/671.

94. 25 May 1885, Charles Grant to Tyndall, William Blackwood and Sons.

95. 10 October 1885, Tyndall to Charles Grant, RI MS JT/1/T/459.

96. 14 October 1885, Tyndall to Charles Grant, RI MS JT/1/T/434.
97. 9 November 1885, Lord Lytton to Tyndall, RI MS JT/1/TYP/2/760–1.
98. 26 November 1885, James Joule to Tyndall, RI MS JT/1/J/143.
99. 6 November 1885, Hirst Journal, RI MS JT/2/32/2219. Tyndall's letter and apology if its publication caused personal offence, was published in *St James's Gazette* on 5 November 1885 and 10 November 1885 respectively (RI MS JT/6/4d/220).
100. 27 May 1885, Tyndall Journal, RI MS JT/2/18.
101. 27 May 1885, Tyndall Journal, RI MS JT/2/18.
102. 28 May 1885, Tyndall Journal, RI MS JT/2/18.
103. 29 May 1885, Tyndall Journal, RI MS JT/2/18.
104. 9 June 1885, Hirst Journal, RI MS JT/2/32e/2198.
105. Krakatoa had erupted in 1883, resulting in spectacular sunrises and sunsets; 24 September 1885, Tyndall to Thomas Hirst, RI MS JT/1/HTYP/672–3; BL Add MS 63092, ff. 131–4.
106. 25 October 1885, Tyndall Journal, RI MS JT/2/18. The thirty-seven acres were given to the National Trust after Louisa's death and are now called 'Tyndall's Wood'.
107. 26 August 1885, Tyndall Journal, RI MS JT/2/18.
108. 21 September 1885, Tyndall Journal, RI MS JT/2/18.
109. 10 November 1885, Hirst Journal, RI MS JT/2/32e/2220.
110. 14 November 1885, Hirst Journal, RI MS JT/2/32e/2222.
111. 13 November 1885, Tyndall Journal, RI MS JT/2/18.
112. Bonney (1919): p. 229.
113. 7 November 1885, Tyndall to Charles Grant, RI MS JT/1/T/436.
114. 18 November 1885, Tyndall Journal, RI MS JT/2/18.
115. Tyndall to *The Times*, 'Professor Tyndall and his Vote', *The Times* (24 November 1885): p. 13.
116. 25 November 1885, Tyndall Journal, RI MS JT/2/18.

Chapter 26: The Final Years (1886–1893)

1. 3 December 1885, Tyndall Journal, RI MS JT/2/18.
2. Tyndall, J., 'Thomas Young', *Proc. Roy. Inst.* 11 (1886): p. 553–88. Also in in Tyndall (1892): p. 248–306.
3. Thompson and Thompson (1920): p. 157.
4. 2 February 1886, Louisa Tyndall to Thomas Hirst, RI MS JT/1/HTYP/675.
5. 27 February 1886, Louisa Tyndall to Thomas Hirst, RI MS JT/1/HTYP/676.
6. 20 April 1888, Tyndall to Mrs Lecky, TCD—143.
7. 5 April 1886, RI Managers' Minutes XIV (1874–1903): p. 43.
8. 21 July 1886, Tyndall to Herbert Spencer, RI MS JT/1/TYP/3/1045.
9. 22 August 1886, Tyndall to Thomas Huxley, RI MS JT/1/TYP/9/3119 (ICHC 8.248).
10. 15 October 1866, Tyndall to Ellen Busk, RI MS JT/1/TYP/1/1/187–8.
11. 21 August 1886, Thomas Hirst to Tyndall, RI MS JT/1/H/316.
12. 18 August 1886, Tyndall to Thomas Hirst, RI MS JT/1/HTYP/681; BL Add MS 63092, ff. 135–6.
13. As Tyndall explained to Hirst: 'The Fischel is 156 Klafter, and each Klafter a square with a side of 6 feet', making the purchase about 575m²: 22 September 1886, Tyndall to Thomas Hirst, RI MS JT/1/HTYP/682–682a; BL Add MS 63092, ff. 138–41; 21 September 1886, Ludwig Salzmann to Tyndall, RI MS JT/1/S/48.
14. 1 November 1886, Hirst Journal, RI MS JT/2/32e/2811.
15. 6 December 1886, Tyndall Journal, RI MS JT/2/19.

16. 7 December 1886, Tyndall to Thomas Hirst, RI MS JT/1/T/747.

17. 30 November 1886, Tyndall Journal, RI MS JT/2/19.

18. 21 December 1886, Tyndall Journal, RI MS JT/2/19.

19. 17 January 1887, Tyndall to Thomas Hirst, RI MS JT/1/T/748.

20. 4 December 1886, Lord Lytton to Tyndall, RI MS JT/1/L/73.

21. 6 February 1887, Tyndall to Henry Pollock, RI MS AD/03/04, RI MS C93L/25; 7 February 1887, RI Managers' Minutes XIV (1874–1903): p. 77.

22. 4 March 1887, Tyndall to Frederick Pollock, RI MS JT/1/TYP/6/2174; Louisa Journal, RI MS JT/2/20.

23. 18 April 1887, Thomas Hirst to Tyndall, RI MS JT/1/H/323.

24. 6 March 1887, Tyndall to Frederick Bramwell, RI MS AD/03/04, RI MS C93L/26.

25. 11 March 1887, Tyndall to Thomas Hirst, RI MS JT/1/T/751.

26. 3 April 1887, Tyndall to Frederick Bramwell, RI MS AD/03/04, RI MS C93L/27.

27. 13 May 1887, Thomas Huxley to Tyndall, RI MS JT/1/TYP/9/3126–7 (ICHC 8.256).

28. 16 May 1887, Tyndall to Thomas Huxley, RI MS JT/1/TYP/9/3129.

29. 'Professor Tyndall and the Scientific Movement', Nature 36 (1887): p. 217–18.

30. 'The Dinner to Professor Tyndall', Nature 36, (1887): p. 222–7.

31. Special General Meeting of the Royal Institution 15 December 1893, Proc. Roy. Inst. 14 (1893): p. 161–8.

32. Tyndall to The Times, 'Lighthouse Illuminants', The Times (7 April 1887): p. 10.

33. Lord Meath to Editor Belfast News-letter (14 April 1887).

34. Tyndall to The Times, 'Professor Tyndall', The Times (9 April 1887): p. 7.

35. Tyndall to The Times, 'Professor Tyndall on the Political Situation', The Times (8 June 1887).

36. Tyndall to The Times, 'Professor Tyndall on the Party Politics', The Times (26 July 1887): p. 11; Tyndall to The Times, 'Professor Tyndall on Mr Gladstone', The Times (9 August 1887): p. 4.

37. 21 August 1887, Tyndall to Thomas Hirst, RI MS JT/1/HTYP/693; BL Add MS 63092, ff. 145–6.

38. 13 November 1887, Emil Du Bois-Reymond to Tyndall, RI MS JT/1/D/175.

39. 20 September 1887, Tyndall to Thomas Hirst, BL Add MS 63092, f. 152.

40. 6 November 1887, Tyndall to Charles Grant, RI MS JT/1/T/449.

41. 21 November 1887, Tyndall to Charles Grant, RI MS JT/1/T/450.

42. 27 December 1887, Tyndall to Thomas Huxley, RI MS JT/1/TYP/9/3132 (ICHC 8.260).

43. 29 December 1887, Joseph Hooker to Tyndall, RI MS JT/1/TYP/8/2764.

44. 1 January 1888, Thomas Huxley to Tyndall, RI MS JT/1/TYP/9/3133 (ICHC 8.262).

45. 5 November 1887, Tyndall to Henry Pollock, RI MS AD/03/04, RI MS C93L/28.

46. 8 November 1887, Frederick Bramwell to Tyndall, RI MS AD/03/04, RI MS C93L/29.

47. 7 December 1887, Tyndall to Thomas Hirst, RI MS JT/1/T/758.

48. 9 December 1887, Tyndall to Lord Rayleigh, Louisa Journal, RI MS JT/2/20.

49. 27 December 1887 Tyndall to Thomas Hirst, RI MS JT/1/T/759.

50. Tyndall to The Times, 'Tory Beacon Light', The Times (3 January 1888): p. 8; Tyndall to The Times, 'Tory Island Light', The Times (6 February 1888): p. 6.

51. 29 April 1888, Tyndall to Thomas Hirst, RI MS JT/1/T/777.

52. 25 June 1888, Tyndall to Thomas Hirst, RI MS JT/1/T/906.

53. Tyndall, J., 'A story of our lighthouses', Nineteenth Century 24 (July 1888): p. 61–80; Tyndall, J., 'A story of the lighthouses (Part I)', Fortnightly Review 44 (December 1888): p. 805–28; Tyndall, J., 'A story of the lighthouses (Part II)', Fortnightly Review 45 (February 1889): p. 198–219.

54. 3 March 1889, Tyndall to Sir John Lubbock, BL Add MS 49658, ff. 47–8.

55. Lighthouse illuminants (south foreland experiments) PP 1890–91 (2).

56. 14 February 1888, Tyndall to Thomas Hirst, RI MS JT/1/HTYP/703.
57. 'The Academy Banquet', *The Times* (7 May 1888): p. 12.
58. 6 June 1888, Tyndall to Thomas Hirst, RI MS JT/1/T/788.
59. 16 May 1888 Tyndall to Lady Welby, YU–018—YU–019 [YU–0016—YU–0017].
60. 'Faraday, Michael', *Dictionary of National Biography* (ed. Stephen, L.), XVIII (London: Smith, Elder & Co., 1889): p. 190–202 (on p. 199).
61. 15 October 1887, Tyndall to Thomas Hirst, BL Add MS 63092, ff. 153–4.
62. 11 August 1889, Wilhelm Weber to Tyndall, RI MS JT/1/W/18.
63. 25 April 1888, Thomas Hirst to Tyndall, RI MS JT/1/H/364.
64. Clausius gave Hirst a gold watch for translating his work on the Mechanical Theory of Heat, published by Van Voorst with a preface by Tyndall.
65. 14 July 1888, Tyndall to Thomas Hirst, RI MS JT/1/T/793.
66. 7 December 1853, Tyndall Journal, RI MS JT/2/13b/645.
67. 16 September 1888, Tyndall to Thomas Hirst, RI MS JT/1/HTYP/731; BL Add MS 63092, ff. 161–2.
68. 18 September 1888, Frederick Pollock jnr to Tyndall, RI MS JT/1/P/141.
69. 30 October 1888, Tyndall to *Nature*, 'Alpine Haze', *Nature* 39 (1 November 1888): p. 7.
70. 9 November 1888, Thomas Hirst to Tyndall, RI MS JT/1/H/393.
71. 11 November 1888, Tyndall to Thomas Hirst, RI MS JT/1/T/798.
72. 6 January 1889, Tyndall to Sir George Stokes, CUL Add 7656/T569.
73. 1 January 1889, Tyndall to Thomas Hirst, RI MS JT/1/T/802.
74. 4 February 1889, Tyndall to Sir John Lubbock, BL Add MS 49658, f. 31.
75. 28 February 1889, Jane Barnard to Tyndall, RI MS JT/1/B/72.
76. 8 April 1889, William Rideing to Tyndall, BL Tyndall Papers, Vol. 1, ff. 59–61.
77. 19 April 1889, Perry Mason & Co to Tyndall, BL Tyndall Papers, Vol. 1, f. 62. The sum was $250 for his article 'Water and its Associates'. Another article, 'The Wonders of Water', appeared on 23 October 1890.
78. 28 February 1889, Tyndall to Thomas Hirst, RI MS JT/1/T/806.
79. 18 April 1889, Tyndall to Sir John Lubbock, BL Add MS 49658, ff.81–2.
80. 12 March 1889, Tyndall Journal, RI MS JT/2/21.
81. 26 May 1889, Hirst Journal, RI MS JT/2/32e/2603.
82. 4 April 1889, Tyndall Journal, RI MS JT/2/21.
83. 3 May 1889, Joseph Hooker to Tyndall, RI MS JT/1/TYP/8/2767.
84. Frederick Abel was elected.
85. 10 March 1889, Lady Welby to Tyndall, YU–031—YU–033.
86. 11 March 1889, Tyndall to Lady Welby, RI MS JT/1/TYP/5/1780.
87. 29 March 1889, Tyndall Journal, RI MS JT/2/21.
88. 2 April 1889, Tyndall Journal, RI MS JT/2/21.
89. 3 April 1889, Tyndall Journal, RI MS JT/2/21.
90. 29 March 1887, Tyndall Journal, RI MS JT/2/20.
91. 19 July 1889, Tyndall to Thomas Hirst, RI MS JT/1/HTYP/745; BL Add MS 63092, ff. 168–9.
92. 11 September 1889, Tyndall to Thomas Hirst, RI MS JT/1/HTYP/747; BL Add MS 63092, ff. 171–2.
93. 22 September 1889, Tyndall to Heinrich Debus, RI MS JT/1/T/288.
94. N.d. September 1889, Tyndall to (probably) Juliet Pollock, RI MS JT/1/TYP/6/2185–6.
95. 25 October 1889, Joseph Hooker to Tyndall, RI MS JT/1/TYP/8/2770.
96. Undated Tyndall notes, RI uncatalogued box 864.
97. 25 November 1889, Tyndall to Herbert Spencer, RI MS JT/1/TYP/3/1073–4.

98. 6 December 1889, Joseph Hooker to Tyndall, RI MS JT/1/TYP/8/2773.
99. 7 December 1889, Tyndall to Joseph Hooker, Louisa Journal RI MS JT/2/21.
100. 5 February 1890, Thomas Hirst to Louisa Tyndall, RI MS JT/1/H/419.
101. 2 October 1890, Hirst Journal, RI MS JT/2/32e/2721.
102. 9 December 1889, Thomas Hamilton to Tyndall, RI MS JT/1/H/14.
103. 20 December 1889, Thomas Hamilton to Tyndall, RI MS JT/1/H/17.
104. 22 December 1889, Tyndall to Thomas Hamilton, RI MS JT/1/T/467/2.
105. 25 December 1889, Tyndall to Thomas Hamilton, RI MS JT/1/T/468.
106. 27 December 1889, Thomas Hamilton to Tyndall, RI MS JT/1/H/19.
107. 5 January 1890, Robert Koch to Tyndall, RI MS JT/1/K/30.
108. Tyndall, J., 'Personal Recollections of Thomas Carlyle', *Fortnightly Review* 50 (January 1890): p. 293–309.
109. 29 January 1890, T. H. Duke to Tyndall, RI MS JT/1/D/187.
110. 3 January 1890, Tyndall to Thomas Hamilton, RI MS JT/1/T/469.
111. RI MS JT/7/6.
112. *Belfast News-Letter* (22 January 1890): p. 5.
113. *Belfast News-Letter* (30 January 1890): p. 5.
114. 24 January 1890, Tyndall to Thomas Hamilton, RI MS JT/1/T/474.
115. 'The Great Unionist Demonstration in Belfast', *Belfast News-Letter* (29 January 1890): p. 6–7. Londonderry had been succeeded by the Earl of Zetland.
116. *Belfast News-Letter* (12 February 1890): p. 7.
117. Tyndall to *The Times*, 'Sir W. Harcourt and Professor Tyndall', *The Times* (13 February 1890): p. 9.
118. Tyndall to *The Times*, 'Sir W. Harcourt and Professor Tyndall', *The Times* (15 February 1890): p. 10.
119. 4 February 1890, Tyndall to William Lecky, TCD—575; 5 February 1890, William Lecky to Tyndall, RI MS JT/1/L/6; 10 March 1890, William Lecky to Tyndall, RI MS JT/1/L/7; Tyndall to *The Times*, 'Professor Tyndall and Mr. Gladstone', *The Times* (10 March 1890): p. 4.
120. 'Mr. Gladstone and Professor Tyndall', *The Standard* (10 March 1890): p. 3.
121. 11 March 1890, Tyndall to William Lecky, TCD—578.
122. 9 March 1890, Tyndall to William Gladstone, BL Add MS 44509, f. 241.
123. 'Professor Tyndall on Politicians of the Day', *Pall Mall Gazette* (29 May 1890): p. 6.
124. 17 April 1890, Hirst Journal, RI MS JT/2/32e/2683.
125. 'Liberal Unionism in Surrey', *The Times* (1 June 1889): p. 15.
126. 18 April 1890, Tyndall to Mrs Steuart, RI MS JT/1/TYP/10/3472.
127. 6 July 1890, Tyndall to Mrs Steuart, RI MS JT/1/TYP/10/3474.
128. 14 May 1890, Hirst Journal, RI MS JT/2/32e/2687–8.
129. 29 June 1890, Tyndall to Thomas Hirst, RI MS JT/1/T/825.
130. 'Sky-Signs in the Country', *Punch* (15 November 1890): p. 238.
131. 23 September 1890, Tyndall to Thomas Hirst, BL Add MS 63092, ff. 178–9.
132. N.d. September 1890, Thomas Hirst to Tyndall, RI MS JT/1/H/428.
133. 5 October 1890, Tyndall to Thomas Hirst, BL Add MS 63092, ff. 181–2.
134. 18 November 1890, Tyndall to Thomas Hirst, RI MS JT/1/T/829.
135. 3 December 1890, Tyndall to Thomas Hirst, RI MS JT/1/T/832.
136. 16 June 1891, Tyndall to Sir Frederick Bramwell, RI MS AD/03/04, RI MS C93L/31.
137. N.d. n.m. 1890, Tyndall to Juliet Pollock, RI MS JT/1/T/1230.
138. Tyndall to *The Times*, 'Professor Tyndall's Health', *The Times* (21 January 1891): p. 6.
139. 17 February 1891, Tyndall to Thomas Hirst, RI MS JT/1/T/839.

140. 10 April 1891, Tyndall to Heinrich Debus, RI MS JT/1/T/292.
141. 17 April 1891, Louisa Tyndall to Thomas Hirst, RI MS JT/1/T/1534a.
142. Tyndall to *The Times*, 'Professor Tyndall's Health', *The Times* (6 May 1891): p. 6.
143. 9 May 1891, Thomas Hirst to Tyndall, RI MS JT/1/H/459.
144. 9 July 1891, Tyndall to Thomas Hirst, RI MS JT/1/T/848.
145. Tyndall, J., 'On the origin, propagation, and prevention of phthisis', *Fortnightly Review* 56, (September 1891): p. 293–309.
146. 11 August 1891, Tyndall to Thomas Hirst, RI MS JT/1/T/850. This was the 1891 International Medical Congress in Berlin, at which Koch described his 'Tuberculin cure', though it proved controversially ineffective.
147. 30 December 1891, Georg Cornet to Tyndall, RI MS JT/1/C/52.
148. Undated Tyndall notes, RI uncatalogued box 864.
149. 'The Other Side of Professor Tyndall's Screens', *Pall Mall Gazette* (28 August 1891). A supportive letter from another neighbour was also published: 'Professor Tyndall and his Screen', *Pall Mall Gazette* (24 August 1891).
150. *The Daily Graphic* (30 October 1891): p. 3, BL MLD22.
151. 4 November 1891, Tyndall to Herbert Spencer, RI MS JT/1/TYP/3/1097–8.
152. 9 September 1891, Tyndall to Herbert Spencer, RI MS JT/1/TYP/3/1094–6.
153. Tennyson (1897): vol. 2, p. 380.
154. *Saturday Review* 73 (9 January 1892): p. 61.
155. 'Reviews: Tyndall's New Fragments', *Saturday Review* 73 (27 February 1892): p. 245–6.
156. Tyndall, J. 'A Morning on Alp Lusgen', *Pall Mall Gazette* (16 August 1881); Tyndall (1892): p. 498.
157. 7 February 1892, Joseph Hooker to Tyndall, RI MS JT/1/TYP/8/2787.
158. 16 February 1892, Joseph Hooker to Tyndall, RI MS JT/1/TYP/8/2789.
159. 18 February 1892, Tyndall to Thomas Huxley, RI MS JT/1/TYP/9/3153 (ICHC 8.276).
160. 26 March 1892, Tyndall to Secretary Royal Academy, Louisa Journal, RI MS JT/2/29.
161. 28 June 1892, Louisa Journal, RI MS JT/2/29.
162. 6 September 1892, Tyndall to Heinrich Debus, RI MS JT/1/T/300.
163. Tyndall to *The Times*, 'Professor Tyndall on the Unionist Prospect', *The Times* (9 July 1892): p. 10.
164. 9 July 1892, Herbert Spencer to Tyndall, RI MS JT/1/S/171.
165. 6 September 1892, Tyndall to Heinrich Debus, RI MS JT/1/T/300.
166. 18 August 1892, Tyndall to Sir John Lubbock, BL Add MS 49658, ff. 152–4.
167. N.d. n.m. 1892?, Tyndall to Herbert Spencer, RI MS JT/1/TYP/3/1101–3.
168. 18 August 1892, Tyndall to Sir John Lubbock, BL Add MS 49658, ff. 152–4.
169. 31 August 1892, Frederick Pollock junior to Tyndall, RI MS JT/1/P/150.
170. 14 November 1891, Joseph Hooker to Tyndall, RI MS JT/1/TYP/8/2782.
171. 24 October 1892, Tyndall to Lady Welby, YU–088—YU–089 [YU–086—YU–087b].
172. 15 February 1893, Tyndall to Édouard Sarasin, RI MS JT/5/15b. Tyndall later wrote his recollections for Hallam Tennyson's memoir of Tennyson: Tyndall, J., 'A Glimpse of Farringford, 1858; and '"The Ancient Sage", 1885', in Tennyson (1897): vol. 2, p. 469–78 (uncorrected at Tyndall's death).
173. 15 October 1892, Thomas Huxley to Tyndall, RI MS JT/1/TYP/9/3160 (ICHC 8.280).
174. 15 May 1893, Thomas Huxley to Tyndall, RI MS JT/1/TYP/9/3163-3164 (ICHC 8.281).
175. 26 August 1893, Tyndall to Heinrich Debus, RI MS JT/1/T/304.
176. 25 November 1893, Tyndall to Herbert Spencer, RI MS JT/1/TYP/3/1112–13.

177. 3 December 1893, Tyndall to William Colles, RI MS JT/2/30; 3 December 1893, Tyndall to Emil Du Bois-Reymond, SB Slg Darmstaedter 1855 Tyndall Be. 1–58, 29–30; 3 December 1893, Tyndall to Gustav Wiedemann, RI MS JT/5/16a.
178. 3 December 1893, Louisa Journal, RI MS JT/2/30.
179. 4 December 1893, Louisa Journal, RI MS JT/2/30.
180. Louisa Tyndall's inquest notes, Elton Hall FAM124.
181. Eve and Creasey (1945): p. 279–80.

Epilogue

1. Huxley, T. H., 'Professor Tyndall', *Nineteenth Century* 35, (1894): p. 1–11.
2. Lord Rayleigh (J. W. Strutt), 'The scientific work of John Tyndall', *Proc. Roy. Inst.* 14 (1894): p. 216–24.
3. Lodge, O., 'Tyndall, John (1820-1893)', *Encyclopaedia Britannica* (1903): vol. 10, p. 517–21.
4. A huge bound tome of rebuttals in the Royal Institution's archive testifies to the offence this biography caused to some: 'Criticism of Oliver Lodge's Biography of John Tyndall', RI MS JT/5/3.
5. Bragg, W. H., 'Tyndall's experiments on magne-crystallic action', *Proc. Roy. Inst.* 25 (1927): p. 161–84.
6. Eve and Creasey (1945).
7. A neat example of this is Stokes's acknowledgement, after Tyndall's death, of Tyndall's experimental discovery that light was refracted far further beyond the red than had been expected from Cauchy's work. It led to theoretical thoughts of his own. 17 February 1896, George Stokes to William Thomson, in Wilson (1990): p. 638.

Bibliography

1. Eve and Creasey (1945): p. 338.

BIBLIOGRAPHY

The bibliography contains works directly referenced in the endnotes, as well as those that have more broadly informed the writing of this biography.

Albritton, V. and F. A. Jonsson, *Green Victorians: The Simple Life in John Ruskin's Lake District* (Chicago: Chicago University Press, 2016).

Allibone, T. E., *The Royal Society and its Dining Clubs* (Oxford: Pergamon, 1976).

Alter, P., *The Reluctant Patron: Science and the State in Britain, 1850–1920* (Oxford, Hamburg, and New York: Berg, 1987).

Andrews, J. H., *A Paper Landscape: The Ordnance Survey in Nineteenth-Century Ireland* (Oxford: Clarendon Press, 1975).

Arapostathis, S. and G. Gooday, *Patently Contestable: Electrical Technologies and Inventor Identities on Trial in Britain* (Cambridge, MA: The MIT Press, 2013).

Ashton, R., *142 Strand: A Radical Address in Victorian England* (London: Chatto & Windus, 2006).

Bade, W. F., *The Life and Letters of John Muir. Volume 1* (Boston, MA: Houghton Mifflin, 1924).

Baldwin, M., 'The Shifting Ground of Nature: Establishing an Organ or Scientific Communication in Britain, 1869–1900', *History of Science* 50 (2012): p. 125–54.

Baldwin, M., *Making 'Nature': The History of a Scientific Journal* (Chicago: University of Chicago Press, 2015).

Baldwin, M. and J. Browne (eds), *Volume 2: The Correspondence, September 1843–December 1849*. In Elwick, J., B. Lightman, and M. Reidy (eds), *The Correspondence of John Tyndall* (Pittsburgh, PA: University of Pittsburgh Press, 2016).

Ball, W. V., *Reminiscences and Letters of Sir Robert Ball* (London: Cassell, 1915).

Barton, R., 'John Tyndall, Pantheist: A Rereading of the Belfast Address', *Osiris* 3 (1987): p. 111–34.

Barton, R., '"An Influential Set of Chaps"': The X-Club and Royal Society Politics 1864–85', *British Journal of the History of Science* 23 (1990): p. 53–81.

Barton, R., '"Huxley, Lubbock, and Half a Dozen Others": Professionals and Gentlemen in the Formation of the X Club, 1851-1864', *Isis* 89 (1998): p. 410–44.

Barton, R., 'Just before *Nature*: The Purposes of Science and the Purposes of Popularization in some English Popular Science Journals of the 1860s', *Annals of Science* 55 (1998): p. 1–33.

Barton, R., '"Men of Science": Language, Identity, and Professionalization in the Mid-Victorian Scientific Community', *History of Science* 41 (2003): p. 73–119.

Barton, R., 'Scientific Authority and Scientific Controversy in *Nature*: North Britain against the X Club', in Henson, L., G. N. Cantor, G. Dawson, R. Noakes, S. Shuttleworth, and J. Topham (eds), *Culture and Science in the Nineteenth Century Media* (Aldershot: Ashgate, 2004): p. 223–35.

Barton, R., *The X Club: Power and Authority in Victorian Science* (Chicago: Chicago University Press, 2018).

Batchelor, J., *Tennyson, To Strive, To Seek, To Find* (London: Chatto and Windus, 2012).

Bateman, C. T., 'Haslemere as a Literary Centre', *Windsor Magazine* 8 (1898): p. 187–95.

Bergman, B. P. and S. A. St J. Miller, 'The Parkes Museum of Hygiene and the Sanitary Institute', *The Journal of The Royal Society for the Promotion of Health* 123 (2003): p. 55–61.

Bibby, C., 'A Victorian Experiment in International Education: The College at Spring Grove', *British Journal of Educational Studies* 5 (1956): p. 25–36.

Blinderman, C., 'John Tyndall and the Victorian New Philosophy', *Bucknell Review* 9 (1961): p. 281–90.

Bonney, T. G., *Annals of the Philosophical Club of the Royal Society, Written from Its Minute Books* (London: Macmillan, 1919).

Bowers, B., *Sir Charles Wheatstone, FRS, 1802–1875* (London: Science Museum/HMSO, 1975).

Braham, T., 'John Tyndall (1820–1893) and Belalp', *Alpine Journal* 98 (1993): p. 196–8.

Braham, T., *When the Alps Cast their Spell: Mountaineers of the Alpine Golden Age* (Glasgow, The In Pinn, 2004).

Braun, C. L. and S. N. Smirnov., 'Why is Water Blue?', *Journal of Chemical Education* 70 (1993): p. 612–14.

Brock, W. H. (ed), *The Atomic Debates* (Leicester: Leicester University Press, 1967).

Brock, W. H. and R. M. MacLeod, 'The Scientists' Declaration: Reflexions on Science and Belief in the Wake of Essays and Reviews', *British Journal of the History of Science* 9 (1976): p. 39–66.

Brock, W. H. and R.M. MacLeod (eds), *Natural Knowledge in Social Context: the Journals of Thomas Arthur Hirst FRS* (London: Mansell, 1980).

Brock, W. H., N. D. McMillan, and R. C. Mollan (eds), *John Tyndall: Essays on a Natural Philosopher* (Dublin: Royal Dublin Society, 1981).

Brock, W. H. and A. J. Meadows, *The Lamp of Learning: Two Centuries of Publishing at Taylor & Francis*, 2nd ed. (London: Taylor & Francis, 1998).

Brooke, J. H., *Science and Religion: Some Historical Perspectives* (Cambridge: Cambridge University Press, 1991).

Brown, A. W., *The Metaphysical Society: Victorian Minds in Crisis, 1869–1880* (New York: Columbia University Press, 1947).

Brown, D., *The Poetry of Victorian Scientists; Style, Science and Nonsense* (Cambridge: Cambridge University Press, 2013).

Browne, J., 'Science and British Culture in the 1830s', *Conference Proceedings, Trinity College, Cambridge* (1994).

Browne, J., *Charles Darwin: The Power of Place* (London: Pimlico, 2003).

Burdon Sanderson, Lady, *Sir John Burdon-Sanderson: A Memoir* (Oxford: Clarendon Press, 1911).

Butt, J. (ed) *Robert Owen: Aspects of his Life and Work* (New York: Humanities Press, 1971).

Cahan, D., 'Helmholtz and the British Scientific Elite: From Force Conservation to Energy Conservation', *Notes Rec.* 66 (2012): p. 55–68.

Cahan, D., 'The Awarding of the Copley Medal and the "Discovery" of the Law of Conservation of Energy: Joule, Mayer and Helmholtz Revisited', *Notes Rec.* 66 (2012): p. 125–39.

Campbell, L. and Garnett, W., *The Life of James Clerk Maxwell* (London: Macmillan, 1882).

Cantor, G., 'John Tyndall's Religion: A Fragment', *Notes Rec.* 69 (2015): p. 419–36.

Cantor, G., G. Dawson, G. Gooday, R. Noakes, S. Shuttleworth, and J. Topham (eds), *Science in the Nineteenth-Century Periodical: Reading the Magazine of Nature* (Cambridge: Cambridge University Press, 2007).

Cantor, G., and G. Dawson (eds), *Volume 1: The Correspondence, May 1840–August 1843*. In Elwick, J., B. Lightman, and M. Reidy (eds), *The Correspondence of John Tyndall* (Pittsburgh, PA: University of Pittsburgh Press, 2016).

Cate, G. A. (ed.), *The Correspondence of Thomas Carlyle and John Ruskin* (Stanford, CA: Stanford University Press, 1982).

Chalmers, D. J., *The Conscious Mind* (Oxford: Oxford University Press, 1996).

Chandler, A., 'The French Universal Exposition of 1855', *World's Fair* 6 (1986) downloaded from http://www.arthurchandler.com/paris-1855-exposition/ (accessed 9 September 2016).

Choi, T. Y., 'Forms of Closure: The First Law of Thermodynamics and Victorian Narrative', *ELH* 74 (2007): p. 301–22.

Clark, R., *The Early Alpine Guides* (London: Phoenix House, 1949).

Clark, R., *The Victorian Mountaineers* (London: Batsford, 1953).

Clark, R., *An Eccentric in the Alps* (London: Museum Press, 1959).

Coley, N. G., 'Henry Bence-Jones, M.D., F.R.S. (1813–1873)', *Notes Rec.* 28 (1973): p. 31–56.

Collini, S., *Public Moralists: Political Thought and Intellectual Life in Britain, 1860–1930* (Oxford, Clarendon Press, 1991).

Cook, E. T. and A. Wedderburn (eds), *The Works of John Ruskin* (39 vols) (London: George Allen, 1903–12).

Conway, M., *Mountain Memories* (New York: Funk and Wagnalls, 1920).

Conway, M. D., *Autobiography: Memories and Experiences of Moncure Daniel Conway* (2 vols) (Boston, MA: Houghton Mifflin Co., 1904).

Crellin, J. K., 'Airborne Particles and the Germ Theory: 1860–1880', *Annals of Science* 22 (1966): p. 49–60.

Crellin, J. K., 'The Problem of Heat Resistance of Micro-organisms in the British Spontaneous Generation Controversies of 1860–1880', *Medical History* 10 (1966): p. 50–9.

Crellin, J. K., 'Antibiosis in the 19th Century', *The History of Antibiotics; a Symposium* (1980): p. 5–13.

Cunningham, F., *James David Forbes; Pioneer Scottish Glaciologist* (Edinburgh: Scottish Academic Press, 1990).

Dangar, D. F. O and T. S. Blakeney, 'The Rise of Modern Mountaineering and the Formation of the Alpine Club 1854–1865', *Alpine Journal* 62 (1957): p. 16–38.

Darwin, C. R., *On the Origin of Species by Means of Natural Selection, or the Preservation of Favoured Races in the Struggle for Life*, 5th ed. (London: John Murray, 1869).

Daston, L., 'Fear & Loathing of the Imagination in Science', *Daedalus* 134 (2005): p. 16–30.

Daub, E. E., 'The Hidden Origins of the Tait-Tyndall Controversy: The Thomson-Tyndall Conflict', in *Proceedings, XIVth International Congress of the History of Science, Tokyo and Kyoto, Japan, 19–27 August, 1974* (Tokyo: Japan Science Council, 1974): p. 241–4.

Daunton. M., *State and Market in Victorian Britain: War, Welfare and Capitalism* (Martlesham: Boydell Press, 2008).

Dawson, G., 'Intrinsic Earthliness: Science, Materialism, and the Fleshly School of Poetry', *Victorian Poetry* 41 (2003): p. 113–29.

Dawson, G., *Darwin, Literature and Victorian Respectability* (Cambridge: Cambridge University Press, 2007).

Dawson, G. and B. Lightman (eds), *Victorian Scientific Naturalism: Community, Identity and Continuity* (Chicago: Chicago University Press, 2014).

DeArce, M., N. McMillan, M. Nevin, and C. Flahaven, 'What Tyndall Read: Provenance, Contents and Significance of the Proby Bequest in the Carlow County Council Library', *Carloviana* 60 (2011): p. 134–41.

DeArce, M., D. McGing, and N. McMillan, 'Two Forgotten Poems by John Tyndall (1841)', *Carloviana* (2013–14).

Desmond, A., *Huxley: The Devil's Disciple* (London: Michael Joseph, 1994).

Desmond, A., *Huxley: Evolution's High Priest* (London: Michael Joseph, 1997).

Desmond, A., 'Redefining the X Axis: "Professionals", "Amateurs" and the Making of Mid-Victorian Biology—A Progress Report', *Journal of the History of Biology* 34 (2001): p. 3–50.

DeYoung, U., *A Vision of Modern Science* (New York: Palgrave Macmillan, 2011).

Donnan, F. G., 'The Scientific Relief Fund and Its Committee', *Notes Rec.* 7 (1950): p. 158–71.

Drayton, R. H., *Nature's Government: Science, Imperial Britain, and the 'Improvement' of the World* (New Haven, CT: Yale University Press, 2000).

Duncan, D., *Life and Letters of Herbert Spencer* (London: Methuen, 1908).

Dyde, S., 'George Combe and Common Sense', *British Journal of the History of Science* 48 (2015): p. 233–59.

Endersby, J., *Imperial Nature: Joseph Hooker and the Practices of Victorian Science* (Chicago: University of Chicago Press, 2008).

Eve, A. S. and C. H. Creasey, *Life and Work of John Tyndall* (London: Macmillan, 1945).

Farley, J. 'The Spontaneous Generation Controversy (1859–1880)', *Journal of the History of Biology* 5 (1972): p. 285–319.

Farley, J. and G. Geison, 'Science, Politics and Spontaneous Generation in Nineteenth Century France: The Pasteur-Pouchet Debate', *Bulletin of the History of Medicine* 48 (1974): p. 161–98.

Feinstein, E., *It Goes with the Territory: Memoirs of a Poet* (Richmond, VA: Alma Books, 2013).

Fielding, K. J., 'Carlyle and John Tyndall: 1', *Carlyle Studies Annual* 18 (1998): p. 43–52.

Finkelstein, G., *Emil du Bois-Reymond; Neuroscience, Self and Society in Nineteenth-Century Germany* (Cambridge, MA: MIT, 2013).

Fisch, M. and S. Schaffer, *William Whewell; A Composite Portrait* (London: Clarendon Press, 1991).

Fiske, J., *Edward Livingston Youmans: Interpreter of Science for the People* (New York: Appleton, 1894).

Fleming, J. R., 'John Tyndall, Svante Arrhenius, and Early Research on Carbon Dioxide and Climate', in *Historical Perspectives on Climate Change* (New York: Oxford University Press, 1998).

Flood, R., M. McCartney, and A. Whitaker (eds), *James Clerk Maxwell: Perspectives on his Life and Work* (Oxford: Oxford University Press, 2014).

Forbes, G. (ed), *Theory of the Glaciers of Savoy* (London: Macmillan, 1874).

Forbes, J. D., *Travels through the Alps of Savoy and other Parts of the Pennine Chain* (Edinburgh: Black, 1843).

Forbes, J. D., *Reply to Professor Tyndall's Remarks, in His Work 'On the Glaciers of the Alps,' Relating to Rendu's 'Théorie Des Glaciers'. Occasional Papers on the Theory of Glaciers* (Edinburgh: A & C Black, 1860).

Forgan, S., 'The Royal Institution of Great Britain, 1840–1873' (PhD thesis: University of London, 1977).

Foster, R. F., *Modern Ireland 1600–1972* (London: Allen Lane, The Penguin Press, 1988).

Foster, S. B., *The Pedigrees of Birkbeck of Mallerstang and Settle, Braithwaite of Kendal, Benson of Stang End* (Private Printing: 1890).

Fowler, W., *Mozley and Tyndall on Miracles* (London: Longmans, 1868).

Francis, M., *Herbert Spencer and the Invention of Modern Life* (Cornell, NY: Cornell University Press, 2007).

French, R. D., *Antivivisection and Medical Science in Victorian Society* (Princeton, NJ: Princeton University Press, 1975).

Friday, J., 'A Microscopic Incident in a Monumental Struggle: Huxley and Antibiosis in 1875', *British Journal for the History of Science* 7 (1974): p. 61–71.

Froude, J. A. (ed), *Reminiscences by Thomas Carlyle* (2 vols) (New York: Harper & Brothers, 1881).

Froude, J. A., *Thomas Carlyle* (4 vols) (London: Longman, 1882–4).

Fyfe, A. and N. Moxham, 'Making Public Ahead of Print: Meetings and Publications at the Royal Society 1752–1892', *Notes Rec.* 70 (2016): p. 361–79.

Galton, F. (ed), *Vacation Tourists and Notes of Travel in 1860* (Cambridge: Macmillan, 1861).

Galton, F., *English Men of Science* (London: Macmillan, 1874).

Gieryn, T. F., 'John Tyndall's Double Boundary-work: Science, Religion and Mechanics in Victorian England', in *Cultural Boundaries of Science; Credibility on the Line* (Chicago: Chicago University Press, 1999), p. 37–64.

Girdlestone, A. G., *The High Alps without Guides* (London: Longmans, 1870).

Golinsky, J., *Science as Public Culture* (Cambridge: Cambridge University Press, 1992).

Gooday, G., 'Ethnicity, Expertise and Authority: The Cases of Lewis Howard Latimer, William Preece and John Tyndall,' in Vandendriessche, J., E. Peeters, and K. Wils (eds) *Scientists' Expertise as Performance* (London: Pickering and Chatto, 2015): p. 15–30.

Gooday, G., *Domesticating Electricity* (Pittsburgh, PA: University of Pittsburgh Press, 2016).

Gooding, D., *Experiment and the Making of Meaning. Science and Philosophy*, Vol. 5 (Dordrecht: Kluwer, 1990).

Gooding, D. and F. A. J. L. James (eds), *Faraday Rediscovered* (London: Macmillan, 1985).

Gos, C., *Alpine Tragedy* (London: Allen and Unwin, 1948).

Haast, J., 'Notes on the Mountains and Glaciers of Canterbury Province, New Zealand', *Journal of the Royal Geographical Society of London* 34 (1864): p. 87–96.

Hall, M. B., *All Scientists Now: The Royal Society in the Nineteenth Century* (Cambridge: Cambridge University Press, 1985).

Hansen, P. H., *The Summits of Modern Man: Mountaineering After the Enlightenment* (Cambridge, MA: Harvard University Press, 2013).

Harman, P. M. (ed), The Scientific Letters and Papers of James Clerk Maxwell (3 vols) (Cambridge: Cambridge University Press, 1990, 1995, 2002).

Hass, J. W., 'The Reverend Dr William Henry Dallinger, F.R.S. (1839–1909)', *Notes Rec.* 54 (2000): p. 53–65.

Haugrud, R. A., 'Tyndall's Interest in Emerson', *American Literature* 41 (1970): p. 507–17.

Hesse, M. B., *Forces and Fields: the Concept of Action at a Distance in the History of Physics* (London: Nelson, 1961).

Hevly, B., 'The Heroic Science of Glacier Motion', *Osiris* 11 (1996): p. 66–86.

Higgitt, R. and C. Withers, 'Science and sociability: Women as audience at the British Association for the Advancement of Science, 1831–1901', *Isis* 99 (2008): p. 1–27.

Hilts, V. L., 'A Guide to Francis Galton's English Men of Science', *Transactions of the American Philosophical Society* 65:5 (1975): p. 3–85.

Holmes, J., 'The X Club: Romanticism and Victorian Science', in Caleb, A. M. (ed) *(Re)Creating Science in Nineteenth-Century Britain* (Newcastle: Cambridge Scholars Publishing, 2007).

Holmes, J., '"The Poet of Science": How Scientists Read Their Tennyson', *Victorian Studies* 54 (2012): p. 655–78.

Howard, J., '"Physics and Fashion": John Tyndall and his Audiences in Mid-Victorian Britain', *Studies in the History and Philosophy of Science* 35 (2004): p. 729–58.

Hutchinson, H. G., *Life of Sir John Lubbock* (2 vols) (London: Macmillan, 1914).

Huxley, L., *Life and Letters of Thomas Henry Huxley* (2 vols) (London: Macmillan, 1900).

Huxley, L., *Life and Letters of Sir Joseph Dalton Hooker* (2 vols) (London: John Murray, 1918).

Huxley, T. H., *Lay Sermons, Addresses and Reviews* (London: Macmillan, 1870).

Jackson, R., 'John Tyndall and the Royal Medal that Was Never Struck', *Notes Rec.* 68 (2014): p. 151–64.

Jackson, R., 'John Tyndall and the Early History of Diamagnetism', *Annals of Science* 72 (2015): p. 435–89.

James, F. A. J. L. (ed), *The Correspondence of Michael Faraday 1849–1855*, vol. 4 (London: Institute of Electrical Engineering, 1996).

James, F. A. J. L. (ed), 'The Common Purposes of Life': Science and Society at the Royal Institution of Great Britain (Ashgate, 2002).

James, F. A. J. L., 'Reporting Royal Institution Lectures 1826–1867', in Cantor, G. and S. Shuttleworth (eds), Science Serialised: Representations of the Sciences in Nineteenth-Century Periodicals (Cambridge, MA: The MIT Press, 2004).

James, F. A. J. L. and A. Peers, 'Constructing Space for Science at the Royal Institution of Great Britain', Physics in Perspective 9 (2007): p. 130–85.

James, F. A. J. L. (ed), The Correspondence of Michael Faraday 1855–1860, vol. 5 (London: Institute of Electrical Engineering, 2008).

James, F. A. J. L., Michael Faraday: A Very Short Introduction (Oxford: Oxford University Press, 2010).

James, F. A. J. L. (ed), The Correspondence of Michael Faraday 1861–1867, vol. 6 (London: Institute of Electrical Engineering, 2011).

Jeans, W. T., Lives of the Electricians: Professors Tyndall, Wheatstone and Morse (London: Whittaker & Co. 1887).

Jenkins, A., 'Spatial Imagery in Nineteenth-Century Representations of Science; Faraday and Tyndall', in Smith, C. and J. Agar (eds) Making Space for Science: Territorial Themes in the Shaping of Knowledge (Basingstoke: Palgrave Macmillan, 1998): p. 181–91.

Jensen, J. V., 'The X Club: Fraternity of Victorian Scientists', The British Journal for the History of Science 5 (1970): p. 63–72.

Jensen, J. V., 'Interrelationships within the Victorian "X" Club', Dalhousie Review 51 (1971): p. 538–52.

Jensen, J. V., Thomas Henry Huxley: Communicating for Science (Newark, DE: University of Delaware Press, 1991).

Jungnickel, C. and R. McCormmach, Intellectual Mastery of Nature, Theoretical Physics from Ohm to Einstein: The Torch of Mathematics 1800–1870 (Chicago: University of Chicago Press, 1986).

Kearney, R., 'John Tyndall and Irish Science', Postnationalist Ireland: Politics, Culture, Philosophy (New York: Routledge, 1997): p. 169–77.

Kent, C., 'The Whittington Club: A Bohemian Experiment in Middle Class Social Reform', Victorian Studies 18 (1974): p. 31–55.

Kurzer, F., 'Chemistry and Chemists at the London Institution 1807–1912', Annals of Science 58 (2001): p. 163–201.

Lambert, R., Sir John Simon, 1816–1904, and English Social Administration (London: MacGibbon & Kee, 1963).

Levine, G., 'Dying to Know', Raritan 21 (2002): p. 100–21.

Libera, S. M., 'John Tyndall and Tennyson's "Lucretius"', Victorian Newsletter 45 (1974): p. 19–22.

Lightman, B., The Origins of Agnosticism: Victorian Unbelief and the Limits of Knowledge (Baltimore, MD: The Johns Hopkins University Press, 1987).

Lightman, B., 'Ideology, Evolution and late-Victorian Agnostic Popularisers', in Moore, J. R. (ed), History, Humanity and Evolution (Cambridge: Cambridge University Press, 1989): p. 285–309.

Lightman, B. (ed), Victorian Science in Context (Chicago: University of Chicago Press, 1997).

Lightman, B., 'Scientists as Materialists in the Periodical Press: Tyndall's Belfast Address', in Cantor, G. and S. Shuttleworth (eds), Science Serialized: Representations of the Sciences in Nineteenth-Century Periodicals (Cambridge, MA: MIT Press, 2004): p. 199–237.

Lightman, B., Victorian Popularisers of Science: Designing Nature for New Audiences (Chicago: University of Chicago Press, 2007).

Lightman, B. and M. Reidy (eds), The Age of Scientific Naturalism (London: Pickering and Chatto, 2014).

Lindquist, J. H., '"The Mightiest Instrument of the Physical Discoverer": The Visual "Imagination" and the Victorian Observer', *Journal of Victorian Culture* 13 (2008): p. 171–99.

Livingstone, D. N., 'Darwinism and Calvinism: The Belfast-Princeton Connection', *Isis* 83 (1992): p. 408–28.

Livingstone, D. N., 'Darwin in Belfast: The Evolution Debate', in Foster, J. W. (ed) *Nature in Ireland: A Scientific and Cultural History* (Dublin: Lilliput Press, 1997): p. 387–408.

Lodge, O., 'Tyndall, John (1820–1893)', *Encyclopaedia Britannica* (10th ed.) 10 (London: Adam and Charles Black, 1903–1904): p. 517–21.

Lunn, A., *A Century of Mountaineering* (London: Allen & Unwin, 1957).

Lyall, A., *The First Descent of the Matterhorn; a Bibliographic Guide to the 1865 Accident and Its Aftermath* (Llandysul: Gomer Press, 1977).

McCabe, I. M., 'Second Best as a Researcher, Second to None as a Populariser? The Atmospheric Science of John Tyndall, FRS (1820–1893)' (PhD thesis: University College London, 2012).

McCarthy, J., *Modern Leaders: Being a Series of Biographical Sketches* (New York: Sheldon and Company, 1872).

Macfarlane, R., *Mountains of the Mind: A History of a Fascination* (London: Granta, 2003).

MacLeod, R. M., 'Science and Government in Victorian England: Lighthouse Illumination and the Board of Trade, 1866–1886', *Isis* 60 (1969): p. 5–38.

MacLeod, R. M., 'The X-Club a Social Network of Science in Late-Victorian England', *Notes Rec.* 24 (1970): p. 305–22.

MacLeod, R. M. and P. Collins (eds), *The Parliament of Science* (Northwood: Science Reviews Ltd, 1981).

MacLeod, R. M. (ed) *Government and Expertise: Specialists, Administrators and Professionals, 1860–1919* (Cambridge: Cambridge University Press, 1988).

McMillan, N. D. and J. Meehan, *John Tyndall: 'X'emplar of Scientific & Technological Education* (Dublin: NCEA, 1980).

Mackowiak, J, R., 'The Poetics of Mid-Victorian Scientific Materialism in the Writings of John Tyndall, W. K. Clifford and Others' (PhD thesis: Cambridge, 2006).

Mallock, W. M., *The New Republic; or Culture, Faith, and Philosophy in an English Country House* (London: Chatto and Windus, 1877).

Marshall, C., Lightman, B. and R. England (eds), *The Papers of the Metaphysical Society, 1869–1880: A Critical Edition* (3 vols) (Oxford: Oxford University Press, 2015).

Martin, R. B., *Tennyson: The Unquiet Heart* (Oxford: Clarendon Press, 1983).

Mason, M., *The Making of Victorian Sexual Attitudes* (Oxford: Oxford University Press, 1994).

Meadows, A. J., *Science and Controversy: a Biography of Sir Norman Locker, Founder Editor of Nature*, 2nd ed. (London: Macmillan, 2008).

Meulders, M., *Helmholtz: from Enlightenment to Neuroscience* (Cambridge, MA: MIT Press, 2010).

Moore, J., 'Deconstructing Darwinism: The Politics of Evolution in the 1860s', *Journal of the History of Biology* 24 (1991): p. 353–408.

Morrell, J. and A. Thackray, *Gentlemen of Science: Early Years of the British Association for the Advancement of Science* (Oxford: Clarendon Press, 1981).

Morus, I. R., *William Robert Grove: Victorian Gentleman of Science* (Cardiff: University of Wales Press, 2017).

Norton, S. and M. A. DeWolfe Howe (eds), *Letters of Charles Eliot Norton* (2 vols) (Boston, MA: Houghton Mifflin Co., 1913).

Nugent, F., *In Search of Peaks, Passes & Glaciers: Irish Alpine Pioneers* (Cork: Collins, 2013).

Numbers, R., J. Hardin, and R. Binzley (eds), *'The Idea that Wouldn't Die': The Warfare between Science and Religion* (Baltimore, MD: Johns Hopkins University Press, forthcoming).

O'Gorman, F., 'Ruskin's "Fors Clavigera" of October 1873: An Unpublished Letter from Carlyle to Tyndall', *Notes and Queries* 43 (1996): p. 430–2.

O'Gorman, F., '"The Eagle and the Whale?": John Ruskin's Argument with John Tyndall', in Wheeler, M. (ed) *Time and Tide: Ruskin and Science* (London: Pilkington Press, 1996): p. 45–64.

O'Gorman, F., 'John Tyndall as Poet: Agnosticism and "A Morning on Alp Lusgen"', *The Review of English Studies* 48 (1997a): p. 353–8.

O'Gorman, F., 'Some Ruskin Annotations of John Tyndall', *Notes and Queries* 44 (1997b): p. 348–9.

O'Gorman, F., '"The Mightiest Evangel of the Alpine Club": Masculinity and Agnosticism in the Alpine Writing of John Tyndall', in Bradstock, A., S. Gill, A. Hogan, and S. Morgan (eds), *Masculinity and Spirituality in Victorian Culture* (New York: Palgrave, 2000): p. 134–48.

O'Gorman, F. (ed), *The Cambridge Companion to Victorian Culture* (Cambridge: Cambridge University Press, 2010).

Odlyzko, A., 'Collective Hallucinations and Inefficient Markets: The British Railway Mania of the 1840s'. (2015) downloaded from http://papers.ssrn.com/sol3/papers.cfm?abstract_id=1537338 (accessed 26 December 2015).

Olohan, M., 'Gate-keeping and Localizing in Scientific Translation Publishing: the Case of Richard Taylor and Scientific Memoirs', *British Journal for the History of Science* 46 (2013): p. 433–50.

Orwin, D. T., 'Why does Levin Read Tyndall?', in Kahla, E. (ed), *The Unlimited Gaze. Essays in Honour of Professor Natalia Baschmakoff* (Helsinki: Aleksanteri Institute, Aleksanteri Series 2, 2009): p. 203–12.

Owen, D. and R. M. Macleod (eds), *The Government of Victorian London, 1855–1889: The Metropolitan Board of Works, the Vestries, and the City Corporation* (Cambridge, MA: Harvard University Press, 1982).

Owen, J., *Darwin's Apprentice; An Archaeological Biography of John Lubbock* (Barnsley: Pen and Sword, 2013).

Parry, J., *The Rise and Fall of Liberal Government in Victorian Britain* (New Haven, CT: Yale University Press, 1993).

Patton, M., *Science, Politics and Business in the Work of Sir John Lubbock* (Aldershot: Ashgate, 2007).

Paul, C. K., *Maria Drummond: A Sketch* (London: Kegan Paul, 1891).

Pollock, W. F., *Personal Remembrances* (2 vols) (London: Macmillan, 1887).

Reid, W., *Memoirs and Correspondence of Lyon Playfair* (London: Cassell, 1900).

Reidy, M., 'Mountaineering, Masculinity and the Male Body in Mid-Victorian Britain', *Osiris* 30 (2015): p. 158–81.

Rey, G., *The Matterhorn* (2nd ed.) (London: Fisher Unwin, 1907).

Ring, J., *How the English Made the Alps* (London: Murray, 2000).

Robertson, D., 'Mid-Victorians amongst the Alps', in Knoepflmacher U. C. and G. B. Tennyson (eds), *Nature and the Victorian Imagination* (Berkeley: University of California Press, 1977): p. 113–36.

Roscoe, H. E., *The Life & Experiences of Sir Henry Enfield Roscoe* (London: Macmillan, 1906).

Ross, S., 'Scientist: The Story of a Word', *Annals of Science* 18 (1962): p. 65–85.

Rowlinson, J. S., 'The Theory of Glaciers', *Notes Rec.* 26 (1971): p. 189–204.

Rowlinson, J. S., *Sir James Dewar, 1842–1923* (Farnham: Ashgate, 2012).

Russell, C. A., *Edward Frankland: Chemistry, Controversy and Conspiracy in Victorian England* (Cambridge: Cambridge University Press, 1996).

Russett, C. E., *Sexual Science: The Victorian Construction of Womanhood* (Cambridge, MA: Harvard University Press, 1989).

Sackmann, W., 'John Tyndall (1820–93) and his relationship to the Alps and Switzerland', [in German] *Gesnerus* 50 (1993): p. 66–78.

Sawyer, P. L., 'Ruskin and Tyndall: The Poetry of Matter and The Poetry of Spirit', in Paradis, J. and T. Postlewaite (eds) *Victorian Science and Victorian Values: Literary Perspectives* (New York: New York Academy of Sciences, 1981): p. 217–46.

Lord Schuster, *Postscript to Adventure* (London: Eyre & Spottiswoode, 1950).

Scrivenor, H., *The Railways of the United Kingdom* (London: Smith, Elder, and Company, 1849).

Secord, J. A., *Victorian Sensation* (Chicago: University of Chicago, 2000).

Semmell, B., 'The Issue of "Race" in the British Reaction to the Morant Bay Uprising of 1865', *Caribbean Studies* 2 (1962): p. 3–15.

Shairp, J. C., P. G. Tait, P. G. and A. Adams-Reilly, *Life and Letters of John David Forbes* (London: Macmillan, 1873).

Sharp, E., *Hertha Ayrton* (London: Edward Arnold, 1926).

Skelton, J., *The Table-Talk of Shirley* (Edinburgh: Blackwood, 1895).

Smith, C. and W. M. Norton, *Energy and Empire: A Biographical Study of Lord Kelvin* (Cambridge: Cambridge University Press, 1989).

Smith, I., *Shadow of the Matterhorn: The Life of Edward Whymper* (Ross-on-Wye: Carreg, 2011).

Smith, M. F. (trans. and ed.) *On the Nature of Things*, Lucretius (Indianapolis, IN: Hackett, 2001).

Sopka, K., 'An Apostle of Science Visits America: John Tyndall's Journey of 1872–1873', *Physics Teacher* 10 (1972): 369–75.

Spencer, H., *An Autobiography* (2 vols) (London: Williams and Norgate, 1904).

Stead, W. T., *The M. P. for Russia: Reminiscences and Correspondence of Madame Olga Novikoff*, vol. 1 (London: Andrew Melrose, 1909).

Stephen, L., *The Playground of Europe* (London: Longmans, 1871).

Stephen, L., *Life of Henry Fawcett* (London: Smith, Elder & Co., 1885).

Stephen, L., *Some Early Impressions* (London: Hogarth Press, 1924).

Stephen, L., *Sir Leslie Stephen's Mausoleum Book*. Introduction by Alan Bell (Oxford: Clarendon Press, 1977).

Strick, J. E., 'Darwinism and the Origin of Life: The Role of H. C. Bastian in the British Spontaneous Generation Debates, 1868–1873', *Journal of the History of Biology* 32 (1999): p. 51–92.

Strick, J. E., *Sparks of Life: Darwinism and the Victorian Debates over Spontaneous Generation* (Cambridge, MA: Harvard University Press, 2000).

Sugiyama, S., 'The Significance of the Particulate Conception of Matter in John Tyndall's Physical Researches', *Historia Scientiarum* 2 (1992): p. 119–38.

Sweet, W. and R. Feist (eds), *Religion and the Challenges of Science* (London: Ashgate, 2007).

Tennyson, H., *Materials for a Life of A. T. Collected for my Children* (4 vols) (Privately printed, 1895).

Tennyson, H., *Alfred, Lord Tennyson, A Memoir* (2 vols) (London: Macmillan, 1897).

Thackray, A. and E. Mendelsohn (eds), *Science and Values* (New York: Humanities Press, 1974).

Thackray, J. C. (ed.), 'To see the Fellows fight: Eye witness accounts of meetings of the Geological Society of London and its Club', 1822–1868, *BSHS Monographs* 12 (2003).

Thompson, D., 'Queenwood College, Hampshire', *Annals of Science* 11 (1955): 246–54.

Thompson, D., 'John Tyndall (1820–1893): A Vocational Enterprise', *Vocational Aspects of Secondary and Further Education* 9 (1957): p. 38–48.

Thompson, J. S. and H. G. Thompson, *Silvanus Phillips Thompson: His Life and Letters* (London: Fisher Unwin, 1920).

Trapp, F. A., 'The Universal Exhibition of 1855', *The Burlington Magazine* 107 (1965): p. 300–5.

Turner, F. M., 'Lucretius Among the Victorians', *Victorian Studies* 16 (1973): p. 329–48.

Turner, F.M., 'Victorian Scientific Naturalism and Thomas Carlyle', *Victorian Studies* 18 (1975): p. 325–43.

Turner, F. M., *Contesting Cultural Authority: Essays in Victorian Intellectual Life* (Cambridge: Cambridge University Press, 1993).

Tyerman, C., *A History of Harrow School* (Oxford: Oxford University Press, 2000).

Tyndall, J., F. Galton, et al., *The Prayer-gauge Debate; By Prof. Tyndall, Francis Galton and others, against Dr Littledale, President McCosh, the Duke of Argyll, Canon Liddon, and 'The Spectator'* (Boston, MA: Congregational Publishing Society, 1876).

Ungureanu, J. C., 'Tyndall and Draper', *Notes and Queries* 64, (2017): p. 1–3.

Wadge, E., 'A Fair Trial for Spiritualism? Fighting Dirty in the Pall Mall Gazette', in Clifford, D., E. Wadge, A. Warwick and M. Willis (eds), *Repositioning Victorian Sciences; Shifting Centres in Nineteenth-Century Scientific Thinking* (London: Anthem Press, 2006): p. 95–106.

Ward, H., *History of the Athenaeum 1824–1925* (London: Clowes and Sons Ltd, 1926).

Whymper, E., *Scrambles amongst the Alps in the Years 1860–1869* (London: John Murray, 1871).

Wilson, A. N., *The Victorians* (London: Hutchinson, 2005).

Wilson, D. B. (ed.), *The Correspondence between Sir George Gabriel Stokes and Sir William Thomson, Baron Kelvin of Largs* (Cambridge: Cambridge University Press, 1990).

Worboys, M., *Spreading Germs; Disease Theories and Medical Practice in Britain 1865–1900* (Cambridge: Cambridge University Press, 2000).

Yamalidou, M., 'John Tyndall, the Rhetorician of Molecularity. Part One. Crossing the Boundary towards the Invisible', *Notes Rec.* 53, (1999a): p. 231–42.

Yamalidou, M., 'John Tyndall, the Rhetorician of Molecularity. Part Two. Questions Put to Nature', *Notes Rec.* 53, (1999b): p. 319–31.

Books by Tyndall

In addition to Tyndall's major books, listed below, he had numerous individual articles and speeches, and notes of lecture courses, separately issued. Many books were published in America, Germany, and France, often in several editions. They were translated into most European languages, and into the languages of India, China, and Japan.[1]

Glaciers of the Alps (London: Murray, 1860).

Mountaineering in 1861: A Vacation Tour (London: Longmans, 1862).

Heat Considered as a Mode of Motion (London: Longmans, 1863).

Sound: A Course of Eight Lectures Delivered at the Royal Institution of Great Britain (London: Longmans, 1867).

Faraday as a Discoverer (London: Longmans, 1868).

Essays on the Use and Limit of the Imagination in Science (London: Longmans, 1870).

Researches on Diamagnetism and Magnecrystallic Action (London: Longmans, 1870a).

Fragments of Science: A Series of Detached Essays, Lectures, and Reviews (London: Longmans, 1871). The 6th edition (1879), in two volumes, contains an extended range of articles.

Hours of Exercise in the Alps (London: Longmans, 1871a).

Contributions to Molecular Physics in the Domain of Radiant Heat (London: Longmans, 1872).

The Forms of Water in Clouds and Rivers, Ice and Glaciers (London: King, 1872a).

Six Lectures on Light. Delivered in America in 1872–1873 (London: Longmans, 1873).

Address Delivered Before the British Association at Belfast 1874 (London: Longmans, 1874).

Lessons in Electricity at the Royal Institution 1875–6 (London: Longmans, 1876).

Essays on the Floating-matter of the Air in Relation to Putrefaction and Infection (London: Longmans, 1881).

New Fragments (London: Longmans, 1892).

See also

Tyndall, J., 'Natural Philosophy', in E. Hughes (ed.), *Reading Lessons: First Book* (London: Longman, 1855), p. 289–338.

Tyndall, J., 'Natural Philosophy', in E. Hughes (ed.), *Reading Lessons: Second Book* (London: Longman, 1855), p. 296–322.

Tyndall, J., 'Natural Philosophy', in E. Hughes (ed.), *Reading Lessons: Third Book* (London: Longman, 1856), p. 238–66.

Tyndall, J., 'Natural Philosophy', in E. Hughes (ed.), *Reading Lessons: Fourth Book* (London: Longman, 1858), p. 229–76.

Tyndall J., *Natural Philosophy in Easy Lessons* (London: Cassell, 1869).

Scientific Papers by Tyndall

This list of eighty-eight scientific papers excludes Tyndall's various articles in *Nature*, which in the nineteenth century was more like a general magazine than the elite scientific journal of today. It also excludes his fifty-five Discourses at the Royal Institution (almost all summarised in the *Proceedings of the Royal Institution*). Individual papers were often published in several journals, including in German (Poggendorff's *Annalen*) and French (*Annales de Chimie*). Tyndall wrote more than fifty articles for a wide range of newspapers and weekly and quarterly magazines. He published a dozen poems in newspapers, all except two—*On the Death of Dean Bernard* and *A Morning on Alp Lusgen*—under pseudonyms. The latter also formed the final piece in his final book, *New Fragments*.

Apart from his doctoral thesis, Tyndall's papers contain little mathematics. They can be read by anyone with a basic knowledge of physics who wishes to give some time to them.

Tyndall, J., 'Die Schraubenfläche mit geneigter Erzeugungslinie und die Bedingungen des Gleichgewichts für solche Schrauben', *Doctoral Thesis*, Marburg (1850).

Tyndall, J. and H. Knoblauch, 'On the Deportment of Crystalline Bodies between the Poles of a Magnet', *Phil. Mag.* 36 (1850a): p. 178–83.

Tyndall, J. and H. Knoblauch, '(First Memoir) On the Magneto-optic Properties of Crystals, and the Relation of Magnetism and Diamagnetism to Molecular Arrangement', *Phil. Mag.* 37 (1850b): p. 1–33.

Tyndall, J., 'On the Magneto-Optical Properties of Crystals', *Brit. Assoc. Rep., Notes and Abstracts of Miscellaneous Communications to the Sections* (1850c), p. 23.

Tyndall, J., 'On the Laws of Magnetism', *Phil. Mag.* 1 (1851a): p. 265–95.

Tyndall, J., 'Phenomena of a Water-jet', *Phil. Mag.* 1 (1851b): p. 105–11.

Tyndall, J., 'On Diamagnetism and Magnecrystallic Action', *Brit. Assoc. Rep., Notes and Abstracts of Miscellaneous Communications to the Sections* (1851c): p. 15–18.

Tyndall, J., '(Second Memoir) On Diamagnetism and Magnecrystallic Action', *Phil. Mag.* 2 (1851d): p. 165–88.

Tyndall, J., 'On the Polarity of Bismuth, Including an Examination of the Magnetic Field', *Phil. Mag.* 2 (1851e): p. 333–44.

Tyndall, J., 'On Molecular Action', *Brit. Assoc. Rep., Notes and Abstracts of Miscellaneous Communications to the Sections* (1852a), p. 20.

Tyndall, J., 'On Poisson's Theoretic Anticipation of Magnecrystallic Action', *Brit. Assoc. Rep., Notes and Abstracts of Miscellaneous Communications to the Sections* (1852b): p. 20–1.

Tyndall, J., 'On Molecular Influences. Part I. Transmission of Heat through Organic Structures', *Phil. Trans. Roy. Soc. Lond.* 143 (1853): p. 217–31.

Tyndall, J., 'On the Vibrations and Tones Produced by the Contact of Bodies Having Different Temperatures', *Phil. Trans. Roy. Soc. Lond.* 144 (1854): p. 1–10.

Tyndall, J., 'On the Nature of the Force by Which Bodies Are Repelled from the Poles of a Magnet; to Which is Prefixed, an Account of Some Experiments on Molecular Influences', *Phil. Trans. Roy. Soc. Lond.* 145 (1855a): p. 1–51.

Tyndall, J., 'On the Existence of a Magnetic Medium in Space', *Phil. Mag.* 9 (1855b): p. 205–9.

Tyndall, J., 'The Polymagnet', *Phil. Mag.* 9 (1855c): p. 425–30.

Tyndall, J., 'On the Currents of the Leyden Battery', *Phil. Mag.* 10 (1855d): p. 226–9.

Tyndall, J., 'Further Researches on the Polarity of the Diamagnetic Force', *Phil. Trans. Roy. Soc. Lond.* 146 (1856a): p. 237–59.

Tyndall, J., 'Further Researches on the Polarity of the Diamagnetic Force', *Phil. Mag.* 12 (1856b), p. 161–84.

Tyndall, J., 'On the Relation of Diamagnetic Polarity to Magnecrystallic Action', *Phil. Mag.* 11 (1856c): p. 125–37.

Tyndall, J., 'On a Peculiar Case of Colour Blindness', *Phil. Mag.* 11 (1856d): p. 329–33.

Tyndall, J. and T. H. Huxley, 'On the Structure and Motion of Glaciers', *Phil. Trans. Roy. Soc. Lond.* 147 (1857): p. 327–46.

Tyndall, J., 'Remarks on Foam and Hail', *Phil. Mag.* 13 (1857a): p. 352–3.

Tyndall, J., 'On the Sounds Produced by the Combustion of Gases in Tubes', *Phil. Mag.* 13 (1857b): p. 473–9.

Tyndall, J., 'On Some Physical Properties of Ice', *Phil. Trans. Roy. Soc. Lond.* 148 (1858): p. 211–29.

Tyndall, J., 'Particulars of an Ascent of Mont Blanc', *Brit. Assoc. Rep., Transactions of the Sections* (1858b): p. 39–40.

Tyndall, J., 'On the Physical Phenomena of Glaciers—Part I. Observations on the Mer de Glace', *Phil. Trans Roy. Soc. Lond.* 149 (1859a): 261–78.

Tyndall, J., 'Remarks on Ice and Glaciers', *Phil. Mag.* 17 (1859b): 91–6.

Tyndall, J., 'On the Veined Structure of Glaciers; with Observations upon White Ice-Seams, Air-Bubbles and Dirt-Bands, and Remarks upon Glacier Theories', *Phil. Trans. Roy. Soc. Lond.* 149 (1859c): p. 279–307.

Tyndall, J., 'Note on the Transmission of Radiant Heat through Gaseous Bodies', *Proc. Roy. Soc. Lond.* 10 (1859d): p. 37–9.

Tyndall, J., 'The Bakerian Lecture: On the Absorption and Radiation of Heat by Gases and Vapours, and on the Physical Connexion of Radiation, Absorption, and Conduction', *Phil. Trans. Roy. Soc. Lond.* 151 (1861a): p. 1–36.

Tyndall, J., 'Remarks on Radiation and Absorption', *Phil. Mag.* 22 (1861b): p. 377–8.

Tyndall, J., 'Observations on Lunar Radiation', *Phil. Mag.* 22 (1861c): p. 470–2.

Tyndall, J., 'On the Absorption and Radiation of Heat by Gaseous Matter. Second Memoir', *Phil. Trans. Roy. Soc. Lond.* 152 (1862a): p. 59–98.

Tyndall, J., 'Remarks on Recent Researches in Radiant Heat', *Phil. Mag.* 23 (1862b): p. 252–66.

Tyndall, J., 'On the Regelation of Snow-granules', *Phil. Mag.* 23 (1862c): p. 312–13.

Tyndall, J., 'On the Conformation of the Alps', *Phil. Mag.* 24 (1862d): p. 169–73.

Tyndall, J., 'On the Relation of Radiant Heat to Aqueous Vapour. Third Memoir', *Phil. Trans. Roy. Soc. Lond.* 153 (1863a): p. 1–12.

Tyndall, J., 'On Radiation through the Earth's Atmosphere', *Phil. Mag.* 25 (1863b): p. 200–6.

Tyndall, J., 'On the Passage of Radiant Heat through Dry and Humid Air', *Phil. Mag.* 26 (1863c): p. 44–54.

Tyndall, J., 'Notes on Scientific History', *Phil. Mag.* 28 (1864a): p. 25–51.

Tyndall, J., 'On the Absorption and Radiation of Heat by Gaseous and Liquid Matter. Fourth Memoir', *Phil. Trans. Roy. Soc. Lond.* 154 (1864b): p. 201–25.

Tyndall, J., 'On the Conformation of the Alps', *Phil. Mag.* 28 (1864c): p. 255–71.

Tyndall, J., 'The Bakerian Lecture: Contributions to Molecular Physics. Being the Fifth Memoir of Researches on Radiant Heat', *Phil. Trans. Roy. Soc. Lond.* 154 (1864d): p. 327–68.

Tyndall, J., 'On Luminous and Obscure Radiation', *Phil. Mag.* 28 (1864e): p. 329–41.

Tyndall, J., 'On the History of Calorescence', *Phil. Mag.* 29 (1865a): p. 218–31.

Tyndall, J., 'On the History of Negative Fluorescence', *Phil. Mag.* 29 (1865b): p. 44–55.

Tyndall, J., 'Note on the Invisible Radiation of the Electric Light', *Proc. Roy. Soc. Lond.* 14 (1865c): 33–5.

Tyndall, J., 'Professor Helmholtz on Ice and Glaciers', *Phil. Mag.* 30 (1865d): p. 393–407.

Tyndall, J., 'On Calorescence', *Phil. Trans. Roy. Soc. Lond.* 156 (1866a): p. 1–24.

Tyndall, J., 'Sixth Memoir on Radiation and Absorption.—Influence of Colour and Mechanical Condition on Radiant Heat', *Phil. Trans. Roy. Soc. Lond.* 156 (1866b): p. 83–96.

Tyndall, J., 'On the Black-bulb Thermometer', *Phil. Mag.* 31 (1866c): p. 191–3.

Tyndall, J., 'Remarks on the Paper of Professor Magnus. The Influence of Heat on the Formation of Dew', *Phil. Mag.* 32 (1866d): p. 118–20.

Tyndall, J., 'On the Action of Sonorous Vibrations on Gaseous and Liquid Jets', *Phil. Mag.* 33 (1867a): p. 375–91.

Tyndall, J., 'Note on Magnus's Paper. The Influence of the Adhesion of Vapour in Experiments on the Absorption of Heat', *Phil. Mag.* 33 (1867b): p. 425.

Tyndall, J., 'On a New Series of Chemical Reactions Produced by Light', *Proc. Roy. Soc. Lond.* 17 (1868a): p. 92–102.

Tyndall, J., 'On the Blue Colour of the Sky, and the Polarization of Light', *Proc. Roy. Soc. Lond.* 17 (1869a): p. 223–33.

Tyndall, J., 'Note on the Formation and Phenomena of Clouds', *Proc. Roy. Soc. Lond.* 17 (1869b): p. 317–19.

Tyndall, J., 'The Generation of Clouds by Actinic Action and the Reaction of such Clouds upon Light', *Proc. Camb. Phil. Soc.* 2 (1869c): p. 136–40.

Tyndall, J., 'On Cometary Theory', *Phil. Mag.* 37 (1869d): p. 241–5.

Tyndall, J., 'On the Action of Rays of High Refrangibility upon Gaseous Matter', *Phil. Trans. Roy. Soc. Lond.* 160 (1870b): p. 333–65.

Tyndall, J., 'On the Polarization of Heat', *Phil. Mag.* 39 (1870c): p. 280–2.

Tyndall, J., 'Preliminary Account of an Investigation on the Transmission of Sound by the Atmosphere', *Proc. Roy. Soc. Lond.* 22 (1874a): p. 58–68.

Tyndall, J., 'On some Recent Experiments with a Fireman's Respirator', *Proc. Roy. Soc. Lond.* 22 (1874b): p. 359–61.

Tyndall, J., 'Further Experiments on the Transmission of Sound', *Proc. Roy. Soc. Lond.* 22 (1874c): p. 359.

Tyndall, J., 'On the Atmosphere as a Vehicle of Sound', *Phil. Trans. Roy. Soc. Lond.* 164 (1874d): p. 183–244.

Tyndall, J., 'On Acoustic Reversibility', *Proc. Roy. Soc. Lond.* 23 (1875): p. 159–65.

Tyndall, J., 'The Optical Deportment of the Atmosphere in Relation to the Phenomenon of Putrefaction and Infection', *Phil. Trans. Roy. Soc. Lond.* 166 (1876a): p. 27–74.

Tyndall, J., 'Note on the Deportment of Alkalized Urine', *Proc. Roy. Soc. Lond.* 25 (1876b): p. 457–8.

Tyndall, J., 'Preliminary Note on the Development of Organisms in Organic Infusions', *Proc. Roy. Soc. Lond.* 25 (1877a): p. 503–6.

Tyndall, J., 'On Heat as a Germicide when Discontinuously Applied', *Proc. Roy. Soc. Lond.* 25 (1877b): p. 569–70.

Tyndall, J., 'Researches on the Deportment and Vital Persistence of Putrefactive and Infective Organisms from a Physical Point of View', *Phil. Trans. Roy. Soc. Lond.* 167 (1877c): p. 149–206.

Tyndall, J., 'Note on Dr. Burdon Sanderson's Latest Views of Ferments and Germs', *Proc. Roy. Soc. Lond.* 26 (1877d): p. 353–6.

Tyndall, J., 'Observations on Hermetically-Sealed Flasks Opened on the Alps', *Proc. R. Soc. Lond.* 26 (1877e): p. 487–8.

Tyndall, J., 'On Schulze's Mode of Intercepting the Germinal Matter of the Air', *Proc. Roy. Soc. Lond.* 27 (1878a): p. 99–100.

Tyndall, J., 'Recent Experiments on Fog-Signals', *Proc. Roy. Soc. Lond.* 27 (1878b): p. 245–58.

Tyndall, J., 'Note on the Influence Exercised by Light on Organic Infusions', *Proc. Roy. Soc. Lond.* 28 (1878c): p. 212–13.

Tyndall, J., 'On Buff's Experiments on the Diathermancy of Air', *Proc. Roy. Soc. Lond.* 30 (1879): p. 10–20.

Tyndall, J., 'Action of an Intermittent Beam of Radiant Heat upon Gaseous Matter', *Proc. Roy. Soc. Lond.* 31 (1881a): p. 307–17.

Tyndall, J., 'Further Experiments on the Action of an Intermittent Beam of Radiant Heat on Gaseous Matter. Thermometric Measurements', *Proc. Roy. Soc. Lond.* 31 (1881b): p. 478–9.

Tyndall, J., 'On the Arrestation of Infusorial Life', *Brit. Assoc. Rep.* (1881c): p. 450–1.

Tyndall, J., 'The Bakerian Lecture: Action of Free Molecules on Radiant Heat, and Its Conversion Thereby into Sound', *Phil. Trans. Roy. Soc. Lond.* 173 (1882a): p. 291–354.

Tyndall, J., 'Note on General Duane's Soundless Zones', *Proc. Roy. Soc. Lond.* 34 (1882b): p. 18–19.

Tyndall, J., 'Note on Terrestrial Radiation', *Proc. Roy. Soc. Lond.* 35 (1883a): p. 21–5.

Tyndall, J., 'On a Hitherto Unobserved Resemblance between Carbonic Acid and Bisulphide of Carbon', *Proc. Roy. Soc. Lond.* 35 (1883b): p. 129–30.

Tyndall, J., 'On Rainbows', *Phil. Mag.* 17 (1884a): p. 61–4.

Tyndall, J., 'Note on the White Rainbow', *Phil. Mag.* 17 (1884b): p. 148–50.

Tyndall, J., 'On Rainbows and Glories', *Phil. Mag.* 17 (1884c): p. 244.

INDEX